Morphogenetic Hormones of Arthropods

*Recent Advances in Compartive Arthropod
Morphology, Physiology, and Development*

Series Editor-in-Chief: Ayodhya P. Gupta

Editorial Board

Morphogenetic Hormones of Arthropods

Discoveries, Syntheses, Metabolism, Evolution, Modes of Action, and Techniques

edited by A. P. GUPTA

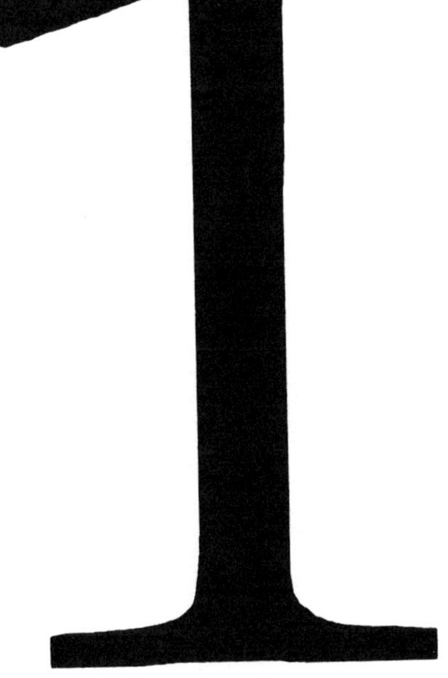

RUTGERS UNIVERSITY PRESS
New Brunswick and London

Library of Congress Cataloging-in-Publication Data

Morphogenetic hormones of arthropods / edited by A.P. Gupta.
     p.   cm. —(Recent advances in comparative arthropod
morphology, physiology, and development)
   Includes bibliographies and indexes.
   Contents: pt. 1. Discoveries, syntheses, metabolism, evolution,
modes of action, and techniques—pt. 2. Embryonic and
postembryonic sources.—pt. 3. Roles in histogenesis, organogenesis,
and morphogenesis
   ISBN 0-8135-1414-2 (pt. 1)
   1. Arthropoda—Morphogenesis.   2. Juvenile hormones.   3. Hormones.
I. Gupta, A. P., 1928–   . II. Series.
   [DNLM: 1. Arthropods. 2. Hormones. 3. Morphogenesis. QL 434.7
M871]
QL434.72.M67 1989
595.2′043—dc19
DNLM/DLC
for Library of Congress   89–6029
British Cataloging-in-Publication information available   CIP

CONTENTS

PREFACE vii

CONTRIBUTORS ix

Chapters:

1. Morphogenetic Hormones and Their Glands in Arthropods: Evolutionary Aspects, Ayodhya P. Gupta 1
2. Methyl Farnesoate: A JH-like Compound in Crustaceans, D. W. Borst and H. Laufer 35
3. Juvenile Hormone–like Compounds in Terrestrial Chelicerata (Arachnida), J. C. Bonaric and C. Juberthie 61
4. Synthesis and Effects on Molting of Juvenile Hormone–like Compounds in Crustacea and Chelicerata, F. Lachaise 73
5. Biosynthesis, Titer Regulation, and Transport of Juvenile Hormones, Walter G. Goodman 83
6. Metabolism of Juvenile Hormones: Degradation and Titer Regulation, R. Michael Roe and Krishnappa Venkatesh 125
7. Modes of Action of Juvenile Hormones at Cellular and Molecular Levels, A. Krishna Kumaran 181
8. Evidence for Ecdysteroids as Molting Hormones in Chelicerata, Crustacea, and Myriapoda, T. C. Jegla 229
9. Synthesis, Metabolism, and Effects on Molting of Ecdysteroids in Crustacea, Chelicerata, and Myriapoda, F. Lachaise 276
10. Metabolism of Insect Molting Hormones: Bioconversion and Titer Regulation, Malcolm J. Thompson, Gunter F. Weirich, and James A. Svoboda 325
11. Mode of Action of Molting Hormones in Insects, G. Kaeuser, J. Koolman, and P. Karlson 361
12. Techniques for Identification and Quantification of Juvenile Hormones and Related Compounds in Arthropods, Fred. C. Baker 389
13. Methods for Ecdysteroid Analysis, René Lafont and Philippe Beydon 455

TAXONOMIC INDEX 513

SUBJECT INDEX 519

Preface

The concept of organizing and publishing a series of comprehensive volumes on comparative arthropod evolution, morphology, physiology, and development occurred to me some 15 years ago and has resulted so far in the publication of volumes on arthropod phylogeny, neurohemal organs, hemocytic and humoral immunity, and brain. This book is the first of three volumes on the morphogenetic hormones of arthropods that will appear during the next two years as part of an international series entitled *Recent Advances in Comparative Arthropod Morphology, Physiology, and Development*. The two recent volumes on insect endocrinology in the Kerkut and Gilbert series are comprehensive treatises on the subject, and the present volumes are not meant to either compete with or duplicate most of the coverage in those two books. Rather, our emphasis here has been on the morphogenetic hormones in Aquatic and Terrestrial Chelicerata, Crustacea, and Myriapoda. Chapters on Insecta have been included to complement the coverage on other arthropod groups. Comparable chapters on other major arthropod taxa could not be produced owing to lack of adequate information in the current literature.

Neurosecretory cells and their hormones have been largely excluded, with the exception of NSCs in Myriapoda, because extensive coverage of them exists elsewhere in books and reviews, and also because most of these topics were covered in the volume on neurohemal organs of arthropods that I edited a few years ago.

Perhaps the single most significant contribution of this series, as indeed of the other volumes on arthropods that I have edited over the past decade, has been the uncovering of the many gaps in our knowledge of the morphogenetic hormones in arthropods in general. Indeed, various contributors have pointed out areas that remain to be explored.

The most conspicuous gaps occur in the Myriapoda, Aquatic Chelicerata, Entomostracan Crustacea, and most of the Terrestrial Chelicerata (Arachnida). Although ecdysteroids have been reported from all the five major arthropod taxa, their glands are largely unknown, except in Insecta and Crustacea. Similarly, juvenile hormone (JH) or JH-like compounds such as methyl farnesoate (MF) have been extracted only from the latter two taxa. MF and 20-hydroxyecdysone (20E) (β-ecdysone) seem to represent the plesiomorphic or ancestral stages in the evolution of these two major morphogenetic hormones in arthropods. MF, like 20E, will most likely eventually be discovered in Aquatic and Terrestrial Chelicerata and Myriapoda as well.

Clearly, comparative studies of all aspects of morphogenetic hormones in all major arthropod taxa, with the exception of those in Insecta, are needed, and I hope this series will provide the impetus and leads for much-needed research in the aforementioned areas.

The three volumes are organized as follows: Part 1—*Discoveries, Syntheses, Metabolism, Evolution, Modes of Action, and Techniques;* Part 2—*Embryonic and Postembryonic Sources;* and Part 3—*Roles in Histogenesis, Organogenesis, and Morphogenesis.*

Although as in any multiauthored treatise, some overlaps between chapters are inevitable, in these volumes we have tried to keep them to a minimum. Wherever relevant, such overlaps and divergences of opinions have been cross-referenced. Because the subject of taxonomic rankings of major arthropod groups is still controversial, each contributor has used his/her preferred taxonomic categories. Furthermore, each contributor has had complete freedom to develop, interpret, and present his/her views. Each chapter presents an in-depth review of the topic it covers.

The endeavor of organizing such a series could not have been successful without the cooperation and assistance of all the authors who responded to my invitation to contribute. To them, I am indebted. To numerous authors, journal publishers, and professional societies who generously allowed reproduction of published or unpublished materials in the original or modified forms, I am grateful. And I greatly appreciate the cooperation and assistance of Karen Reeds, Marilyn Campbell, Barbara Kopel, and Dina Bednarczyk at Rutgers University Press, and the meticulous copy editing by Bert N. Zelman of Publishers Workshop Inc. I gratefully acknowledge the financial assistance of the Rutgers Research Council in establishing the Series and in the preparation of Volume 1. As always in the past, the ungrudging support of my wife and children during the preparation of the book has been enormous and has allowed me to devote to this project considerable time that rightfully belonged to them. To them, I am as ever grateful.

Ayodhya P. Gupta
New Brunswick, New Jersey
December 8, 1988

Contributors

Fred. C. Baker
Zoecon Research Institute
975 California Avenue
Palo Alto, CA 94304–1104

Philippe Beydon
CNRS U. A. 686
Biochemistry and Physiology
  of Development
46 Ulm Street
75230 Paris Cedex 05
France

J. C. Bonaric
Laboratoire Souterrain du CNRS
Moulis
09200 Saint Girons
France

David W. Borst
Department of Biological Sciences
Felmley Hall
Illinois State University
Normal, IL 61761–6901

Walter G. Goodman
Department of Entomology
237 Russell Laboratories
University of Wisconsin–Madison
Madison, WI 53706

Ayodhya P. Gupta
Department of Entomology
Rutgers University
New Brunswick, NJ 08903

Thomas C. Jegla
Department of Biology
Kenyon College
Gambier, OH 43022

Christian Juberthie
Laboratoire Souterrain du CNRS
Moulis
09200 Saint Girons
France

G. Kaeuser
Institute for Physiological Chemistry
Karl von Frisch Street
Philipps University
D-3550 Marburg (Lahn)
West Germany

Peter Karlson
Institute for Physiological Chemistry
Karl von Frisch Street
Philipps University
D-3550 Marburg (Lahn)
West Germany

Jan Koolman
Institute for Physiological Chemistry
Karl von Frisch Street
Philipps University
D-3550 Marburg (Lahn)
West Germany

A. Krishna Kumaran
Department of Biology
Wehr Life Sciences Building
Marquette University
Milwaukee, WI 53233

Fabienne Lachaise
Biochemistry and Physiology
  of Development
CNRS U. A. 686
46 Ulm Street
75230 Paris Cedex 05
France

René Lafont
Biochemistry & Physiology
  of Development
CNRS U. A. 686
46 Ulm Street
75230 Paris Cedex
France

Hans Laufer
Department of Molecular
  and Cell Biology
University of Connecticut
Storrs, CT 06268

R. Michael Roe
Department of Entomology
School of Agriculture & Life Sciences
North Carolina State University
Raleigh, NC 27607–7613

James A. Svoboda
Insect and Nematode Hormone
  Laboratory
ARS-USDA
Beltsville, MD 20705

Malcolm J. Thompson
Insect and Nematode Hormone
  Laboratory
ARS-USDA
Beltsville, MD 20705

Krishnappa Venkatesh
Department of Entomology
School of Agriculture & Life Sciences
North Carolina State University
Raleigh, NC 27607–7613

Gunter F. Weinrich
Insect & Nematode Hormone
  Laboratory
ARS-USDA
Beltsville, MD 20705

Morphogenic Hormones of Arthropods

Morphogenetic Hormones and Their Glands in Arthropods: Evolutionary Aspects

1

AYODHYA P. GUPTA

1.1. Introduction   3
1.2. Growth and Differentiation in Arthropods   3
1.3. Morphogenetic Hormones and Their Glands in
      Arthropods   4
    1.3.1. Molting Hormones, or Ecdysteroids, and
      Their Glands   4
        1.3.1.1. Aquatic Chelicerata   6
        1.3.1.2. Crustacea   7
        1.3.1.3. Terrestrial Chelicerata (Arachnida)   8
        1.3.1.4. Myriapoda   9
        1.3.1.5. Insecta (Hexapoda)   9
    1.3.2. Hormones of the Inhibitory System (JH or
      Its Counterparts) and Their Glands   10
        1.3.2.1. Aquatic Chelicerata   11
        1.3.2.2. Crustacea   12
        1.3.2.3. Terrestrial Chelicerata (Arachnida)   12
        1.3.2.4. Myriapoda   13
        1.3.2.5. Insecta (Hexapoda)   13
    1.3.3. Morphogenetic Hormones Other Than
      Ecdysteroids and JHs and MF   14
        1.3.3.1. Androgenic Hormones in Crustacea
      and Insecta   14
        1.3.3.2. Other Morphogenetic Hormones   15
1.4. Evolution of Arthropods and Their Morphogenetic
      Hormones   15
    1.4.1. Evolution of Arthropods   15
    1.4.2. Evolution of Molting Hormones or
      Ecdysteroids   16

1.4.2.1.   Plesiomorphic or Ancestral
Ecdysteroid   17
1.4.2.2.   Apomorphic or Newly Evolved
Ecdysteroids   18
1.4.3.   Evolution of the Hormones of the
Inhibitory System   18
1.4.3.1.   Evolution of the Arthropod Brain and
Its Neurosecretory Cells That Secrete
Hormones of the Inhibitory Mechanisms in
Some Arthropods   19
1.4.3.2.   Evolution of the Neurohemal Organs
That Secrete the Inhibitory Hormones in
Some Arthropods   21
1.4.3.3.   Evolution of the JH or Its
Counterparts   24
1.5.   Summary   25
Acknowledgments   26
References   27

1.1. Introduction

Organisms and their organs have a defined, genetically determined form or final state. The developmental process that gradually unfolds this form or state and eventually completes it is called *morphogenesis.* In other words, morphogenesis may be referred to as the realization of a predetermined three-dimensional morphology of the organism or the final state of any of its organs or structures over a period of time, the latter termed as the fourth dimension by Waddington (1965). He also makes a distinction between morphogenesis and "pattern-formation." According to him, "pattern" refers to morphologies whose most characteristic features are a number of discrete "elements arranged in some form of order" (e.g., venation or color patches on an insect wing), the units within the pattern having some epigenetic relation with one another.

Morphogenesis, of necessity, involves cellular differentiation, histogenesis, organogenesis, and formation of organ systems or structures. Each part of an organism follows a specific pattern of morphogenesis, bears a specific relation to other parts in terms of size and cellular content, and is strictly regulated or controlled by hormones and some external environmental factors, such as gravity, temperature, and light intensity. Morphogenesis is accomplished by cell movement and growth.

1.2. Growth and Differentiation in Arthropods

All arthropods have a rigid body covering or cuticle that must be periodically shed to accomplish growth. Growth is caused by molting and involves mitosis, deposition of new cuticle, protein synthesis for new cells, synthesis of specific enzymes, and new tissue formation by differentiation. The cuticle is deposited by the underlying epidermis and is firmly attached to it. Once tanned or sclerotized, following its deposition, the cuticle prevents any further epidermal growth by limiting cellular multiplication. For any further growth to occur, the epidermis must first be detached from the old cuticle; this separation of the epidermis from the old cuticle, termed *apolysis,* normally occurs at predetermined intervals. Following apolysis, cellular multiplication and differentiation continue until sufficient growth has occurred to define the shape and size of the next developmental stage. Once this is accomplished, any further epidermal growth ceases and the new cuticle is deposited, the last stage of the above chain of events being the eventual shedding, molting, or ecdysis of the old cuticle and the emergence of the newly molted morph. Eventually, as a result of several such overall structural changes, the animal metamorphoses into the adult form (e.g., in insects) or continues to grow as adult (e.g., in some apterygote insects and decapod Crustacea).

Metamorphosis is generally associated with postembryonic growth, although Snodgrass (1956) defines it to include structural changes at "any time during the life history of an animal," including the embryonic stage. Most terrestrial arthropods are *epimorphic*, that is, they hatch with full complement of body segments and appendages and are capable of readily adapting to the environment upon hatching; others (e.g., the majority of Crustacea and aquatic Chelicerata, Myriapoda, and some insects, e.g., Protura), however, are *anamorphic*, that is, they hatch in a much earlier stage of development and generally lack (Diplopoda and Chilopoda being exceptions) a full complement of body segments and appendages; the latter are added later during postembryonic or adult stages by teloblastic proliferation in a subterminal growth zone. Since anamorphic growth occurs also in annelids, Snodgrass (1956) regarded it as the "primitive method of growth in annulate animals."

Some adult insects, chelicerates, crustaceans, and myriapods maintain morphogenetic or regenerative capabilities throughout life.

1.3. Morphogenetic Hormones and Their Glands in Arthropods

1.3.1. *Molting Hormones, or Ecdysteroids, and Their Glands*

Apolysis and molting are brought about by molting hormones (MHs), or ecdysteroids, which are some of the principal morphogenetic hormones in arthropods. Several major ecdysteroids have been reported from arthropods, the 20-hydroxyecdysone (20E) ($= \beta$-ecdysone), often formed by hydroxylation of a precursor (α-ecdysone), is the most commonly found ecdysteroid in these animals. It was first extracted from the spiny lobster, *Jasus lalandei* (Hampshire and Horn, 1966), and is the major ecdysteroid in Aquatic Chelicerata (*Limulus*), Crustacea, Terrestrial Chelicerata (ticks and spiders), and pterygote insects (Fig. 1.1). Crustaceans also have 25-deoxyecdysone (25De), 2-deoxy-20-hydroxyecdysone (2De-20E) (Galbraith et al., 1968), makisterone A (MaA),

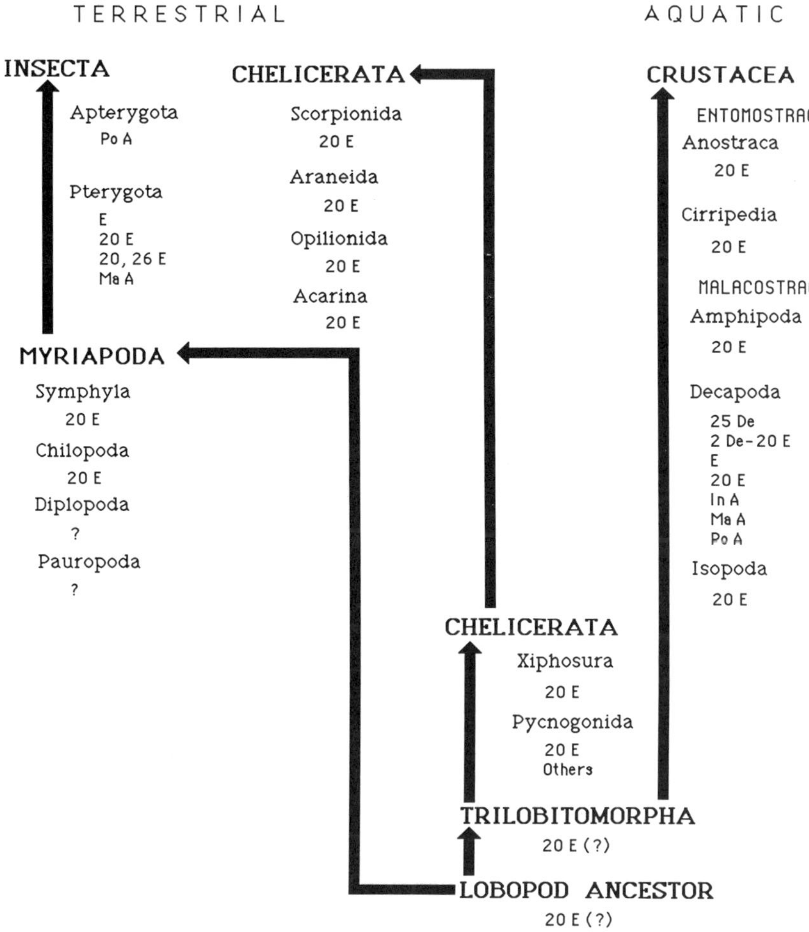

FIGURE 1.1. Monophyletic arrangement of the five arthropod taxa and list of major ecdysteroids that have been found in represetatives of each taxon: 25 De = 25-deoxyecdysone; 2De-20 E = 2-deoxy-20-hydroxyecdysone; E = α-ecdysone; 20E = 20-hydroxyecdysone (β-ecdysone); 20,26-dihydroxyecdysone (= 20-hydroxy-β-ecdysone); InA = inokosterone A (25-deoxy-20, 26-dihydroxyecdysone); MaA = makisterone A (20-hydroxy-24-methylecdysone; others = seven other ecdysteroids have been reported, but these presumably do not induce molting; PoA = ponasterone A (25-deoxy-20-hydroxyecdysone); ? = Not extracted; putative molting gland known or unknown; (?) = unknown, but postulated to be present.

inokosterone A (InA) (Faux et al., 1969), and ponasterone A (PoA) (McCarthy, 1979; Lachaise et al., 1981) (see also Section 1.3.1.2. and Chapter 9 by Lachaise herein). According to Lachaise, PoA is also found in the apterygote insect orders Collembola and Thysanura but is absent in pterygote insects as well as in ticks (Acarina) (Terrestrial Chelicerata).

Although the presence of various ecdysteroids has been experimentally demonstrated or suspected in some members of all the major arthropod groups, the various glands that produce these hormones are mostly unknown, except in insects and some crustaceans and myriapods. In addition to the prothoracic or ecdysial glands of insects, the Y-organs of crustaceans, and the ecdysial, head, and collar glands of myriapods, there are other tissues or cells that are known to produce ecdysteroids. For example, the ovaries of some insects and crabs (Lachaise et al., 1981; Lachaise and Hoffmann, 1982) and the oenocytes of insects (Locke, 1969) and opilionids (Terrestrial Chelicerata) (Romer and Gantzy, 1981) produce ecdysteroids. And according to Seifert (Chapter 6 in Part 2), the so-called lymphatic tissue and the nephrocytes of some chilopods (Myriapoda) are suspected to produce MF.

It appears that some form of MH must have evolved prior to the evolution of the arthropods from the annelid ancestors, because MH is present not only in annelids (see Chapter 5 by Spindler in Part 3) but also in nematodes (Rogers, 1973; Horn et al., 1974; Matsuda, 1986). It is reasonable to assume that the trilobites also must have possessed this hormone. Accounts of these hormones and their glands in various arthropod taxa follow.

1.3.1.1.   AQUATIC CHELICERATA

This group consists of the Xiphosura (king or horseshoe crabs), Pycnogonida (sea spiders) (Fig. 1.1), and Tardigrada (water bears), and is the most neglected group in terms of the information on the MH and its glands. 20-Hydroxyecdysone has been reported in the horseshoe crab, *Limulus polyphemus* (Jegla and Costlow, 1979; Winget and Herman, 1976) and in the sea spider, *Pycnogonum litorale* (Behrens and Bückmann, 1983; Bückmann et al., 1986). According to these latter authors, *P. litorale* also has several other ecdysteroids, but presumably these do not induce molt (see Chapter 8 by Jegla herein).

The molting glands in *L. polyphemus,* one of the most studied representatives of Xiphosura, are unknown (Tombes, 1979; Solomon

et al., 1982). Jegla et al. (1972) were able to induce molting by injecting
α- and β-ecdysones, which indicates that the tissues of this primitive
arthropod are sensitive to both forms of ecdysone. The molting glands
of *P. litorale* also are unknown.

1.3.1.2. CRUSTACEA

This group consists of the primitive Entomostraca (Anostraca, Cirripedia) and the more advanced Malacostraca (Amphipoda, Decapoda,
Isopoda). In addition to α-ecdysone, at least six other ecdysteroids
have been reported in this group (Fig. 1.1): 25 De; 25-deoxy-20-
hydroxyecdysone (= ponasterone A, or PoA); 20E; 2De-20E (= 2-
deoxyecdysterone) (Galbraith et al., 1968); 20-hydroxy-24-methyl-
ecdysone (= makisterone A, or MaA), and 25-deoxy-20, 26-dihydroxy-
ecdysone (= inokosterone A, or InA) (Faux et al., 1969). According to
Lachaise and Lafont (1984), InA is a metabolite of PoA. Of these
ecdysteroids, 20E has been reported from at least two entomostracans,
Balanus balanoides (Cirripedia) (Bebbington et al., 1977) and *Artemia
salina* (Anostraca) (Spindler et al., 1980; Walgraeve et al., 1986), and
many representatives of the more advanced Malacostraca, the most
commonly studied subgroup being the Decapoda; 20E has been
reported in the amphipod *Orchestia gammarella* (Blanchet et al., 1976)
and one isopod, *Ligia oceanica* (see Chapter 9 by Lachaise herein).
   In addition to its presence in Crustacea (McCarthy, 1979; Lachaise
et al., 1981), PoA is also found in Insecta (Collembola and Thysanura
(see Chapter 13 by Lafont and Beydon herein). According to Lachaise
(Chapter 9 herein), PoA is responsible for adult molt in crustaceans.
   Thus, on the basis of the information available to date, the crustaceans (as well as pycnogonids, if we include those other reported ecdysteroids that presumably do not induce molting) show greater diversity
in their ecdysteroid repertoire than any other arthropod taxon.
   Although considered monophyletic (Baccetti, 1979; Bergström, 1979;
Paulus, 1979), the Crustacea is a heterogeneous group. This heterogeneity is reflected in the four types of neurohemal/neurohemal-
endocrine organs (Gupta, 1983a) and might eventually be true also in
terms of the diversity of the MH-producing glands when these are
discovered in most subgroups of entomostracan and malacostracan
crustaceans. At present, however, molting glands are unknown in
Entomostraca (Matsuda, 1979). The Y-organs (also known as rostral or
ventral glands) in most malacostracans are known to produce the MH
and thus may be regarded as analogous to the prothoracic or ecdysial

glands of Insecta. Unlike such glands in most insects, however, these Y-organs do not disappear in adults, and this accounts for the continuance of molting in adult crustaceans.

(For details of the ultrastructure of the Y-organ, see Chapter 4 by Spaziani in Part 2).

1.3.1.3. TERRESTRIAL CHELICERATA (ARACHNIDA)

This group includes the pulmonate (with book lungs) taxa, such as Scorpionida (scorpions), Araneida (= Araneae) (spiders), (Fig. 1.1), and Pedipalpida (whip scorpions) (Uropygi and Amblypygi), and the apulmonate (without book lungs, but with well-developed tracheal system) taxa, including Opilionida (= Phalangida) (harvestmen), Acarina (ticks and mites), Pseudoscorpionida, Ricinulei (= Podogona), and Solifugae (= Solpugida) (sun spiders).

It is generally agreed that Terrestrial Chelicerata originated from aquatic progenitors (Merostomata) (Kraus, 1976; Gupta, 1979a) and is monophyletic (with the exception of Scorpionida) (Bergström, 1979). Considering the diversity of the subgroups of Terrestrial Chelicerata, comparable to that of Crustacea, it is surprising that ecdysteroids other than 20E have not been discovered in this group to date; 20E has so far been found in the pulmonate subgroups, Scorpionida and Araneida, and in the apulmonate subgroups, Opilionida and Acarina. Obviously, other subgroups listed above await studies.

Terrestrial Chelicerata is a heterogeneous group in terms of the types of neurohemal organs (Gupta, 1983a) and might show similar diversity when MH-producing organs will be known in many more representatives of this group.

In general, molting glands of Terrestrial Chelicerata are poorly known. According to Meewis and Naisse (1960), in Opilionida, the putative molting gland is located in the cephalothorax. In *Heterometrus swammerdami* (Scorpionida), the ''blind anterior organ'' (Habibulla, 1970) may be the site of MH synthesis (Matsuda, 1979). The presence of MH (not the gland) in *Ornithodoros turicata* (Acarina) has been suggested by Cox (1960).

Because ecdysteroids are present in some spiders (Araneida) (Bonaric and De Reggi, 1977; Solomon et al., 1982), in ticks (Acarina) (Delbecque et al., 1978), and in some scorpions (Scorpionida) (El Bakary et al., 1987), their glands should be investigated.

1.3.1.4. MYRIAPODA

This group includes the Symphyla, Chilopoda, Diplopoda, and Pauropoda (Fig. 1.1). Although it is debatable whether the Myriapoda as a group is mono- or polyphyletic, its origin from a lobopod ancestor seems more plausible (Gupta, 1979c). Like the Terrestrial Chelicerata, Myriapoda is a poorly studied group in terms of the presence of ecdysteroids, and so far only the 20E has been extracted from *Lithobius forficatus* (Chilopoda) (Joly et al., 1979; Leubert et al., 1979) and *Hanseniella ivorensis* (Symphyla) (Juberthie-Jupeau et al., 1979). Ecdysteroids of Diplopoda and Pauropoda await investigation.

Myriapoda is a heterogeneous group in terms of the diversity of neurohemal organs (NHOs); there are six types of NHOs in this group (Gupta, 1983a). And it may be that, when more species of this group are studied, similar diversity in the molting glands will be revealed. Indeed, at least three kinds of molting glands have been reported. The most commonly recognized molting glands in this group are the ecdysial (*Lithobius*) (Lithobiomorpha) and head glands (*Scutigera*) (Scutigeromorpha) in Chilopoda and the collar glands in Diplopoda (Seifert and Rosenberg, 1974; Seifert, Chapter 6 in Part 2). According to Seifert, the so-called lymphatic tissue and the nephrocytes of some chilopods are thought to be the molting glands. No molting glands have so far been reported in Pauropoda and Symphyla.

1.3.1.5. INSECTA (HEXAPODA)

This group includes the Apterygota (consisting of the entognathous orders Protura, Diplura, and Collembola and the ectognathous Thysanura) and the Pterygota (Fig. 1.1). Both mono- and polyphyletic origins of Insecta have been proposed (Gupta, 1978, 1979c). Insecta is certainly a diverse group, and this diversity is also reflected in the various kinds of morphogenetic hormones found in insects. Five major types of MHs (ecdysones) are presently known in insects: α-ecdysone (E); β-ecdysone (20E); 20,26-dihydroxyecdysone (= 26-hydroxy-β-ecdysone) (20,26E); PoA; and MaA. Of these, the β-ecdysone is the true MH hormone in insects, because it alone induces molting. It seems to be nonspecific inasmuch as it causes molting also in crustaceans, spiders, and ticks.

MaA has been reported in hemipteran nymphs by Kelly et al. (1984; see also Lafont and Beydon, Chapter 13 herein), and PoA has been found in the apterygote orders Collembola and Thysanura (see Chapter 9 by Lachaise herein).

Both the MHs and the various ecdysial or prothoracic glands that produce them have been extensively studied in insects. According to Wigglesworth (1965), Toyama (1902) described the origin of the prothoracic glands in the silkworm long before their function was identified. Depending on their locations and morphologies, the ecdysial glands are variously known as prothoracic or thoracic, ventral (in the head), pericardial, peritracheal, and ring (Weismann's) glands (Novak, 1975). These glands are juvenile tissue; they are usually present in immature stages and disappear in adults, with a few exceptions (e.g., Thysanura, the preimago of Ephemeroptera, and the solitary phase of locusts). Hazarika and Gupta (1986) reported the persistence of these glands in the adult German cockroach, *Blattella germanica,* for considerable time.

(For the ultrastructural and other details of the ecdysial gland in Insecta, see Chapter 7 by Beaulaton in Part 2).

1.3.2. *Hormones of the Inhibitory System (JH or Its Counterparts) and Their Glands*

From what we know of their mode of action in insects and other arthropods, the MHs (ecdysteroids) do not act alone but rather in concert—either synergistically or antagonistically—with another morphogenetic hormone, *neotenin,* or juvenile (inhibitory or *status quo*) hormone (JH), in insects or its counterparts in other arthropods.

Molting is regulated or controlled by JH in insects and its counterparts in other arthropods (e.g., the hormone secreted by the neurosecretory (NS) cells in the eyestalk X-organ and stored in the sinus gland in Crustacea; it inhibits the Y-organ that secretes the MH). These hormones of the regulatory mechanism are secreted and stored in and released from glands and/or neurohemal organs (e.g., corpora allata and corpora cardiaca in insects and the sinus gland in Crustacea) prior to their action on the MH-producing glands or interactions with the MH.

Note that many of these compounds also regulate morphogenesis of various morphotypes in social insects and some crustaceans (see several pertinent chapters in Part 3).

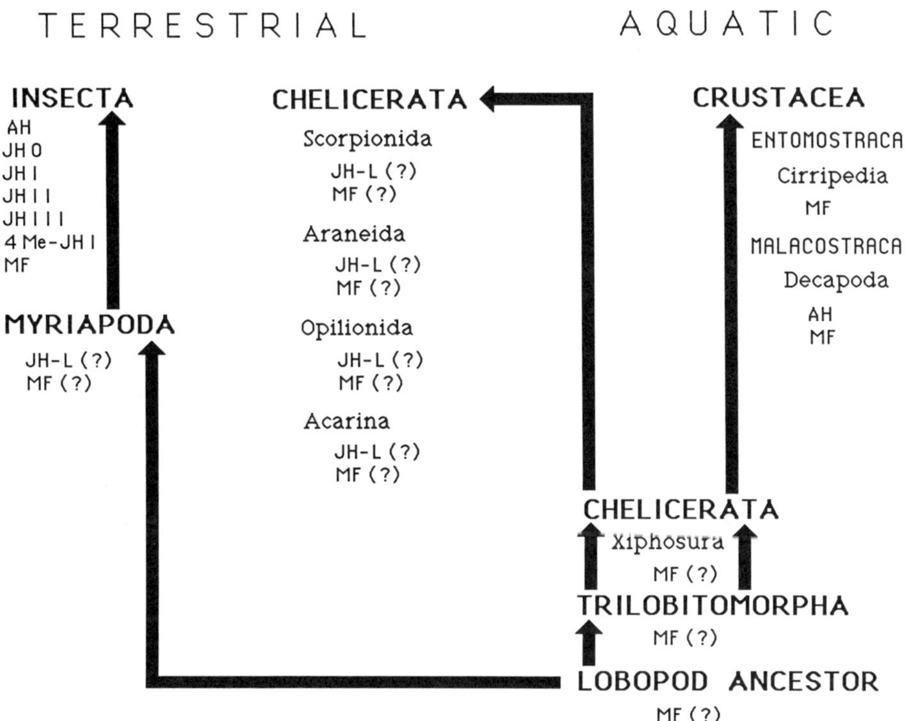

FIGURE 1.2. Monophyletic arrangement of five arthropod taxa and list of juvenile hormone (JH) or JH-like compounds in each taxon: AH = androgenic hormone; JH I, II, II = juvenile hormones I, II, and III; JH-L = juvenile hormone–like substance of unknown identity and unlike MF in structure; JH 0 = Juvenile hormone zero; 4Me–JH I = 4-methyl–JH I (an isomer of JH I); MF = methyl farnesoate; ? = not extracted; putative JH-like hormone-producing gland known or unknown; (?) = unknown, but postulated to be present.

1.3.2.1. AQUATIC CHELICERATA

The inhibitory JH-like hormone of *Limulus* is unknown (Fig. 1.2). Schneiderman and Gilbert (1958) were unable to show JH activity in this animal. According to Jegla (1982), JH-like compounds do not play any role in morphogenesis of *Limulus*.

The rudimentary lateral eye of *Limulus* is comparable to the sinus gland–X-organ system in Crustacea (Gabe, 1966, p. 151) and according to Solomon et al. (1982), the latter system is comparable to the NSC–CC–CA (neurosecretory cells–corpora cardiaca–corpora allata) complex in Insecta. However, this remains to be confirmed in Aquatic Chelicerata.

1.3.2.2. CRUSTACEA

A JH-like hormone, methyl farnesoate (MF), was recently discovered in the spiny crab, *Libinia emarginata* (Laufer et al., 1987a,b). At least in one species, *Balanus balanus* (Cirripedia), belonging to the primitive Entomostraca, and several other decapod species belonging to the more advanced Malacostraca are now known to possess MF (see Chapter 2 by Borst and Laufer herein) (Fig. 1.2). To date, no other JH-like hormone has been reported in Crustacea, and these authors suggest that MF may not be the only JH-like substance in this group. According to Ferezou et al. (1977) and Borst and Laufer (Chapter 2 herein), the androgenic gland in Crustacea produces farnesylacetone, a terpenoid, closely related to MF.

For ultrastructural details of androgenic glands, see Chapter 8 by Payen in Part 2).

The mandibular glands that produce MF in Crustacea have been identified only in some decapods and may be regarded as equivalent to insect CA. Chaigneau (1983) suggested that the molt-inhibiting hormone in Crustacea is produced in the X-organ and stored in the sinus gland. Thus, the sinus gland–X-organ system may be likened to the NSC–CC–CA complex in insects (Matsuda, 1979).

For ultrastructural details of the mandibular gland in Crustacea, see Chapter 10 by Hinsch in Part 2.

1.3.2.3. TERRESTRIAL CHELICERATA (ARACHNIDA)

Based on the information available to date, insect JH-like compounds may be present in some apulmonate Acarina (*Ornithodoros moubata*, *Amblyomma hebraeum*, and *Boophilus microplus*) and some pulmonate Araneida (*Pisaura mirabilis*) (Connat, 1987). And according to El Bakary et al. (1987), JH-like compounds are absent in the pulmonate scorpion *Leiurus quinquestriatus*. However, topical applications of JH or JH analogues suggest that such compounds may be present.

According to Gabe (1955), the Schneider's organs in Araneida have connections with the NSC in the brain and thus may be comparable to the NSC–CC–CA complex in insects (Matsuda, 1979); the same may be said of the sympathetic nervous system (Habibulla, 1970) in Scorpionida (Matsuda, 1979). Apparently, ticks (Acarina) do not have CC or CA (Solomon et al., 1982). According to these authors, the retrocerebral portions of the NS system in ticks resemble the sinus gland–X-organ system of Crustacea and the NSC–CC–CA complex of insects.

1.3.2.4. MYRIAPODA

The inhibitory hormone in Myriapoda, whose identity is unknown, is apparently released from the cerebral glands (neurohemal organs) (Joly, 1962; Matsuda, 1979). In fact, Scheffel (1969) suggested that the inhibitory hormone influences the release of MH in *Lithobius*. Because the cerebral glands are innvervated by the NS axons from cells in the frontal lobes of the protocerebrum, Solomon et al. (1982) regarded them as comparable to the NSC–CC–CA complex in insects.

For the ultrastructural and other details of these glands, see Gupta (1983b) and Chapter 3 by Descamps et al. in Part 2.

1.3.2.5. INSECTA (HEXAPODA)

The inhibitory hormone of the regulatory mechanism in insects is called juvenile hormone (JH), or *neotenin,* and is secreted by the corpora allata (CA), which, like the ecdysial glands, are ectodermal in origin. Six kinds of JHs have been reported in insects (Fig. 1.2): JH 0 (in eggs; see Bergot et al., 1980); 4Me–JH I (an isomer of JH 0) (Bergot et al., 1981); JH I (= Cercropia I or Röller's compound (Röller et al., 1967); JH II (Meyer et al., 1968); JH III (Judy et al., 1973); and MF (methyl farnesoate), which is closely similar to JH III and shows JH activity (Wakabayashi et al., 1971). MF has been reported in *Dysdercus fasciatus* (Feldlaufer et al., 1982), *Nauphoeta cinerea* (Lanzrein et al., 1984), and *Periplaneta americana* (Pratt et al., 1984). Of these, JH I, JH II, and JH III occur only in Lepidoptera; only JH III has been reported in other orders, such as Orthoptera, Dictyoptera, Hemiptera, Isoptera, Coleoptera, Hymenoptera, and Diptera.

Because the androgenic hormone reported in the beetle *Lampyris noctiluca* (Naisse, 1963, 1965, 1966) acts as an inhibitory hormone in female differentiation, it has been included in the list of inhibitory hormones of Insecta in Fig. 1.2.

For the ultrastructural details of the JH-producing glands in Insecta, see Chapter 2 by Cassier in Part 2.

1.3.3. *Morphogenetic Hormones Other Than Ecdysteroids and JHs and MF*

1.3.3.1. ANDROGENIC HORMONES OF CRUSTACEA AND INSECTA

Crustacea

The following account of the androgenic gland and its hormone(s) in Crustacea is based on the information in Chapters 8 and 14 by Payen in Parts 2 and 3, respectively, and should be consulted for details.

Male sex hormone has been found in malacostracan crustaceans and is produced in the androgenic gland (AG), which may be located in the last thoracic segment apposed to the ejaculatory part of the sperm duct or along the testicular utriculus, against the seminal vesicle, and/or against the vas deferens. These glands were first discovered by Charniaux-Cotton (1954) in the amphipod *Orchestia gammarella* and have since been found in many other malacostracans. Note that part of the testis does not function as the androgenic gland in these crustaceans, unlike the androgenic gland in the insect *Lampyris noctiluca* (see the discussion under *Insecta*, below), and the hormone does not control secondary sexual characters of the female.

In *O. gammarella*, several isopods, and the decapod *Macrobrachium rosenbergi*, the androgenic hormone (AH) controls the differentiation of both the genital apparatus, including spermatogenesis, and the secondary sexual characters. It is proteinaceous in nature, and its action is regulated (inhibited) by a hormone produced in the NSC of the protocerebrum. In addition to the AH, the AGs produce two C_{18} isoprenoid ketones: hexahydrofarnesylacetone and farnesylacetone. These compounds do not induce spermatogenesis or secondary male sexual characters.

The activity of farnesylacetone seems to vary with the sexual cycle of these animals and inhibits ovarian protein synthesis.

Insecta

Sex hormones, although common in vertebrates, are unknown in insects, with the sole exception of the glowworm *Lampyris noctiluca* (Doane, 1973). The two sexes in this insect are dimorphic. The winged males are smaller than the females and have larger eyes and poorly developed luminescent organs; the apterous females have smaller eyes and complex luminescent organs. Curiously, the sex determination occurs late in the larval life under the influence of the androgenic hormone (AH), produced in the apical tissue of the testis (Naisse, 1963, 1965, 1966). This hormone controls the differentiation of all primary and secondary sexual characters in the male. The female is initially neuter, or—more precisely—of "neutral sex," and its primary and secondary sexual characters develop in the female direction only if the AH does not prevent this sexual differentiation. Whether the AHs of Crustacea and Insecta are structurally similar is unconfirmed.

1.3.3.2. OTHER MORPHOGENETIC HORMONES

In addition to the AHs in the Crustacea and Insecta, there are other morphogenetic hormones that influence a number of morphogenetic processes and/or pattern formation (see Section 1.1), such as sclerotization, color change, caste determination in social insects and different morphs in locusts and aphids, morphogenesis of eyes in insects, wing polymorphism in Blattaria, regeneration, differentiation of silk glands in *Bombyx*, spermatogenesis, and pupation. All these topics have been discussed in detail in various relevant chapters in Part 3.

1.4. Evolution of Arthropods and Their Morphogenetic Hormones

1.4.1. *Evolution of Arthropods*

For any discussion of the evolution of the morphogenetic hormones, it is necessary to first consider the evolution and phylogeny of the arthropods themselves, because both monophyletic and polyphyletic origins have been suggested. In fact, there is so much differing phylogenetic evidence in the literature that one can make a case for either the mono- or polyphyletic origin of arthropods. Because I subscribe to the monophyletic origin, I will discuss the evolution of the morphogenetic hormones in that context.

It is generally accepted that arthropods originated from a lobopod annelid ancestor (Gupta, 1979a, 1983a, 1986, 1987). The extant arthropods can be divided into aquatic (Aquatic Chelicerata and Crustacea) and terrestrial (Terrestrial Chelicerata, Myriapoda, Insecta, and some decapod and isopod crustaceans) groups. The phylogenetic arrangement of the principal arthropod taxa and their major morphogenetic hormones are shown in Figs. 1.1 and 1.2. Note that according to this scheme the arthropods are monophyletic and the lobopod ancestor gave rise to both Myriapoda and Trilobitomorpha; the latter were the progenitors of both the Crustacea and the Aquatic Chelicerata. Exactly when the Aquatic Chelicerata and the ancestors of Myriapoda colonized land is controversial and shrouded in mystery.

According to Størmer (1977), the arthropod invasion of land occurred during the late Silurian and Devonian times, some 400 million years ago. It is also believed that aquatic arthropods colonized land independently several times. And, finally, I have suggested that both the apterygote and pterygote insects originated from myriapodan ancestors (Gupta, 1979c). All the phylogenetic controversies notwithstanding, land colonization had profound effects in bringing about important anatomic and physiological changes that may have had some bearing on the evolution of the major morphogenetic hormones. I have suggested comparable effects of land colonization on the evolution of neurohemal organs (see Section 1.4.3.2).

1.4.2. *Evolution of Molting Hormones or Ecdysteroids*

Whether the exoskeleton in arthropods developed before or after land colonization is a controversial question. Contrary to Manton's (1973) view that uniramians (Onychophora, Myriapoda, and Insecta or Hexapoda) colonized land in a soft-bodied condition, there is paleontological evidence that some myriapod-like creatures with well-sclerotized exoskeletons and articulated uniramous legs were already present in the Lower and Middle Cambrian (Bergström, 1979). This suggests that some form of a MH must have been present in arthropods some 400–500 million years ago and that it must have originated even earlier during evolution from the lobopod ancestor to an arthropodan body form (i.e., an animal whose exoskeleton must, of necessity, be shed to accomplish any growth). Indeed, it seems that the origins of MH and JH or its counterparts predate those of the arthropods and even annelids; Rogers (1973) has reported the presence of these hormones even in nematodes, as has Spindler (Chapter 5 in Part 3) in Platyhelminthes, Annelida, and Mollusca. Since the actual source of the MH has not been identified in nematodes, and since epidermal molting hormones are present in annelids, it is reasonable to assume that perhaps the epidermal (ectodermal) cells themselves originally produced MH and later evolved into various types of molting glands. Indeed, the prothoracic or ecdysial glands of Insecta, Y-organs of Crustacea, and the ecdysial, head, and collar glands of Myriapoda are ectodermal in origin.

1.4.2.1. PLESIOMORPHIC OR ANCESTRAL ECDYSTEROID

Figure 1.1 shows the various types of ecdysteroids that have been reported in the five major arthropod taxa. It appears that 20E (= β-ecdysone) represents the most primitive (ancestral) or plesiomorphic stage in the evolution of the molting hormones in arthropods and is highly conserved in the entire phylum, since all five taxa possess it. It can be further postulated that trilobites, the progenitors of both the Aquatic Chelicerata and Crustacea, also possessed this molting hormone, and its existence even in the lobopod ancestor of arthropods cannot be ruled out. Indeed, according to Sandor and Mehdi (1979), steroids are present throughout the biosphere and thus are universal biomolecules. These authors go on to suggest that steroids were originally bioregulators (i.e., they exerted their message within the same cells in which they were synthesized or where they found their way through ingestion or diffusion) and that their hormonal role ''came about by target organ specialization and not by the evolution of the steroids themselves'' (p. 3). At any rate, 20E is the only ecdysteroid reported to be present in the primitive entomostracan Crustacea and in the Terrestrial Chelicerata that supposedly evolved from their aquatic cousins, as well as in myriapods which share a common lobopod ancestry with the trilobites (Gupta, 1979b,c). The antiquity of 20E is also indicated by the fact that, of all the known ecdysteroids, it is the only one that induces molting in crustaceans, insects, spiders, and ticks, and thus appears to be an ancestral molecule.

But is 20E the only ecdysteroid of antiquity in arthropods? The answer to this question must await the discovery of any other ecdysteroid that proves to be as ubiquitous as is 20E throughout the entire phylum.

1.4.2.2. APOMORPHIC OR NEWLY EVOLVED ECDYSTEROIDS

Of the several other ecdysteroids found in various arthropods, ponasterone A (PoA) and makisterone A (MaA) have been reported to be present in some terrestrial crustaceans and in insects. Did they evolve concomitantly with a terrestrial mode of life? If this were the case, one would have to postulate their presence in Myriapoda, especially in Symphyla, the progenitors of Insecta (Gupta, 1979c), as well as in Terrestrial Chelicerata. The same might be said of several other ecdysteroids that have been reported from terrestrial crustaceans. Thus, some of these ecdysteroids that appear to be of recent origin and are confined to only certain recent, more advanced arthropod taxa may be tentatively regarded as the apomorphic (derived or newly evolved) stages in the evolution of ecdysteroids.

1.4.3. *Evolution of the Hormones of the Inhibitory System*

Concomitant with the origin of the MH, some sort of endogenous regulatory or control system must have evolved to control the action of MH by triggering its action at precise intervals and inhibiting it at other times. Indeed, inhibitory factors have been reported in a few polychaetes. For example, in *Platynereis dumerilii,* a factor from the cerebral ganglia inhibits oocyte and sperm maturation, and Huenschild (1956) called it "juvenile hormone." The same factor supposedly inhibits epitoky or epigamy. The hormones from the NS cells in the pars intercerebralis (of the protocerebrum) of the brain in insects and those from the NS cells in the X-organ in Crustacea (Bliss, 1951; Bliss and Walsh, 1952; Passano, 1953; Highnam and Hill, 1977) provide this regulatory or control mechanism. Similar inhibitory control mechanisms perhaps occur in other arthropods but are either unknown or poorly understood. Because the brain is involved in the regulatory mechanism, it is useful to consider how it evolved in its present form in arthropods.

1.4.3.1. EVOLUTION OF THE ARTHROPOD BRAIN AND
          ITS NEUROSECRETORY CELLS THAT SECRETE
          HORMONES OF THE INHIBITORY SYSTEM IN
          SOME ARTHROPODS

I have stated elsewhere (Gupta, 1983a) that from whatever is known
about the evolution of the brain in invertebrates (Bullock, 1965), one
can reconstruct the following scenario of the evolution of the "brain":
(1) In some ancestral metazoan, some of the epidermal cells became
differentiated into neurons and neurosecretory cells (NSCs). This
stage is comparable to Ramón y Cajal's (1909–1911, as cited by Bullock, 1965) hypothetical invertebrate. It is likely that at this stage, the
neurons were just sensory. It is unknown whether effector cells (e.g.,
nematocysts of *Hydra*) also evolved at this time. (2) In possibly the
next stage, the sensory neurons and the NSCs became basiepithelial,
as is found in some present-day coelenterates (e.g., *Hydra*). Because
motor neurons are not found in coelenterates (Bullock, 1965), it is conceivable that during this stage also only sensory neurons were
present. This stage is comparable to Raymón y Cajal's (1909–1911)
second stage and, in certain features, to the present-day coelenterate
medusae (Bullock, 1965). (3) During the subsequent third stage, some
of the dispersed sensory neurons became differentiated into motor
neurons and eventually became grouped into ganglia (comparable to
Raymón y Cajal's third stage and to the present-day platyhelminths or
their ancestors). Apparently, the NS cells followed the same course
and became aggregated and/or associated with the ganglia. (4) Finally,
during the process of cephalization, the migration of the ganglia,
along with the associated NS cells, into the cephalic region established
the primitive "brain." It is in the present-day primitive flatworms
(Turbellaria) that a distinct "brain" is encountered for the first time in
evolution. The annelids have a well-defined brain.

   Tombes (1970) has stated that the nervous system of the primitive
arthropod is comparable to that of a polychaete annelid. Based primarily on the theories of Holmgren (1916) and Hanström (1927, 1928,
1930), Snodgrass (1935) hypothesized that the arthropod body, and
concomitantly the brain, evolved in several stages from the ancestral
worm-like annelid form. He considered the unsegmented preoral
prostomium (= acron) of the annelids as the primitive head, or archicephalon, with its ganglion, the archicerebrum. In the ancestral lobopod ancestor, the mouth was located between the first segment (the
unsegmented preoral prostomium) and the first postoral segment; the

prostomium secondarily divided into the preantennal or precheliceral and the antennal segments, the neuropil masses of the two segments constituting the proto- and deutocerebrum, respectively. Then the first postoral segment (also known as the intercalary, second antennal, postantennal, or premandibular segment) became fused with the pre- and postantennal segments, producing the protocephalon and its neuropil mass forming the tritocerebrum of the brain.

The protocephalon, according to Snodgrass (1935), represents the first stage in the development of the definitive arthropodan head or brain, forming the procephalic part of the insect head; in his view, the tritocerebrum represents the first paired ganglia of the primitive ventral nerve cord. Together, the proto-, deuto-, and the tritocerebrum constitute the supraesophageal ganglion. Note, however, that in insects, according to Chaudonneret (1987), a super numerary neuropil mass exists between the tritocerebrum and the mandibular neuropil masses, and he regards it as the fourth constituent (tetrocerebrum) of the supraesophageal brain mass.

The neuropils of the next three postoral segments (mandibular, first maxillary, and the second maxillary, or labial) fused to form the subesophageal ganglion. The supra- and the subesophageal ganglia together constitute the *brain*.

The NSCs are ectodermal in origin and are present not only in the proto-, deuto-, and tritocerebrum of the present-day arthropods but are distributed throughout the nervous system. According to Gabe (1966), the protocerebral NS cells in all arthropods are similar histochemically and ultrastructurally.

Earlier in this section I stated that the NSCs perhaps had the same fate during cephalization as the primitive ganglia, which migrated to the cephalic region and formed the primitive brain. Indeed, NSCs, whether within the brain or associated with it, have been reported in platyhelminths (Scharrer and Scharrer, 1954; Ude, 1962, 1964; Battaglini, 1964; Lender, 1964) as well as in Polychaeta, Oligochaeta (Gabe, 1966; Tombes, 1970; Highnam and Hill, 1977), and Hirudinea (Gabe, 1966; Tombes, 1970).

Clark (1956a,b), on the basis of his studies of the polychaetes *Nephtys* and *Nereis,* suggested that the NSCs in these annelids constitute the most primitive part of the brain and retain, to a certain degree, the secretory characteristics of the original epidermal (ectodermal) cells from which they have been derived.

The question as to whether the NSCs were originally true neurons (Hanström, 1954; Gersch, 1957, 1958, 1960) and later became glandular or vice versa (Clark, 1956a,b) is controversial and has been discussed by Novak (1964) and Highnam and Hill (1977).

There is no evidence as to what led to the grouping and location of NSCs in various parts of the nervous system in various arthropods, although it is conceivable that once the regulatory mechanism (the brain or activation hormone) became established, the need for efficient transport, storage, and release could have necessitated the formation of the neurohemal organs that act as storage and release sites for the hormones of the regulatory mechanisms in many arthropods.

1.4.3.2. EVOLUTION OF THE NEUROHEMAL ORGANS THAT SECRETE THE INHIBITORY HORMONES IN SOME ARTHROPODS

Neurohemal organs (NHOs) serve as storage and release sites for the hormones of the inhibitory mechanism in some arthropod groups, and I have discussed their evolution in detail elsewhere (Gupta, 1983a). It is unlikely that the platyhelminths or the nematodes should possess any type of NHO, because the evolution of the NHO seems to be linked with that of the circulatory system, which appears for the first time in the phylum Nemertinea, or ribbon worms. The system in these animals consists of three longitudinal blood vessels (Storer and Usinger, 1957). According to Bianchi (1969), axons from the NS cells of the cerebral ganglion in the nemertine *Cerebratulus marginatus,* running singly or in groups, terminate in close contact with the lateral blood vessel. Thus, this association of the NS axons with the lateral blood vessels in nemertines should be considered the very beginning of a NHO of the most primitive kind. The so-called glandular cerebral organ of nemertines (Scharrer, 1941; Lechenault, 1962; Meglitsch, 1967), although fused with the cerebral ganglion in higher forms and in proximity to the circulatory system, apparently is a sensory organ (Ling, 1970; Bianchi et al., 1972) and not a NHO. Its role in osmoregulation (Lechenault, 1965), however, supports its neurohemal nature.

It is easy to imagine that once a simple cephalic endocrine system of some sort became established, the need for an efficient transport of the NS material from the NS cells in the brain, its storage, and its release could have necessitated further grouping and consolidation of the axon terminals of the NSCs in discrete organs, the *neurohemal organs,* associated with the circulatory system. Such aggregation supposedly produced the so-called NS organs (Olsson, 1963).

If the beginnings of a simple NHO are present in Nemertinea, it should not be surprising at all that the annelids possess them too. In fact, such a NHO is present in the polychaete annelids (Clark, 1959; Knowles, 1963; Tombes, 1970). According to Clark (1959), in *Nephtys*, the axon of at least the B-type of NS cells in the brain is closely associated with the dorsal vessel, comparable to the axon terminals adjacent to the sinusoids of the circulatory system in *Limulus* (Fahrenbach, 1973). Golding and Whittle (1978) recognized four types of NS elements associated with the supraesophageal ganglion of the polychaetes. Of these, the "neurohemal complexes," formed by the axons of the NS cells, and the so-called secretory end foot (SEF) system, whose axon terminals contribute to the neurohemal areas, clearly suggest that some sort of NHO had already evolved in the annelids prior to the evolution of the arthropod. For detailed information on the NHOs in arthropods, the reader is referred to various reviews cited in Gupta (1983b).

I have suggested elsewhere (Gupta, 1983a) that an efficient circulatory system is crucial for the transport of the regulatory hormones from the NHOs and thus a prerequisite for the survival of the arthropods. Aquatic arthropods possess a comparatively elaborate circulatory system (as compared with that in the terrestrial arthropods) and thus have a more effective internal transport system than their terrestrial cousins, in which the tubular dorsal vessel is the only blood vessel present and the hemolymph flow is relatively slow.

It is likely that the simplification of the circulatory system in terrestrial arthropods may have been caused by the evolution of tracheae for the transport of oxygen to the tissues, which in aquatic arthropods is accomplished by the hemocyanin in the blood (Gupta, 1983a). With the evolution of tracheae in terrestrial arthropods, hemocyanin and its transport through the elaborate circulatory system were no longer needed. Consequently, the former disappeared and the latter became simpler and less efficient through the loss of arteries, veins, and diaphragms, because rapid hemolymph circulation was no longer an important factor for survival.

The need for transport of the NHO hormones, the regulatory hormones, as well as those involved in such important functions as cardiac stimulation, color regulation for defense, and osmoregulation to avoid desiccation, still remained crucial. But with the less efficient circulatory system in the terrestrial arthropods, rapid transport of the most crucial hormones could not be effectively carried out. This factor was perhaps one of the major forces in the evolution of the more dispersed system of NHOs, located in the thoracoabdominal region, to compensate for the lack of effective hemolymph circulation. It seems that a centralized cephalic NHO was no longer sufficient to cope with the demands of a terrestrial mode of life and prompted the evolution of a more dispersed system of neurohemal–endocrine organs (NH–EOs), the culmination of this evolutionary trend being the appearance of the NH–EOs that are close to or in contact with the target organs [e.g., the "distal" perisympathetic organs (PSOs) in *Carausius, Roscius,* and *Glossina* (Baudry-Partiaoglou, 1983)] and the synaptoid junctions (Raabe, 1983) between the NS axons and the target organs in some insects. The distal PSOs are thin networks of NHOs located far from the ganglia and in proximity to their target organs. The synaptoid junctions clearly represent a situation where transport of a hormone is no longer dependent on the circulatory system.

The survival value of a decentralized, dispersed system of NH–EOs becomes readily apparent when one considers that in a terrestrial environment, where the animal may be forced to conserve water or seek defense against an enemy at a moment's notice, the prompt availability, ideally locally, of certain hormones (e.g., diuretic and antidiuretic hormones in case of a sudden need for water conservation, or pigment-effector hormones needed for physiological color change for defense) would be of paramount importance. Furthermore, it is reasonable to assume that the effectiveness of NH–EOs would be considerably enhanced if the quantitative and qualitative output of these organs were augmented by the evolution of additional secretory cells, multiplicity of these organs, and their dispersed, segmental arrangement.

This need for increased effectiveness could have been the cause of the evolution of the intrinsic secretory cells very early in the evolution of the terrestrial arthropods and later of the proliferation of the dispersed NH–EOs. The evolution of the lateral cephalic nerve plexus in some terrestrial Crustacea and of the PSOs, the distal organs, and the synaptoid junctions in Insecta represents the culmination of that process. The last specialization (synaptoid junctions) would not only obviate the necessity for transport of hormones but would even prevent their dilution.

1.4.3.3. EVOLUTION OF THE JH OR ITS COUNTERPARTS

Plesiomorphic JH

Unfortunately, only in Insecta and Crustacea, do we have information on the existence of JH or JH-like compounds (Fig. 1.2). In Aquatic and Terrestrial Chelicerata and Myriapoda, the existence of such compounds have been suspected and in some cases experimentally demonstrated, but they have not yet been extracted and characterized. Figure 1.2 lists the JH or JH-like compounds reported so far in the five major arthropod taxa.

On the basis of the information available to date, methyl farnesoate (MF), although considered a ''prohormone'' that is epoxidated to JH III (see Chapter 12 by Baker herein), appears to be the most primitive (ancestral) or plesiomorphic hormone of the inhibitory mechanisms in arthropods. Its existence both in Crustacea and Insecta, the most advanced among arthropods, suggests that it has been conserved throughout the Arthropoda. On this basis, one may postulate that it probably occurs in both Aquatic and Terrestrial Chelicerata and Myriapoda. Among Crustacea, it has been reported in at least one species, *Balanus balanus* (Cirripedia), belonging to the primitive Entomostraca, and in several terrestrial decapods belonging to the more advanced Malacostraca. Among insects, it has been reported in *Dysdercus fasciatus* (Feldlaufer et al., 1982), *Periplaneta americana* (Pratt et al., 1984), and *Nauphoeta cinerea* (Lanzrein et al., 1984).

Apomorphic or Newly Evolved JHs

All insect JHs except MF seem to have developed relatively recently in the course of evolution, probably derived from MF or similar precursors. It may not be just a coincidence that JH III, the most commonly occurring JH in insects, is chemically similar to MF (a nonepoxidated form of JH III) (Borst and Laufer, Chapter 2 herein) and that radiolabeled JH III was produced by nonbiological oxidation of MF (Laufer et al., 1987a).

1.5. Summary

Organisms and their organs have a defined, genetically determined form or final state. The developmental process that gradually unfolds this form or state and eventually completes it is called *morphogenesis*. It involves cellular differentiation, histogenesis, organogenesis, and formation of organ systems or structures. Growth and differentiation in arthropods is accomplished through moltings and involves mitosis, deposition of new cuticle, protein synthesis for new cells, synthesis of specific enzymes, and new tissue formation by differentiation.

Apolysis and molting in arthropods are brought about by molting hormones (MHs), or ecdysteroids, which are some of the principal morphogenetic hormones found in these animals.

Because myriapod-like creatures with well-sclerotized exoskeletons and articulated uniramous legs were already present in the Lower and Middle Cambrian, some form of MH not only must have been present some 400–500 million years ago in arthropods, but must have originated even earlier during evolution of arthropods from lobopod annelid ancestors. Indeed, it seems that the origins of MH (as well as of JH-like compounds) must have predated those of the arthropods themselves and even annelids. Furthermore, because epidermal MHs are present in annelids, it is reasonable to assume that perhaps the epidermal (ectodermal) cells themselves produced MH, and later evolved into various types of molting glands. Indeed, the prothoracic or ecdysial glands of Insecta, Y-organs of Crustacea, and the ecdysial, head, and collar glands of Myriapoda are ectodermal in origin.

Several major ecdysteroids have been reported from arthropods; the 20-hydroxyecdysone (β-ecdysone, or 20E), often formed by hydroxylation of its precursor, α-ecdysone (E), is the most commonly occurring ecdysteroid in these animals. It is the major ecdysteroid in Aquatic and Terrestrial Chelicerata, Crustacea, and pterygote Insecta. Crustaceans also have 25-deoxyecdysone (25De), 2-deoxy-20-hydroxyecdysone (2De-20E), makisterone A (MaA), inokosterone A (InA), and ponasterone A (PoA); PoA is also found in the apterygote orders Collembola and Thysanura. β-ecdysone represents the most primitive (ancestral) or plesiomorphic stage in the evolution of the MHs in the arthropods and is highly conserved in the entire phylum, since all five taxa possess it. It can be further postulated that trilobites, the progenitors of both the Aquatic Chelicerata and Crustacea, also possessed this MH, and its existence in the lobopod ancestor cannot be ruled out. Indeed steroids are present throughout the biosphere and thus are universal biomolecules, and probably originally served as bioregulators. The evolutionary antiquity of 20E is also indicated by the fact that it is nonspecific and induces molting in individuals of several arthropod taxa.

Other ecdysteroids, such as PoA and MaA, which are confined to certain terrestrial taxa, may have evolved concomitantly with terrestrial mode of life and may be tentatively regarded as the apomorphic (derived or newly evolved) stages in the evolution of ecdysteroids.

Unfortunately, the ecdysteroid-producing glands are largely unknown, except in insects and some crustaceans and myriapods. In addition to the prothoracic or ecdysial glands of insecta, the Y-organs of crustaceans, and the ecdysial, head, and collar glands of myriapods, oenocytes of insects and opilionids and the ovaries of some insects and crabs also produce ecdysteroids.

From what we know of their mode of action in insects and other arthropods, the MHs or ecdysteroids do not act alone but are influenced by another morphogenetic hormone, *neotenin* [or juvenile (inhibitory or *status quo*) hormone (JH)] in insects or its counterparts in other arthropods. Unfortunately, JHs or JH-like compounds (e.g., methyl farnesoate, MF) have been extracted from only Insecta and Crustacea. In Aquatic and Terrestrial Chelicerata and Myriapoda, the existence of such compounds have been suspected and in some cases experimentally demonstrated, but they have not yet been extracted and characterized. Of the several known JHs and MF, the latter appears to be the most primitive (ancestral) or plesiomorphic hormone of the inhibitory mechanisms in arthropods. Its existence both in Crustacea and Insecta, the most advanced among arthropods, suggests that it has been conserved throughout the phylum. On this basis, it can be postulated that it probably also occurs in both Aquatic and Terrestrial Chelicerata as well as in Myriapoda. Also on this basis, all insect JHs except MF seem to be recent evolutionary developments, probably derived from MF or similar precursors.

Acknowledgments

New Jersey Agricultural Experiment Station Publication No. F-08125-01-88, supported by State Funds, U.S. Hatch Funds, and a grant from the Rutgers Research Council.

References

Baccetti, B. 1979. Ultrastructure of sperm and its bearing on arthropod phylogeny. Pp. 609–644. *in* A. P. Gupta (ed.), *Arthropod Phylogeny.* Van Nostrand Reinhold, New York.

Battaglini, P. 1964. Attivita di neurosecrezione nel sistema nervosa di un Turbellaria rabdocelo: *Mesostoma linqua.* Experientia (Basel) 20: 150–151.

Baudry-Partiaoglou, N. 1983. Ultrastructure of perisympathetic organs in insects. Pp. 513–551 *in* A. P. Gupta (ed.) *Neurohemal Organs of Arthropods.* C. Thomas, Springfield.

Bebbington, P. M., E. D. Morgan, and C. F. Poole. 1977. Detection and identification of moulting hormones (ecdysones) in the barnacle *Balanus balanoides.* Comp. Biochem. Physiol. 56B: 77–79.

Behrens, W. and D. Bückmann. 1983. Ecdysteroids in the pycnogonid *Pycnogonum litorale* (Strom) (Arthropoda, Pantopoda). Gen. Comp. Endocrinol. 51: 8–14.

Bergot, B. J., G. C. Jamieson, M. A. Ratcliff, and D. A. Schooley. 1980. JH Zero: new naturally occurring insect juvenile hormone from developing embryos of the tobacco hornworm. Science (Wash., DC) 210: 336–338.

Bergot, B. J., F. C. Baker, D. C. Cerf, G. C. Jamieson, and D. A. Schooley. 1981. Qualitative and quantitative aspects of juvenile hormone titers in developing embryos of several insect species: discovery of a new JH-like substance extracted from eggs of *Manduca sexta.* Pp. 33–45 *in* G. E. Pratt and G. T. Brooks (eds.), *Juvenile Hormone Biochemistry.* Elsevier, Amsterdam and New York.

Bergström, J. 1979. Morphology of fossil arthropods as a guide to phylogenetic relationship. Pp. 3–56 *in* A. P. Gupta (ed.), *Arthropod Phylogeny.* Van Nostrand Reinhold, New York.

Bianchi, S. 1969. On the neurosecretory system of *Cerebratulus marginatus* (Heteronemertini). Gen. Comp. Endocrinol. 12: 541–548.

Bianchi, S., V. Esposito, and N. Granata. 1972. On the cerebral organs of *Cerebratulus marginatus* (Heteronemertini). Gen. Comp. Endocrinol. 18: 5–11.

Blanchet, M. F., P. Porcheron, and F. Dray. 1976. Etude des variations due taux des ecdysones au cours du cycle d'intermue chez le mâle d'*Orchestia gammarella* Pallas (Crustace Amphipode). C. R. Acad. Sci. Paris 283: 651–654.

Bliss, D. E. 1951. Metabolic effects of sinus gland or eyestalk removal in the land crab, *Gecarcinus lateralis.* Anat. Rec. 111: 502.

Bliss, D. E. and J. H. Walsh. 1952. The neurosecretory system of brachyuran Crustacea. Biol. Bull. (Woods Hole) 103: 157–169.

Bonaric, J. C. and M. De Reggi. 1977. Changes in ecdysone levels in the spider *Pisaura mirabilis* nymph (Araneae, Pisauridae). Experientia (Basel) 33: 1664–1665.

Bückmann, D., G. Starnecker, K. H. Tomaschko, E. Wilhelm, R. Lafont, and J. P. Gerault, 1986. Isolation and identification of major ecdysteroids from the pycnogonid *Pycnogonum litorale* (Strom) (Arthropoda, Pantopoda). J. Comp. Physiol. 156B: 759–765.

Bullock, T. H. 1965. Introduction. Pp. 4–34. *in* T. H. Bullock and G. A. Horridge (eds.), *Structure and Function in the Nervous System of Invertebrates*, Vol. 1. Freeman, San Francisco.

Chaigneau, J. 1983. Neurohemal organs in Crustacea. Pp. 53–89. *in* A. P. Gupta (ed.), *Neurohemal Organs of Arthropods*. Thomas, Springfield, Illinois.

Charniaux-Cotton, H. 1954. Découverte chez un Crustacé Amphipode (*Orchestia gammarella*) d'une glande endocrine responsable de la différenciation des caractères sexuels primaires et secondaires mâles. C. R. Acad. Sci. Paris 239: 780–782.

Chaudonneret, J. 1987. Evolution of the insect brain with special reference to the so-called tritocerebrum. Pp. 3–26 *in* A. P. Gupta (ed.), *Arthropod Brain*. Wiley, New York.

Clark, R. B. 1956a. On the origin of neurosecretory cells. Ann. Sci. Nat. Zool. Biol. Anim. 18: 199–207.

Clark, R. B. 1956b. On the transformation of neurosecretory cells into ordinary nerve cells. K. Fysiogr Sällsk. Lund Förh. 26: 1–8.

Clark, R. B. 1959. The neurosecretory system of the supraoesophageal ganglion of *Nephtys*. Zool. Jahrb. Physiol. 68: 395–424.

Connat, J. L. 1987. Aspects endocrinologiques de la physiologie du développement et de la reproduction chez les tiques. Doctoral thesis, University of Dijon, France.

Cox, B. L. 1960. Hormonal involvement in the molting process in the soft-tick *Ornithodoros turicata* Duges. Ph.D. dissertation, University of Oklahoma, Oklahoma City.

Delbecque, J. P., P. A. Diehl, and J. D. O'Connor. 1978. Presence of ecdysone and ecdysteroid in the tick, *Amblyomma hebraeum* Koch. Experientia (Basel) 34: 1377–1380.

Doane, W. W. 1973. Role of hormones in insect development. Pp. 291–497 *in* S. J. Counce and C. H. Waddington (eds.), *Development Systems: Insects*, Vol. 2. Academic Press, Orlando, Florida.

El Bakary, Z., P. Porcheron, M. Moriniere, and S. Fuzeau-Braesch. 1987. Mise en évidence et dosage d'ecdystéroïdes chez le Scorpion *Leiurus quinqestraiatus*. C. R. Acad. Sci. Paris 304: 453–456.

Fahrenbach, W. H. 1973. The morphology of the *Limulus* visual system. V. Protocerebral neurosecretion and ocular innervation. Z. Zellforsch. 144: 153–166.

Faux, A., D. H. S. Horn, E. J. Middleton, H. M. Fales, and M. E. Lowe. 1969. Moulting hormones of a crab during ecdysis. Chem. Commun. 4: 175–176.

Feldlaufer, M. F., W. S. Bowers, D. M. Soderlund and P. H. Evans. 1982. Biosynthesis of the sesquiterpenoid skeleton of juvenile hormone 3 by *Dysdercus fasciatus* corpora allata *in vitro*. J. Exp. Zool. 223: 295–298.

Ferezou, J. P., J. Berraur-Bonnenfant, J. J. Meusy, M. Barbier, M. Suchy, and H. K. Wipf. 1977. 6,10,14-trimethylpentadecan-2-one and 6,10,14-trimethyl-5-*trans*, 9-*trans*, 13-pentadecatrien-2-one from the androgenic glands of the male crab *Carcinus maenas*. Experientia (Basel) 33: 290.

Gabe, M. 1955. Données histologiques sur la neurosécrétion chez les arachnides. Arch. Anat. Microsc. Morphol. Exp. 44: 351–383.

Gabe, M. 1966. *Neurosecretion.* Pergamon Press, Oxford and Elmsford, New York.

Galbraith, M. N., D. H. S. Horn, E. J. Middleton, and R. J. Hackney. 1968. Structure of deoxycrustecdysone, a second crustacean moulting hormone. Chem. Commun. 3: 83–85.

Gersch, M. 1957. Wesen under Wirkungsweise von Neurohormonen im Tierreich. Naturwissenschaften 20: 525–532.

Gersch, M. 1958. Neurohormone bei wirbellosen Tieren. Verh. Dtsch. Zool. Ges. Zool. Anz. (Suppl.) 20: 40–76.

Gersch, M. 1960. Neurosekretion und Neurohormone bei wirbellosen Tieren. Symp. Biol. Hung. 1: 153–180.

Golding, D. W. and A. C. Whittle. 1978. Neurosecretion and related phenomena in Annelida. Int. Rev. Cytol. Suppl. 5: 189 302.

Gupta, A. P. 1978. Origin of Apterygota. Detta Fisiocrit. Accad. Sci. Siena (Italy) pp. 179–181.

Gupta, A. P. (Ed.) 1979a. *Arthropod Phylogeny.* Van Nostrand Reinhold, New York.

Gupta, A. P. 1979b. Arthropod hemocytes and phylogeny. Pp. 669–735 *in* A. P. Gupta (ed.), *Arthropod Phylogeny.* Van Nostrand Reinhold, New York.

Gupta, A. P. 1979c. Origin and affinities of Myriapoda. Pp. 373–390 *in* M. Camatini (ed.), *Myriapod Biology.* Academic Press, Orlando, Florida.

Gupta, A. P. 1983a. Neurohemal and neurohemal–endocrine organs and their evolution in arthropods. Pp. 17–50 *in* A. P. Gupta (ed.), *Neurohemal Organs of Arthropods.* Thomas, Springfield, Illinois.

Gupta, A. P. (ed.) 1983b. *Neurohemal Organs of Arthropods.* Thomas, Springfield, Illinois.

Gupta, A. P. 1986. Arthropod immunocytes: identification, structure, functions, and analogies to the functions of vertebrate B- and T-lymphocytes. Pp. 3–59 *in* A. P. Gupta (ed.), *Hemocytic and Humoral Immunity in Arthropods.* Wiley, New York.

Gupta, A. P. 1987. Evolutionary trends in the central and mushroom bodies of the arthropod brain: a dilemma. Pp. 27–42 *in* A. P. Gupta (ed.), *Arthropod Brain.* Wiley, New York.

Habibulla, M. 1970. Neurosecretion in the scorpion, *Heterometrus swammerdami.* J. Morphol. 131: 1–18.

Hampshire, F. and D. H. S. Horn. 1966. Structure of crustecdysone, a crustacean moulting hormone. Chem. Commun. 1: 37–38.

Hanström, B. 1927. Das zentrale und periphere Nervensystem des Kopflappens einiger Polychaeten. Z. Morphol. Oekol. Tiere 7: 543–596.

Hanström, B. 1928. Vergleichende Anatomie des Nervensystems der wirbellosen Tiere unter Berücksichtigung seiner Funktion. Springer-Verlag, Berlin.

Hanström, B. 1930. Über das Gehirn von *Termopsis nevadensis* u. *Phyllium pulchrifolium* nebst Beitragen zur Phylogenie der Corpora Pedunculata der Arthropoden. Z. Morphol. Oekol. Tiere 19: 732–773.

Hanström, B. 1954. On the transformation of neurosecretory cells into ordinary nerve cells. K. Fysiogr. Sällsk. Lund Förh. 26: 1–8.

Hauenschild, C. 1956. Hormonale Hemmung der Geschlechtsreife und Metamorphose bei dem Polychaeten *Platynereis dumerilii*. Z. Naturforsch. 11B: 125–132.

Hazarika, L. K. and A. P. Gupta. 1986. Structure, innervation, persistence, and effects of juvenile hormone on the prothoracic glands in adult *Blattella germanica* (L.) (Dictyoptera: Blattellidae). Zool. Sci. 3: 145–150.

Highnam, K. C. and L. Hill. 1977. *The Comparative Endocrinology of the Invertebrates.* University Park Press, Baltimore.

Holmgren, N. 1916. Zur vergleichenden Anatomie des Gehirns von Polychaeten, Onychophoren, Xiphosuren, Arachniden, Crustaceen, Myriapoden, und Insekten. K. Sven. Vetenskapsakad. Handl. 56(1): 1–303.

Horn, D. H. S., J. S. Wilkie, and J. A. Thompson. 1974. Isolation of β-ecdysone (20-hydroxyecdysone) fom the parasitic nematode *Ascaris lumbricoides.* Experientia (Basel) 30: 1109–1110.

Jegla, T. C. 1982. A review of the molting physiology of the trilobite larva of *Limulus.* Pp. 83–111 *in* J. Bonaventura, C. Bonaventura, and S. Tesh (eds.), *Physiology and Biology of Horseshoe Crabs: Studies on Normal and Environmentally Stressed Animals.* Liss, New York.

Jegla, T. C. and J. D. Costlow. 1979. Ecdysteroids in *Limulus* larvae. Experientia (Basel) 35: 554–555.

Jegla, T. C., J. D. Costlow, and J. Alspaugh. 1972. Effects of ecdysones and some synthetic analogs on horseshoe crab larvae. Gen. Comp. Endocrinol. 19: 159–166.

Joly, R. 1962. Les glandes cérébrales, organes inhibiteurs de la mue chez les Myriapodes. C. R. Acad. Sci. Paris 254D: 1679–1681.

Joly, R., P. Porcheron, and F. Dray. 1979. Etude des variations du taux d'ecdystéroïdes au cours du cycle de mue dans l'hémolymphe de *Lithobius forficatus* (Myriapode Chilopode). C. R. Acad. Sci. Paris 288: 243–246.

Juberthie-Jupeau, L., A. Strambi, M. De Reggi, and C. Juberthie. 1979. The presence of ecdysteroids and variations of their levels during the first adult stage of the myriapod *Hanseniella ivorensis* Juberthie-Jupeau and Kehe (Symphyla). Experientia (Basel) 35: 1406–1408.

Judy, K. J., D. A. Schooley, L. L. Dunham, M. S. Hall, B. J. Bergot, and J. B. Siddall. 1973. Isolation, structure, and absolute configuration of a new natural insect juvenile hormone from *Manduca sexta.* Proc. Nat. Acad. Sci. USA 70: 1509–1513.

Kelly, T. J., J. R. Aldrich, C. W. Woods and A. J. Borkovec. 1984. Makisterone-A: its distribution and physiological role as the molting hormone of true bugs. Experientia (Basel) 40: 996–997.

Knowles, F. G. W. 1963. The structure of neurosecretory systems in invertebrates. Pp. 47–62 *in* U. S. von Euler and H. Heller (eds.), *Comparative Endocrinology*, Vol. 2. Academic Press, Orlando, Florida.

Kraus, O. 1976. Zur phylogenetischen Stellung und Evolution der Chelicerata. Entomol. Ger. 3: 1–12.

Lachaise, F. and J. A. Hoffmann. 1982. Ecdysteroids and embryonic development in the shore crab, *Carcinus maenas*. Hoppe-Seyler's Z. Physiol. Chem. 362: 521–529.

Lachaise, F. and R. Lafont. 1984. Ecdysteroid metabolism in a crab: *Carcinus maenas* L. Steroids 43: 243–260.

Lachaise, F., M. Goudeau, C. Hetru, C. Kappler, and J. A. Hoffmann. 1981. Ecdysteroids and ovarian development in the shore crab, *Carcinus maenas*. Hoppe-Seyler's Z. Physiol. Chem. 362: 521–529.

Lanzrein, B., H. Imboden, C. Burgin, E. Brunning, and H. Gfeller. 1984. On titers, origin and functions of juvenile hormone III, methyl farnesoate, and ecdysteroids in embryonic development of the ovoviviparous cockroach *Nauphoeta cinerea*. Pp. 454–464 *in* J. Hoffmann and M. Porchet (eds.), *Biosynthesis, Metabolism and Mode of Action of Invertebrate Hormones.* Springer-Verlag, Berlin and New York.

Laufer, H., D. Borst, F. C. Baker, C. Carrasco, M. Sinkus, C. C. Reuter, L. W. Tsai, and D. A. Schooley. 1987a. Identification of a juvenile hormone-like compound in a crustacean. Science (Wash., DC) 235: 202–205.

Laufer, H., M. Landau, D. Borst, and E. Homola. 1987b. The synthesis and regulation of methyl farnesoate, a new juvenile hormone for crustacean reproduction. Pp. 135–143 *in* M. Porchet, J. C. Andries, and D. Dhainault (eds.), *Advances in Invertebrate Reproduction*, Vol. 4. Elsevier, Amsterdam and New York.

Lechenault, H. 1962. Sur l'existence de cellules neurosécrétices dans les ganglions cérébroïdes des Lineidae (Hétéronémertes). C. R. Acad. Sci. Paris 255D: 194–196.

Lecheneault, H. 1965. Neurosécrétion et osmorégulation chez les Lineidae (Heteronemertes). C. R. Acad. Sci. Paris 261D: 4868–4871.

Lender, T. 1964. Mise en évidence et rôle de la neurosécrétion chez les Planaires d'eau douce (Turbellaires, Triclades). Ann. Endocrinol. 25(Suppl.): 61–65.

Leubert, F., H. Eibisch, and H. Scheffel. 1979. Isolierung und Chrakterisierung des Hautungshormons von *Lithobius forficatus* (L.) (Chilopoda). Zool. Jahrb. Physiol. 83: 334–339.

Ling, E. A. 1970. Further investigations on the structure and function of cephalic organs of a nemertine *Lineus ruber*. Tissue Cell 2: 569–588.

Locke, M. 1969. The ultrastructure of the oenocytes in the molt/intermolt cycle of an insect (*Calpodes ethlius* Stoll). Tissue Cell 1: 103–154.

Manton, S. M. 1973. Arthropod phylogeny: a modern synthesis. J. Zool. (Lond.) 171: 111–130.

Matsuda, R. 1979. Abnormal metamorphosis and arthropod evolution. Pp. 137–256 *in* A. P. Gupta (ed.), *Arthropod Phylogeny.* Van Nostrand Reinhold, New York.

Matsuda, R. 1986. *Animal Evolution in Changing Environments.* Wiley, New York.

McCarthy, J. F. 1979. Ponosterone A: a new ecdysteroid from embryos and serum of brachyuran crustaceans. Steroids 34: 799–806.

Meewis, A. S. and J. Naisse 1960. Pheromones neurosécrétrices et glandes endocrines chez les Opilions. C. R. Acad. Sci. Paris 245D: 858–860.

Meglitsch, P. A. 1967. *Invertebrate Zoology.* Oxford University Press, London and New York.

Meyer, A. S., H. A. Schneiderman, E. Hanzmann, and J. H. Ko. 1968. The two juvenile hormones from the Cecropia silk moth. Proc. Nat. Acad. Sci. USA 60: 853–860.

Naisse, J. 1963. Cellule neurosécrétrices chez *Lampyris noctiluca* L. C. R. Acad. Sci. Paris 256: 3895–3897.

Naisse, J. 1965. Contrôl endocrinien de la différenciation sexuelle chez les insectes. Arch Anat. Microsc. Morphol. Exp. 54: 417–428.

Naisse, J. 1966. Contrôl endocrinien de la différenciation sexuelle chez l'insecte *Lampyris noctiluca.* I. Rôle androgene des testicule. Arch. Biol. 77: 139–201.

Novak, V. J. A. 1964. The phylogenetic origin of neurosecretion. Gen. Comp. Endocrinol. 4: 696–703.

Novak, V. J. A. 1975. *Insect Hormones.* Wiley, New York.

Olsson, R. 1963. The evolution of neurosecretory cells and systems. Proc. 16th Int. Congr. Zool. 3: 38–43.

Passano, L. M. 1953. Neurosecretory control of molting in crabs by X-organ-sinus gland complex. Physiol. Comp. Oecol. 3: 155–189.

Paulus, H. F. 1979. Eye structurew and the monophyly of the Arthropoda. Pp. 299–383 *in* A. P. Gupta (ed.), *Arthropod Phylogeny.* Van Nostrand Reinhold, New York.

Pratt, G. E., R. C. Jennings, and R. J. Weaver. 1984. The influence of a P_{450} inhibitor on methyl farnesoate levels in cockroach corpora allata, *in vitro.* Insect Biochem. 14: 609–614.

Raabe, M. 1983. Role of perisympathetic organs and other neurohemal organs in the neurosecretory system of insects. Pp. 552–580 *in* A. P. Gupta (ed.), *Neurohemal Organs of Arthropods.* Thomas, Springfield, Illinois.

Ramón y Cajal, S. 1909–1911. *Histologie du système nerveaux de l'homme et des vertébrés.* Paris. (Republished in 1952 by Inst. Ramón y Cajal, Madrid.)

Rogers, W. P. 1973. Juvenile and moulting hormones from nematodes. Parasitology 67: 105–113.

Röller, H., K. H. Dahm, C. C. Sweeley, and B. M. Trost. 1967. The structure of the juvenile hormone. Angew. Chem. Int. Ed. Engl. 6: 179–180.

Romer, F. and W. Gantzy. 1981. Arachnid oenocytes: ecdysone synthesis in the legs of harvestment (Opilionidae). Cell Tissue Res. 216: 449–453.

Sandor, T. and A. Z. Mehdi. 1979. Steroids and evolution. Pp. 1–72 *in* E. J.

W. Barrington (ed.), *Hormones and Evolution*, Vol. 1. Academic Press, Orlando, Florida.

Scharrer, B. 1941. Neurosecretion. 3. The cerebral organ of the nemerteans. J. Comp. Neurol. 74: 109–130.

Scharrer, E. and B. Scharrer. 1954. Neurosecretion. Pp. 5, 6, and 953–1066 *in* W. H. W. Mollendorff (ed.), *Handbuch der Mikroskopisch Anatomie.* Springer-Verlag, Berlin and New York.

Scheffel, H. 1969. Untersuchungen über die hormonale Regulation von Hatung und Anamorphose von *Lithobius forficatus* (L.) (Myriapoda, Chilopoda). Zool. Jahrb. Physiol. 74: 436–505.

Schneiderman, H. A. and L. I. Gilbert. 1958. Substances with juvenile hormone activity in crustaceans and other invertebrates. Biol. Bull. (Woods Hole) 115: 530–535.

Seifert, G. and J. Rosenberg. 1974. Elektronmikroskopische Untersuchungen der Hautungsdrusen Lymphstrange von *Lithobius forficatus* L. (Chilopoda). Z. Morphol. Tiere 78: 263–279.

Snodgrass, R. E. 1935. *Principles of Insect Morphology.* McGraw-Hill, New York.

Snodgrass, R. E. 1956. Crustacean metamorphosis. Smithson. Misc. Collect, 131(10): 1–78.

Solomon, K. R., C. K. A. Mano, and F. D. Obenchain. 1982. Endocine mechanisms in ticks: effects of insect hormones and their mimics on development and reproduction. Pp. 339–438 *in* F. D. Obenchain and R. Galun (eds.), *Physiology of Ticks.* Pergamon Press, Oxford and Elmsford, New York.

Spindler, K.-D., R. Keller, and J. D. O'Connor. 1980. The role of ecdysteroids in the crustacean molting cycle. Pp. 247–280 *in* J. A. Hoffmann (ed.), *Progress in Hormone Research.* Elsevier, Amsterdam and New York.

Storer, T. I. and R. L. Usinger. 1957. *General Zoology.* McGraw-Hill, New York.

Størmer, L. 1977. Arthropod invasion of land during Silurian and Devonian times. Science (Wash., DC) 197: 1362–1364.

Tombes, A. S. 1970. *An Introduction to Invertebrate Endocrinology.* Academic Press, Orlando, Florida.

Tombes, A. S. 1979. Comparison of arthropod neuroendocrine structures and their evolutionary significance. Pp. 645–668 *in* A. P. Gupta (ed.), *Arthropod Phylogeny.* Van Nostrand Reinhold, New York.

Toyama, K. 1902. Development of prothoracic glands in *Bombyx* embryo. Bull. Coll. Agric. Tokyo 5: 73–117.

Ude, J. 1962. Neurosekretorische Zellen im Cerebralganglion von *Dictocoelium lanceatum* St. u. H. (Trematoda: Digena). Zool. Anz. 169: 455–457.

Ude, J. 1964. Untersuchungen zur neurosekretion bei *Dendrocoelum lacteum* Oerst. (Platyhelminthes–Turbellaria). Z. Wiss. Zool. 170: 455–457.

Waddington, C. H. 1965. The morphogenesis of patterns in *Drosophila.* Pp. 499–535 *in* S. J. Counce and C. H. Waddington (eds.), *Development Systems: Insects*, Vol. 2. Academic Press, Orlando, Florida.

Wakabayashi, N., M. Schwartz, P. E. Sonnet, R. M. Waters, R. E. Redfern, and M. Jacobson. 1971. Compounds related to insect juvenile hormones. Bull. Soc. Entomol. Suisse 44: 131–139.

Walgraeve, H., E. van Beek, G. Criel, K. van Brussel, and A. de Leenheer. 1986. Comparison of three separation systems for radioimmunoassay of ecdysteroids in *Artemia* (Crustacea: Branchipoda). Insect Biochem. 16: 41–44.

Wigglesworth, V. B. 1965. *The Principles of Insect Physiology*, 6th ed. Methuen, London.

Winget, R. and W. S. Herman. 1976. Occurrence of ecdysone in the blood of chelicerate arthropod, *Limulus polyphemus*. Experientia (Basel) 32: 1345–1346.

Methyl Farnesoate: A JH-like Compound in Crustaceans

2

D. W. BORST AND H. LAUFER

2.1.   Introduction   37
2.2.   The Roles of JH in Insect Development and Reproduction   38
2.3.   The Case for a Crustacean JH   38
    2.3.1.   Larval Development and Metamorphosis   38
    2.3.2.   Postmetamorphic Morphogenesis   40
    2.3.3.   Reproduction   41
2.4.   The Effects of JH and JH Analogues on Crustaceans   42
    2.4.1.   Effects on Larval Development   42
    2.4.2.   Effects on Reproduction   43
2.5.   JH Bioactivity in Crustaceans   43
2.6.   The Mandibular Organ   44
2.7.   Methyl Farnesoate: A Product of the Mandibular Organ   45
    2.7.1.   Technical Procedures   45
    2.7.2.   MF Synthesis in *Libinia emarginata*   45
    2.7.3.   MF Synthesis in Other Species   47
    2.7.4.   Hemolymph Levels of MF   47
2.8.   The Roles of MF in Crustacean Reproduction and Development   48
    2.8.1.   Reproduction   49
    2.8.2.   Development   51
2.9.   The Regulation of MF Synthesis   51
2.10.   Other Considerations   54
2.11.   Summary   54
Acknowledgments   55
References   55
    55

2.1. Introduction

In spite of recent research efforts, our knowledge regarding the regulation of development and reproduction in crustaceans is still rather rudimentary. It is clear that these events are regulated by a variety of endocrine factors, including peptide and steroid hormones (Skinner, 1985; Quackenbush, 1986). However, many of these compounds are either unknown or are poorly characterized.

In attempting to identify new hormones, it is useful to consider endocrine compounds that are found in related species. During the last decade, many studies have demonstrated that the same hormones are often present in species from widely separated phylogenetic groups. For example, peptides antigenically related to "vertebrate" hormones, such as insulin, glucagon, and somatostatin, have been detected in many different invertebrate phyla. In some cases, the mammalian forms of these hormones have been shown to have effects on invertebrates (Niall, 1982; Kramer, 1985).

Thus, it is hardly surprising that major arthropod groups, such as insects and crustaceans, utilize the same endocrine compounds to regulate common physiological and developmental processes. A good example of the common endocrine heritage of these two arthropod classes is ecdysterone. Though first identified as a steroid that regulates molting in insects (Huber and Hoppe, 1965), this steroid was subsequently identified as the regulator of the same process in other arthropods, including crustaceans (Hampshire and Horn, 1986).

Similar logic has led many investigators, including ourselves, to look in crustaceans for another important insect hormone, juvenile hormone (JH). In this chapter, we review previous studies in this area. In addition, recent work from our laboratories identifying a JH-like compound and its site of synthesis in these animals will be discussed. Some initial studies regarding the physiological and developmental roles of this compound and the regulation of its synthesis will be presented. These data suggest that the functions and the regulation of this crustacean JH-like compound are at least partly similar to that observed for JH in insects. Although our studies are largely confined to work with decapod crustaceans, we suspect that these regulatory mechanisms are present in most, if not all, orders of this arthropod class.

2.2. The Roles of JH in Insect Development and Reproduction

JH is a particularly interesting endocrine compound because of its diverse endocrine functions. First identified (and named) for its regulation of metamorphosis, JH is now known to have a wide variety of other regulatory roles (Laufer and Borst, 1983; Wiggleworth, 1985). For example, it regulates the development of different morphotypes in a number of insects, such as termites, bees, and aphids (for reviews, see Lüscher, 1963; Hardie and Lees, 1985; also Chapters 8–12 and 15 in Part 3). It has important roles in reproduction, regulating vitellogenesis in the female fat body (Engelmann, 1983) and the function of the male accessory glands (Chen, 1984; Tobe and Stay, 1985). JH even appears to have a role in the regulation of molting by affecting the synthesis of ecdysteroids in the prothoracic gland (Watson et al., 1987).

JH refers to a family of epoxidated sesquiterpenoids secreted by the corpus allatum. JH I (methyl 10,11-epoxy-7-ethyl-3,11-dimethyl-2,6-tridecadienoate), the first member of this family to be chemically identified, was isolated by Roller and colleagues (1967) from the silk moth *Hyalophora cecropia*. The other naturally occurring JHs differ only slightly from this basic structure (see Fig. 2.1). JH III appears to be the most common form of his hormone. Other JH homologues (JH I, JH II, JH 0, 4Me–JH I) have been found to date only in the Lepidoptera (Schooley and Baker, 1985). Methyl farnesoate (MF), a nonepoxidated form of JH III, has been detected in some insect species and is known to be biologically active in insects (Sláma, 1971; Lanzrein et al., 1984; Brüning et al., 1985). However, the physiological importance of MF in insect development and reproduction is not clear.

2.3. The Case for a Crustacean JH

2.3.1. *Larval Development and Metamorphosis*

The processes regulated by JH are not, of course, unique to insects. Metamorphosis is common to many invertebrates, though its regulation in most species is poorly understood. Many invertebrates have the ability to delay this event until conditions are appropriate, suggesting that some hormonal mechanism(s) is involved (Highnam, 1981).

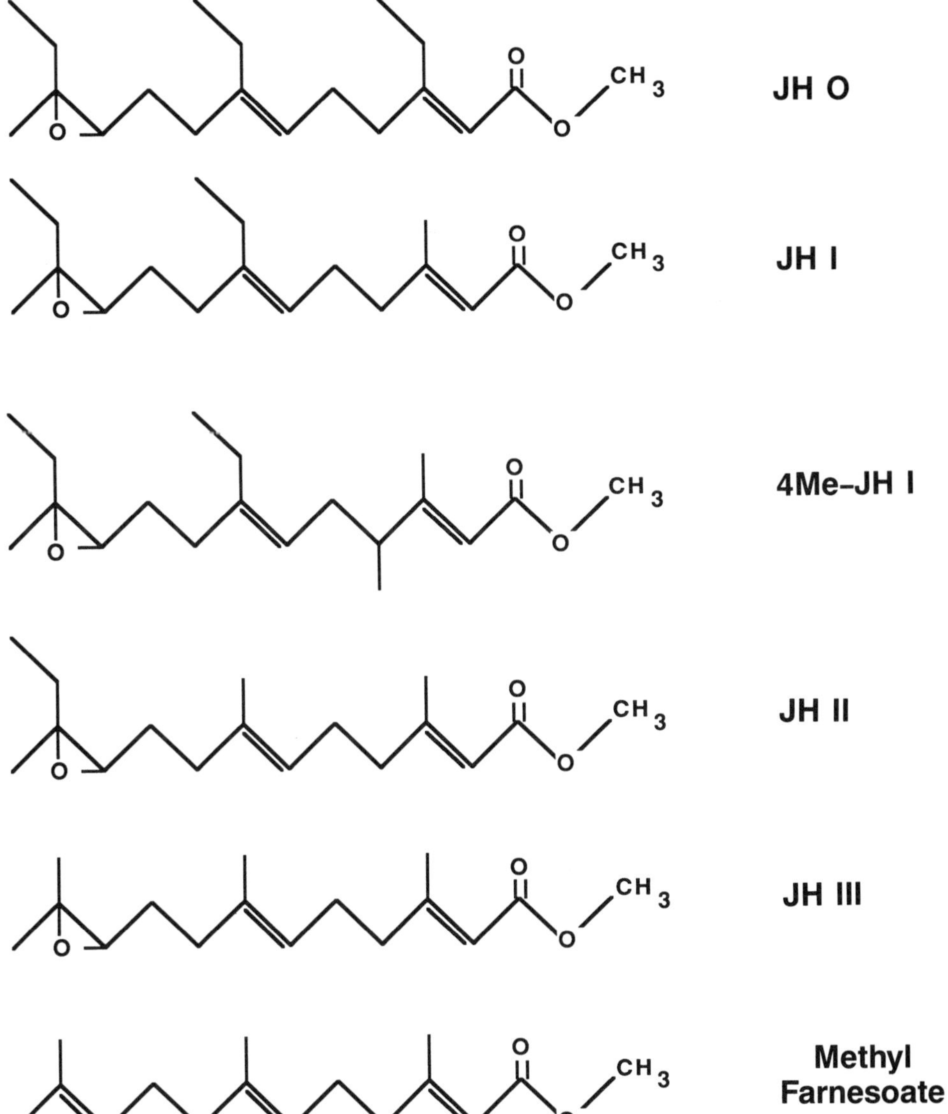

FIGURE 2.1. Juvenile hormone homologues.

Metamorphosis in crustaceans is similar to that observed in hemimetabolous insects and involves the gradual accumulation of morphological changes until the first juvenile stage is reached (Snyder and Chang, 1986). In many crustacean species, larval development proceeds through a fixed number of stages. For example, the lobster *Homarus americanus* normally has three larval stages. Intermediate larval stages are occasionally observed, usually as a result of adverse environmental conditions (Snyder and Chang, 1986). In insects that have a fixed number of stages, intermediate larval stages can be produced by experimental manipulations that affect JH levels (Wigglesworth, 1985; Sehnal, 1983).

Considerable evidence implicates the eyestalk, a potent source of neuroendocrine compounds, as a regulatory center for larval development. A number of studies have shown that eyestalk ablation causes the formation of extra zoeal stages and intermediates in the larvae of the mud crab *Rhithropanopeus harrisii* (Costlow, 1966), the blue crab *Callinectes sapidus* (Costlow, 1968), the shrimp *Palaemon macrodactylus* (Little, 1969), the anomuran *Pisidia longicornis* (Le Roux, 1979), and the lobster *H. americanus* (Charmantier et al., 1985; Charmantier and Aiken, 1987). One interpretation of these studies is that eyestalk ablation removes factors that directly or indirectly inhibit larval growth. Alternately, it has been suggested that the formation of intermediates in eyestalk-ablated larvae may reflect their inadequate nutrition because of a reduced ability to capture prey (Snyder and Chang, 1986). In either case, larval intermediates can occur, suggesting that development and metamorphosis are regulated by some hormonal mechanism. By analogy to insects, it seems likely that such regulation may involve a compound similar to JH.

2.3.2. *Postmetamorphic Morphogenesis*

Another JH-regulated process in insects, the formation of morphotypes, also has its parallel in crustaceans. Many crustacean species have juvenile and adult morphotypes, and these forms are usually related to reproductive activity. For example, the crayfish *Procambarus clarkii* changes from a rapidly growing juvenile to a slow-growing adult (form I) at sexual maturity. After reproduction, form I animals molt to a quasi-juvenile morphotype (form II) until the next reproductive cycle (Avault and Huner, 1985). Thus, the regulation of reproduction appears to affect morphogenesis in this species.

Likewise, males of the freshwater prawn *Macrobrachium rosenbergii* have several morphotypes. These vary in size, behavior, color, reproductive activity, and growth rates (Ra'anan and Sagi, 1985; Sandifer and Smith, 1985). Removal of dominant males allows the smaller morphotypes to grow to the next larger morphotype. Thus, there appear to be mechanisms in these animals that inhibit morphogenesis in response to environment.

Females of the spider crab *Libinia emarginata* also display two morphotypes. The juvenile (sexually immature) form continues to grow and molt, whereas the adult (sexually mature) form is anecdysic. These morphotypes are easily distinguished by their brood pouch apron, which is wider in adults than in juveniles (Hinsch, 1972). Although the juvenile forms are usually smaller than the adults, there is considerable overlap in their sizes. There is a high probability that small juveniles will molt to larger juvenile forms and that large juveniles will molt to adult forms. However, removal of eyestalks from large juvenile animals results in the formation of abnormally large juveniles (Hinsch, 1972). These giant immature animals are reminiscent of the supernumerary larvae produced in insects in response to JH treatment (Sehnal, 1983; Wigglesworth, 1985).

2.3.3. *Reproduction*

Many aspects of insect and crustacean reproduction are similar. For example, vitellogenins (yolk protein precursors) are produced in extraovarian sites in many of these arthropods. The secreted proteins are then transported to the ovary in the hemolymph (Engelmann, 1983; Paulus and Laufer, 1987). In many insect, vitellogenesis is stimulated by JH. In contrast, this process in crustaceans has been shown to be regulated by at least two peptides, the gonad-inhibiting and gonad-stimulating hormones (Adiyodi and Subramoniam, 1983; Charniaux-Cotton, 1985; Quakenbush, 1986). However, these peptides may be stimulating vitellogenesis indirectly by regulating another endocrine tissue. Thus, eyestalk peptides could be analogous to insect neuroendocrine peptides that regulate JH synthesis by the corpus allatum.

2.4. The Effects of JH and JH Analogues
on Crustaceans

2.4.1. *Effects on Larval Development*

The suggestion that larval development in crustaceans is regulated by a JH-related compound has been investigated by treating larvae with JH or JH analogues (synthetic compounds with JH activity). The developmental effects of these compounds on larval development and metamorphosis have been particularly well documented in barnacles. Barnacle nauplii molt into cyprids, a motile, nonfeeding stage that normally lasts about 30 days until they metamorphose and attach to a substrate. When cyprids of the acorn barnacle *Balanus galeatus* were treated with the JH analogue hydroprene (ZR-512) or JH I (Gomez et al., 1973; Ramenofsky et al., 1974), metamorphosis occurred prematurely (within 24 h) without attachment of the cyprid to a substrate. The response of cyprids is fairly selective, since methoprene (ZR-515), a JH analogue related to hydroprene, did not affect cyprid metamorphosis (Gomez et al., 1973).

In another barnacle, *Eliminus modestus,* similar effects were observed when cyprids were treated with the JH analogues farnesyl methyl ether and RO-8-4314 (Tighe-Ford, 1977). In addition, treatment of stage VI nauplii (the developmental stage preceding the cyprid stage) with these analogues resulted in intermediate larval forms with characteristics of both cyprids and nauplii. These developmental abnormalities are not unlike those observed in some insects after treatment with JH or JH analogues (Sehnal, 1983).

Larvae of decapods are also affected by JH and JH analogues. Treatment of *Homarus americanus* larvae with JH III increased the duration of larval development and affected carapace and appendage size (Hertz and Chang, 1986). Treatment with methyl farnesoate had a smaller but still significant effect on the time required for metamorphosis (Borst et al., 1987b). Farnesol, a compound structurally related to JHs but without activity in insects, had no effect on lobster larvae (Hertz and Chang, 1986). The JH analogues methoprene and hydroprene had effects on larvae of *Rhithropanopeus harrisii*. Low doses of these compounds lowered larval survival (the first zoeal stage was most sensitive) and increased the duration of larval development. Methoprene also caused abnormal development of the megalopa stage in this species, possibly by increasing the stress on treated animals (Christiansen et al., 1977a,b).

2.4.2. *Effects on Reproduction*

JH and JH analogues have also been shown to affect crustacean reproduction. For example, injection of JH I into the amphipod *Orchestia
gammarellus* inhibited vitellogenesis (Charniaux-Cotton, 1974). Treatment of *Libinia emarginata* females with JH analogues inhibited ovarian
development and had deleterious effects on oocyte morphology
(Hinsch, 1981a). In *R. harrisii,* methoprene had no effect on early
stages of oogenesis, but it did block vitellogenesis, leaving oocytes in
a previtellogenic stage (Payen and Costlow, 1977). Methoprene also
inhibited spermatogenesis in this species and increased the overall
mortality of treated males and females. In *Daphnia magna,* methoprene
interrupted its reproductive cycle and caused sterility (Templeton and
Laufer, 1983).

Although JH and JH analogues affect reproduction in crustaceans, it
has been difficult to determine if these inhibitory effects result from a
true physiological action of these compounds or from a pharmaco-
logical/toxic effect. Furthermore, these actions contrast with the ability
of JH and JH analogues to stimulate reproductive events in insects.
However, little is known about the turnover of these compounds in
crustaceans or the optimum treatment schedule. Thus, the inhibitory
effects of these compounds may reflect inappropriate doses or treatment protocols.

2.5. JH Bioactivity in Crustaceans

The parallels between insect and crustacean development and reproduction suggest that these processes may be controlled by a similar
endocrine compound. In addition, the specificity and sensitivity of the
response to JH and JH analogues suggests that these compounds are
interacting with an endogenous endocrine system. Early attempts to
detect an endogenous crustacean JH appeared to vindicate this conclusion. When extracts of lobster eyestalks were tested in a JH bioassay
using the moth *Antheraea polyphemus,* they showed potent JH bioactivity (Schneiderman and Gilbert, 1958). However, the nonspecific
response of JH bioassays to many common compounds made these
observations difficult to interpret (Schneiderman and Gilbert, 1964).
For this and other reasons, the chemical basis for such bioactivity has
never been identified.

2.6.   The Mandibular Organ

One potential site for the synthesis of a crustacean JH is the mandibular organ (MO), a tissue named for its location on the posterior abductor muscle tendon of the mandibles (Le Roux, 1968). Electron micrographs of the MO disclose an extensive smooth endoplasmic reticulum and numerous mitochondria (Bazin, 1976; Hinsch and Al Hajj, 1975; Yudin et al., 1980; see also Chapter 10 by Hinsch in Part 2), features that are typical of steroid-secreting cells and of the corpus allatum (Tobe and Stay, 1985; Sedlack, 1985).

The function of the MO has not been easy to identify, partly because it has been frequently mistaken for the Y-organ, the crustacean molting gland (Sochasky et al., 1972; Hinsch and Al Hajj, 1975). However, a number of observations suggest that the MO may have a role in development and molting. For example, the ultrastructure of the MO changes during the molt cycle (Aoto et al., 1974; Byard et al., 1975; Hinsch, 1981b), showing an apparent increase in tissue activity during the premolt stage. In addition, eyestalk removal causes hypertropy of MO cells in larvae (Le Roux, 1983) and adults (Byard et al., 1975; Bazin, 1976; Hinsch, 1977), often resulting in a distorted ultrastructural appearance, including the presence of bizarrely shaped mitochondria. Transplantation of MOs from *Callinectes sapidus* into the white shrimp, *Penaeus setiferus*, shortens the length of its molt cycle (Yudin et al., 1980). However, no ecdysteroids were detected by radioimmunoassay (RIA) after *in vitro* incubation of MOs from either *C. sapidus* (Yudin et al., 1980) or the crayfish *Orconectes limosus* (Keller and Schmid, 1979). Thus, it appears that some compound other than the molting hormone is responsible for this effect (Yudin et al., 1980). The detection of progesterone in the MO of *Homarus americanus* is of interest in this regard (Couch et al., 1987a).

Other observations link the MO with reproduction in females. In the green crab, *Carcinus maenas*, the histological appearance of the MO varies in females, assuming a more active appearance during oogenesis (Le Roux, 1968). Likewise, the ultrastructure of the MO changes during the vitellogenic cycle (Hinsch, 1981b). Furthermore, transplantation of MOs from *Libinia emarginata* adult males into juvenile females stimulated precocious vitellogenesis (Hinsch, 1980). In contrast to these activities in females, MOs in male *C. maenas* and *L. emarginata* have a uniform appearance that does not change appreciably during reproduction.

These data suggest that the MO is an endocrine organ involved in the regulation of molting and reproduction. Indeed, it has been speculated that the MO might be the crustacean homologue of the insect corpus allatum (Byard et al., 1975). However, until recently, no information existed to support this view.

2.7. Methyl Farnesoate: A Product of the Mandibular Organ

2.7.1. *Technical Procedures*

Speculation that the MO might be the crustacean homologue of the corpus allatum in insects (Le Roux, 1968; Byard et al., 1975) encouraged us to analyze this tissue for the production of JH-like materials (Borst et al., 1987a,b; Laufer et al., 1987a,b). For this purpose, we adapted a method used to study the synthesis and secretion of JH by corpora allata *in vitro* (Judy et al., 1973; Pratt and Tobe, 1974). This method measures the transfer of radioactive methyl groups from methionine (via S-adenosylmethionine and a specific methyl transferase) to JH acid to form the methyl ester. Radioactive products can be quantified after separation and identification by high-pressure liquid chromatography (HPLC). Since methionine is rarely used in the biosynthesis of lipids, this method produces few contaminating products (Schooley and Baker, 1985).

2.7.2 *MF Synthesis in* Libinia emarginata

For initial studies, MOs from the spider crab, *L. emarginata*, were incubated in modified Pantin's saline (Laufer et al., 1987a) supplemented with [^{14}C- or ^{3}H-*methyl*]methionine. At the end of the incubation, the tissue and the culture medium were extracted with pentane. Organosoluble materials were analyzed by normal phase HPLC (npHPLC) for radioactive compounds co-eluting with JH standards (Laufer et al., 1987a).

A major radiolabeled product was found in both the culture media and the tissue homogenates. This product comigrated with methyl farnesoate (MF) on npHPLC (see Fig. 2.2). In most samples, this product contained more than 90% of the pentane-extractable radioactivity in the sample. About 1% of the radiolabeled material from these samples comigrated with JH III. Each of these products was collected from npHPLC and analyzed further by reversed phase HPLC (rpHPLC). More than 90% of the radioactivity in each sample migrated with the

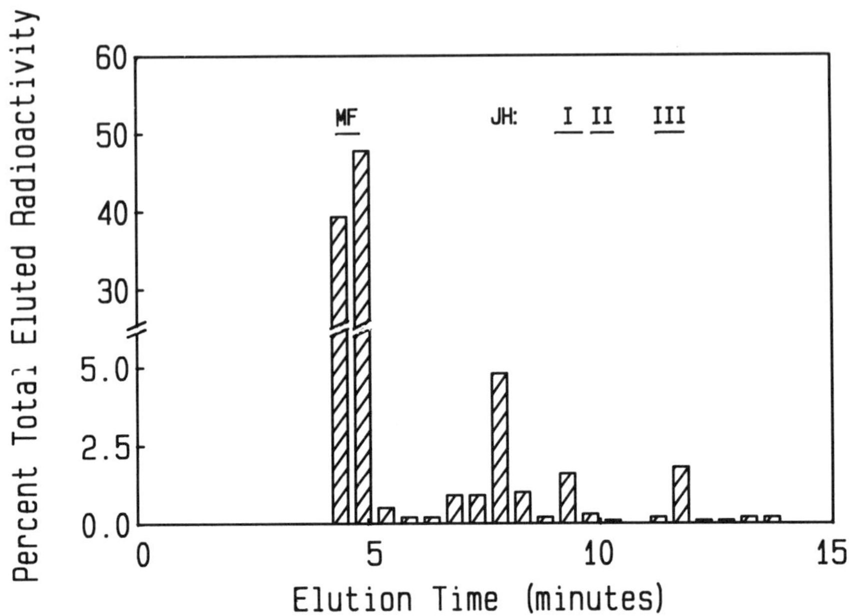

FIGURE 2.2. HPLC separation of radiolabeled products secreted *in vitro* by MOs from male *L. emarginata*. MOs were incubated in modified Pantin's saline supplemented with [*methyl-³H*]methionine. Hexane extractable products in the culture medium were analyzed by HPLC. (From Laufer and Borst, 1988; reprinted with permission.)

appropriate standard on rpHPLC (Laufer et al., 1987a). By gas chromatographic/mass spectroscopic (GC/MS) analysis, we confirmed the identity of the minor product as JH III and the major product as MF, using the methods of Bergot et al. (1981) and Borst et al. (1987b), respectively.

Because JH III is the major JH found in insects, our work was initially focused on this MO product. However, stereoisomer analysis of the radiolabeled JH III showed the presence of both enantiomers, suggesting that the radiolabeled JH III was produced by the nonbiological oxidation of some other material, such as methyl farnesoate (Laufer et al., 1987a). Furthermore, the secreted products from MOs of other decapods did not contain detectable JH III, suggesting that its presence in *L. emarginata* may be a characteristic restricted to that species.

Several other tissues, including epidermis, muscle, eyestalk, and hepatopancreas were also analyzed for the production of JH-like materials. None of the tissues tested synthesized or secreted products that comigrated with methyl farnesoate or JH III on both npHPLC and rpHPLC (Laufer and Borst, 1988). Thus, methyl farnesoate appears to be a product that is characteristic of the MO.

2.7.3. *MF Synthesis in Other Species*

We have used the *in vitro* procedure described in Section 2.7.2 to investigate MF synthesis in a number of crustaceans. In Table 2.1 are listed 14 species that have been tested, including several crabs, a lobster, a crayfish, a penaeid shrimp, and a barnacle. In each species, radiolabeled products were obtained that comigrated with MF on npHPLC and rpHPLC (Borst et al., 1987b; unpublished results). These studies suggest that MF synthesis is a process common to many crustaceans.

TABLE 2.1. Crustacean Species in Which Methyl Farnesoate has Been Detected[a]

Libinia emarginata	*Libinia dubia*
Cancer borealis	*Cancer irroratus*
Callinectes sapidus	*Carcinus maenas*
Uca pugnax	*Uca pugilator*
Procambarus clarkii	*Procambarus bartonni*
Ovalipes ocellatus	*Homarus americanus*
Penaeus vannamei	*Balanus nubilus*

[a] *MF was detected by incubating MOs or tissue from a similar anatomical location with methyl-labeled methionine. Products were analyzed by both normal and reversed phase HPLC.*

2.7.4. *Hemolymph Levels of MF*

To determine the circulating levels of MF, we analyzed hemolymph samples using GC/MS with selective ion monitoring (SIM) (Borst et al., 1987b; Laufer et al., 1987a). Hemolymph from all of the species tested (*L. emarginata, H. americanus, C. maenas,* and the fiddler crab *Uca pugilator*) had material with the same retention time as MF that exhibited peaks at m/z 69, 114, and 121 in the ratios characteristic for MF. Using ethyl farnesoate as an internal standard (Borst et al., 1987b), we detected substantial amounts of MF in their hemolymph, ranging from about 1 ng/ml for the lobster to 10–60 ng/ml for the three species of crabs. The difference in MF levels of the lobster and crabs is similar to the difference in MF synthesis by MOs from these species. Hemolymph was also analyzed for the presence of JHs using GC/MS–SIM. In *L. emarginata,* both JH III and JH III acid were detected in the hemolymph. The amount (3–30 pg/ml) of JH III was low, less than 0.1% of the level of methyl farnesoate in each hemolymph sample. Therefore, it is unlikely that this material has much physiological significance. Furthermore, methyl farnesoate standards are converted to JH III at a similar level (about 0.1%) during analysis by this procedure (Baker and Schooley, 1986). Thus, it seems likely that the JH III detected in these samples is due to the artifactual oxidation of endogenously produced methyl farnesoate. The level of JH III acid (0.1–0.8 ng/ml) was higher than that of JH III but still substantially lower than that of MF. JH III acid may reflect a similar oxidation of farnesoic acid during analysis (Laufer et al., 1987a). In *H. americanus,* no JH III was detected in the hemolymph (Borst et al., 1987b).

2.8. The Roles of MF in Crustacean Reproduction and Development

The studies described in Section 2.7 indicate that MF is a major, if not the sole, product synthesized and secreted by the MO. Furthermore, MF is the major JH-like compound found in the hemolymph. Of course, these studies do not indicate the active form of this compound, since MF could be a prohormone that is converted by peripheral tissues to the active form. Although it is hard to rule out such conversion, our data suggest that MF is not converted to JH III. The only JH III found in the hemolymph appears to be due to oxidation of MF during sample preparation. In addition, when tissue (hepatopancreas, epidermis, muscle, or green gland) was incubated with MF, no detectable JH III was produced (unpublished observations).

These studies also do not clarify the role(s) that the MO and its secretory product(s) have in the physiological and developmental regulation of crustaceans. The classical method for demonstrating an endocrine role for a tissue involves removal of the tissue and the subsequent detection of some physiological or developmental change. Thereafter, these endocrine-compromised animals can be injected with tissue extracts or purified compounds to identify the active agent secreted by the tissue. Unfortunately, the location of the MO in many crustaceans makes surgical ablation difficult. Furthermore, in many species the MO is small (often less than 1 mg) and has multiple lobes, complicating its complete removal. Probably more than anything else, difficulty in removing the MO has retarded studies of its function. In spite of such difficulties, recent studies from our laboratories support suggestions that the MO has regulatory functions similar to those of the corpus allatum in insects.

2.8.1. *Reproduction*

As mentioned in Secetion 2.6, the histological and ultrastructural appearance of the MO changes during the reproductive cycle of female crustaceans. Furthermore, JH analogues are known to affect reproduction. Such observations, plus the knowledge that JH III was an important regulator of insect reproduction, led us to investigate the relationship of MF synthesis and reproduction in the crabs *L. emarginata*, *U. pugilator*, and *C. maenas* (Borst et al., 1987b; Laufer et al., 1987a).

Libinia emarginata has been particularly useful for these studies because it has multiple ovarian cycles during the summer. Stages in the ovarian cycle (which lasts about 20 days) can be estimated from the color of the embryos being held in the brood pouch from the previous cycle (Hinsch, 1981b). Embryos are initially orange, and then become red, brown, and finally gray. They remain gray for the last few days before hatching. As shown in Fig. 2.3, MF synthesis by MOs from adult females is low at the beginning of the ovarian cycle, with levels of synthesis similar to those observed in males and juvenile (sexually immature) females. In animals with older (brown and gray) embryos, MF synthesis rises significantly (about 250%). This rise is also correlated with oocyte growth (mean oocyte length increased from 0.39 mm in females with orange embryos to 0.53 mm in females with gray embryos).

MF synthesis and ovarian development (measured as the ovarian index, OI = ovarian weight/total body weight) were also correlated in *C. maenas*. MF synthesis was 250% greater in MOs from animals with the largest ovarian indexes (OI = 9.5) than in animals with the smal-

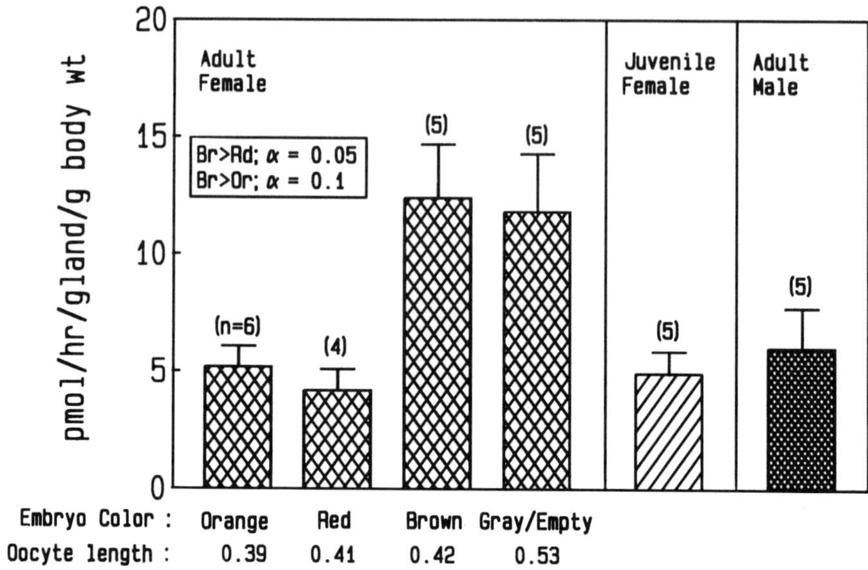

FIGURE 2.3. MF synthesis by MOs from *L. emarginata*. MOs were removed from males, juvenile females, and adult females. The ovarian cycle stage of the adult females was estimated from the color of the embryos in the brood pouch from the previous cycle (Hinsch, 1981b). Oocyte diameter was measured to confirm the stage of oocyte development. MOs were removed at the indicated stages and MF synthesis was determined (see Fig. 2.2.) MF synthesis [mean $+/-$ SEM (standard error of mean)] by MOs from females with brown embryos and those with gray embryos or an empty brood pouch was significantly higher [a = 0.05 and 0.10, respectively, by comparison of rank sums (Dunn, 1964)] than those from females with orange embryos.

lest or intermediate ovarian indexes (OIs of 0.7 and 2.7, respectively). No difference was observed in MF synthesis in MOs from these latter two groups, suggesting that MF may not be involved in early stages of ovarian development. Increased MF synthesis and ovarian development were also correlated in *U. pugilator* (Borst et al., 1987a).

These studies indicate that MF synthesis increases at the same time as ovarian maturation occurs. Thus, the data are consistent with a regulatory role for this compound in ovarian development. MF is also synthesized by MOs of male crustaceans, but the role of this compound in male reproductive physiology has not yet been investigated.

2.8.2. *Development*

Several reports in the literature suggest an involvement of an endocrine compound in postlarval development. Perhaps most persuasive are studies with *L. emarginata*. As discussed in Section 2.3.2, this species has two female morphotypes. Of particular interest is the observation that eyestalk removal from large juveniles causes a molt to a giant juvenile, rather than adult, form (Hinsch, 1972). As shown in Table 2.2, eyestalk removal stimulates MF synthesis by the MOs from juvenile *L. emarginata* (Laufer et al., 1987b). This observation suggests that elevated levels of MF might be a factor in the retention of the juvenile morphotype in smaller juvenile animals. Likewise, lower levels of MF at the molt might result in the formation of the adult morphotype. Preliminary measurements of MF levels in hemolymph from small and large juvenile animals by GC/MS support this view.

2.9. The Regulation of MF Synthesis

As mentioned in Section 2.6, eyestalk removal affects the histological and ultrastructural appearance of the MO in several species (Le Roux, 1968; Bazin, 1976; Hinsch, 1977). These changes in MO appearance

---

TABLE 2.2.   Effect of Eyestalk Ablation of MF Production[a,b]

Species	Treatment	MF synthesis[c] (pmol/2 h per gland)	MF level[d] (ng/ml hemolymph)
Libinia emarginata	Intact	116 +/ − 30	ND
juvenile female	Ablated	251 +/ − 119[e]	ND
Homarus americanus	Intact	75 +/ − 15	0.8 +/ − 0.1
male	Ablated	1407 +/ − 417[f]	7.8 +/ − 2.1[e]

[a] Laufer et al. (1987b).

[b] Couch et al. (1987b).

[c] MF synthesis: determined by incubation of MOs *in vitro.*

[d] MF levels: determined by GC/MS–SIM analysis. (ND = not determined.)

[e] $P < 0.01$ (Student's *t*-test) for intact vs. ablated.

[f] $P < 0.02$ (Student's *t*-test) for intact vs. ablated.

(hypertrophy of the cells) suggest that MO activity is stimulated by eyestalk ablation, implying the presence of an MO-inhibiting hormone (MOIH) in the eystalk.

To investigate this possibility, we removed eyestalks from several species of decapods. Eyestalk ablation stimulated MF synthesis about 250% in MOs from juvenile female *L. emarginata* (Table 2.2) and about 200% in MOs from adult female *U. pugilator*. In the latter study, eyestalk removal also stimulated the ovarian index (Borst et al., 1987a), further supporting the correlation between MF synthesis and reproduction. Likewise, eyestalk ablation caused a 20-fold stimulation in the level of MF synthesis by MOs from male *H. americanus* (Couch et al., 1987b) and a 10-fold increase in circulating MF levels (see Table 2.2). Eyestalk ablation also increased MF synthesis by MOs from male crayfish *Procambarus clarkii* (Laufer et al., 1987c). In contrast, the ablation of eyestalks from males of *L. emarginata* and *U. pugilator* had no detectable effect on MF synthesis.

These results indicate that the eyestalk inhibits MO function. Such inhibition might involve a direct effect of an MO-inhibiting factor on this tissue. However, the lack of response by male *L. emarginata* and *U. pugilator* suggests that removal of eyestalk inhibition may not be alone sufficient to increase MO function, implying that a stimulatory factor may also be required.

To test for the presence of a MO-inhibiting factor, juvenile *L. emarginata* were eyestalk-ablated and one MO from each animal was incubated with eyestalk extract. The other gland was incubated in saline alone. MF synthesis in MOs incubated with eyestalk extract was about 40% of that observed in the contralateral gland (Fig. 2.4), suggesting that eyestalk factors can act directly on this tissue (Laufer et al., 1987b).

The nature of such eyestalk factors was investigated further, using two synthetic crustacean eyestalk peptides, crab pigment dispersing hormone [PDH (Rao et al., 1985)] and red pigment concentrating hormone [RPCH (Fernlund and Josefsson, 1972)]. When MOs from eyestalk-ablated *P. clarkii* were incubated for 6 h with 10^{-7} M PDH, MF synthesis was inhibited 80% compared with MOs incubated in saline. This result suggests that PDH may be similar to the MO-inhibiting factor of the eyestalk. Incubation of MOs with 10^{-6} M RPCH resulted in a two-fold stimulation in MF synthesis, compared with MOs incubated in saline, suggesting that RPCH may be similar to the MO-stimulating factor *in vivo* (Laufer et al., 1987c).

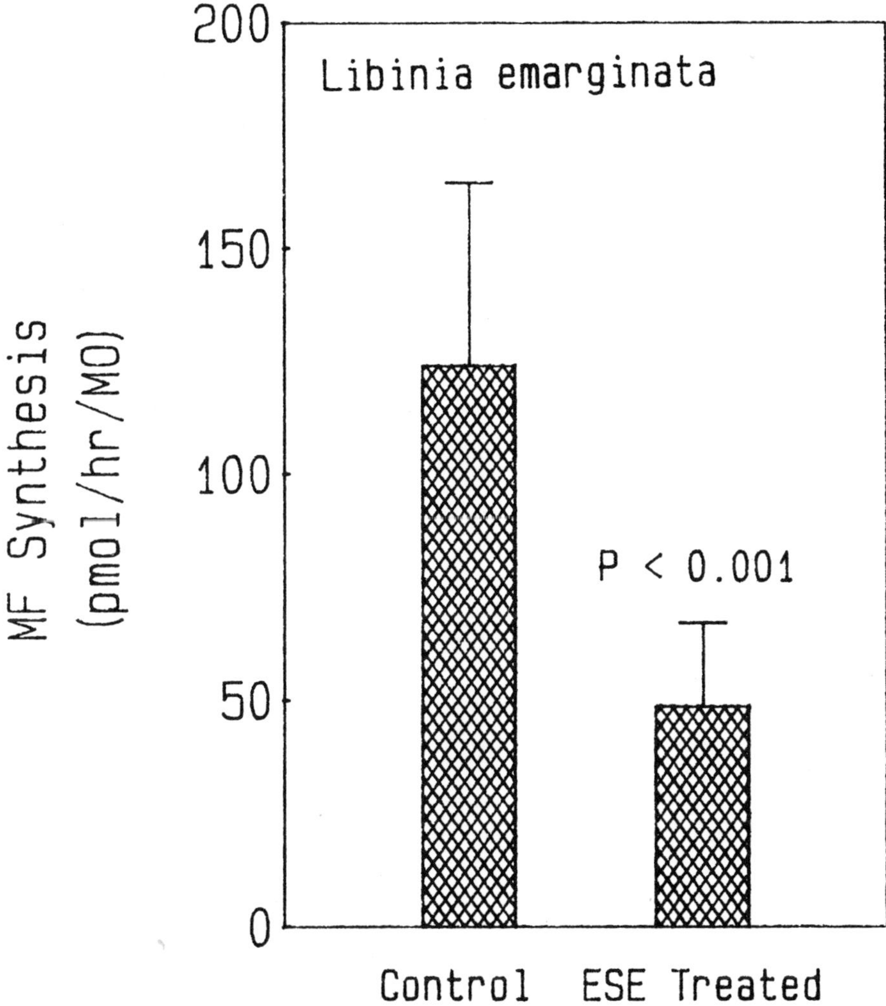

FIGURE 2.4. The effect of eyestalk extracts on MF synthesis by MOs from juvenile *L. emarginata*. Eyestalks were ablated from animals, and MOs were removed 8 days later. MOs were incubated with or without eyestalk extract (ESE treated: 0.75 eyestalk equivalents), and MF synthesis was determined (see Fig. 2.2.). Significance of differences between the two groups [mean +/− SD (standard deviation)] was determined by Student's *t* test (Laufer et al., 1987b).

2.10.   Other Considerations

From a broader perspective, MF appears to belong to a class of terpenoid growth regulators about which our knowledge is rapidly increasing. Although the JHs were initially thought to be a unique class of hormones because of their terpenoid structure, other terpenoids have been identified that regulate development and reproduction. In crustaceans, it is of special interest that the androgenic gland produces farnesylacetone (Ferezou et al., 1977), a terpenoid closely related to MF. Low concentrations of farnesylacetone inhibit protein and RNA synthesis in ovarian tissue incubated *in vitro* (Berreur-Bonnenfant and Lawrence, 1984).

In addition, several recent studies have focused on the retinoids as important regulator of vertebrate differentiation. For example, it has recently been shown that gradients of retinoic acid direct the anterior–posterior digit pattern of the chick wing (Thaller and Eichele, 1987). These and other data suggest that terpenes may be widely used as regulatory agents.

2.11.   Summary

Considerable progress has been made during the last few years in identifying a crustacean JH. Currently, the best candidate for this role is methyl farnesoate (MF), a compound that is chemically related to JH III, the JH homologue most widely distributed among insects. This compound has been detected in all crustaceans tested to date, suggesting that it is a major—if not the sole—JH-like compound in this class of arthropods. In decapods, the site of MF synthesis is the mandibular organ, identifying this organ as the functional equivalent of the insect corpus allatum. These studies provide the first chemical identification of a JH-related compound in an animal other than insects, and suggest that MF or other JH-like compounds may be present in species from other classes of arthropods and possibly other invertebrate phyla.

Because JH has multiple actions in insects, it seems likely that MF may also .be involved in a many aspects of crustacean development and reproduction. Some of these possible roles of MF have been investigated. The data from these studies are consistent with a role for MF in the regulation of vitellogenesis and postmetamorphic morphogenesis. MF may also regulate larval metamorphosis.

Acknowledgments

The research from our laboratories described in this review was supported in part by research grants from the NSF (PCM 84-0213), the NIH (1-R15-HD22677-01), and the Sea Grant College Program (NA 85AA-083N-RA2) to D. W. B. and from the Sea Grant College Program (NA 82 AA-D-0001 and NA 85-D-SO 101-84-86) and the U.S. Department of Agriculture (85-CRCR-1-1839) to H. L. We thank both the Marine Biological Laboratory, Woods Hole, Massachusetts, and the Environmental Laboratory, Millstone Power Plant, Waterbury, Connecticut, for supplying crustacean species. We gratefully acknowledge the collaboration of Drs. D. A. Schooley and F. C. Baker of the Zoecon Research Institute (Palo Alto, California), who did many of the GC/MS analyses and provided us with HPLC standards for our studies. In addition, we appreciate the collaboration of Drs. E. Chang (University of California, Davis), E. F. Couch (Texas Christian University, Fort Worth), and M. Landau (Stockton State Colleage, Pomona, New Jersey). Finally, our special thanks to Ms. E. Homola (University of Connecticut, Storrs), Ms. L. Kissee (Illinois State University, Normal), and Mr. D. Konzelman (Eastern Illinois University, Charleston) for their technical help in these studies.

References

Adiyodi, R. G. and T. Subramoniam. 1983. Arthropoda: Crustacea. Pp. 443–495 *in* K. G. Adiyodi and R. G. Adiyodi (eds.), *Reproductive Biology of Invertebrates*, Vol. I. Wiley, New York.

Aoto, T., Y. Kamiguchi, and S. Hisano. 1974. Histological and ultrastructural studies on the Y-organ and the mandibular organ of the freshwater prawn, *Palaemon paucidens*, with special reference to their reactions with the molting cycle. J. Fac. Sci. Hokkaido Univ. Ser. VI Zool. 19: 295–308.

Avault, J. W. and J. V. Huner. 1985. Crawfish culture in the United States. Pp. 1–61 *in* J. V. Huner and E. E. Brown (eds.), *Crustacean and Mollusk Aquaculture in the United States*. AVI, Westport, Connecticut.

Baker, F. C. and D. A. Schooley. 1986. Preparation of (10R,S)-[10-^{3}H]juvenile hormone III and (10R,S,-11S,R)-[10-^{3}H]juvenile hormone 0; conversion of [10-^{3}H]juvenile hormone III to methyl (2E,6E)-[10-^{3}H]farnesoate and (2E,6E)-[10-^{3}H]farnesol. J. Labelled Compd. & Radiopharm. 23: 533–543.

Bazin, F. 1976. Mise en évidence des caractères cytologiques des glandes stéroïdogènes dans les glandes mandibulaires et les glandes Y du crabe

Carcinus maenas (L.) normal et épédonculé. C. R. Acad. Sci. Paris 282D: 739–741.

Bergot, B. J., M. Ratcliff, and D. A. Schooley. 1981. Method for quantitative determination of the four known juvenile hormones in insect tissue using gas chromatography–mass spectroscopy. J. Chromatogr. 204: 231–244.

Berreur-Bonnenfant, J. and F. Lawrence. 1984. Comparative effect of farnesylacetone on macromolecular synthesis in gonads of crustaceans. Gen. Comp. Endocrinol. 54: 462–468.

Borst, D. W., L. Kissee, and D. Ramlose. 1987a. The synthesis of methyl farnesoate (MF) by mandibular organs (MO) of two crabs. Am. Zool. 27: 69a (abstract).

Borst, D. W., H. Laufer, M. Landau, E. S. Chang, W. A. Hertz, F. C. Baker, and D. A. Schooley. 1987b. Methyl farnesoate and its role in crustacean reproduction and development. Insect Biochem. 17:1123–1127.

Brüning, E., A. Saxer, and B. Lanzrein. 1985. Methyl farnesoate and juvenile hormone III in normal and precocene-treated embryos of the ovoviviparous cockroach *Nauphaeta cinerea*. Int. J. Invertebr. Reprod. Dev. 8: 269–278.

Byard, E. H., R. R. Shivers, and D. E. Aiken. 1975. The mandibular organ of the lobster, *Homarus americanus*. Cell Tissue Res. 162: 13–22.

Charniaux-Cotton, H. 1974. Données nouvelles concernant la vitellogenèse des Crustacés Malacostracés obtenues chez l'Amphipode *Orchestia gammarellus* (Pallas): folliculogenèse à partir d'un tissu permanent; action du busulfan; action inhibitrice de l'hormone juvénile. C. R. Acad. Sci. Paris 279D: 563–566.

Charniaux-Cotton H. 1985. Vitellogenesis and its control in malacostracan Crustacea. Am. Zool. 25: 197–206.

Charmantier, G. and D. E. Aiken. 1987. Intermediate larval and postlarval stages of *Homarus americanus* Edwards, H. Milne, 1837. J. Crustacean Biol. 7: 525–535.

Charmantier, G., M. Charmantier-Dkaures, and D. E. Aiken. 1985. Intervention des pédoncules oculaires dan le contrôle de la métamorphose chez *Homarus americanus* H. Milne-Edwards, 1837 (Crustacea, Decapoda). C. R. Acad. Sci. Paris 300D: 271–276.

Chen, P. S. 1984. The functional morphology and biochemistry of insect male accessory glands and their secretions. Annu. Rev. Entomol. 29: 233–255.

Christiansen, M. E., J. D. Costlow, and R. J. Monroe. 1977a. Effect of the juvenile hormone mimic ZR-514 (Altozar) on larval development of the mud crab *Rhithropanopeus harrisii* at various cyclic temperatures. Mar. Biol. (NY) 39: 281–288.

Christiansen, M. E., J. D. Costlow, and R. J. Monroe. 1977b. Effects of the juvenile hormone mimic ZR-515 (Altosid) on larval development of the mud crab *Rhithropanopeus harrisii* at various salinities and cyclic temperatures. Mar. Biol. (NY) 39: 269–279.

Costlow, J. D., Jr. 1966. The effect of eyestalk extirpation on larval development of the mud crab, *Rhithropanopeus harrisii* (Gould). Gen. Comp. Endocrinol. 7: 255–274.

Costlow, J. D., Jr. 1968. Metamorphosis in crustaceans. Pp. 3–41 *in* W. Etkin and L. I. Gilbert (eds.), *Metamorphosis.* Appleton-Century-Crofts, New York.

Couch, E. F., N. Hagino, and J. W. Lee. 1987a. Changes in estradiol and progesterone immunoreactivity in tissues of the lobster, *Homarus americanus,* with developing and immature ovaries. Comp. Biochem. Physiol. 87A: 765–770.

Couch, E. F., D. W. Borst, H. Laufer, and J. K. Butler. 1987b. Regional differences in methyl farnesoate (MF) production by the mandibular organ (MO) of the lobster. Am. Zool. 27: 68A (abstract).

Dunn, O. J. 1964. Multiple comparisons using rank sums. Technometrics 6: 241–252.

Engelmann, F. 1983. Vitellogenesis controlled by juvenile hormone. Pp. 259–270 *in* R. G. H. Downer and H. Laufer (eds.), *Endocrinology of Insects.* Liss, New York.

Ferezou, J. P., J. Berreur-Bonnenfant, J. J. Mausy, M. Barbier, M. Suchyý, and H. K. Wipf. 1977. 6,10,14-Trimethylpentadecan-2-one and 6,10,14-trimethyl-5-*trans*,9-*trans*,13-pentadecatrien-2-one from the androgenic glands of the male crab *Carcinus maenas.* Experientia (Basel) 33: 290.

Fernlund, P. and L. Josefsson. 1972. Crustacean color-change hormone: amino acid sequence and chemical synthesis. Science (Wash., DC) 177: 173–175.

Gomez, E. D., D. J. Faulkner, W. A. Newman, and C. Ireland. 1973. Juvenile hormone mimics: effects on cirriped crustacean metamorphosis. Science (Wash., DC) 179: 813–814.

Hampshire, R. and D. H. S. Horn. 1966. Structure of crustecdysone, a crustacean moulting hormone. Chem. Commun. 1966: 37–38.

Hardie, J. and A. D. Lees. 1985. Endocrine control of polymorphism and polyphenism. Pp. 441–490 *in* G. A. Kerkut and L. I. Gilbert (eds.), *Comprehensive Insect Physiology, Biochemistry and Pharmacology,* Vol. 8. Pergamon Press, Oxford and Elmsford, New York.

Hertz, W. A. and E. S. Chang. 1986. Juvenile hormone effects on metamorphosis of lobster larvae. Int. J. Invertebr. Reprod. Dev. 10: 71–77.

Highnam, K. C. 1981. A survey of invertebrate metamorphosis. Pp. 43–73 *in* L. I. Gilbert and E. Frieden (eds.), *Metamorphosis: A Problem in Developmental Biology.* Plenum Press, New York.

Hinsch, G. W. 1972. Some factors controlling reproduction in the spider crab, *Libinia emarginata.* Biol. Bull. (Woods Hole) 143: 358–366.

Hinsch, G. W. 1977. Fine structural changes in the mandibular gland of the male spider crab, *Libinia emarginata* (L.), following eyestalk ablation. J. Morphol. 154: 307–316.

Hinsch, G. W. 1980. Effect of mandibular organ implants upon the spider crab ovary. Trans. Am. Microsc. Soc. 99: 317–322.

Hinsch, G. W. 1981a. Effects of juvenile hormone mimics on the ovary in the immature spider crab, *Libinia emarginata*. Int. J. Invertebr. Reprod. 3: 237–244.

Hinsch, G. W. 1981b. The mandibular organ of the female spider crab, *L. emarginata*, in immature, mature, and ovigerous crabs. J. Morphol. 168: 181–187.

Hinsch, G. W. and H. Al Hajj. 1975. The ecdysial gland of the spider crab, *Libinia emarginata* (L). I. Ultrastructure of the gland in the male. J. Morphol. 145: 179–187.

Huber, R. and W. Hoppe. 1965. Die Kirstall- und Molekülstrukturanalyse des Insektenverpuppungshormons Ecdyson mit der automatisierten Faltmolekülmethode. Chem. Ber. 98: 2403–2424.

Judy, K. J., D. A. Schooley, L. L. Durham, M. S. Hall, B. J. Bergot, and J. B. Siddall. 1973. Isolation, structure, and absolute configuration of a new natural insect juvenile hormone from *Manduca sexta*. Proc. Nat. Acad. Sci. USA 70: 1509–1513.

Keller, R. and E. Schmid. 1979. *In vitro* secretion of ecdysteroids by Y-organs and lack of secretion by mandibular organs of the crayfish following molt induction. J. Comp. Physiol. 130: 347–353.

Kramer, K. J. 1985. Vertebrate hormones in insects. Pp. 511–536 *in* G. A. Kerkut and L. I. Gilbert (eds.), *Comprehensive Insect Physiology Biochemistry and Pharmacology*, Vol. 7. Pergamon Press, Oxford and Elmsford, New York.

Lanzrein, B., H. Imboden, C. Bürgin, E. Brüning, and H. Gfeller. 1984. On titers, origin, and functions of juvenile hormone III, methylfarnesoate, and ecdysteroids in embryonic development of the ovoviviparous cockroach *Nauphoeta cinerea*. Pp. 454–465 *in* J. Hoffmann and M. Porchet (eds.), *Biosynthesis, Metabolism and Mode of Action of Invertebrate Hormones*. Springer-Verlag, Berlin and New York.

Laufer, H. and D. W. Borst. 1983. Juvenile hormone and its mechanisms of action. Pp. 203–216 *in* R. G. H. Downer and H. Laufer (eds.), *Insect Endocrinology*. Liss, New York.

Laufer, H. and D. W. Borst. 1988. Juvenile hormone in Crustacea. Pp. 305–313 *in* H. Laufer and R. G. H. Downer (eds.), *Endocrinology of Selected Invertebrate Types*. Liss, New York.

Laufer, H., D. Borst, F. C. Baker, C. Carrasco, M. Sinkus, C. C. Reuter, L. W. Tsai, and D. A. Schooley. 1987a. Identification of a juvenile hormone–like compound in a crustacean. Science (Wash., DC) 235: 202–205.

Laufer, H., M. Landau, D. Borst, and E. Homola. 1987b. The synthesis and regulation of methyl farnesoate, a new juvenile hormone for crustacean reproduction. Pp. 135–143 *in* M. Porchet, J. C. Andries, and A. Dhainaut (eds.), *Advances in Invertebrate Reproduction*, Vol. 4. Elsevier, Amsterdam and New York.

Laufer, H., E. Homola, and M. Landau. 1987c. Control of methyl farnesoate synthesis in crustacean mandibular organs. Am. Zool. 27: 69a (abstract).

Le Roux, A. 1968. Description d'organes mandibularies nouveaux chez les Crustacés Décapodes. C. R. Acad. Sc. Paris 266D: 1414–1417.

Le Roux, A. 1979. Effects de l'ablation des pédoncules oculaires sur le développement de *Pisidia longicornis* (Crustacé, Décapode, Anomoure). Ann. Endocrinol. 40: 79–80.

Le Roux, A. 1983. Réactions de l'organe mandibulaire à l'ablation des pédoncules oculaires chez les larves et les juvéniles de *Palaemonetes varians* (Leach) (Decapoda, Natantia). C. R. Acad. Sci. Paris 296D: 697–700.

Little, G. 1969. The larval development of the shrimp, *Palaeomon macrodactylus* Ratbun, reared in the laboratory, and the effect of eyestalk extirpation on development. Crustaceana (Leiden) 17: 69–87.

Lüscher, M. (ed.) 1963. *Phase and Caste Determination in Insects: Endocrine Aspects.* Proc. 15th Int. Congr. Entomol. Pergamon Press, Oxford and Elmsford, New York.

Niall, H. D. 1982. The evolution of peptide hormones. Annu. Rev. Physiol. 44: 615–624.

Paulus, J. E. and H. Laufer. 1987. Vitellogenocytes in the hepatopancreas of *Carcinus muenus* and *Libinia emarginata* (Decapoda: Brachyura). Int. J. Invertebr. Reprod. Dev. 11: 29–44.

Payen, G. C. and J. D. Costlow. 1977. Effects of a juvenile hormone mimic on male and female gametogenesis of the mud crab *Rhithropanopeus harrisii* (Gould) (Brachyura: Xanthidae). Biol. Bull. (Woods Hole) 152: 199–208.

Pratt, G. E. and S. S. Tobe. 1974. Juvenile hormones radiobiosynthesised by corpora allata of adult female locusts *in vitro.* Life Sci. 14: 575–586.

Quackenbush, L. S. 1986. Crustacean endrocinology: a review. Can. J. Fish. Aquat. Sci. 43: 2271–2282.

Ra'anan, Z. and A. Sagi. 1985. Alternative mating strategies in male morphotypes of the freshwater prawn *Macrobrachium rosenbergii* (de Man). Biol. Bull. (Woods Hole) 169: 592–601.

Ramenofsky, M., D. J. Faulkner, and C. Ireland. 1974. Effect of juvenile hormone on cirriped metamorphosis. Biochem. Biophys. Res. Commun. 60: 172–178.

Rao, K. R., J. P. Riehm, C. A. Zahnow, L. H. Kleinholz, G. E. Tarr, L. Johnson, S. Norton, M. Landau, O. J. Semmes, R. M. Sattelberg, W. H. Jorenby, and M. F. Hintz. 1985. Characterization of a pigment-dispersing hormone in eyestalks of the fiddler crab *Uca pugilator.* Proc. Nat. Acad. Sci. USA 82: 5319–5322.

Roller, H., K. H. Dahm, C. C. Sweeley, and B. M. Trost. 1967. The structures of the juvenile hormone. Angew. Chem. Int. Ed. Engl. 6: 179–180.

Sandifer, P. A. and T. I. J. Smith. 1985. Freshwater prawns. Pp. 63–125 *in* J. V. Huner and E. E. Brown (eds.), *Crustacean and Mollusk Aquaculture in the United States.* AVI, Westport, Connecticut.

Schneiderman, H. A. and L. I. Gilbert. 1958. Substances with juvenile hormone activity in Crustacea and other invertebrates. Biol. Bull. (Woods Hole) 115: 530–535.

Schneiderman, H. A. and L. I. Gilbert. 1964. Control of growth and development in insects. Science (Wash., DC) 143: 325–333.

Schooley, D. A., and F. C. Baker 1985. Juvenile hormone biosynthesis. Pp. 363–389 *in* G. A. Kerkut and L. I. Gilbert (eds.), *Comprehensive Insect Physiology Biochemistry and Pharmacology*, Vol. 7. Pergamon Press, Oxford and Elmsford, New York.

Sedlack, B. J. 1985. Structure of endocrine glands. Pp. 25–60 *in* G. A. Kerkut and L. I. Gilbert (eds.), *Comprehensive Insect Physiology Biochemistry and Pharmacology*, Vol. 7. Pergamon Press, Oxford and Elmsford, New York.

Sehnal, F. 1983. Juvenile hormone analogues. Pp. 657–671 *in* R. G. H. Downer and H. Laufer (eds.), *Endocrinology of Insects*. Liss, New York.

Skinner, D. M. 1985. Molting and regeneration. Pp. 43–146 *in* D. E. Bliss and L. H. Mantel (eds.), *The Biology of Crustacea*, Vol. 9. Academic Press, Orlando, Florida.

Sláma, K. 1971. Insect juvenile hormone analogues. Annu. Rev. Biochem. 40: 1079–1102.

Snyder, M. J. and E. S. Chang. 1986. Effects of eyestalk ablation on larval molting rates and morphological development of the american lobster, *Homarus americanus*. Biol. Bull (Woods Hole) 170: 232–243.

Sochasky, J. B., D. E. Aiken, and N. H. F. Watson. 1972. Y-organs, molting gland, and mandibular organ: a problem in decapod crustacea. Can. J. Zool. 50: 993–997.

Templeton, N. S. and H. Laufer. 1983. The effects of a juvenile hormone analog (Altosid ZR-515) on the reproduction and development of *Daphnia magna* (Crustacea: Cladocera). Int. J. Invertebr. Reprod. 6: 99–110.

Thaller, C. and G. Eichele. 1987. Identification and spatial distribution of retinoids in the developing chick limb bud. Nature (Lond.) 327: 625–628.

Tighe-Ford, J. D. 1977. Effects of juvenile hormone analogues on larval metamorphosis in the barnacle *Eliminus modestus* Darwin (Crustacea: Cirripedia). J. Exp. Mar. Biol. Ecol. 25: 163–176.

Tobe, S. S. and B. Stay. 1985. Structure and regulation of the corpus allatum. Adv. Insect Physiol. 18: 305–432.

Watson, R. D., N. Agui, M. E. Haire, and W. E. Bollenbacher. 1987. Juvenile hormone coordinates the regulation of the hemolymph ecdysteroid titer during pupal commitment in *Manduca sexta*. Insect Biochem. 17: 955–959.

Wigglesworth, V. B. 1985. Historical perspectives. Pp. 1–24 *In* G. A. Kerkut and L. I. Gilbert (eds.), *Comprehensive Insect Physiology Biochemistry and Pharmacology*, Vol. 7. Pergamon Press, Oxford and Elmsford, New York.

Yudin, A. I., R. A. Diener, W. H. Clark, Jr., and E. S. Chang. 1980. Mandibular gland of the blue crab, *Callinectes sapidus*. Biol. Bull. (Woods Hole) 159: 760–772.

Juvenile Hormone–like Compounds in Terrestrial Chelicerata (Arachnida)

3

J. C. BONARIC AND C. JUBERTHIE

3.1.	Introduction	63
3.2.	Evidence of JH or JH-like Compounds in Arachnida	63
	3.2.1. Acari	63
	3.2.2. Scorpionida	64
	3.2.3. Araneae	64
3.3.	Effect of JH or JH-like Compounds	65
	3.3.1. Acari	65
	3.3.2. Araneae	68
	3.3.3. Scorpionida and Other Arachnida	69
3.4.	Concluding Remarks	69
3.5.	Summary	70
	References	70

3.1. Introduction

In Arthropoda, development, growth and reproduction are dependent on morphogenetic hormones. In insects, two types of morphogenetic hormones, ecdysteroids and juvenile hormones (JHs), are known. The numerous structural and physiological similarities among arthropods suggest that Arachnida, Crustacea, and Myriapoda also possess hormonal systems similar to those in insects.

The importance of ecdysteroids in the control of molting is now well known in insects and crustaceans. In Chelicerata and Myriapoda, however, the research on hormonal physiology began later than in insects, and therefore only a few contributions are available so far.

The ecdysteroids were detected and measured for the first time in spiders (Bonaric and De Reggi, 1977), and later in Acari (Delbecque et al., 1978) and in Scorpionida (El Bakary, 1986).

Assays to detect JHs or JH-like compounds are scarce and quite recent in other Arthropods. For example, although Schneiderman and Gilbert (1958), using bioassays, reported JH activity in ocular peduncle extracts of *Homarus americanus* (Crustacea), it took another 27 years to demonstrate the existence of "JH-like compound" in Crustacea (Laufer et al., 1985, 1986) by radioimmunoassay (RIA) technique used in insects. Some attempts to detect JHs in several classes of Arachnida have been made (Connat, 1987a; Bonaric, 1988), but the results are preliminary and difficult to interpret.

3.2. Evidence of JH or JH-like Compounds in Arachnida

3.2.1. *Acari*

Using crude hexane extracts of larvae and females of *Ornithodoros moubata*, Connat (1987a) investigated compounds that have JH activity in bioassays, by RIA of JH, according to the protocol described by Strambi et al. (1981). He also used hemolymph extracts of *O. moubata* and *Boophilus microplus* to reduce the large amounts of contaminating lipids contained in the whole-body extracts, and performed RIA with three different antibodies (anti-JHI, anti-JHII, anti-JHIII) according to the protocol of Baehr et al. (1976).

The results of the first method indicate that hexane extracts from vitellogenic females (5 days after a blood meal) and larvae contained substances that cross-reacted with anti-JH diols of Strambi et al.

(1981). High-pressure liquid chromatographic (HPLC) analysis of hemolymph extracts from vitellogenic *B. microplus* females and the use of the second RIA method on the final fraction indicated that several HPLC fractions of the extract cross-reacted with antibodies directed against JH I, JH II, and JH III. However, the JH activity in this tick does not appear to correspond to these hormones, since neither the retention times nor the dilution curves corresponded to those of the three hormones. Since with each screening two of the three antibodies used cross-reacted with the same fractions, Connat assumed that this RIA method could be a good means of screening for the presence of JH-like compounds in the tick.

Further analysis of each of these immunopositive fractions with gas chromatography and mass spectrometry showed that none of the fractions contained JH 0, JH I, JH II, JH III, or FM (farnesyl methyl ether); however, some spectra corresponded to molecules that could be analogous to JHs, but that is not determined yet.

According to Connat (1987a), JHs of insects or MF (methyl farnesoate) recently found in Crustacea (Laufer et al., 1985, 1986) have not been detected in the Acari that have been investigated (*Ornithodoros moubata, Amblyomma hebraeum,* and *Boophilus microplus*); however, the products present in Acari may have homologies with JHs. (See Table 3.1.)

3.2.2.   *Scorpionida*

El Bakary (1986) investigated JHs in the scorpion, *Leiurus quinquestriatus,* using RIA protocol of Baehr et al. (1976, 1979), but found no immunoreactive products and concluded that JHs were absent in this species.

3.2.3.   *Araneae*

Radioimmunoassay of JHs was performed with total extracts of the spider *Pisaura mirabilis* by J. C. Bonaric, A. Strambi, and M. De Reggi (unpublished data), using the protocol of Strambi et al. (1981, 1984). During the molting cycle of the eighth instar, the total amount of the immunoreactive products fluctuates from 20 to 80 pmol/g fresh weight. Equivalent JH diol. RIA, performed on each purified fraction obtained after HPLC separation, does not permit determination of the immunoreactive substances, which comigrate with the JH I, JH II, JH III of insects (Bonaric, 1988).

Based on these data, two hypotheses are proposed: (1) the amount of immunoreactive compounds obtained could represent JH-like substances not clearly identified; or (2) these immunoreactive products could be the result of nonspecific binding with lipids of the hexane extract of crushed spiders.

3.3. Effect of JH or JH-like Compounds

3.3.1. *Acari*

In insects, vitellogenesis is dependent on JH, secreted by the corpora allata (CA), and JH functions gonadotropically in adults. Allatectomy or ligations, or chemical blockage of CA by means of precocene, prevents the secretion of JH and blocks oocyte maturation. Maturation is restored by injection of JH, topical application of JH, or juvenile hormone analogues (JHA), or implantation of active CA from other individuals.

Several reports (Bassal and Roshdy, 1974; Pound and Oliver, 1979; Obenchain and Mango, 1980; Hayes and Oliver, 1981; Solomon et al., 1982; Connat et al., 1983; Oliver et al., 1985; Connat, 1987a,b) suggest a role for JH and JH-like compounds in the reproduction of ticks, similar to that reported in insects; these authors assumed that these hormones or hormone analogs induced vitellogenesis and oviposition in ticks.

According to Pound and Oliver (1979), topical application of precocene 2, an antiallototropic substance in insects, on the soft tick *Ornithodoros parkeri*, blocks vitellogenesis, ovarian development, and embryogenesis. Topical application of JH III reverses the action of precocene, and oogenesis is restored. In *O. moubata*, topical application with different JHAs or with an isomeric mixture of JH I, JH II, JH III, farnesol, and FM induced vitellogenesis and egg laying in up to 70% of treated females; in contrast, only 13.7% of control females oviposited (Connat et al., 1983; Connat, 1987a,b).

Topical application of hexane extracts of vitellogenic females of the cattle tick, *Boophilus microplus*, or virgin engorged females of *O. moubata*, indicated a gonadotropic effect of the extracts. These results suggest the presence of JH-like substances, involved in the hormonal control of vitellogenesis. In contrast, in some experiments, JHA inhibited oogenesis and oviposition in *B. microplus*, according to Mansingh and Rawling (1977). In *Dermanyssus gallinae*, collected from chicken houses, fed females exposed to the antiallatotropic compound precocene 2 (P2)

TABLE 3.1. Summary of Effects of Exogenous Juvenile Hormones (JH and JHA) in Arachnida

Authors	Species, families[a]	Effect
Acari:		
Wright (1969)	*Dermacentor albipictus* (Ix) D. L.)	Rupture larval diapause (D. L.)
Bassal and Roshdy (1974)	*Argas arboreus* (Arg) DR	Rupture reproductive diapause (DR)
Bassal and Roshdy (1974)	*Argas arboreus* (Arg) (V; O)	
Pound and Oliver (1979)	*Ornithodoros parkeri* (Arg) (V)	↑ Vitellogenesis (V)
Obenchain and Mango (1980)	*Ornithodoros porcinus* (Arg) (V)	
Hayes and Oliver (1981)	*Dermacentor variabilis* (Ix) (O; V)	↑ Oviposition (O)
Connat et al. (1983)	*Ornithodoros moubata* (Arg) (O; V)	Gonadotropic effect (G)
Oliver et al. (1985)	*Dermanyssus gallinae* (Derm) (G; OL)	↑ Egg-laying (OL)
Connat (1987b)	*Boophilus microplus* (Ix) (O)	
Mansingh and Rawling (1977)	*Boophilus microplus* (Ix)	↘ Oogenesis
		↘ Embryogenesis antigonotropic effect

Bassal (1974)	*Hyalomma dromedarii* (Ix) (E)	
Klein-Koch (1975, 1976)	*Tetranychus urticae* (Tet) (PE; E)	↘ Embryogenesis (E)
	Phytoseiulus persimilis (Phyt) (PE)	↘ Post-embryonic (PE) development
El-Banhawy (1977)	*Amblyseius brazilli* (Phyt) (PE)	↘ Percent hatching (Ec)
Solomon and Evans (1977)	*Boophilus microplus* (Ix) and *B. decoloratus* (Ix) (E; EC)	
	Amblyomma hebraeum (Ix) (E; Ec)	
McDaniel and Oliver (1978)	*Dermacentor variabilis* (Ix) (E;Ec)	
Gothe and Morawietz (1979)	*Argas walkerae* (Arg) (E; PE)	
Khalil et al. (1984)	*Hyalomma dromedarii* (Ix) (E)	

Araneae:

| Bonaric (1979) | *Pisaura mirabilis* (Pis) | Postembryonic development |

[a] *Key:* Arg, Argasidae; Dem, Dermanyssidae; Ix, Ixodidae; Phyt, Phytoseiidae; Pis, Pisauridae; Tet, Tetranychidae.

produced fewer progreny than did untreated females; reproductive capacity was restored by application of JH III, but not to the level of untreated control females.

Blockage of embryonic development by JHAs has been noted in the mites *Tetranychus urticae* and *T. persimilis* (Klein-Koch, 1976), *Amblyseius brasilli* (El-Banhawy, 1977), and also in Ixodidae, *Hyalomma dromedarii* (Bassal, 1974; Khalil et al., 1984) and *Dermacentor variabilis* (McDaniel and Oliver, 1978), and in Argasidae, *Argas walkerae* (Gothe and Morawietz, 1979). In Ixodidae, according to Solomon and Evans (1977) and McDaniel and Oliver (1978), reduced hatchability in eggs of *B. microplus, B. decoloratus, Amblyomma hebraeum,* and *Dermacentor variabilis* was attributed to desiccation of eggs and not to a direct effect on embryogenesis. It seems possible that the mode of action of JHAs on embryogenesis might be to affect the female's ability to waterproof her eggs, and this would provide indirect evidence to suggest hormonal interference with Gene's organ, which functions to waterproof tick eggs. According to Pound and Oliver (1979), some effects could be the result of toxicity more than endocrine action.

Topical application of JHs or JHAs to the postembryonic instars of mites, Ixodidae, and Argasidae retarded the development of postembryonic stages and affected the molting cycle, which was retarded or inhibited (Klein-Koch, 1976; Gothe and Morawietz, 1979; McDaniel and Oliver, 1978; Solomon et al., 1982).

None of the JHAs *trans-trans*-10- and -11-epoxyfarnesenic acid methyl ester applied topically during larval diapause gave significant results in *Dermacentor albipictus* (Wright, 1969) and *Rhipicephalus sanguineus* (Sannasi and Subramoniam, 1972). However, when the treatment was performed by means of immersion, Wright (1969) obtained recovery of the activity in diapausing larvae of *D. albipictus*.

3.3.2. *Araneae*

JHAs applied to the seventh instar of *Pisaura mirabilis* (Pisauridae) induced postembryonic development period longer than that of the controls; the duration of each intermolt period was increased (Bonaric, 1979). A morphological examination of the affected individuals did not show any abnormalities in the secondary sexual characters, particularly in the bulb of the male palps, which develops normally.

The technical difficulties of injecting anything, specially JH and JHA, in spiders limits the range of experiments.

3.3.3. *Scorpionida and Other Arachnida*

No JH or JHA have been reported in scorpions and other Arachnida.

3.4. Concluding Remarks

Topical applications of insect JH or JHAs on spiders and Acari, bio-assays, and dosages by means of RIA suggested that JH-like products are present and play similar roles in the gonadotropic cycle and growth, at least in Acari.

Data suggest the presence of a JH-like substance in ticks (Connat, 1987a,b), because the effects of JH, JHAs, and anti-JH are similar to those obtained in insects. Furthermore, hemolymph extracts of vitellogenic females of *B. microplus* applied to virgin females of *O. moubata* or to young nymphs of *Tenebrio molitor* produced JH-like effects. And inhibitors of the biosynthesis of JH in insects also act in ticks to prevent maturation, reduce number of eggs, block oviposition, or induce dissociation of mature oocytes. Presence of immunoreactive products detected by RIA with antibodies of insect JH or JH diols also suggest the presence of JH like compounds in Arachnids.

In *O. moubata* and *B. microplus*, the insect JHs are converted into compounds identical to those of insects (JH diol, JH acid, JH acid–diol). These data demonstrate that ticks possess enzymes (hydrolases, esterases) which produce the same metabolites than in insects. These compounds are transformed into more polar molecules, and some of those extracted from *B. microplus* comigrate in HPLC, as the anti–JH III antibodies do (Connat et al., 1988).

RIA applied to spiders, Acari, and Scorpionida does not detect compounds identical to those of insect JH I, JH II or JH III. The JH-like hormones found in ticks have some fragments in common with insect JHs, but they are not identical to insect JHs. Their exact nature may well be established in the near future (see Connat et al., 1986). In spiders, the results are similar (Bonaric, 1987).

Comparison with data reported on JH-like substances in Crustacea (Laufer et al., 1985) suggests some similarities (see also Borst and Laufer, Chapter 2 herein).

It may well be that ancestors of arthropods possessed gene encoding for one or several prohormones (JH-like compounds) and that certain groups, such as Crustacea and Arachnida, have retained these prohormones. However, in insects, the prohormones are not the terminal products of synthesis; rather, the final products are JH I, JH II, and JH III.

3.5. Summary

Data in Acari and Araneae demonstrated that these Arachnida possess JH-like compounds but not JH I, JH II, JH III. By contrast to insects, Arachnida have retained these prohormones as terminal products.

Endocrine cells that secrete JH-like compounds are unknown in all the Arachnida.

References

Baehr, J. C., P. Pradelles, and F. Dray. 1979. A radioimmunological assay for naturally occuring insect juvenile hormones using iodinated tracers: its use in the analysis of biological samples. Ann. Biol. Anim. Biochim. Biophys. 19: 1827–1832.

Baehr, J. C., P. Pradelles, C. Lebreux, P. Cassier, and F. Dray. 1976. A simple and sensitive radioimmunoassay of insect juvenile hormones using and iodinated tracer. FEBS (Fed. Eura. Biochem. Soc.) Lett. 69: 123–128.

Bassal, T. T. M. 1974. Biochemical studies of certain ticks (Ixodoidae): activity of juvenile hormone analogs during embryogenesis in *Hyalomma dromedarius* Koch (Ixodidae). Z. Parasitenkd. 45: 85–89.

Bassal, T. T. and M. A. Roshdy. 1974. *Argas (Persicargas) arboreus:* juvenile hormone analog termination if diapause and oviposition control. Exp. Parasitol. 36: 34–39.

Bonaric, J. C. 1799. Effets des ecdysones et de l'hormone juvénile sur la durée du cycle de mue chez l'araignée *Pisaura mirabilis* (Araneae, Pisauridae). Rev. Arachnol. 2: 205–207.

Bonaric, J. C. 1987. Moulting hormones. Pp. 111–118 *in* W. Nentwig (ed.), *Ecophysiology of Spiders.* Springer-Verlag, Berlin and New York.

Bonaric, J. C. 1988. Présence de composés apparentés aux hormones juvéniles (JH) d'insectes chez les arachnides: détection et difficultés d'interprétation. Bull. Soc. Sci. Bretagne 59: 27–36.

Bonaric, J. C. and M. De Reggi. 1977. Changes in ecdysone levels in the spider *Pisaura mirabilis* nymphs (Araneae, Pisauridae). Experientia (Basel) 33: 1664–1665.

Connat, J. L. 1987a. Aspect endocrinologique de la physiologie du développement et de la reproduction chez les tiques. Doctoral thesis, University of Dijon, France, 131 pp. (+ annexes).

Connat, J. L. 1987b. Effects of different anti–juvenile hormone agents on the fecundity of females of the cattle tick *Boophilus microplus*. Pestic. Biochem. Physiol. (in press).

Connat, J. L., J. Ducommun, and P. A. Diehl. 1983. Juvenile hormone-like substances can induce vitellogenesis in the tick *Ornithodoros moubata* (Acarina: Argasidae). Int. J. Invertebr. Reprod. 6: 285–294.

Connat, J. L., B. Mauchamps, and J. C. Baehr. 1988. An attempt to identify juvenile hormone in the hemolymph of the cattle tick *Boophilus microplus* during vitellogenesis. (In preparation.)

Connat, J. L., J. Ducommun, P. A. Diehl, and A. Aeschlimann. 1986. Some aspects of the control of the gonotrophic cycle in the tick *Ornithodoros moubata* (Ixodoidae, Argasidae). Pp. 194–216 *in* J. R. Sauer and J. A. Hair (eds.), *Morphology, Physiology and Behavioral Biology of Ticks.* Ellis Horwood, Chichester, England.

Delbecque, J. P., P. A. Diehl, and J. D. O'Connor. 1978. Presence of ecdysone and ecdysterone in the tick *Amblyomma hebraeum* Koch. Experientia (Basel) 34: 1379–1380.

El Bakary, J. 1986. Contribution à l'étude biologique du scorpion *Leiurus quinquestriatus* (Buthidae). Doctoral thesis, University of Orsay, France, 198 pp.

El-Banhawy, E. M. 1977. Growth inhibition of the predacious mite *Amblyseius brazilli* (Mesostigmata: Phytoseiidae) by a synthetic juvenile hormone under laboratory conditions. Entomophaga 22: 429–434.

Gothe, R. and Morawietz. 1979. Zum Wirkungsspecktrum von Juvenoiden in postembryonal Phasen von *Argas* (*Persiargas* *walkerae* Kaiser und Hoogstraal, 1969. Zeutralbl. Veterinaermed. 26: 779–797.

Hayes, M. J. and J. H. Oliver. 1981. Immediate and latent effects induced by precocene 2 on embryonic *Dermacentor variabilis* (Acari: Ixodidae). J. Parasitol. 67: 923–927.

Khalil, G. M., D. E. Sonenshine, H. A. Hanafi, and A. E. Abdelmonem. 1984. Juvenile hormone effects on the camel tick *Hyalomma dromedarii* (Acari: Ixodidae). J. Med. Entomol. 21: 561–566.

Klein-Koch, C. 1975. Wirking von Insektenwachstumsregulatoren (IWR) auf Adulte und Eier von Spinn und Raubmilben. Z. Pflanzenkr. Pflanzenschutz 82: 406–409.

Klein-Koch, C. 1976. Die Wirkung von Juvenoiden und ähnlichen Substanzen auf Spinnmilben. Z. Angew. Entomol. 82: 193–199.

Laufer, H., D. W. Borst, F. C. Baker, and D. A. Schooley. 1985. Juvenile hormone-like compounds in *Libinia emarginata.* Am. Zool. 25: 103.

Laufer, H., M. Landau, D. Borst, and E. Homola. 1986. The synthesis and regulation of methyl farnesoate, a new juvenile hormone for crustacean reproduction. Pp. 135–143 *in* M. Porchet, J. C. Andries, and A. Dhainaut (eds.), *Advances in Invertebrate Reproduction. Elsevier, Amsterdam and New York.*

Laufer, H., D. Borst, F. C. Baker, C. Carrasco, M. Sinkus, C. C. Reuter, L. W. Tsai, and D. A. Schooley. 1987. Identification of a juvenile hormone-like compound in a crustacean. Science (Wash., DC) 235: 202–205.

Mansingh, A. and S. C. Rawlins. 1977. Antigonadotropic action of insect hormone analogues on the cattle tick *Boophilus microplus*. Naturwissenchaften 64: 41.

McDaniel, R. S. and J. H. Oliver, Jr. 1978. Effects of two juvenile hormone analogs and β-ecdysone on nymphal development, spermatogenesis and embryogenesis in *Dermacentor variabilis* (Say) (Acari: Ixodidae). J. Parasitol. 64: 571–573.

Obenchain, F. D. and C. Mango. 1980. Effects of exogenous ecdysteroids and juvenile hormone on female reproductive development in *Ornithodoros porcinus*. Am. Zool. 20: 936.

Oliver, J. H., Jr., J. M. Pound, and G. Severino. 1985. Evidence of a juvenile-hormone-like compound in the reproduction of *Dermanyssus gallinae* (Acari: Dermanyssidae). J. Med. Entomol. 22: 281–286.

Payen, G. C. and J. D. Costlow. 1977. Effects of a juvenile hormone mimic on male and female gametogenesis of the mud crab *Rhithropanopeus harrisii* (Gould) (Brachyura: Xanthidae). Biol. Bull. (Woods Hole) 152: 199–208.

Pound, J. M. and J. H. Oliver, Jr. 1979. Juvenile hormone: evidence of its role in the reproduction of ticks. Science (Wash., DC) 206: 355–357.

Ramenofsky, M., D. J. Faulkner, and C. Ireland. 1974. Effect of juvenile hormone on cirriped metamorphosis. Biochem. Biophys. Res. Commun. 60: 172–178.

Sannasi, A. and T. Subramoniam. 1972. Hormonal rupture of larval diapause in the tick *Rhipicephalus sanguineus* (Lat.). Experentia (Basel) 28: 666–667.

Schneiderman, H. A. and L. I. Gilbert. 1958. Substances with juvenile hormone activity in Crustacea and other invertebrates. Biol. Bull. 115: 530–535.

Solomon, K. R. and A. A. Evans. 1977. Activity of juvenile hormone mimics in egg-laying ticks. J. Med. Entomol. 14: 433–436.

Solomon, K. R., C. K. A. Mango, and F. D. Obenchain. 1982. Endocrine mechanisms in ticks: effects of insect hormones and their mimics on development and reproduction. Pp. 399–438 *in* F. Obenchain and R. Galun (eds.), *Physiology of Ticks*. Pergamon Press, Oxford and Elmsford, New York.

Strambi, C., A. Strambi, M. De Reggi, and M. Delaage. 1984. Radioimmunoassays of juvenile hormones: state of the methods and recent data on validation. Pp. 355–362 *in* J. Hoffmann and M. Porcher (eds.), *Biosynthesis, Metabolism and Mode of Action of Invertebrates Hormones*. Springer-Verlag, Berlin and New York.

Strambi, C., A. Strambi, M. De Reggi, M. Hirn, and M. Delaage, 1981. Radioimmunoassay of insect juvenile hormones and their diol derivatives. Eur. J. Biochem. 118: 401–406.

Wright, J. E. 1969. Hormonal termination of larval diapause in *Dermacentor albipictus*. Science (Wash., DC) 163: 390–391.

Synthesis and Effects on Molting of Juvenile Hormone–like Compounds in Crustacea and Chelicerata

4

F. LACHAISE

4.1.	Introduction	75
4.2.	Various JH-like Compounds	75
4.3.	Synthesis	75
4.4.	Effects on Molting and Metamorphosis	76
	4.1.1. Chelicerates	76
	4.4.1.1. Action on Molt	76
	4.4.1.2. Diapause Termination	77
	4.4.1.3. Juvenilizing Effect	77
	4.4.2. Crustaceans	78
	4.4.2.1. Cirripeds	78
	4.4.2.2. Decapods	78
4.5.	Summary	79
	Acknowledgments	79
	References	79

4.1. Introduction

Regarding juvenile hormone–like compounds (JH-like compounds) compared to ecdysteroids, few studies have been devoted to their identification in arthropods other than insects, although the existence of compounds with JH activity has been shown in crustaceans since the late 1950s (Schneiderman and Gilbert, 1958). The insect juvenile hormones (JH 0, JH I, JH II and JH III) belong to the terpenoid group and are classically recognized as hormones inducing larval molt in the immature insect and as a gonadotropic hormone in the adult. Sections 4.2 and 4.3, respectively, will deal with the identification and synthesis of JH-like compounds in crustaceans. Section 4.4 will be devoted to the effects of these hormones on metamorphosis and molt of chelicerate and crustacean larvae.

4.2. Various JH-like Compounds

In 1958, Schneiderman and Gilbert provided evidence that there exist, in crustaceans, compounds with activity similar to that of insect JH. Both the hemolymph and the androgenic glands of the crab *Carcinus maenas* contain two compounds that are closely related to JH: hexahydrofarnesylacetone and farnesylacetone, the latter being a JH-like compound (Berreur-Bonnenfant et al., 1973; Ferezou et al., 1976, 1977). Most recently, Laufer et al. (1987) identified in the hemolymph of the crab *Libinia emarginata* methyl farnesoate (Fig. 4.1) and two artifactually derived compounds: JH III (Fig. 4.1) and JH III acid. Adult females of the tick *Ornithodoros moubata* possess compounds with juvenilizing and gonadotropic effects similar to those of JH (Connat, 1987).

4.3. Synthesis

Synthesis of such products in arthropods other than insects has been reported in a single study. Laufer et al. (1987) provided evidence of a synthesis of methyl farnesoate (MF) (Fig. 4.1) in the mandibular organs of the spider crab *Libinia emarginata*. These mandibular organs, of hypodermic origin, were described in decapod crustaceans by Le Roux (1968), who had suggested their role in oogenesis. Synthesis of MF by mandibular organs seems to occur via an inhibiting hormone located in the eyestalks (Laufer et al., 1986; see also Borst and Laufer, Chapter 2 herein).

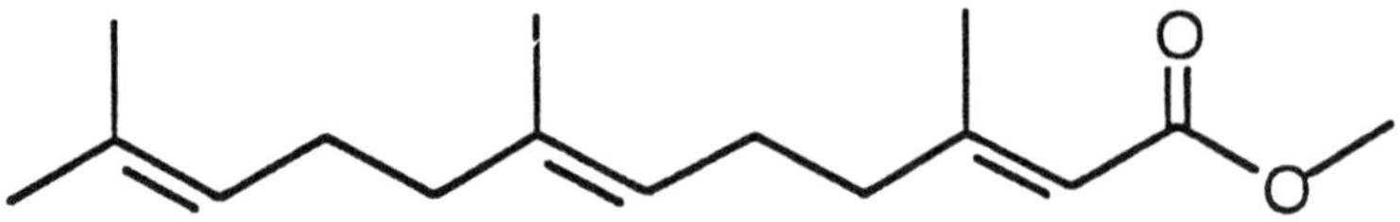

METHYL FARNESOATE

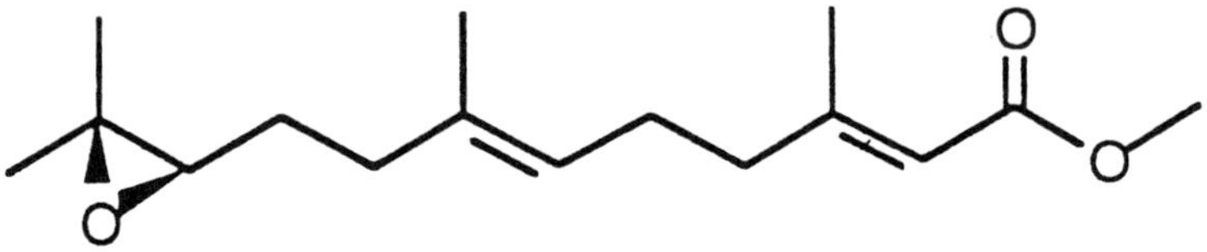

JH III

FIGURE 4.1. Juvenile hormone-like compounds found in crustaceans and major juvenile hormone of insects.

4.4. Effects on Molting and Metamorphosis

In arthropods other than insects, endogenous JH-like compounds have been identified only in crustaceans. Therefore, their role has so far only been studied with regard to exogenous supply. A number of studies have focused attention on JH-like compounds owing to their toxic effect on undesirable species of crustaceans (e.g., barnacles) and on ticks, mite pests, and predatory mite species. Their action was tested in the larva to provide evidence of their role in the larval molt, and in the adult to analyze their role in reproduction.

4.4.1. *Chelicerates*

4.4.1.1. ACTION ON MOLT

Addition of JH-like compounds generally seems to induce a delay in both molt and metamorphosis of larvae and results in high mortality. In ticks, when JH-like compounds are applied to *Dermacentor variabilis* and *Haemaphysalis cancinna* nymphs immediately after engorgement, there is a disruption of molting, development, and occurrence of high mortality (McDaniel and Oliver, 1978; Ioffe et al., 1977). The effects of

JH-like compounds vary according to their nature and the animal used: Solomon and Evans (1977, 1978) have determined the 50% effective concentrations for induction of molt inhibition in engorged *Amblyomma hebraeum* larvae by a number of JH-like compounds (Solomon and Evans, 1978) and for induction of the desiccation of laid eggs by topically treated female *Boophilus decoloratus* (Solomon and Evans, 1977). According to these studies, the same compound can have an effect on molt in ticks and none on reproduction. Moreover, the effects of JH-like compounds vary according to the time of application to the experimental animal: engorged females of *Argas habraeum* topically treated with the JH-like compounds, molt normally (Solomon and Evans, 1978), and females obtained from treated nymphs laid a normal number of eggs. However, treatment of female *Boophilus decoloratus* on the day engorgement ended resulted in desiccation of the eggs (Solomon and Evans, 1977). A hypothesis was developed for ticks based on the antagonistic effect of JH-like compounds and ecdysteroids. Dose-dependent reversal of 20-hydroxyecdysone (20E)-induced supermolting among mated *Ornithodoros porcinus*, with promotion of normal oviposition, occurred in response to topical application of JH III (Obenchain and Mango, 1980). Fed fourth-stage nymphal *Ornithodoros porcinus* also exhibited a molting delay after exposure to precocene (a compound with ''antiallatal'' activities), but that delay could be antagonized if the nymphs were fed on blood containing 20E (Solomon et al., 1982). These results suggest that endogenous JH-like compounds may be involved in the ecdysteroid-dependent control of molting and developmental processes.

4.4.1.2. DIAPAUSE TERMINATION

In the tick *Argas arboreus*, reproductive diapause was broken in females by the application of 10 μg of JH-like compounds (Bassal and Roshdy, 1974). In nondiapausing *Ixodes ricinus* nymphs, topical application of Altozar, a JH-like compound, delayed ecdysis. In contrast, the molting period is reduced in nymphs undergoing a developmental diapause.

4.4.1.3. JUVENILIZING EFFECT

The juvenilizing effect of JH-like compounds on the development of insect larvae consists of inducing a larval molt and producing supernumerary larval stages. This juvenilizing effect does not seem to exist in ticks (Obenchain, 1979), other arachnids (Matsuda, 1979), or *Limulus* (Schneiderman and Gilbert, 1958; Jegla, 1982). In *Limulus*, Jegla (1982) concluded "present data suggest that JH-like compounds have no role in the morphogenesis of *Limulus*."

4.4.2. *Crustaceans*

In crustaceans, JH-like compounds have been tested more particularly for their action on the metamorphosis of larvae and their gonadotropic effect (Gomez et al., 1973; Cheung and Nigrelli, 1973; Ramenofsky et al., 1974; Christiansen et al., 1977a,b; Landau and Finney, 1977; Tighe-Ford, 1977; Mortlock et al., 1984; Hertz and Chang, 1986).

4.4.2.1. CIRRIPEDS

Barnacles are ship-bottom-fouling animals, and so their economic impact is considerable; therefore a number of studies have been devoted to means of controlling their development. The action of JH-like compounds has been analyzed in these animals in order to better understand the mechanism of their control and to test their possible toxic effect. The addition of exogenous JH-like compounds provokes mostly precocious metamorphosis in this crustacean. In the cirriped *Balanus galeatus*, larval development seems unaffected by the presence of ZR 512 and ZR 515 (JH-like compounds) used at nonlethal doses. With ZR 512, nonfeeding cyprid stage barnacles metamorphosed prematurely without attaching to a substrate, whereas ZR 515 did not cause premature metamorphosis (Gomez et al., 1973). The metamorphosis of cirriped larvae is also affected by other JH-like compounds (farnesyl methyl ester and farnesol, among others). These compounds caused precocious metamorphosis and abnormalities in cyprid larvae (Cheung and Nigrelli, 1973; Ramenofsky et al., 1974; Landau and Finney, 1977; Tighe-Ford, 1977; Mortlock et al., 1984).

4.4.2.2. DECAPODS

ZR 512 and ZR 515 do not inhibit metamorphosis of larvae of the mud crab *Rhithropanopeus harrisii*; in contrast, duration of zoeal development was significantly delayed with an increase in concentration of

ZR 512 or ZR 515 (Christiansen et al., 1977a,b). In the lobster, JH III delayed metamorphosis whereas farnesol was not effective (Hertz and Chang, 1986). The apparent contradiction between the two results induced by JH III or JH-like compounds, i.e., precocious metamorphosis in *Balanus galeatus* and delayed metamorphosis in lobster larvae, may possibly be explained by the interaction of JH-like compounds with ecdysteroids (Cheung, 1974).

4.5. Summary

Crustaceans possess JH-like compounds that are synthesized by mandibular organs. The role of JH-like compounds in the molting process of arthropods other than insects has so far received little attention compared to that of ecdysteroids. However, owing to their toxic effect, their action was tested on undesirable crustaceans or ticks. Addition of JH-like compounds induced a delay in both molt and metamorphosis of decapod and tick larvae. In contrast JH-like compounds caused precocious metamorphosis in cyprid larvae.

Acknowledgments

I am indebted to P. Diehl for having supported this project. I wish to express deep gratitude to P. Harry, D. Lachaise, and G. Somme for help during the preparation of the manuscript. I also thank J. L. Connat, R. Lafont, and F. Xavier for helpful comments on the chapter. I gratefully acknowledge the work of G. Carpentier in preparing the final version of the figures.

References

Bassal, T. T. M. and M. A. Roshdy. 1974. *Argas (Persicargas) arboreus:* juvenile hormone analog termination of diapause and oviposition control. Exp. Parasitol. 36: 34–39.

Berreur-Bonnenfant, J., J. J. Meusy, J. P. Ferezou, M. Devys, A. Quesneau-Thierry, and M. Barbier. 1973. Recherches sur la sécrétion de la glande androgène des crustacés malacostracés: purification d'une substance à activité androgène. C. R. Acad. Sci. Paris 277D: 971–974.

Cheung, P. J. 1974. The effect of ecdysterone on cyprids of *Balanus eburneus* Gould. J. Exp. Mar. Biol. Ecol. 15: 223–229.

Cheung, P. J. and R. F. Nigrelli. 1973. The development of barnacles from cyprids in pre-heated seawaters, with and without farnesol. Am. Zool. 13(4): 1339–1340 (abstract).

Christiansen, M. E., J. D. Costlow, Jr., and R. J. Monroe. 1977a. Effects of the juvenile hormone mimic ZR-514 (Altosid®) on larval development of the mud crab *Rhithropanopeus harrisii* in various cyclic temperatures. Mar. Biol. (NY) 39(3): 269–280.

Christiansen, M. E., J. D. Costlow, and R. J. Monroe. 1977b. Effects of the juvenile hormone mimic ZR-515 (Altozar®) on larval development of the mud crab *Rhithropanopeus harrisii* at various cyclic temperatures. Mar. Biol. (NY) 39(3): 269–280.

Connat, J. L. 1987. Aspects endocrinologiques de la physiologie du développement et de la reproduction chez les tiques. Doctoral thesis, University of Dijon, France.

Ferezou, J. P., J. Berreur-Bonnenfant, J. J. Mausy, M. Barbier, M. Suchy, and H. K. Wipf. 1976. 6,10,14-Trimethylpentadecan-2-one and 2,10,14-tri-methyl-5-*trans*,9-*trans*,13-pentadecatrien-2-one from the androgenic glands of the male crab *Carcinus maenas*. Experientia (Basel) 33: 3–4.

Ferezou, J. P., J. Berreur-Bonnenfant, A. Tekitek, M. Rojas, M. Barbier, M. Suchy, H. K. Wipf, and J. J. Meusy. 1977. Biologically-active lipids from the androgenic gland of the male crab *Carcinus maenas*. *Pp. 361–366 in* D. J. Faulkner and W. H. Fenical (eds.), *Marine Natural Products Chemistry.* Plenum Press, New York.

Gomez, E. D., D. J. Faulkner, W. A. Newman, and C. Ireland. 1973. Juvenile hormone mimics: effect on cirriped crustacean metamorphosis. Science (Wash., DC) 179: 813–814.

Hertz, W. A. and E. S. Chang. 1986. Juvenile hormone effects on metamorphosis of lobster larvae. Int. J. Invertebr. Reprod. 6:99–110.

Ioffe, E. D., I. V. Uspensky, I. K. Bessmertnaya, and N. I. Kachanko. 1977. Effect of synthetic juvenile hormone analogue on Ixodoidea. J. Gen. Biol. 38: 885–892.

Jegla, T. C. 1982. A review of the molting physiology of the trilobite larva of *Limulus*. Pp. 83–101 *in* J. Bonaventura, C. Bonaventura, and S. Tesh (eds.), *Physiology and Biology of Horseshoe Crabs: Studies on Normal and Environmentally Stressed Animals.* Liss, New York.

Landau, M. and C. Finney. 1977. Insect juvenile hormones and their mimics: application in crustacean research. Fl. Sci. 40(Suppl.): 14.

Laufer, H., M. Landau, D. Borst, and E. Homola. 1986. The synthesis and regulation of methyl farnesoate, a new juvenile hormone for crustacean reproduction. Pp. 135–143 *in* M. Porchet, J. C. Andries, and A. Dhainaut (eds.), *Advances in Invertebrate Reproduction. Elsevier/North-Holland, Publ., Amsterdam and New York.*

Laufer, H., D. Borst, F. C. Baker, C. Carrasco, M. Sinkus, C. C. Reuter, L. W. Tsai, and D. A. Schooley. 1987. Identification of a juvenile hormone–like compound in a crustacean. Science (Wash., DC) 235: 202–205.

Le Roux, A. 1968. Description d'organes mandibulaires nouveaux chez les Crustacés Décapodes. C. R. Acad. Sc. Paris 266D: 1414–1417.

Matsuda, R. 1979. Abnormal metamorphosis and arthropod evolution. Pp. 137–256 *in* A. P. Gupta (ed.), *Arthropod Phylogeny.* Van Nostrand Reinhold, New York.

McDaniel, R. S. and J. H. Oliver, Jr. 1978. Effects of two juvenile hormone analogs and β-ecdysone on nymphal development, spermatogenesis and embryogenesis in *Dermacentor variabilis* (Say) (Acari: Ixodidae). J. Parasitol. 64: 571–573.

Mortlock, A. M., J. T. R. Fitzsimons, and G. A. Kerkut. 1984. The effects of farnesol on the late stage nauplius and free swimming cypris larvae of *Elminius modestus* (Darwin). Comp. Biochem. Physiol. 78A: 345–357.

Obenchain, F. D. 1979. Non-acaricidal chemicals for management of Acari of medical and veterinary importance. Pp. 35–43 *in* J. G. Rodiquez (ed.), *Recent Advances in Acarology.* Academic Press, Orlando, Florida.

Obenchain, F. D. and C. Mango. 1980. Effects of exogenous ecdysteroids and juvenile hormone on female reproductive development in *Ornithodoros porcinus*. Am. Zool. 20: 936.

Ramenofsky, M., D. J. Faulkner, and C. Ireland. 1974. Effect of juvenile hormone on cirriped metamorphosis. Biochem. Biophys. Res. Commun. 60: 172–178.

Schneiderman, H. A. and L. I. Gilbert. 1958. Substances with juvenile hormone activity in Crustacea and other invertebrates. Biol. Bull. (Woods Hole) 115: 530–535.

Solomon, K. R. and A. A. Evans. 1977. Activity of juvenile hormone mimics in egglaying ticks. J. Med. Entomol. 14: 433–436.

Solomon, K. R. and A. A. Evans. 1978. The activity of insect juvenile mimics in larval *Amblyomma hebraeum* Koch (Acarina: Metastriata: Ixodidae). Onderstepoort J. Vet. Res. 45: 39–42.

Solomon, K. R., C. K. A. Mango, and F. D. Obenchain. 1982. Endocrine mechanisms in ticks: effects of insect hormones and their mimics on development and reproduction. Pp. 399–438 *in* F. D. Obenchain and R. Galun (eds.), *Physiology of Ticks.* Pergamon Press, Oxford and Elmsford, New York.

Tighe-Ford, J. D. 1977. Effects of juvenile hormone analogues on larval metamorphosis in the barnacle *Eliminus modestus* Darwin (Crustacea: Cirripedia). J. Exp. Mar. Biol. Ecol. 26: 163–176.

Biosynthesis, Titer Regulation, and Transport of Juvenile Hormones   5

WALTER G. GOODMAN

5.1. Introduction   85
5.2. Biosynthesis of the Juvenile Hormones   85
    5.2.1. Juvenile Hormone-Producing Tissues   86
      5.2.1.1. Corpora Allata   86
      5.2.1.2. Accessory Sex Glands   87
      5.2.1.3. Imaginal Disks   88
    5.2.2. Synthesis of JH   89
      5.2.2.1. Formation of the Carbon Skeleton
      (JH III)   89
      5.2.2.2. Formation of the Methylester and the
      Epoxide Moieties   91
      5.2.2.3. Formation of the Carbon Skeleton
      (Higher Homologues)   92
5.3. Regulation of the CA During Larval Development
and Metamorphosis   93
    5.3.1. Innervation of the CA   93
    5.3.2. Brain Regulation of CA Activity   94
      5.3.2.1. Neural Regulation   94
      5.3.2.2. Evidence for Allatotropins   96
      5.3.2.3. Chemical Nature of the ATF   97
      5.3.2.4. Evidence for Allatostatins   97
      5.3.2.5. Chemical Nature of the ASF   98
      5.3.2.6. Mode of Action of the Allato-Active
      Factors   99
      5.3.2.7. Ecdysteroid Regulation of CA
      Activity   101
5.4. Hemolymph JH Binding   102
    5.4.1. Low-Affinity JH–Protein Associations   103
      5.4.1.1. Hemolymph Storage Protein Binding   103
      5.4.1.2. Hemolymph Lipophorin Binding   104

5.4.2.   High-Affinity, Low-Molecular-Weight
         JH-Binding Proteins   105
         5.4.2.1.   Purification and Characterization   105
         5.4.2.2.   Specificity   107
         5.4.2.3.   Biosynthesis and Titers   108
5.4.3.   High-Affinity, High-Molecular-Weight
         JH-Binding Proteins   109
         5.4.3.1.   Purification and Characterization   109
         5.4.3.2.   Specificity   110
         5.4.3.3.   Biosynthesis and Titers   111
5.4.4.   Functions of the JH-Binding Proteins   112
5.5.   Summary   113
Acknowledgments   114
References   115

5.1. Introduction

The juvenile hormones (JHs) are a group of acyclic sesquiterpenoids that regulate insect development and reproduction. During the immature stages, ecdysteroids, in concert with the JHs, initiate apolysis when the growth of the insect is inhibited by the confining exoskeleton. The nature of the internal organs and new integument is a manifestation of the JH titers during the preceding stadium. Insect development as well as insect evolution are thus dependent upon regulation of biosynthesis and peripheral distribution of the JHs. It has become increasingly evident that regulation of JH cannot be easily dissociated from that of other known insect growth hormones. The endocrine systems are regulated by their own products via a complex arrangement of hormonal interactions and feedback mechanisms that modulate endocrine activity both temporally and quantitatively (Watson et al., 1986). Thus, in the half-century since the discovery of JH, the field has matured sufficiently so that no one review can encompass all aspects of this very important group of developmental hormones. Accordingly, this review will focus on JH biosynthesis, titer regulation, and transport during development.

5.2. Biosynthesis of the Juvenile Hormones

Although it has long been recognized that the corpora allata (CA) are responsible for the biosynthesis of the JHs, this view has been modified in recent years to include extra-allatal sites. The accessory sex glands of *Hyalophora cecropia* (Shirk et al., 1983) and the imaginal disks of *Manduca sexta* (Sparagana et al., 1985) synthesize JH by enzymatically converting JH acids to JH via the enzyme epoxyfarnesoic acid methyltransferase [= JH acid methyltransferase (see Weirich, 1985)]. Although they do not manufacture the JH carbon skeleton, these tissues nevertheless produce JH and therefore must be considered in any evaluation of total JH in the insect.

5.2.1. *Juvenile Hormone-Producing Tissues*

5.2.1.1. CORPORA ALLATA

The evolutionary divergence in the structure and location of the CA has been well documented by Cassier (1979; see also Chapter 2 in Part 2) and reviewed more recently by Sedlak (1985). As noted by these authors, there is a great deal of variation in size between species as well as variation within the same species, due to age, sex, polymorphism, and activity cycle, making generalizations difficult. In a number of insect orders, the glands are paired and lie within the head capsule dorsolateral to the esophagus. In some orders, the glands become fused to form a single body either ventral to the aorta, as in the Embioptera, Dermaptera, and Psocoptera, or dorsal as in the Diptera.

In addition to external morphology and relative position of the CA, there are also order and species differences with regards to cell size and arrangement within the gland. The CA of *Manduca sexta* offer an example of the macrocell type (Sedlak, 1985), which is thought to be the most highly evolved of the four types (Cassier, 1979). The large cells are surrounded by a noncellular basal lamina, which is invaded by neurosecretory cell axons. The neurosecretory cells also penetrate the gland itself with numerous terminal vericosities observed (Sedlak, 1985). The larval CA cell shape is irregular with numerous projections that interdigitate with neighboring cells, and in actively secreting cells intercellular spaces are observed.

At the ultrastructural level, the CA cells of *Manduca* are connected to the basal lamina by hemidesmosomes, while junctions between cells appear to be of the maculae adherens, septate desmosome, or gap junction types. The most distinguishing feature of the active CA is the abundance of smooth endoplasmic reticulum (SER) during periods of high JH biosynthesis. In Lepidoptera, the SER becomes less abundant as the glandular activity is reduced and the concentric whorls of SER, characteristic of active glands during the fourth and early fifth stadia, are absent in the pupal glands. Another distinguishing feature of active glands are the numerous mitochondria and free ribosomes. Conversely, there is little rough endoplasmic reticulum and few Golgi complexes, indicating a reduced protein synthetic and secretory capacity (Sedlak, 1985).

The use of CA ontogeny, JH titer studies, and gross and ultrastructural correlates to assess synthetic capacity of the CA has been the focus of a number of investigators (see Tobe and Stay, 1985, for a review). The CA are of ectodermal origin arising from the maxillary segment at the base of the mandibular segment or from a region between the two segments, depending upon the species (Johannsen and Butt, 1941). The glands appear to become functional early in embryonic life, as determined by changing JH titers and by the use of inhibitors of JH synthesis (Bergot et al., 1981). The addition of the inhibitor fluoromevalonate to embryos reduced JH titers, thus suggesting that the JH present is embryonic in origin rather than maternal. The resulting larvae also displayed symptoms of JH deficiency. Therefore, the embryo appears capable of synthesizing JH, presumably through CA activity, and the hormone is required for normal larval development.

As previously noted, the CA of many insects undergo changes in volume depending upon the physiological state of the insect, and these observations have, in turn, been used to describe the putative activity of the gland (see Tobe and Stay, 1985, for a review). However, in the immature insect, the change in CA size is due to intrinsic growth rather than increases in JH biosynthesis (Sedlak, 1985). Gland volume–activity correlates are further challenged by the observations that JH biosynthesis in nymphal *Schistocerca gregaria* and *Diploptera punctata* is lower in the later part of the stadium when the glands are actually larger (see Tobe and Stay, 1985, for a review). Moreover, storage of JH within the glands has not been detected (see Tobe and Stay, 1985, for a review; Granger et al., 1979, 1982a). Thus, a positive correlation between volume and activity can be made only on a case-by-case basis. It is expected that these generalizations will be eliminated by increased use of JH radioimmunoassays (Granger et al., 1979) or radiochemical assays (see Feyereisen, 1985b) to measure JH biosynthesis.

5.2.1.2. ACCESSORY SEX GLANDS

One of the more interesting discoveries in the field of JH biochemistry was the detection of extra-allatal tissue, which could support limited JH biosynthesis. Since the mid-1950s, it was recognized that abdomens of male *Hyalophora cecropia* contained large amounts of JH; however, nearly two decades passed before it was revealed that the accessory sex glands (ASGs) were the source of the hormone (Dahm et al., 1976). Following up on this discovery, it was found that the pharate male CA synthesized JH I acid as a precursor, which was converted to

JH I in the ASGs (Shirk et al., 1983) by the enzyme JH acid methyl-transferase (JHAMT).

ASGs are relatively long (30–60 mm), paired tubular organs that extend distally from and open into the seminal vesicles. The structures are invested with a thin layer of longitudinal transverse muscle, and beneath this sheath is a single layer of glandular cells, which face a central lumen. The lumen contains a granular secretory material that first appears at day 16 and is secreted until day 19. The composition of this material is unknown but may contain JH-binding proteins that prevent extensive diffusion of the hormone from the gland.

JHAMT appears in homogenates of the ASGs on day 14 of adult development, but the glands do not become competent to convert JH acid to JH until day 16. High levels of JH are not sequestered until just prior to adult eclosion some 8 days later (Shirk et al., 1983). Although the enzyme is found in both the lumen and the glands, the newly synthesized JH is localized in the lumen. Parabiosis studies have indicated that despite hormone sequestration, a portion of the JH is liberated from the ASGs into the hemolymph.

5.2.1.3. IMAGINAL DISKS

The recent discovery that the imaginal disks of *Manduca sexta* are capable of producing JH presents a singularly intriguing idea (Sparagana et al., 1985). As in the case of the ASGs, the biosynthetic capabilities of the imaginal disks are limited to enzymatic conversion of JH acid to JH. The crucial role for this biosynthetic site in normal pupal development was demonstrated in a series of transplantation experiments based on observations that allatectomized fifth instars undergo precocious adult development (Kiguchi and Riddiford, 1978). Sparagana et al. (1985) removed CA from fourth or early fifth (last) instars and reimplanted them into allatectomized fourth instar hosts. The hosts subsequently molted to the fifth instar and pupated successfully, showing no precocious adult development. CA taken from wandering larvae, pupae, or early pharate adults and tested in the same bioassay were incapable of promoting larval-to-larval molting but were able to prevent precocious adult development. CA taken from adult males and tested in the same manner resulted in a larval-to-pupal molt, the animals displaying precocious development.

Sparagana et al. (1985) interpreted these observations as follows: (1) CA taken from fourth and early fifth instars secrete JH I and II, which rescue the allatectomized animals; (2) CA from wandering-stage animals secrete JH I and II acids, which cannot promote the fourth-to-fifth stadium molt but can rescue allatectomized fifth instars owing to the presence of a JHAMT in the imaginal disks; and (3) CA from adult males produce only JH III acid, which is biologically inactive in this assay. These conclusions were supported by the observation that injection of JH I acid into allatectomized fifth-instar animals could be recovered 24 h later as JH I (also see Baker et al., 1987, for JH and JH acid titers). Analysis of JHAMT activity of various tissue homogenates demonstrated that JH I acid could be converted to JH I by imaginal disks; however, other tissues, including fat body and epidermis, were inactive.

These findings have opened several very fertile areas for future research. First, the CA may synthesize JH or JH acid, depending upon the stage-specific presence of a JHAMT (see Section 5.2.2.2, below). Second, JH acids, heretofore thought of as inactive metabolites, may actually be prohormones for JH-dependent tissue, such as the imaginal disks, during prepupal development (Bhaskaran et al., 1986). Third, the imaginal disks, whose development is suppressed by JH, can convert the precursor JH acid to the active hormone, which is released back into the hemolymph, presumably to prevent precocious imaginal disk development.

5.2.2. *Synthesis of JH*

5.2.2.1. FORMATION OF THE CARBON SKELETON
(JH III)

Virtually all of our knowledge of the biosynthesis of the JH carbon skeleton comes from *in vitro* studies pioneered in the early 1970s and continued to the present by the Zoecon group (see Schooley and Baker, 1985, for a review). By employing *in vitro* techniques, these investigators were able to overcome problems such as massive precursor dilution engendered by *in vivo* analysis. Coupled with cell-free homogenate studies, *in vitro* analysis has enabled these investigators to elucidate the synthetic pathway for each of the homologues. For a comprehensive report, the reader is directed to the review by Schooley and Baker (1985).

The synthesis of JH III, the homologue lacking branched side chains, is now known to follow the pathway prescribed for the initial steps in cholesterol synthesis. The potential precursors for the carbon skeleton of JH III are two-carbon (C_2) units that have arisen from the metabolism of glucose, leucine, isoleucine, and threonine as well as the immediate precursors of acetate (Brindle et al., 1987). Identification of the precursors' immediacy was determined by its degree of dilution in the final product. Three C_2 units comprising acetyl–coenzyme A (acetyl-CoA) undergo enzymatic condensation to yield a six-carbon intermediate, 3-hydroxy-3-methylglutaryl-CoA (HMG-CoA). The condensation reaction is under the control of several enzymes, including a thiolase, which is responsible for activation of the acetyl and acetoacetyl groups and the enzyme HMG-CoA synthetase.

HMG-CoA is reduced to mevalonate by the microsomal-associated enzyme HMG-CoA reductase, which uses NADPH (nicotinamide adenine dinucleotide phosphate, reduced) as an electron donor. This enzyme is known to play a central role in mammalian steroid biosynthesis and has been well characterized. It is negative feedback controlled by cholesterol and other isoprenoids. Phosphorylation inhibits the enzyme, whereas dephosphorylation activates it (Kennely and Rodwell, 1985). Based on enzyme requirements and inhibitor studies, the process of phosphorylation/dephosphorylation has been proposed as a possible regulatory control point for JH biosynthesis (Feyereisen et al., 1981a; Monger et al., 1982). Because so little is known about the CA HMG-CoA reductase it cannot be distinguished from other phosphorylated proteins in a CA homogenate.

A series of studies of HMG-CoA reductase in the *Drosophila* K_c cell line (Watson et al., 1985; Havel et al., 1986) proposed that a rapid regulatory event linked to pool size of a nonsterol, post–isopentenyl pyrophosphate ligand serves as a negative signal for the irreversible modulation of HMG-CoA reductase. According to the aforementioned authors, the regulatory molecule is generated from shunted mevalonate carbons that end up in the nonisoprenoid end products. They suggest that when the concentration of the regulatory signal molecule is increased, the HMG-CoA reductase rate constant for irreversible inactivation is rapidly enhanced. This inactivation, coupled with repression of HMG-CoA reductase gene transcription, would give rise to a new steady state commensurate with the increase in the pool of the regulatory molecule. If parallels can be drawn between isoprenoid synthesis in K_c cells and the CA, the existence of regulatable post–isopentenyl pyrophosphate carbon shunting suggests that caution should be exercised in attributing changes in the rate of isopentenoid end product synthesis exclusively to modulation of HMG-CoA reductase activity. Further evidence that HMG-CoA reductase is not involved in regulating levels of JH has been presented by Bhaskaran et al. (1987). It was noted that the reductase activity of the

CA does not follow the expected titers of JH or JH acid, since high levels of the enzyme are observed in late prepupal and pupal glands when titers are very low (see Granger et al., 1984, for a contrasting viewpoint).

The next step in JH biosynthesis is the conversion of mevalonate to 3-isopentenyl pyrophosphate (IPP). In insects, this reaction is not well characterized; however, assuming the pathway to be similar to that observed in mammals, three enzymes appear to be involved: (1) mevalonate kinase, which phosphorylates mevalonate to form 5-phosphomevalonate; (2) 5-phosphomevalonate kinase, which converts its substrate to the intermediate 3-phospho-5-pyrophosphome-valonate; and (3) adenosine triphosphate(ATP)–dependent mevalonate 5-pyrophosphate decarboxylase, responsible for the synthesis of IPP, which isomerizes to 3,3'-dimethylallyl pyrophosphate (DMAPP). Both of these precursors have been isolated from the CA of *Manduca* (Baker et al., 1981). The last step in the synthesis of the carbon skeleton is the condensation of two units of IPP and one unit of DMAPP to form the basic farnesyl pyrophosphate unit. The reaction is catalyzed by the enzyme prenyl transferase. Farnesyl pyrophosphate is converted to farnesol enzymatically, presumably by a phosphatase or pyrophospha-tase (Reibstein et al., 1976). Farnesol is then oxidized to farnesal via a dehydrogenase that requires NAD and then further oxidized to far-nesoic acid by another dehydrogenase also requiring NAD (Baker et al., 1983).

5.2.2.2. FORMATION OF THE METHYLESTER AND THE EPOXIDE MOIETIES

The terminal steps in biosynthesis are methylester formation of C-1 and epoxidation of the C-10/C-11 position. Early studies examining the origins of the branched side chains found that radiolabeled methionine was incorporated into the C-1 position (Metzler et al., 1971). Reibstein and Law (1973) identified *S*-adenosyl-L-methionine (SAM) as the methyl donor by supplementing homogenates with radiolabeled SAM to yield labeled JH. At the present time, very little is known about the physicochemical characteristics of the CA JHAMT.

The little that is known about methyltransferase pertains to its sub-strate specificity and developmental profile. Reibstein et al. (1976) demonstrated that the enzyme from homogenates of *Manduca sexta* CA preferred the epoxidized substrate; in contrast, the methyl-transferase from Orthoptera and Blattodea CA homogenates does not require the epoxide moiety to perform the *O*-methylation. In *Manduca,* the enzyme's activity appears to be regulated; the CA of the fifth instar switch from JH to JH acid secretion just before the onset of wandering (Sparagana et al., 1985; Bhaskaran et al., 1986, 1987), as is reflected in the hemolymph JH acid titers (Baker et al., 1987). JH esterase (see Chapter 6 herein by Roe and Venkatesh) titers in *Manduca* (Schooley and Baker, 1985; Bhaskaran et al., 1986) and *Galleria* (Wiśnewski et al., 1986) CA begin to rise during the same time period, and one might suggest that the rapidly synthesized JH is metabolized to JH acid before leaving the gland. However, use of potent JH esterase inhibitors in the gland homogenate dispels this idea and demonstrates that the CA methyltransferase activity is indeed regulated.

The formation of the epoxide moiety on the JH carbon chain has received far less attention. This enzymatic process, depending upon the species, is either the last or penultimate step in the biosynthetic pathway. The enzyme 10,11-epoxidase appears to be a microsomal enzyme, with many properties of a cytochrome P-450–linked enzyme, including inhibition of carbon monoxide and reversal of this inhibition by white light (Feyereisen et al., 1981b), inhibition by methylene blue and piperonyl butoxide, and a requirement for NADPH as a cofactor (Hammock, 1975).

5.2.2.3. FORMATION OF THE CARBON SKELETON (HIGHER HOMOLOGUES)

The origin of the branched side chains of JH 0, JH I, JH II and 4-methyl–JH I was also elucidated by the Zoecon group. Employing a series of elegant microchemical analyses, Schooley et al. (1976) demonstrated that branched precursors to the higher homologs originated from propionyl-CoA. The homoisoprenoid units, which give rise to the branched chains, are formed from condensation of one propionyl-CoA unit and two acetyl-CoA units. This enzymatic reaction, similar to that described for JH III, results in the synthesis of a homomevalonate intermediate. Thus, JH 0 is composed of three homomevalonate units, whereas JH I is composed of two homomevalonate and one mevalonate units and JH II is composed of one homomevalonate and two mevalonate units. For a discussion of 4-methyl–JH I, the reader is referred to Schooley and Baker (1985).

5.3. Regulation of the CA During Larval Development and Metamorphosis

The well-defined orchestration of CA activity during development implies the existence of glandular regulating mechanisms. Two sources of CA regulatory input have long been recognized but are only now being extensively investigated. These regulatory sources include nervous input from the brain, both negative and positive, and the allatotropins and allatostatins, which until recently have been inferred and not actually demonstrated. In conjunction with the allato-active factors, there may be peripheral factors, such as key substrates, and metabolites, which regulate CA activity via feedback loops (Feyereisen, 1985a).

5.3.1. *Innervation of the CA*

Innervation from the brain has been shown to be a vital component in regulation of CA activity, and because of its importance, especially in light of recent advances, a brief overview of the neural connectives is justified. The evolutionary diversity of insects has led to a number of variations in the innervation of the CA; however, the prevalent pattern appears to be a series of inputs from the brain and from the subesophageal ganglion (see Tobe and Stay, 1985, and Feyereisen, 1985a, for comprehensive reviews). Three nerves emanate from the brain and enter the corpus cardiacum (CC), nervi corporis cardiaci (NCC) I, NCC II, and NCC III. Axons originating in the medial neurosecretory cells of the protocerebral lobe contribute to the contralateral NCC I, whereas axons arising from the lateral neurosecretory cells of the protocerebrum contribute to the ipsilateral NCC II. Axons originating in the tritocerebral neurosecretory cells reach the CC via NCC III. In some species, the neural tracts, which appear to innervate the CC, pass through the gland and terminate in or on the CA. These nerves, which exit from the CC and enter the CA, are collectively termed the nervi corporis allati (NCA) I, whereas tracts from the subesophageal ganglion are termed NCA II. While this is the basic plan, there appear to be significant differences between species.

Using *Manduca* as an example, it has been observed that NCC I and II, which innervate the CA, fuse outside the brain and have been labeled NCC I + II (Nijhout, 1975). Upon entering the brain, the fused neural tract splits, with one branch running along the ventral side of the brain and bending sharply upward upon crossing into the contralateral lobe. This bundle of axons connects with the lateral neurosecretory cells L-NSC III (= IIa of Carrow et al., 1984; see also Agui et al., 1979) of the brain, which are responsible for the synthesis of prothoracicotropic hormone (PTTH). The CA are also innervated by fibers from the contralateral medial cells M-NSC Ia, which travel to the surface of the gland; this may explain the innervation seen in ultrastructural examination of the basal lamina (Sedlak, 1985). Terminal varicosities in M-NSC I suggest that neurosecretory material may be released from this area. Fibers leading away from the group of cells labeled IIb (Carrow et al., 1984) run ipsilaterally to the CA via NCC II.

Using monoclonal antibodies to big PTTH, O'Brien et al. (1988) immunologically identified an extensive dendritic field in the protocerebrum in addition to the long fibers that extend into and arborize in the CA. Given this extensive field and probable overlap with dendritic fields from different neurosecretory axons, it becomes evident that neuronal intercommunication is quite likely.

5.3.2. *Brain Regulation of CA Activity*

The extensive innervation of the CA supports the fact that the glands are subject to regulation by the brain. Until recently, attention has centered on brain regulation of the CA during reproduction and only a few studies addressed the problem covered in this review, i.e., insect development. Advances in *in vitro* technology have opened the door for studies of CA regulation, which have clearly identified the brain of the immature insect as a primary regulatory agent. Brain regulatory mechanisms may be purely neural, neurosecretory, or a combination of both; as noted by Feyereisen (1985a), these neurosecretory factors may be released into neurohemal organs, such as the CC, and act via the hemolymph, and/or may be released inside the CA. To provide a comprehensive overview of this process in the immature insect, attention will be primarily focused on two species, *Manduca sexta* and *Diploptera punctata*, for which a substantial body of information on allatotropins and allatostatins as well as JH has been generated.

5.3.2.1. NEURAL REGULATION

In surveying the literature, it becomes apparent that neural regulation of the CA has been loosely defined as control of JH synthesis or release by either neurosecretory factors or by depolarization of the CA cell.

In *Diploptera*, Szibbo and Tobe (1983) demonstrated that denervation of the CA in the last nymphal instar would lead to supernumerary instars and suggested that nerve tracts from the brain supplied an inhibitory response. A radiochemical assay (RCA), which monitors the incorporation of [^{3}H]methionine into JH as a measurement of *in vitro* hormone synthesis, revealed a significant increase in JH synthesis in denervated glands (see Feyereisen, 1985b, for details of this assay). The inhibitory role of innervation was further revealed by studies in which the intact complex of adult brain–CC–CA cultured *in vitro* displayed a low JH biosynthetic rate in comparison to either isolated CA or CA coincubated with detached brain–CC (Rankin et al., 1986). The inhibition could be intensified by treatment in medium containing high potassium, presumably by depolarization of nerves innervating the CA. It was suggested that allatostatic factors stored in the brain as neurosecretory granules are released directly within the CA upon depolarization. To test the neurosecretion hypothesis, glands were incubated in Mg^{2+}, a known calcium antagonist. As extracellular Ca^{2+} is required for neurosecretion, the presence of Mg^{2+} would be expected to have an antagonistic or inhibitory effect on potassium-induced secretion of the allatostatin. As expected, incubation in medium containing a high Mg^{2+} concentration abolished the potassium-induced response. Although these findings strongly suggest neurosecretory regulation of JH biosynthesis, a high potassium concentration may act directly on the CA cells to induce JH release. A dissociated cell preparation of CA, free of nerve terminals and challenged in the manner described above, should yield definitive results.

Based on biological activity, it has been postulated that JH is still present in *Manduca* at day 3 of the fifth instar (Nijhout and Williams, 1974; however, see Baker et al., 1987). Bhaskaran (1981) demonstrated that an intact brain–CC–CA shuts off biosynthesis and release of JH; however, this inactivation can be overcome by severing NCA I prior to day 3. Severing the nerve after day 3 could not reverse the inhibition. As observed in *Diploptera*, it is difficult to determine whether the inhibition is purely neural or is derived from neurosecretory molecules transported directly to the gland.

5.3.2.2. EVIDENCE FOR ALLATOTROPINS

Allatotropins are regulatory molecules that stimulate an increase in the biosynthetic activity of the CA. With the exception of *Manduca,* where the putative allatotropic molecules have been partially characterized, these factors have yet to be unequivocally identified in other immature insects. Williams (1976) noted that under certain experimental conditions, denervated *Manduca* CA lose their activity upon transplantation and suggested that the glands require a factor to maintain biosynthetic activity. Soon after this report, Bhaskaran and Jones (1980), Bhaskaran (1981), and Granger et al. (1981, 1982a,b) demonstrated the presence of allatotropins in *M. sexta.*

Capitalizing on the observation that prolonged starvation in *Manduca* during the early part of the last larval instar induces a significant elevation in hemolymph JH titers and leads to a supernumerary stadium, Bhaskaran and Jones (1980) and Bhaskaran (1981) suggested that a starvation-induced allatotropin could be involved. Removing the CA from the starved animal and replacing them with another gland resulted in the host molting to the sixth instar, indicating that the factor could act humorally. The developmental profile of the allatotropic factor (ATF) was followed by cauterizing the median neurosecretory cells (M-NSC) on day 0 of the fifth instar, starving the animals for 3 days, refeeding for 1 day, then implanting test brains into the donor. Since the median neurosecretory cells are responsible for allatotropic factor (ATF) synthesis, the destruction of the region by cautery left only the donor brain ATF source intact. Brains removed from different stages of larval development and tested in the starved animal paradigm exhibited varying degrees of ATF activity, with the early fourth instar displaying the highest activity. Brains taken from later stages had no effect in this particular assay; however, the shutdown in ATF synthesis or release during the fifth stadium does not appear to be irreversible. This point was demonstrated by cauterizing the median neurosecretory cells on day 0 of the fifth instar, implanting donor brains, starving the animals for 3 days, and then refeeding. If the donor brain produced a humorally released ATF, then a supernumerary molt should occur. It was observed that both ''active'' and ''inactive'' brains were induced by starvation conditions to produce sufficient ATF to promote a sixth instar. It would be of considerable interest to determine if ATF activity can be observed in this same manner under nonstarvation conditions.

Granger et al. (1981), using radioimmunoassay to detect JH biosynthesis, compared the *in vitro* activity of isolated *Manduca* CA with that of the intact brain–CC–CA complex. These studies revealed that ATF(s) from the brain stimulate biosynthesis of JH III acid during the later part of the fifth larval stadium and the very early part of the pupal stage (Granger et al., 1984; also Granger, personal communication). Although neural connectives appear important in the activation process, it is unclear whether the ATF reaches the CA by neurosecretory tracts or by humoral transport; until specific histological probes become available, the delivery process will remain unclear. Unfortunately, JH titers (Baker et al., 1987) do not rise during this period, and it is suggested that ATF may be synthesized and stored in the brain but not necessarily released for activation of the CA. Tissue distribution studies indicate that ATF is limited to the brain and the first three abdominal ganglia. Remarkably, the ATF discovered by Granger et al. (1981, 1984) is homologue specific and activates only JH III acid synthesis.

5.3.2.3. CHEMICAL NATURE OF THE ATF

At present very little is known about the chemical nature of the ATF from any insect species, and what little is known comes from *Manduca*. Building on the observation that early pupal brains contained ATF, Granger et al. (1984) found that an aqueous postmicrosomal supernatant of pupal brain induced *in vitro* a selective, dose-dependent activation of JH III acid biosynthesis in fifth instars (day 7). ATF does not appear to be heat stable but is resistant to freeze-thawing. Determination of apparent molecular weight by gel filtration revealed a macromolecule of 40 kD. The molecule is sensitive to pronase and has an isoelectric point of 5.5.

5.3.2.4. EVIDENCE FOR ALLATOSTATINS

Factors that curtail biosynthesis and secretion of JH have been termed *allatohibins* (Williams, 1976) but more recently have been retermed *allatostatins* to be consistent with the vertebrate literature (Tobe and Stay, 1985). Szibbo and Tobe (1983) noted that CA removed from nymphal *Diploptera* and implanted into an adult undergo a significant increase in JH III biosynthesis as determined by the RCA. They concluded that a humoral allatostatic factor (ASF) in the nymphal hemolymph inactivates the CA.

Paulson and Stay (1985) confirmed this observation and suggested that the humoral environment of the CA during the last part of the stadium was inhibitory. In addition, they were able to demonstrate that JH biosynthesis of the adult female CA was inhibited following implantation into last nymphal instar males. Inhibition could be overcome by decapitation prior to the head critical period but not afterward, whereas implantation of a nymphal protocerebrum restored the inhibition. From these observations it was concluded that the nymphal brain releases an inhibitory factor.

An ASF was obtained from aqueous extracts of *Diploptera* protocerebra (Paulson et al., 1987). A comparison of brain extracts from males and females at day 0 and day 14 of the final stadium and from day 0 adult males and females demonstrated an equivalent ability to inhibit the JH biosynthetic activity of CA from day-2 virgin females. This observation reveals that animals of different ages contain similar quantities of extractable inhibitory material and hence the differences between brain titers of ASF and CA activity may be attributable to the storage of this allato-active factor.

In their *in vitro* coincubation studies of brain–CC–CA or brain–CC coincubated with CA, Granger et al. (1981) observed allatostatic as well as allatotropic activity originating from brains of *Manduca*. The allatostatic activity was specific for JH I and inhibited CA activity on days 2–4 of the fifth stadium (Granger and Janzen, 1987), a finding that correlates well with the drop in hemolymph JH titers (Baker et al., 1987).

5.3.2.5. CHEMICAL NATURE OF THE ASF

As in the case of allatotropins, attempts to characterize the ASF are in their infancy owing to their paucity and the laborious bioassays necessary to characterize them. Granger and Janzen (1987) have partially characterized the JH I–specific ASF in *Manduca* neural tissue. The factor appears to be a peptide on the basis of its heat lability and pronase sensitivity. They report two forms of the factor, with apparent molecular weights of 6.8 and 13 kD. The molecular weight relationship suggests that the larger form may be a dimer of the smaller form. Inhibition of JH I synthesis occurs within 1 min of exposure of the CA to the factor and is reversible by 6 h after this exposure.

An allatostatic factor from the CC of adult *Locusta migratoria migratorioides* has now been purified and polyclonal antiserum developed (Girardie et al., 1987b). The protein was purified from approximately 400 CC by a combination of centrifugation (to remove cellular debris) and polyacrylamide gel electrophoresis. Immunohistochemical studies indicated that the peptide is synthesized in the pars intercerebralis region of the brain in M-NSC of the A1 type. Earlier work on this cerebral peptide, termed neuroparsin A (Girardie et al., 1987a), indicated that it was a trimer with an aggregate apparent molecular weight of 54 kD and a pI of 5.6 (Girardie et al., 1985). Interestingly, neuroparsin A breaks down to a smaller molecule, neuroparsin B, whose biological activity is yet unclear. Injecting antiserum developed against neuroparsin A (and presumably cross-reacting with neuroparsin B) into immature locusts resulted in green pigmentation, intermediate morphological forms, and precocious sexual maturation, all characteristics of increased JH titers. Unfortunately, a key test for neuroparsin's allatostatic activity, curtailment of JH synthesis, could not be demonstrated.

5.3.2.6. MODE OF ACTION OF THE ALLATO-ACTIVE FACTORS

Having established that there are allato-active factors and that crude but biologically active preparations can be obtained, several laboratories have begun to explore the biochemical mode of action of these cerebral peptides. Although it was postulated in the early 1970s (Vedeckis and Gilbert, 1973) that certain insect endocrine glands were activated via a second messenger system (see Bodnaryk, 1983; Smith and Combest, 1985; Smith and Gilbert, 1986), only recently have these observations been extended to the CA. Virtually all of the work to date on the mode of action of the allato-active factors is derived from studies on adults, primarily *Diploptera punctata*. How the factors act at the biochemical level during nymphal development is still unknown for even a single species. On the basis of work by Paulson et al. (1987) demonstrating that extracts from the nymphal brain inhibit JH biosynthesis in both nymphs and adults, it may be surmised that ASF(s) act similarily in both stages.

Meller et al. (1985) observed that cyclic AMP (i.e., adenosine 3′,5′-cyclic monophosphate) levels were relatively high in the CA of newly emerged females, which corresponds to a period when JH synthesis is low. Further developmental evaluation of CA cyclic AMP titers suggested a reciprocal relationship between cyclic AMP titers and JH biosynthesis. JH biosynthsis was shown to be significantly depressed by 8-bromo–cyclic AMP, a membrane-permeable analogue of cyclic AMP. Rankin et al. (1986) pieced together two important components of the problem. First, they noted that aqueous extracts of the cerebral lobes, CC, and CA inhibited JH biosynthesis in a dose-dependent manner. Secondly, incubating the glands in a high-potassium medium led to an increase in cyclic AMP. They reasoned that if the allotostatin were present in the terminal vericosities of the nervous tract innervating the glands, the high potassium would depolarize the neurons, thus leading to a discharge of the neurosecretory factors. Incubating CA with increasing levels of brain extract led to a dose-dependent increase in glandular cyclic AMP, with a maximal response (120–140% over control) induced with less than 0.25 brain equivalents.

In many endocrine-driven systems, one of the initial effects of hormone on the target cell is the increase in cytoplasmic free calcium ion (Ca^{2+}) concentration (Gershengorn, 1985). Kikukawa et al. (1987), using adult virgin female CA, demonstrated that, in the absence of Ca^{2+} in the medium, the CA did not release JH; however, upon increasing Ca^{2+} concentrations (up to 5 mM), a dose-dependent increase in JH release was noted. Moreover, analysis of the final step in JH biosynthesis, i.e., conversion of methyl farnesoate to JH III, indicated that this process was impaired by the lack of calcium in the medium. The pharmacological agent verapamil, a molecule known to inhibit Ca^{2+} uptake, was ineffective in modifying JH release. The role of calcium was further examined by employing the ionophore A23187, which creates a divalent cation channel in the plasma membrane and permits Ca^{2+} to flow down a gradient. In this experiment, the CA were incubated with the ionophore in the presence of high calcium. JH biosynthesis as well as its release into the medium was shown to be significantly depressed. Unfortunately, it is not entirely clear if the event is strictly calcium dependent, as inhibition was not reversible upon removal of the ionophore. Thus, JH synthesis and release require extracellular calcium, but an increase in cytosolic Ca^{2+} levels leads to curtailment of JH synthesis and release.

Studying the role of Ca^{2+} in relation to brain-induced inhibition, Aucoin et al. (1987) demonstrated that extracellular Ca^{2+} altered the ability of *Diploptera* CA to respond to brain extracts. In medium containing low concentrations of Ca^{2+}, brain extract could inhibit JH biosynthesis within minutes, yet if the allata were incubated with brain extract in medium containing 10 mM Ca^{2+}, JH release was not inhibited (Aucoin et al., 1987). Conversely, the effect of high extracellular Ca^{2+} was abolished in the presence of the phosphodiesterase inhibitor 3-isobutyl-1-methylxanthine, and Ca^{2+} had no effect on the phosphodiesterase activity or on the brain-induced increases in cyclic AMP. On the basis of these rather confusing results, Aucoin et al. (1987) have suggested that there may be two pathways in reducing JH biosynthesis: a calcium-dependent route, as observed when glands were incubated in calcium-free medium, and a second route that is calcium independent and involves cyclic AMP.

5.3.2.7. ECDYSTEROID REGULATION OF CA ACTIVITY

The regulation of ATF and ASF release into the hemolymph and/or within the CA itself is dictated by biochemical changes in tissues peripheral to the brain–CC–CA axis and by environmental cues. In the larval insect, response to nutritional intake (Bhaskaran, 1981; Cymborowski et al., 1982), parasitism (see Beckage, 1985, for review), temperature (Bogus and Cymborowski, 1981, 1984), caste status in social insects (Nijhout and Wheeler, 1982; Rembold, 1987), injury (Bogus et al., 1986), and JH feedback (Feyereisen, 1985a; Tobe and Stay, 1985) all influence the regulation of the CA.

One of the most intriguing aspects of brain–CA regulation is the recent discovery that ecdysteroids may control the synthesis and/or release of allato-active factors. From what little is known, it appears that the ecdysteroids act on the brain in a species-specific manner, with 20-hydroxyecdysone (20HE) stimulating release of allatostatins in *Diploptera* and allatotropins in *Manduca*.

Paulson and Stay (1987) reexamined an early observation that adult female CA were inhibited following implantation into final instar male *Diploptera*. It was observed that this inhibition could be overcome by decapitating the nymphal host prior to the head critical period for molting; however, if decapitation were performed afterward, the gland remained inhibited. Implantation of one nymphal protocerebrum into the decapitated host prior to the head critical period reestablished the inhibition, which suggested that CA activity may be inhibited by elevated ecdysteroid levels that are normally present prior to ecdysis. This hypothesis was confirmed by the observation that adult CA implanted into decapitated nymphs prior to the head critical period were inhibited by injections of 20HE (1 μg/ml). Yet when adult CA were incubated *in vitro* in 20HE at widely varying concentrations, no inhibitory effect was detected. These investigators suggest that 20HE acts indirectly on the CA by eliciting secretion of inhibitory factors from other sources, such as the ventral ganglia. Thus, the CA inhibitory activity regulated by ecdysteroids originates from two different sources—the brain and a yet unidentified source, possibly the ventral ganglia.

In contrast to *Diplotera*, the temporal relationship between JH titers (Baker et al., 1987) and 20HE (Bollenbacher et al., 1981; Baker et al., 1987) in *Manduca* suggests that 20HE may stimulate CA biosynthetic activity (Watson et al., 1986; Whisenton et al., 1985, 1987a,b; Granger et al., 1987). Ecdysteroids in physiological concentrations stimulated *in vitro* JH release in both the fourth and precommitment fifth instars via the brain–CC complex. In the fourth instar, ecdysteroid-increased CA activity led to JH I release, whereas in the fifth stadium, during the precommitment period, ecdysteroids induced release of JH I and JH III acids. Incubation of the CA in 20HE did not activate the glands above basal synthetic levels, indicating that the molting hormones act via the brain. One hour's incubation of the brain–CC–CA complex in the presence of 20HE induces the maximal activity; however, the response was not observed until 6 h later. Once stimulated by 20HE, the glands continued to produce JH for at least 10 h. According to these investigators, the lag time between ecdysteroid challenge and the increase in JH biosynthesis represents the time needed for (1) release of allatotropin or (2) inhibition of release of an allatostatin or (3) a change in the nervous stimulation of the gland. To establish an explanation for the lag period, Granger et al. (1987) incubated brain–CC–CA with and without 20HE and compared the results with preparations containing brain–CC coincubated with CA. As expected, the brain–CC–CA complex responded to the 20HE by producing JH I and JH III acids. When the brain–CC preparation was coincubated with CA, only JH I acid biosynthesis was increased, presumably via the allatotropin diffusing from the severed ends of NCC I + II. JH III acid synthesis was not increased and appears to be controlled by direct neural activation arising from the ecdysteroid action on the brain.

In contrast to the positive regulation afforded by the ecdysteroids, the postcommitment period appears to be negatively regulated. Incubation of day 6 brain–CC–CA complexes with physiological concentrations of 20HE suppresses the biosynthetic activity of the CA. Granger et al. (1987) concluded that higher levels of 20HE had no effect, or had several simultaneous effects, which, when summed, led to no stimulation. Thus, the controlling mechanisms in the brain undergo a switchover at the time of commitment, which permits the insect to begin the metamorphic process.

5.4.   Hemolymph JH Binding

The regulation of the CA represents but one aspect of the overall control of JH titers within the insect. Peripherally, factors such as degradation, excretion, and uptake also regulate hormone titers. These aspects have been reviewed elsewhere, and attention will be directed toward the hemolymph juvenile hormone binding proteins (JHBPs).

Upon release into the hemolymph, JH associates with various macromolecules in a noncovalent fashion, forming a dissociable hormone–protein complex. Although the hormones are water soluble at physiological concentrations (Kramer et al., 1976a), their amphiphilic nature promotes surface binding that *in vivo* would include both nonspecific as well as specific sites (Law, 1980). The surface-active nature of the hormone will not permit the unbound molecule to develop sufficient titers at the peripheral sites unless it is associated with a hemolymph protein. The reversible interaction between hormone and binding protein allows the hormone to be dispersed in an aqueous medium and yet be available for interaction at the target site.

This assertion was tested in *Manduca* by rapidly isolating various fractions of hemolymph using preparative polyacrylamide gel electrophoresis and then bioassaying the organic solvent extracts of each of the fractions (Goodman et al., 1978a). It was assumed that detectable levels of the hormone–protein complex would remain associated during the process of electrophoresis. Although two areas of the gel showed activity, one area, now known to contain a specific JHBP, displayed maximal JH bioassay activity. It was concluded that *M. sexta* hemolymph contains a binding protein capable of transporting JH; since then, a number of insects have been shown to bind radiolabeled JH *in vitro*.

5.4.1. *Low-Affinity JH–Protein Associations*

Two classes of JH-binding macromolecules are present in the hemolymph of most insects: nonspecific binding macromolecules characterized by a low affinity constant ($<K_a = 10^6 M^{-1}$), and specific binding macromolecules exhibiting a high association constant ($>K_a = 10^6 M^{-1}$) (Goodman, 1983; Goodman and Chang, 1985). The nonspecific binding of JH to various hemolymph proteins is well documented by studies using JH photoaffinity labels (see Prestwich et al., 1985). For example, in *Manduca* and other Lepidoptera, there are apparently at least three to five JH-binding molecules in addition to a high-affinity binding protein (Koeppe et al., 1984; Prestwich, 1987).

5.4.1.1. HEMOLYMPH STORAGE PROTEIN BINDING

One well-studied group of hemolymph proteins, the storage proteins, are likely candidates for nonspecific JH binding. This type of nonspecific binding is thought to be similar to albumin–fatty acid binding where hydrophobic amino acids have formed a nonspecific, low polarity binding site. Photoaffinity labeling of hemolymph from larval *Manduca* indicates that the storage protein manducin may be a nonspecific binder. As noted by Law (1980), various lipids associated with the protein constitute approximately 2% of the protein's mass. The unusually high content of aromatic amino acids and its high titer during much of larval development (Kramer et al., 1980) makes manducin a very likely candidate for nonspecific JH binding. For further information about the storage proteins, the reader is referred to a recent review by Levenbook (1985).

5.4.1.2. HEMOLYMPH LIPOPHORIN BINDING

The other group of JH binders are the lipoproteins. In this case, the hormone appears to associate with or dissolve into the nonpolar mass of lipid present on the surface of the protein. The lipophorins are high-density lipoproteins that serve as transport molecules for diacylglycerols, phospholipids, cholesterol, hydrocarbons, carotenoids (see Chino, 1985, and Ryan et al., 1987, for reviews), and JH (de Kort and Koopmanschap, 1986). Because they act as reusable shuttles between the fat body where they are synthesized (Prasad et al., 1986, 1987) and the energy utilization sites, they have been termed lipophorins (Chino et al., 1981). Remarkably, in certain species of insects, these proteins bind the hormone specifically (see Section 5.4.3, below), and it must be concluded that separate binding sites exist for JH and the other lipid ligands.

These high-molecular-weight proteins (approximately 600 kD) are composed of two glycosylated apolipoproteins, apolipoprotein I ($M_r = 250$ kD) and apolipoprotein II ($M_r = 80$ kD). The difference between the native molecular weight and the apoproteins approaches the total lipid content of lipophorin (between 40% and 50% of the total mass). The molecular weight and subunit composition is conserved among widely divergent orders (Ryan et al., 1984), and although no immunological cross-reactivity was observed for apolipoprotein I, a number of species displayed cross-reactivity for epitopes on apolipoprotein II.

The total concentration of lipophorins in larval *Manduca* hemolymph is astoundingly high (Prasad et al., 1987; compare with total hemolymph protein for the same stages, given by Trost and Goodman, 1986), and considering its lipophilic nature it would seem a certain candidate for nonspecific JH binding. This point has been borne out recently with the discovery that lipophorins from the hemolymph of *Leptinotarsa decemlineata* and *Periplaneta americana* bind JH (de Kort and Koopmanschap, 1986; see Section 5.4.3, below). Thus, as our information about hemolymph JH binding from different species grows, categorization of these proteins into the previously defined classes becomes more difficult.

Despite binding affinities that are several orders of magnitude lower than that of their high-affinity counterparts, the far greater concentrations of lipophorins and storage proteins makes them important in any consideration of JH transport from the CA to the target tissue.

5.4.2. *High-Affinity, Low-Molecular-Weight JH-Binding Proteins*

Following the discovery of low-affinity, high-molecular-weight JH-associating hemolymph proteins in the early 1970s (Whitmore and Gilbert, 1972), another form of hemolymph protein was characterized that displayed a relatively high affinity and a lower molecular weight. These low-molecular-weight proteins are now known to be characteristic of many lepidopteran species. Since Lepidoptera is the only insect order to exhibit the JH homologue array and the lower-molecular-weight binding proteins, there may be some evolutionary relationship between the hormone and the binding protein. Since it is the most thoroughly studied to date, our attention will focus on the JHBP from *Manduca sexta* (Goodman, 1983; Goodman et al., 1984; Goodman and Chang, 1985).

5.4.2.1. PURIFICATION AND CHARACTERIZATION

For good reason, the only low-molecular-weight JHBP that has been purified to homogeneity is that of *M. sexta*. The combination of low hemolymph concentrations of JHBP (20–40 μg/ml) and use of merely conventional purification methods has made successful purification difficult. Despite these problems, JHBP work on purification from late fourth and mid-fifth instars managed to proceed by conventional methods (Kramer et al., 1976b; Goodman et al., 1978a; Peterson et al., 1982; Peterson, 1985). But, owing to poor recovery, purification by conventional methods was eventually abandoned and an alternative method, affinity chromatography, was developed (Goodman and Goodman, 1981). With JH II used as the immobilized ligand, the protein was purified approximately 150-fold with a recovery of better than 75% of the starting material. Unfortunately, the affinity purified protein was still contaminated with several other proteins. As these proteins do not bind JH, as determined by photoaffinity labeling, it is assumed that they bind nonspecifically to the hydrophobic matrix.

The physicochemical characteristics of *Manduca* JHBP have been extensively reviewed elsewhere (Goodman and Chang, 1985); however, recent evidence sheds new light on certain aspects of JHBP. The apparent molecular weight of JHBP in whole hemolymph was reported to be 34 kD by gel filtration (Kramer et al., 1974), yet upon conventional purification the molecular weight was some 5% lower (Kramer et al., 1976b; Goodman et al., 1978a). However, the molecular weight of affinity-purified JHBP and photoaffinity-labeled protein in hemolymph, as determined by sodium dodecyl sulfate–polyacrylamide gel electrophoresis (SDS–PAGE), corresponded to the early molecular-weight estimates (Prestwich et al., 1987). It may be that during conventional purification of JHBP, a portion of the protein, away from the binding site, was hydrolyzed, since the affinity of the protein was apparently unaffected. Although less likely, it is also possible that both affinity-purified and photoaffinity-labeled JHBP, having all binding sites totally occupied, migrate differently from the partially loaded conventionally purified protein.

Several groups of researchers have now sequenced a portion of the protein from the N-terminal [35 of approximately 260 residues (Peterson et al., 1982; Kulcsar and Prestwich, 1988)]. The sequence is as follows:

ASP-GLN-GLY-ALA-LEU-PHE-GLU-PRO-CYS-SER-THR-GLN-
ASP-ILE-ALA-CYS-LEU-SER-THR-ALA-THR-GLN-GLN-PHE-
LEU-ASP-LYS-ALA-CYS-ARG-GLY-VAL-PRO-ASN-ILE-

The proportion of cysteine residues in this N-terminal fragment is relatively high, since amino acid compositions indicate that there are less than 10 cysteine residues in the entire protein (Kramer et al., 1976b; Goodman et al., 1978a). Further characterization of the protein indicates that the JHBP is neither a glycoprotein ($<1\%$) nor a true lipoprotein.

There have been several reports of multiple forms of JHBP in the hemolymph. During purification of the JHBP from the fifth instar, Kramer et al. (1976b) noted two forms of the protein. Peterson (1977) suggested that ampholytes bound to JHBP may have given rise to the multiple forms and found that, if the protein was purified without the use of ampholytes, a second form did not arise (Peterson et al., 1982). Goodman et al. (1978a), using nondenaturing PAGE, also found several forms (see also Prestwich, 1987); however, prerunning the gels with thioglycolic acid eliminated the artifact (Goodman et al., 1978a). Kulscar and Prestwich (1988) have reported isoforms of the protein taken from fifth instars. Multiple forms deserve further investigation since they may be the result of developmental changes in expression of different genes, action of proteolytic enzymes that may remove a portion of the protein, or previously described artifacts.

5.4.2.2. SPECIFICITY

Since JH homologues exhibit varying degrees of biological activity, specificity studies were performed to correlate biological activity with relative binding affinities. It was demonstrated that binding interactions are based primarily on lipophilicity; binding affinity decreases with increasing homologue polarity in the order JH 0 > JH I > JH II > JH III and gives rise to the "polarity rule" for homologue binding. This homologue binding order has been described in another lepidopteran, *Diatrea grandiosella* (Lenz et al., 1986); yet this binding pattern is not order specific, because the *Trichoplusia ni* JHBP shows a preference for JH II over JH I. None of the biologically active, low-polarity JH analogues have been demonstrated to interact with JHBP. It should be emphasized that binding preference based on polarity was developed to describe the binding of the JH homologues only (Goodman et al., 1978b); the polarity rule does not explain deviations in the pattern of binding based on changes in the epoxide and methylester moieties. Moreover, there are a number of factors that go into the assessment of relative binding affinities, such as the radiolabeled homologue probe, the type of assay, the degree of JHBP purity, and the mathematical transformation of the data. Comparisons of JH binding between laboratories will be difficult to make until these parameters are standardized.

In combination with the homologues and analogues, geometrical isomers and stereoisomers of JH I (Goodman et al., 1978b) and III (Peterson et al., 1977; Schooley et al., 1978) have been used to probe the dimensions and conformation of the binding site. These studies envision the binding site as a hydrophobic groove or pocket on the surface of the protein. The primary interactions between the hormone and the binding site occur along the alkyl side chain and the methyl of the ester group, which together constitute a distinct hydrophobic "face." Hydrogen bonding, which involves electrophilic groups in the binding domain, anchors the ligand at the epoxide and carbomethoxy group, with the distance between these substituents fixed.

Recent evidence indicates that the binding site domain accommodating the epoxide moiety may be less rigid than first anticipated. Studies using JH III 10R and JH III 10S indicated that a change in stereochemistry leads to a highly significant (100-fold) difference in relative binding affinity (Schooley et al., 1978). In contrast, Prestwich and Wawrzenczyk (1985) found that JH I stereoisomers did not display a large difference in binding avidity (two- to threefold), suggesting that the binding site could accommodate an epoxide moiety of unnatural configuration. The contrasting binding affinities between JH I and JH III enantiomers supports the concept that lipophilicity is the most important factor in binding for JH I while less so for JH III. The significant binding energy derived from hydrogen bonding of the epoxide to JHBP in the natural enantiomer position is probably disrupted in JH III 10S-JHBP interactions; without the hydophobic side chain–JHBP interaction, the loss of the hydrogen bonding is magnified for JH III 10S.

5.4.2.3. BIOSYNTHESIS AND TITERS

Immunochemical analyses demonstrated that *M. sexta* JHBP is present in the hemolymph, fat body, and body wall [cuticle with attached epidermal cells and muscle (Nowock et al., 1975)]. *In vitro* synthesis of the protein was detected in the fat body only; the newly synthesized protein is released rather than stored. The fat body has been confirmed as the site of synthesis in a number of insects (see Goodman and Chang, 1985; Koeppe and Kovalick, 1986).

It has been recognized for some time that the number of JH-binding sites in the hemolymph is not constant through even a single stadium. Goodman and Gilbert (1978) demonstrated that it is the relative concentration of JH-specific binding protein that changes rather than the affinity of the protein. More recently, Goodman (1985) titered the JHBP through the fourth and fifth larval stadia, the pupal stage, and adult development, focusing on the binding of just one homologue, JH I. When JH titers are relatively high, during the early part of each stadium, the concentration of JHBP is relatively low. Midway through the stadium, the JHBP titers rise while JH levels fall. Since this inverse relationship between JHBP titers and hormone suggested that JH may control the levels of binding protein, preliminary studies were performed to determine the effect of exogenous hormone on JHBP concentration (Goodman et al., 1984). Dose–response studies conducted on head-ligated fourth instars (68 h, second gate) did not reveal a significant difference between hormone-treated and control animals. It may be that JHBP is stage-specifically controlled and the time of hormone treatment did not coincide with a hormone-sensitive window.

Although the relationship between JH and JHBP titers remains unclear, there is a direct correlation between the JHBP and total hemolymph protein concentration. This observation has been confirmed in other lepidoptera (Ożyhar et al., 1983; Wing et al., 1981). Interestingly, female hemolymph contained significantly greater concentrations of JHBP at nearly every developmental time point during pupation and adult development. Adult females exhibit the highest concentration of JHBP found at any time during the life cycle, nearly double the JHBP present in males of the same age (Goodman, 1985).

5.4.3. *High-Affinity, High-Molecular-Weight JH-Binding Proteins*

It has become evident that many insects outside the order Lepidoptera have high-affinity hemolymph JHBP as well. For the most part, these proteins are characterized by their high molecular weight and their greater affinity for JH III than for JH I. Although it has been suggested that those insects which synthesize only JH III (see Schooley et al., 1984, for a listing) have high-affinity, high-molecular-weight binding proteins (de Kort et al., 1984), several orders display a low-molecular-weight, high-affinity JHBP (see Koeppe and Kovalick, 1986, for a listing). Because of limited studies on the immature stages, most of the information reported here is concerned with JHBP from adult hemolymph.

5.4.3.1. PURIFICATION AND CHARACTERIZATION

The first of the high-affinity, high-molecular-weight binding proteins to be purified was that from the grasshopper *Gomphocerus rufus* (Hartmann, 1978). Purification was effected by a combination of gel filtration and PAGE, with the isolated protein having a relative molecular weight of approximately 200 kD. The protein is composed of three subunits (M_r = 116, 63, and 40 kD). Based on staining properties, the JHBP appears to be a lipoglycoprotein.

The JHBP from *Leucophaea maderae* has now been isolated and demonstrated to be a lipophorin (J. K. Koeppe, personal communication). Hybridoma technology was used to generate monoclonal antibodies against the protein without extensive prepurification (Koeppe et al., 1987). The resulting monoclonal antibodies were attached to an immobilized matrix and immunoaffinity chromatography performed to purify the protein. The *Leucophaea* JHBP is a lipophorin displaying the 220- and 87-kD peptide subunits characteristic of this class of proteins. Studies on the molecular weight of the native protein indicate a mass of approximately 400 kD. Koeppe's group has demonstrated that the 220-kD subunit is photoaffinity labeled by 10,11-epoxyfarnesyl diazoacetate (EFDA). Other JH-binding lipophorins have been purified from *Leptinotarsa decemlineata* and *Periplaneta americana* (de Kort et al., 1978) by KBr gradient centrifugation (Shapiro et al., 1984). Considering the overwhelming lipid composition of the lipophorins, one might expect that the proteins would exhibit an unlimited number of binding sites; however, according to de Kort et al. (1987), "on every two molecules of lipophorin about one JH-binding site occurs."

5.4.3.2. SPECIFICITY

Surprisingly, JH III is bound with higher affinity than the more biologically active JH I (see Koeppe and Kovalick, 1986). *Leucophaea* JHBP displays a preference for JH III, followed in order by the less polar homologues. Curiously, the photoaffinity label EFDA, the diazoacetate derivative of JH III, binds with the least affinity even though it is the most polar of the compounds tested (Koeppe et al., 1984). In *L. decemlineata*, an even more striking disparity exists between JH I and JH III binding, with >100-fold difference observed. The increased affinity for the more polar homologues does not extend to EFDA (de Kort et al., 1987) or the JH acids (de Kort et al., 1983) and indicates that the methylester binding domain is highly selective. The high-molecular-weight binding proteins also discriminate between JH III enantiomers JH III 10*R* and 10*S*. Peter et al. (1979) and de Kort and Koopmanschap (1987) demonstrated that JHBP from *Locusta* bound the naturally occurring 10*R* form more avidly than its racemate. The same enantioselectivity has been reported for *L. decemlineata* (de Kort and Koopmanschap, 1987).

In comparing dissociation constants for the high-molecular-weight proteins and their low-molecular-weight counterparts, it is evident that the high-molecular-weight forms bind JH III more avidly than the low-molecular-weight forms bind JH I. Dissociation constants for the low-molecular-weight JHBPs average about 10^{-7} M, whereas the high-molecular-weight proteins display dissociation constants nearly a magnitude higher. The greater binding energy of these proteins may be derived from specific interactions between the amino acids of the binding site and the functional groups of the hormone. In contrast, the low-molecular-weight binding proteins may rely more on lipophilic interactions along the alkyl side chains.

5.4.3.3. BIOSYNTHESIS AND TITERS

Koeppe et al. (1987) demonstrated that the fat body synthesizes and secretes the lipophorin responsible for JH binding in the hemolymph of *Leucophaea maderae*. The fat body is also responsible for synthesis of JHBP in the grasshopper *Gomphocerus rufus* (Hartmann, 1978).

Although circumstantial evidence indicates that the low-molecular-weight JHBP from *Manduca* may be regulated, the evidence is more substantial in insects exhibiting the high-molecular-weight form of JHBP. The hemolymph levels of JHBP fluctuate during the reproductive cycle of *L. maderae* (Koeppe et al., 1987), peaking at the time of oocyte maturation. Analysis of fat body JHBP synthesis indicates that newly emerged adult female fat body synthesizes significantly less JHBP than does the fat body of the mated female. As these differences suggested hormonal control, changes in JH titers were experimentally manipulated in an attempt to induce *in vivo* JHBP synthesis. It was found that fat body from vitellogenic females or from females injected with JH actively synthesized more JHBP than fat bodies from decapitated controls or virgins. Since the lipophorins have functions other than JH binding, the observed increase may not be directly related to the need for increased JH transport or protection.

Mohamed and Prestwich (1987) have presented preliminary data on regulation of JHBP levels in their study on caste differentiation in the termite *Reticulitermes flavipes*. Regulation of caste differentiation is mediated by a primer pheromone, an orally transmitted substance that in this case initiates development of the worker into a presoldier. According to these investigators, the worker hemolymph contains two proteins (with molecular weights of 210 and 220 kD) that bind JH as determined by photoaffinity labeling. In the hemolymph of workers induced to differentiate, the original JHBPs disappear and two lower-molecular-weight JHBPs of 134 and 102 kD appear. If this case is analogous to that of *L. maderae*, the worker JHBP protein is a lipophorin with only the high-molecular-weight subunit binding the photoaffinity label.

5.4.4. *Functions of the JH-Binding Proteins*

The functions of the hemolymph JHBP are more complex than is conveyed by the terms carrier or transport protein. Indeed, the JHBPs are thought to have evolved several important functions in the regulation of JH titer in addition to that of transport. Although these functions have been extensively described (Goodman and Chang, 1985; see Westphal, 1986, for an overview), they bear restatement because they provide the insect endocrinologist with an eminently sound foundation for future experimental investigations. (1) The hormone–protein complex prevents JH from adsorbing to membranes and other structures that may compete for the hydrophobic domains of the hormone. This ''solubilization'' effect could enhance the rate of release from the CA, enhance dispersion, and lower the excretion rate of hormone. (2) The association of JH with a hemolymph protein reduces enzymatic metabolism of the hormone. In the insect hemolymph, circulating esterases would otherwise rapidly hydrolyze JH. (3) The hormone–portein complex provides a reservoir of JH that is readily available to the target tissue. This pool is important in maintaining hemolymph titers throughout larval development to prevent precocious adult development. (4) The hormone–protein complex may aid or be requisite in the transfer of hormone from the hemolymph into the target cell.

Binding protein–mediated uptake has stimulated considerable discussion among the vertebrate endocrinologists (Westphal, 1986). Until recently, many steroid endocrinologists ascribed cellular entry of lipophilic hormones to free diffusion; however, with the development of more sensitive probes, this premise has been challenged. There are three basic mechanisms by which lipophilic hormones may enter into the cell: (1) free diffusion, (2) internalization via plasma membrane receptors for the unbound hormone, and (3) internalization via plasma membrane receptors for the specific hormone-binding proteins. Westphal (1986) has recently reviewed the first two mechanisms, and attention is here focused on recent evidence suggesting that serum hormone-binding proteins are required in the process of hormone uptake at the target site.

Corticosteroid-binding globulin (CBG), a vertebrate serum hormone-binding protein, has been found in the cytoplasm of target tissue (see Westphal, 1986, for a review), but it was unclear whether its presence signaled mediated steroid transport into target cells. Siiteri et al. (1982) found that uptake of labeled CBG is enhanced by coadministration of corticosterone, suggesting that the serum hormone-binding proteins are central to the endocrine-mediated event (see also Hsu et al., 1986). Additional evidence that the binding proteins for the lipophilic steroids are important in the process of hormone transfer has been presented by Hyrb et al. (1985) and Rosner et al. (1986). These authors demonstrated that sex steroid-binding protein (SBP) binds to specific receptors on the plasma membrane of hormone-sensitive cells. The binding is characterized as high affinity and displays considerable specificity. In a similar system, Strel'chyonok et al. (1984) showed that SBP would not bind to membranes unless the hormone-binding site was occupied. The results presented by these authors suggest that the conformation of SBP changes upon hormone loading to permit interaction with the specific membrane receptor. In all serum or hemolymph hormone–protein interactions there is an overwhelming number of unoccupied binding proteins, and if these unoccupied proteins were allowed to bind to the specific receptors for hormone transfer, very little transfer would occur.

Based on the vertebrate studies, investigations should be initiated on the interaction of JHBP with target tissue. Indeed, Goodman and Chang (1985) cited evidence that the *Manduca* JHBP may be interacting with the cell surface to effect a transfer of the hormone. Until such time as significant amounts of JHBP are isolated and a well-defined JH mode of action can be established *in vitro,* the biological consequences of interaction between JHBP and plasma membrane will remain unclear.

5.5. Summary

The JHs play a central role in insect growth and development, and as such are tightly regulated by a number of physiological and biochemical processes. Although there are a number of components in the regulatory process, this review focuses on three important mechanisms: biosynthesis, central titer regulation, and hemolymph transport. JH biosynthesis, primarily the function of the CA, also involves peripheral tissues such as the imaginal disks and accessory sex glands. Although these extra-allatal sites are responsible only for *O*-methylation of JH acids, they are nevertheless important in preventing precocious adult development. Considering that CA synthesize JH acids exclusively during parts of larval and pupal development, the methylation of these acids at specific developmental stages suggests that the JH acids can be considered prohormones.

The biosynthetic pathways for each of the JH homologues has now been elucidated, and attention has turned to the enzymes responsible for synthesis. There are some indications that the enzymes responsible for the terminal steps in biosynthesis may be regulated.

Brain regulation of CA activity is controlled by neural input and neurosecretory factors that may be released into the hemolymph or may be delivered directly to the CA by neurosecretory axons. Both allatotropins and allatostatins have been isolated from the brains of several insect species. All appear to be peptides, but little else is known about their physical characteristics.

Regulation of JH cannot be easily dissociated from that of the other known insect developmental hormones. It has now been demonstrated that the ecdysteroids act via the brain to elicit the release of allatotropins in *Manduca sexta;* however, the molting hormones induce the release of allatostatins in *Diploptera punctata.* The allato-active factors may act through a second messenger system utilizing cyclic AMP and/or calcium.

Three types of proteins responsible for hemolymph transport of JH have been identified. Nonspecific JH binding can be attributed to storage proteins and certain lipoproteins. Specific JH-binding proteins are divided into two categories based on their molecular weights: the low-molecular-weight proteins, predominantly from Lepidoptera, preferentially bind the higher JH homologues, whereas the high-molecular-weight binding proteins bind JH III preferentially over JH I. The proteins appear to have several important functions in the maintenance of peripheral JH titers. In addition, there is some indication that JHBP may be important in the process of cellular uptake.

Acknowledgments

I thank Ms. H. L. Goodman for her excellent editorial assistance. Portions of the work reviewed here were supported in part by the College of Agricultural and Life Sciences at the University of Wisconsin, Madison, the Johnson Wax Fund, and the National Institutes of Health. The literature search for this review was completed in November 1987.

References

Agui, N., N. A. Granger, L. I. Gilbert, and W. E. Bollenbacher. 1979. Cellular localization of the insect prothoracicotropic hormone: *in vitro* assay of a single neurosecretory cell. Proc. Nat. Acad. Sci. USA 76: 5694–5698.

Aucoin, R. R., S. M. Rankin, B. Stay, and S. S. Tobe. 1987. Calcium and cyclic AMP involvement in the regulation of juvenile hormone biosynthesis in *Diploptera punctata*. Insect Biochem. 17: 965–969.

Baker, F. C., E. Lee, B. J. Bergot, and D. A. Schooley. 1981. Isomerization of isopentenyl pyrophosphate and homoisopentenyl pyrophosphate by *Manduca sexta* corpora cardiaca–corpora allata homogenates. Pp. 67–80 *in* G. E. Pratt and G. T. Brooks (eds.), *Juvenile Hormone Biochemistry.* Elsevier/North-Holland Publ., Amsterdam and New York.

Baker, F. C., B. Mauchamp, L. W. Tsai, and D. A. Schooley. 1983. Farnesol and farnesal dehydrogenase(s) in the corpora allata of the tobacco hornworm moth *Manduca sexta*. J. Lipid Res. 24: 1586–1594.

Baker, F. C., L. W. Tsai, C. C. Reuter, and D. A. Schooley. 1987. *In vivo* fluctuation of JH, JH acid, and ecdysteroid, titer and JH esterase activity, during development of fifth stadium, *Manduca sexta*. Insect Biochem. 17: 989–996.

Beckage, N. E. 1985. Endocrine interactions between endoparasitic insects and their hosts. Annu. Rev. Entomol. 30: 371–414.

Bergot, B. J., F. C. Baker, D. C. Cerf, G. Jamieson, and D. A. Schooley. 1981. Qualitative and quantitative aspects of juvenile hormone titers in developing embryos of several insect species: discovery of a new JH-like substance extracted from eggs of *Manduca sexta*. Pp. 35–45 *in* G. E. Pratt and G. T. Brooks (eds.), *Juvenile Hormone Biochemistry.* Elsevier/North-Holland Publ., Amsterdam and New York.

Bhaskaran, G. 1981. Regulation of corpus allatum activity in last instar *Manduca sexta*. Pp. 53–81 *in* G. Bhaskaran, S. Friedman, and J. G. Rodriguez (eds.), *Current Topics in Insect Endocrinology and Nutrition.* Plenum Press, New York.

Bhaskaran, G. and G. L. Jones. 1980. Neuroendocrine regulation of corpus allatum activity in *Manduca sexta*: the endocrine basis for starvation induced supernumerary larval moult. J. Insect Physiol. 26: 431–440.

Bhaskaran, G., S. P. Sparagana, P. Barrera, and K. H. Dahm. 1986. Change in corpus allatum function during metamorphosis of the tobacco hornworm, *Manduca sexta*: regulation at the terminal step in juvenile hormone biosynthesis. Arch. Insect Biochem. Physiol. 3: 321–338.

Bhaskaran, G., K. H. Dahm, G. L. Jones, K. Peck, and S. Faught. 1987. Juvenile hormone acid synthesis and HMG-CoA reductase activity in corpora allata of *Manduca sexta* prepupae. Insect Biochem. 17: 933–938.

Bodnaryk, R. P. 1983. Cyclic nucleotides. Pp 567–614 *in* R. G. H. Downer and H. Laufer (eds.), *Endocrinology of Insects.* Liss, New York.

Bogus, M. I. and B. Cymborowski. 1981. Chilled *Galleria mellonella* larvae: mechanism of supernumerary moulting. Physiol. Entomol. 6: 343–348.

Bogus, M. I. and B. Cymborowski. 1984. Induction of supernumerary moults in *Galleria mellonella:* evidence for an allatotropic function of the brain. J. Insect Physiol. 22: 669–672.

Bogus, M. I., J. R. Wisniewski, and B. Cymborowski. 1986. Effects of injury to the neuroendocrine system of last-instar larvae of *Galleria mellonella.* J. Insect Physiol. 32: 1011–1017.

Bollenbacher, W. E., S. L. Smith, W. Goodman, and L. I. Gilbert. 1981. Ecdysteroid titer during larval–pupal–adult development of the tobacco hornworm, *Manduca sexta.* Gen. Comp. Endocrinol. 44: 302–306.

Brindle, P. A., F. C. Baker, L. W. Tsai, C. C. Reuter, and D. A. Schooley. 1987. Sources of propionate for the biogenesis of ethyl-branched insect juvenile hormones: role of isoleucine and valine. Proc. Nat. Acad. Sci. USA 84: 7906–7910.

Carrow, G. M., R. L. Calabrese, and C. M. Williams. 1984. Architecture and physiology of insect cerebral neurosecretory cells. J. Neurosci. 4: 1034–1044.

Cassier, P. 1979. The corpora allata of insects. Int. Rev. Cytol. 57: 1–73.

Chino, H. 1985. Lipid transport: Biochemistry of hemolymph lipophorin. Pp. 115–135 *in* G. A. Kerkut and L. I. Gilbert (eds.), *Comprehensive Insect Physiology Biochemistry and Pharmacology,* Vol. 10. Pergamon Press, Oxford and Elmsford, New York.

Chino, H., R. G. H. Downer, G. R. Wyatt, and L. I. Gilbert. 1981. Lipophorins, a major class of lipoprotein of insect hemolymph. Insect Biochem. 11: 491.

Cymborowski, B., M. Bogus, N. E. Beckage, C. M. Williams, and L. M. Riddiford. 1982. Juvenile hormone titres and metabolism during starvation-induced supernumerary larval moulting of the tobacco hornworm, *Manduca sexta.* L. J. Insect Physiol. 28: 129–135.

Dahm, K. H., G. Bhaskaran, M. G. Peter, P. D. Shirk, K. R. Seshan, and H. Röller. 1976. On the identity of the juvenile hormones in insects. Pp. 19–47 *in* L. I. Gilbert (ed.), *The Juvenile Hormones.* Plenum Press, New York.

de Kort, C. A. D. and A. B. Koopmanschap. 1987. Specificity of binding of juvenile hormone III to hemolymph proteins of *Leptinotarsa decemlineata* and *Locusta migratoria.* Experentia (Basel) 43: 904–905.

de Kort, C. A. D. and A. B. Koopmanschap. 1986. High molecular transport proteins from *Locusta migratoria.* Experientia 42: 834–836.

de Kort, C. A. D., M. G. Peter, and A. B. Koopmanschap. 1983. Binding and degradation of juvenile hormone III by haemolymph proteins of the Colorado potato beetle: a re-examination. Insect Biochem. 13: 481–487.

de Kort, C. A. D., A. B. Koopmanschap, and A. A. Ermens. 1984. A new class of juvenile hormone binding proteins in insect haemolymph. Insect Biochem. 14: 619–623.

de Kort, C. A. D., M. A. Khan, and A. B. Koopmanschap. 1987. Juvenile hormone and control of adult diapause in the Colorado potato beetle, *Leptinotarsa decemlineata.* Insect Biochem. 17: 985–988.

Feyereisen, R. 1985a. Regulation of juvenile hormone titer: synthesis. Pp. 391–429 *in* G. A. Kerkut and L. I. Gilbert (eds.), *Comprehensive Insect Physiology Biochemistry and Pharmacology,* Vol. 7. Pergamon Press, Oxford and Elmsford, New York.

Feyereisen, R. 1985b. Radiochemical assay for juvenile hormone III. Methods Enzymol. 111: 530–539.

Feyereisen, R., J. Koener, and S. S. Tobe. 1981a. *In vitro* studies with C_2, C_6, C_{15} precursors of C_{16} JH biosynthesis in the corpora allata of adult *Diploptera punctata.* Pp. 81–92 *in* G. E. Pratt and G. T. Brooks (eds.), *Juvenile Hormone Biochemistry.* Elsevier/North-Holland Publ., Amsterdam and New York.

Feyereisen, R., G. E. Pratt, and A. F. Hamnett. 1981b. Enzymatic synthesis of juvenile hormone in locust corpora allata: evidence for a microsomal cytochrome P-450 linked methyl farnesoate epoxidase. Eur. J. Biochem. 118: 231–238.

Gershengorn, M. C. 1985. Thyrotropin-releasing hormone action: mechanism of calcium mediated stimulation of prolactin secretion. Recent Prog. Endocrinol. 41: 607–645.

Girardie, J., A. Faddoul, and A. Girardie. 1985. Characterization of three neurosecretory proteins from the A median neurosecretory cells of *Locusta migratoria* by coupled chromatographic, electrophoretic and isoelectrofocusing methods. Insect Biochem. 15: 85–92.

Girardie, J., D. Boureme, and A. Girardie. 1987a. Production sites of the three neurosecretory proteins characterized in the corpora cardiaca of the migratory locust. Insect Biochem. 17: 29–36.

Girardie, J., D. Boureme, F. Couillaud, M. Tamarelle, and A. Girardie. 1987b. Anti-juvenile effect of neuroparsin A, a neuroprotein isolated from locust corpora cardiaca. Insect Biochem. 17: 977–983.

Goodman, W. G. 1983. Hemolymph transport of ecdysteroids and juvenile hormones. Pp. 49–57 *in* R. G. H. Downer and H. Laufer (eds.), *Endocrinology of Insects.* Liss, New York.

Goodman, W. G. 1985. Relative hemolymph juvenile hormone binding capacity during larval, pupal and adult development of *Manduca sexta.* Insect Biochem. 15: 557–564.

Goodman, W. G. and E. S. Chang. 1985. Juvenile hormone cellular and hemolymph binding proteins. Pp. 491–510 *in* G. A. Kerkut and L. I. Gilbert (eds.), *Comprehensive Insect Physiology Biochemistry and Pharmacology,* Vol. 7. Pergamon Press, Oxford and Elmsford, New York.

Goodman, W. G. and L. I. Gilbert. 1978. The hemolymph titer of juvenile hormone binding protein and binding sites during the fourth instar of *Manduca sexta.* Gen. Comp. Endocrinol. 35: 27–34.

Goodman, W. G. and H. L. Goodman. 1981. Development of affinity chromatography for juvenile hormone binding proteins. Pp. 365–374 *in* G. E. Pratt and G. T. Brooks (eds.), *Juvenile Hormone Biochemistry.* Elsevier/ North-Holland Publ., Amsterdam and New York.

Goodman, W. G., D. A. Schooley, and L. I. Gilbert. 1978b. Specificity of the juvenile hormone binding protein: the geometrical isomers of juvenile hormone I. Proc. Nat. Acad. Sci. USA 75: 185–189.

Goodman, W. G., P. A. O'Hern, R. H. Zaugg, and L. I. Gilbert. 1978a. Purification and characterization of a juvenile hormone binding protein from the hemolymph of the fourth instar tobacco hornworm, *Manduca sexta*. Mol. Cell. Endocrinol. 11: 225–242.

Goodman, W. G., J. T. Trost, C. T. Reiter, D. W. Bean, and R. O. Carlson. 1984. Hemolymph transport of the juvenile hormones. Pp. 426–437 *in* J. Hoffman and M. Porchet (eds.), *Biosynthesis, Metabolism and Mode of Action of Invertebrate Hormones*. Springer-Verlag, Berlin and New York.

Granger, N. A. and W. P. Jansen. 1987. Inhibition of *Manduca sexta* corpora allata *in vitro* by a cerebral allatostatic neuropeptide. Mol. Cell. Endocrinol. 49: 237–248.

Granger, N. A., W. E. Bollenbacher, and L. I. Gilbert. 1981. An *in vitro* approach for investigating the corpora allata during larval–pupal metamorphosis. Pp. 83–105 *in* G. Bhaskaran, S. Friedman, and J. G. Rodriguez (eds.), *Current Topics in Insect Endocrinology and Nutrition*. Plenum Press, New York.

Granger, N. A., S. A. Niemiec, L. I. Gilbert, and W. E. Bollenbacher. 1982a. Juvenile hormone III biosynthesis by the larval corpora allata of *Manduca sexta*. J. Insect Physiol. 18: 385–391.

Granger, N. A., S. M. Niemiec, L. I. Gilbert, and W. E. Bollenbacher. 1982b. Juvenile hormone III biosynthesis by the larval corpora allata of *Manduca sexta*. Mol. Cell. Endocrinol. 28: 587–604.

Granger, N. A., L. J. Mitchell, W. P. Janzen, and W. E. Bollenbacher. 1984. Activation of *Manduca sexta* corpora allata *in vitro* by a cerebral neuropetide. Mol. Cell. Endocrinol. 37: 349–358.

Granger, N. A., L. R. Whisenton, W. P. Janzen, and W. E. Bollenbacher. 1987. Interendocrine control by 20-hydroxyecdysone of the corpora allata of *Manduca sexta*. Insect Biochem. 17: 949–953.

Granger, N. A., W. E. Bollenbacher, R. Vince, L. I. Gilbert, J. C. Baehr, and F. Dray. 1979. *In vitro* biosynthesis of juvenile hormone by the larval corpora allata of *Manduca sexta*. Mol. Cell. Endocrinol. 16: 1–17.

Hammock, B. D. 1975. NADPH dependent epoxidation of methyl farnesoate to juvenile hormone in the cockroach *Blaberus giganteus* (L.) Life Sci. 17: 323–328.

Hartmann, R. 1978. The juvenile hormone carrier in haemolymph of the acrine grasshopper, *Gomphocerus rufus* L.: blocking of the juvenile hormone's action by means of repeated injections of antibody to the carrier. Wilhelm Roux' Arch. Dev. Biol. 184: 301–324.

Havel, C. M., E. R. Rector, and J. A. Watson, 1986. Isopentenoid synthesis in isolated embryonic *Drosophila* cells. J. Biol. Chem. 261: 10150–10156.

Hsu, B. S. R., P. K. Siiteri, and R. W. Kuhn. 1986. Interactions between corticosteroid binding globulin and target tissues. Pp. 577-592 *in* M. G. Forest and M. Pugat (eds.), *Binding Proteins of Steroid Hormones*, Vol. 7. Libbey Eurotext, Montrouge, France.

Hyrb, D. J., S. Khan, and W. Rosner. 1985. Testosterone-estradiol-binding globulin binds to human prostatic cell membranes. Biochem. Biophys. Res. Commun. 128: 432–440.

Johannsen, O. A. and F. H. Butt. 1941. *Embryology of Insects and Myriapods.* University Microfilms International, publ., Ann Arbor, Michigan.

Kennely, P. J. and V. W. Rodwell. 1985. Regulation of 3-hydroxy-3-methylglutaryl coenzyme A reductase by reversible phosphorylation-dephosphorylation. J. Lipid Res. 26: 903–914.

Kigushi, K. and L. M. Riddiford. 1978. A role of juvenile hormone in pupal development of the tobacco hornworm, *Manduca sexta.* J. Insect Physiol. 24: 673–680.

Kikukawa, S., S. S. Tobe, S. Solowiej, S. M. Rankin, and B. Stay. 1987. Calcium as a regulator of juvenile hormone biosynthesis and release in the cockroach *Diploptera punctata.* Insect Biochem. 17: 179–187.

Koeppe, J. K. and G. E. Kovalick. 1986. Juvenile hormone binding proteins. Pp. 266–304 *in* G. Litwack (ed.), *Biochemical Actions of Hormones,* Vol. 13. Academic Press, Orlando, Florida.

Koeppe, J. K., R. C. Rayne, E. A. Whitsel, and M. E. Pooler. 1987. Synthesis and secretion of juvenile hormone binding protein by fat body from the cockroach *L. maderae:* protein A immunoassay. Insect Biochem. 17: 1027–1032.

Koeppe, J. K., G. D. Prestwich, J. J. Brown, W. G. Goodman, G. E. Kovalick, T. Briers, M. D. Pak, and L. I. Gilbert. 1984. Photoaffinity labeling of the hemolymph juvenile hormone binding protein of *Manduca sexta.* Biochemistry 23: 6674–6679.

Kramer, K. J., L. L. Sanburg, F. Keźdy, and J. H. Law. 1974. The juvenile hormone binding protein in the hemolymph of *Manduca sexta* (Johannson) (Lepidoptera: Sphingidae). Proc. Nat. Acad. Sci. USA 71: 493–497.

Kramer, K. J., P. E. Dunn, R. Peterson, and J. H. Law. 1976a. Interaction of juvenile hormone with binding proteins of the insect hemolymph. Pp. 327–341 *in* L. I. Gilbert (ed.), *The Juvenile Hormones.* Plenum Press, New York.

Kramer, K. J., P. E. Dunn, R. C. Peterson, H. Seballos, and J. H. Law. 1976b. Purification and characterization of the carrier protein for juvenile hormone from the hemolymph of the tobacco hornworm *Manduca sexta* Johannson (Lepidoptera: Sphingidae). J. Biol. Chem. 251: 4979–4985.

Kramer, S. J., E. C. Mundall, and J. H. Law. 1980. Purification and properties of manducin, an amino acid storage protein of the hemolymph of larval and pupal *Manduca sexta.* Insect Biochem. 10: 279–288.

Kulcsar, P. and G. D. Prestwich. 1988. Detection and microsequencing of juvenile hormone binding proteins of an insect using an iodinated JH analog. FEBS (Fed. Euro. Biochem. Sci.) Lett. (in press).

Law, J. H. 1980. Lipid–protein interactions in insects. Pp. 295–310 *in* M. Locke and D. S. Smith (eds.), *Insect Biology in the Future.* Academic Press, Orlando, Florida.

Lenz, C. J., J. W. Dillwith, and G. M. Chippendale. 1986. Comparison of some properties of the high affinity juvenile hormone binding protein from the larval hemolymph of pyralid moths. Arch. Insect Biochem. Physiol. 3: 61–73.

Levenbook, L. 1985. Insect storage proteins. Pp. 307–346 *in* G. A. Kerkut and L. I. Gilbert (eds.), *Comprehensive Insect Physiology Biochemistry and Pharmacology*, Vol. 10. Pergamon Press, Oxford and Elmsford, New York.

Meller, V., R. R. Aucoin, S. S. Tobe, and R. Feyereisen. 1985. Evidence for an inhibitory role of cyclic AMP in the control of juvenile hormone biosynthesis by cockroach corpora allata. Mol. Cell Endocrinol. 43: 155–163.

Metzler, M., K. H. Dahm, D. Meyer, and H. Röller. 1971. On the biosynthesis of juvenile hormone in the adult cecropia moth. Z. Naturforsch. 26B: 1270–1276.

Mohamed, M. A. and G. D. Prestwich. 1987. The pheromonal suppression of predifferentiation hemolymph juvenile hormone binding proteins in the termite *Reticulitermes flavipes*. Am. Zool. 27: 67 (abstract).

Monger, D. J., W. A. Lim, F. J. Keździ, and J. H. Law. 1982. Compactin inhibits insect HMG-CoA reductase and juvenile hormone biosynthesis. Biochem. Biophys. Res. Commun. 105: 1374–1380.

Nijhout, H. F. 1975. Axonal pathways in the brain–retrocerebral neuroendocrine complex of *Manduca sexta*. Int. J. Insect Morphol. Embryol. 4: 529–538.

Nijhout, H. F. and C. M. Williams. 1974. Control of moulting and metamorphosis in the tobacco hornworm, *Manduca sexta*: cessation of juvenile hormone secretion as a trigger for pupation. J. Exp. Biol. 61: 493–501.

Nijhout, H. F. and D. E. Wheeler. 1982. Juvenile hormone and the physiological basis of insect polymorphisms. Q. Rev. Biol. 57: 109–133.

Nowock, A., W. Goodman, W. E. Bollenbacher, and L. I. Gilbert. 1975. Synthesis of juvenile hormone binding proteins by the fat body of *Manduca sexta*. Gen. Comp. Endocrinol. 27: 230–239.

O'Brien, M. A., E. J. Katahira, T. R. Flanagan, L. W. Arnold, G. Haughton, and W. E. Bollenbacher. 1988. A monoclonal antibody to the insect prothoracicotropic hormone. J. Neurosci. 8:3247–3257.

Ożyhar, A., J. R. Wiśniewski, F. Sehnal, and M. Kochman. 1983. Age dependent changes in the binding and hydrolysis of juvenile hormone in the hemolymph of last instar larvae of *Galleria mellonella*. Insect Biochem. 13: 435–441.

Paulson, C. R. and B. Stay. 1985. Inhibition of the corpora allata of last-instar male larvae of the cockroach, *Diploptera punctata*. J. Insect Physiol. 31: 625–630.

Paulson, C. R. and B. Stay. 1987. Humoral inhibition of the corpora allata in larvae of *Diploptera punctata*: role of the brain and ecdysteroids. J. Insect Physiol. 33: 613–622.

Paulson, C. R., B. Stay, S. Kikukawa, and S. S. Tobe. 1987. *In vitro* inhibition of the corpora allata by a larval brain factor in the cockroach *Diploptera punctata*. Insect Biochem. 17: 961–964.

Peter, M. G., S. Gunawan, G. Gellissen, and H. Emmerich. 1979. Differences in hydrolysis and binding of homologous juvenile hormones in *Locusta migratoria* hemolymph. Z. Naturforsch. 34C: 588–598.

Peterson, R. C. 1977. Juvenile hormone carrier protein from *Manduca sexta* (L.): an improved purification, determination of binding specificity and selected modification. Ph.D. thesis, University of Chicago.

Peterson, R. C. 1985. Purification and properties of the juvenile hormone carrier protein from the hemolymph of *Manduca sexta*. Methods Enzymol. 111: 482–487.

Peterson, R. C., M. F. Reich, P. E. Dunn, J. H. Law, and J. A. Katzenellenbogen. 1977. Binding specificity of the juvenile hormone carrier protein from the hemolymph of the tobacco hornworm *Manduca sexta* Johannson (Lepidoptera: Sphingidae). Biochemistry 16: 2305–2311.

Peterson, R. C., P. E. Dunn, H. L. Seballos, B. K. Barbeau, P. S. Keim, C. T. Riley, R. L. Heinrickson, and J. H. Law. 1982. Juvenile hormone carrier protein of *Manduca sexta* haemolymph: improved purification procedure; protein modification studies and sequence of the amino terminus of the protein. Insect Biochem. 12: 643–650.

Prasad, S. V., G. P. Fernando-Warnakulasuriya, M. Sumida, J. H. Law, and M. A. Wells. 1986. Lipoprotein biosynthesis in the larvae of the tobacco hornworm, *Manduca sexta*. J. Biol. Chem. 261: 17174–17176.

Prasad, S. V., K. Tsuchida, K. D. Cole, and M. A. Wells. 1987. Lipoprotein biosynthesis during the life cycle of the tobacco hornworm, *Manduca sexta*. Pp. 267–273 *in* J. H. Law (ed.), *Molecular Entomology*. Liss, New York.

Prestwich, G. D. 1987. Chemistry of pheromone and hormone metabolism in insects. Science (Wash., DC) 237: 999–1006.

Prestwich, G. D. and C. Wawrzenczyk. 1985. High specific activity enantiomerically enriched juvenile hormones: synthesis and binding assay. Proc. Nat. Acad. Sci. USA 82: 5290–5294.

Prestwich, G. D., S. Robles, C. Wawrzenczyk, and A. Buhler. 1987. Hemolymph juvenile hormone binding proteins of lepidopterous larvae: enantiomeric selectivity and photoaffinity labeling. Insect Biochem. 17: 551–560.

Prestwich, G. D., J. K. Koeppe, G. E. Kovalick, J. J. Brown, E. S. Chang, and A. K. Singh. 1985. Methods Enzymol. 111: 509–530.

Rankin, S. M., B. Stay, R. R. Aucoin, and S. S. Tobe. 1986. *In vitro* inhibition of juvenile hormone synthesis by corpora allata of the viviparous cockroach *Diploptera punctata*. J. Insect Physiol. 32: 151–156.

Reibstein, D. and J. H. Law. 1973. Enzymatic synthesis of juvenile hormones. Biophys. Biochem. Res. Commun. 55: 266–272.

Reibstein, D., J. H. Law, S. B. Bowlus, and J. A. Katzenellenbogen. 1976. Enzymatic conversion of juvenile hormones in *Manduca sexta*. Pp. 131–146 *in* L. I. Gilbert (ed.), *The Juvenile Hormones*. Plenum Press, New York.

Rembold, H. 1987. Caste specific modulation of juvenile hormone titers in *Apis mellifera*. Insect Biochem. 17: 1003–1006.

Rosner, W., M. S. Khan, N. A. Romas, and D. J. Hyrb. 1986. Interactions of plasma steroid-binding proteins with cell membranes. Pp. 567–576 *in* M. G. Forest and M. Pugat (eds.), *Binding Proteins of Steroid Hormones*, Vol. 149. Libbey Eurotext, Montrouge, France.

Ryan, R. O., J. O. Schmidt, and J. H. Law. 1984. Chemical and immunological properties of lipophorins from seven insect orders. Arch. Insect Biochem. Physiol. 1: 375–383.

Ryan, R. O., M. A. Wells, and J. H. Law. 1987. The dynamics of lipophorin interconversions in *Manduca sexta*. Pp. 257–266 *in* J. H. Law (ed.), *Molecular Entomology*. Liss, New York.

Schooley, D. A. and F. C. Baker. 1985. Juvenile hormone biosynthesis. Pp. 363–390 *in* G. A. Kerkut and L. I. Gilbert (eds.), *Comprehensive Insect Physiology, Biochemistry and Pharmacology*, Vol. 7. Pergamon Press, Oxford and Elmsford, New York.

Schooley, D. A., B. J. Bergot, W. Goodman, and L. I. Gilbert. 1978. Synthesis of both optical isomers of insect juvenile hormone III and their affinity for the JH-specific binding protein of *Manduca sexta*. Biochem. Biophys. Res. Commun. 81: 743–749.

Schooley, D. A., K. J. Judy, B. J. Bergot, M. S. Hall, and R. C. Jennings. 1976. Determination of the physiological levels of juvenile hormones in several insects and biosynthesis of the carbon skeleton of the juvenile hormones. Pp. 101–117 *in* L. I. Gilbert (ed.), *The Juvenile Hormones*. Plenum Press, New York.

Schooley, D. A., F. C. Baker, L. W. Tsai, C. A. Miller, and G. C. Jamieson. 1984. Juvenile hormones 0, I, II exist only in Lepidoptera. Pp. 373–383 *in* J. Hoffmann and M. Porchet (eds.), *Biosynthesis, Metabolism and Mode of Action of Invertebrate Hormones*. Springer-Verlag, Berlin and New York.

Sedlak, B. J. 1985. Structure of endocrine glands. Pp. 25–60 *in* G. A. Kerkut and L. I. Gilbert (eds.), *Comprehensive Insect Physiology, Biochemistry and Pharmacology*, Vol. 7. Pergamon Press, Oxford and Elmsford, New York.

Shapiro, J. P., P. S. Keim, and J. H. Law. 1984. Structural studies on lipophorin, an insect lipoprotein. J. Biol. Chem. 259: 3680–3685.

Shirk, P. D., G. Bhaskaran, and H. Röller. 1983. Developmental physiology of corpora allata and accessory sex glands in the cecropia silkmoth. J. Exp. Zool. 227: 69–79.

Siiteri, P. K., J. T. Murai, G. L. Hammond, J. A. Nisker, W. J. Raymoure, and R. W. Kuhn. 1982. The serum transport of steroid hormones. Recent Prog. Horm. Res. 38: 457–510.

Smith, W. A. and W. L. Combest. 1985. Role of cyclic nucleotides in hormone action. Pp. 263–299 *in* G. A. Kerkut and L. I. Gilbert (eds.), *Comprehensive Insect Physiology, Biochemistry and Pharmacology*, Vol. 8. Pergamon Press, Oxford and Elmsford, New York.

Smith, W. A. and L. I. Gilbert. 1986. Cellular regulation of ecdysone synthesis by the prothoracic gland of *Manduca sexta*. Insect Biochem. 16: 143–147.

Sparagana, S. P., G. Bhaskaran, and P. Barrera. 1985. Juvenile hormone acid methyltransferase activity in imaginal discs of *Manduca sexta* prepupae. Arch. Insect Biochem. Physiol. 2: 191–202.

Strel'chyonok, O. A., G. V. Avvakumov, and L. I. Survivlo. 1984. A recognition system for sex-hormone-binding protein–estradiol complex in human decidual endometrium plasma membranes. Biochem. Biophys. Acta 802: 459–466.

Szibbo, C. M. and S. S. Tobe. 1983. Nervous humoral inhibition of C_{16} juvenile hormone synthesis in last instar females of the viviparous cockroach, *Diploptera punctata*. Gen. Comp. Endocrinol. 49: 437–445.

Tobe, S. S. and B. Stay. 1985. Structure and regulation of the corpus allatum. Adv. Insect Physiol. 18: 305–432.

Trost, J. T. and W. G. Goodman. 1986. Hemolymph titers of the biliprotein, insecticyanin, during development of *Manduca sexta*. Insect Biochem. 16: 353–358.

Vedeckis, W. V. and L. I. Gilbert. 1973. Production of cyclic AMP and adenosine by the brain and prothoracic glands of *Manduca sexta*. J. Insect Physiol. 19: 2445–2457.

Watson, J. A., C. M. Havel, D. V. Lobos, F. C. Baker, and C. J. Marrow. 1985. Isoprenoid synthesis in isolated embryonic *Drosophila* cells. J. Biol. Chem. 260: 14083–14091.

Watson, R. D., L. R. Whisenton, W. E. Bollenbacher, and N. A. Granger. 1986. Interendocrine regulation of the corpora allata and prothoracic glands of *Manduca sexta*. Insect Biochem. 16: 149–155.

Weirich, G. F. 1985. Epoxyfarnesoic acid methyltransferase. Methods Enzymol. 111: 540–544.

Westphal, U. 1986. *Steroid–Protein Interactions II.* Springer-Verlag, Berlin and New York.

Whisenton, L. R., M. F. Bowen, N. A. Granger, L. I. Gilbert, and W. E. Bollenbacher. 1985. Brain-mediated 20-hydroxyecdysone regulation of juvenile hormone synthesis by the corpora allata of the tobacco hornworm, *Manduca sexta*. Gen. Comp. Endocrinol. 58: 311–318.

Whisenton, L. R., N. A. Granger, and W. E. Bollenbacher. 1987a. A kinetics analysis of brain-mediated 20-hydroxyecdysone stimulation of the corpora allata of *Manduca sexta*. Mol. Cell. Endocrinol. 54: 171–178.

Whisenton, L. R., R. D. Watson, N. A. Granger, and W. E. Bollenbacher. 1987b. Regulation of JH biosynthesis by 20-hydroxyecdysone during the fourth instar of the tobacco hornworm, *Manduca sexta*. Gen. Comp. Endocrinol. 66: 62–70.

Whitmore, E. and L. I. Gilbert. 1972. Haemolymph lipoprotein transport of juvenile hormone. J. Insect Physiol. 18: 1153–1168.

Williams, C. M. 1976. Juvenile hormone: in retrospect and in prospect. Pp. 1–14 *in* L. I. Gilbert (ed.), *The Juvenile Hormones*. Plenum Press, New York.

Wing, K. D., T. C. Sparks, V. M. Lovell, S. O. Levinson, and B. D. Hammock. 1981. The distribution of juvenile hormone esterase and its interrelationship with other proteins influencing juvenile hormone

metabolism in the cabbage looper, *Trichoplusia ni*. Insect Biochem. 11: 473–485.
Wiśnewski, J. R., M. Muszynska-Pytel, and M. Kochman. 1986. Juvenile hormone degradation in brain and corpora cardiaca–corpora allata complex during the last larval instar of *Galleria mellonella*. Experientia (Basel) 42: 167–168.

Metabolism of Juvenile Hormones: Degradation and Titer Regulation

6

R. MICHAEL ROE

AND KRISHNAPPA VENKATESH

6.1. Introduction 127
6.2. Routes of JH Metabolism 128
    6.2.1. Assay Methods for JH Metabolic Activity 129
    6.2.2. Oxidation Reactions 130
    6.2.3. Conjugation Reactions 131
    6.2.4. Primary Route of *In Vivo* JH Metabolism 131
6.3. Role of JH Metabolism in Development 132
    6.3.1. Larval Growth 132
    6.3.2. Metamorphosis 134
    6.3.3. Pupal Development 139
    6.3.4. Reproduction 139
    6.3.5. Embryogenesis 143
    6.3.6. Diapause 145
6.4. Characterization of Ester Hydrolysis 146
    6.4.1. JH-Specific Esterase 146
    6.4.2. JH Esterase Inhibition 150
    6.4.3. JH Esterase Purification 152
    6.4.4. Distribution and Source of JH Esterase 154
6.5. Regulation of JH Esterase 157
    6.5.1. Role of JH 157
    6.5.2. Role of Head Factors 160
    6.5.3. Role of Environmental Factors 162
6.6. Characterization of Epoxide Hydrolase 163
6.7. Prospectus 165
6.8. Summary 166
Acknowledgment 167
References 167

6.1. Introduction

The insect juvenile hormones are methyl esters of farnesoic acid 10,11-epoxide (JH III); 11-ethyl (JH II), 7,11-diethyl (JH I), or 3,7,11-triethyl (JH 0) methyl farnesoate 10,11-epoxide; and 4-methyl–JH I (iso-JH 0) (see Fig. 6.6 in Section 6.4.1). Studies suggest that JH regulates embryogenesis, larval and adult development, metamorphosis, reproduction, diapause, migration, polymorphism, ecdysone regulation, and metabolism (Denlinger, 1985; Hoffmann and Lagueux, 1985; Koeppe et al., 1985; Pener, 1985). It is this wide spectrum of biological activity, much of which has not yet been comprehended, and the uniqueness of JH to the Insecta, that has made it a major focus of entomological research. Much of this research has centered on the mechanisms that regulate JH titer.

Several primary factors may be important in determining the concentration of JH in insects: *de novo* biosynthesis of JH acid and JH by the corpora allata (CA); JH biosynthesis from JH acid by other tissues; release and distribution rates for JH acid and JH; the interaction of JH and JH-metabolizing enzymes with hemolymph JH-binding proteins; tissue sequestration and receptor binding; excretion; and degradation by hemolymph and tissue. These possible interactions are illustrated in Fig. 6.1. There are, in addition, secondary regulatory processes that control these primary elements.

In insects, it is clear that JH titer is regulated by a balance between the rate of its synthesis and degradation. What is not clear, however, is the relative importance of the many possible factors in the maintenance of JH titer, i.e., the dynamics of the interactions outlined in Fig. 6.1. This is clouded even further by the variability within the Insecta, which makes generalizations difficult. Caution is also needed in the assignment of importance to one factor or another without the development of whole animal and specific tissue models or without considering the importance of interactions that may occur under field conditions.

This review focuses on the role of JH metabolism in JH regulation, including some discussion of other factors involved in JH regulation. For greater detail on the latter, the reader is referred to reviews on the role of JH synthesis (de Kort and Granger, 1981; Feyereisen, 1985; Tobe and Stay, 1985) and binding protein (de Kort and Granger, 1981; Goodman and Chang, 1985; Tobe and Stay, 1985) on JH titer regulation. The presence and function of JH-like materials in arthropods other than insects (Solomon et al., 1982; Laufer et al., 1987a,b; see also other relevant chapters herein) is not as well known; and even

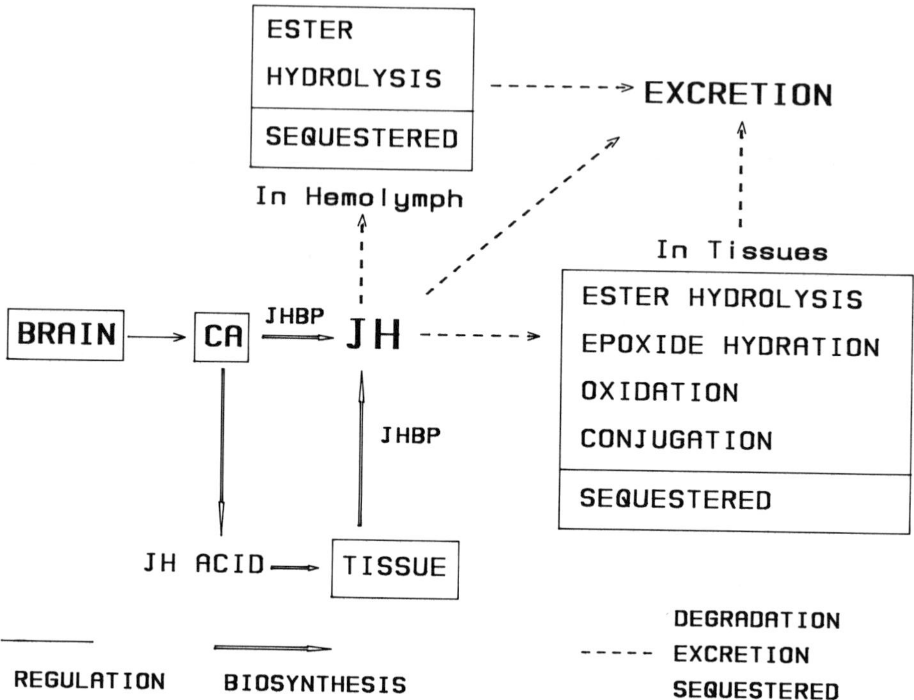

FIGURE 6.1. Possible factors regulating juvenile hormone concentration in insects. Key: CA = corpora allata; JH = juvenile hormone; JHBP = hemolymph JH = hemolymph JH = binding protein.

less so on the possible existence of developmentally regulated, specific degradation processes (Hammock, 1985). Recent unpublished investigations on the tick in our laboratory will be discussed in this respect.

6.2. Routes of JH Metabolism

Several potential routes of JH metabolism exist: ester hydrolysis, epoxide hydration, oxidation, and conjugation (Fig. 6.1). The primary route of metabolism in the hemolymph is ester hydrolysis, while multiple primary and secondary routes are possible in the tissues. These reactions can occur alone or in several combinations and have recently been reviewed extensively by Hammock (1985). A number of reviews on JH metabolism have been published in the 1975–1985 period (Akamatsu et al., 1975; Hammock and Quistad, 1976, 1981; Kramer and Law, 1980; Hammock, 1981, 1985; Hammock et al., 1981, 1984b; de Kort and Granger, 1981; Tobe and Stay, 1985). More recently, research on the characterization of JH esterase from Lepidoptera was reviewed by Hammock et al. (1987).

The major research emphasis on JH degradation, currently an active area of interest for a number of laboratories, is ester hydrolysis. JH metabolism by JH esterases has been demonstrated in approximately 60 insect species. Most of these studies have involved isolated insect hemolymph and tissue homogenate, with less attention given to *in vivo* metabolism and whole animal modeling. JH epoxide hydrolase activity has been demonstrated in more than 30 insect species. Much of the published work was comparative, however, and represents a relatively small number of studies. Ester hydrolysis and epoxide hydration will be discussed in more detail later. In comparison with the studies of JH esterase and epoxide hydrolase, studies demonstrating oxidation and conjugation reactions with JH are almost nonexistent.

6.2.1. *Assay Methods for JH Metabolic Activity*

One reason for the emphasis on JH esterase activity has been the availability of sensitive assay methods. Esterase metabolism is most frequently monitored with a basic aqueous methanol–isooctane partition assay utilizing commercially available ^{3}H-labeled JH I or JH III (Hammock and Sparks, 1977; Hammock and Roe, 1985). Other available methods include a charcoal precipitation technique that monitors the release of radioactive methanol from methoxy-labeled JH (Sanburg et al., 1975), an extraction by a Folch distribution of unmetabolized JH (Nowock et al., 1976), and an electrophoretic (Pratt, 1975) and thin-layer chromatographic (TLC) (Hammock et al., 1975; Hammock and Roe, 1985) separation of JH from JH acid. The partition assay (Hammock and Sparks, 1977; Hammock and Roe, 1985) is the most desirable method because it is rapid, inexpensive, uses commercially available substrates, and does not require extensive controls. The common commercially available substrates used in partition assays, JH I and JH III, have a high degree of isomeric purity (greater than 98% 2*E*,6*E*,10*Z*) but are racemic at C-10/C-11 and C-10, respectively. Recent studies have shown that the hemolymph JH esterase of some insect species hydrolyzes the (10*R*,11*S*)-JH II enantiomer, which is believed to be the natural form, faster than the (10*S*,11*R*)-enantiomer at substrate concentrations normally used in JH esterase partition assays (see Section 6.4.1, below). Specific spectrophotometric substrates for JH esterase would be most desirable for routine assays and kinetic studies, but they are not yet available.

TLC methods on silica gel are frequently used to simultaneously monitor ester hydrolysis, epoxide hydration, oxidation and conjugation (Slade and Zibitt, 1972; Ajami and Riddiford, 1973; Erley et al., 1975; Hammock et al., 1975; Yu and Terriere, 1978; Hammock and Roe, 1985). The TLC system, most useful in our laboratory for baseline separation of JH, JH acid, JH diol, JH acid–diol, and JH conjugates in the absence of oxidative products, is silica gel developed in hexane : ethyl acetate : glacial acetic acid (66:33:1). Oxidative products can be visualized with this system by incubating microsomes with and without NADPH (nicotinamide adenine dinucleotide phosphate) or by two-dimensional TLC (Hammock et al., 1975). Reverse-phase high-pressure liquid chromatography (HPLC) and TLC techniques are useful in some cases but often lack the resolution of normal-phase chromatography. Procedures for the synthesis of standards and for metabolite identification are available from several sources (Slade and Zibitt, 1972; Ajami and Riddiford, 1973; Slade and Wilkinson, 1974; Hammock et al., 1975; Yu and Terriere, 1978; Hammock and Roe, 1985; Share and Roe, 1988). Artificial substrates specific for epoxide hydrolase activity, for example, styrene oxide and cyclodiene, are available, but studies indicate that these model substrates may be hydrated by enzymes different from those involved in JH metabolism (Slade et al., 1976). Isooctane–methanol partition assays are available for JH epoxide hydrolase activity in the absence of JH esterase (Mumby and Hammock, 1979) and for the simultaneous assay of both JH esterase and epoxide hydrolase (Share and Roe, 1988). Both of these partition assays assume no oxidation or conjugation.

6.2.2.   *Oxidation Reactions*

Probably the most complete work on oxidative reactions was that of Yu and Terriere (1978). They found that microsomal monooxygenases were a component in the primary metabolism of JH I in the flesh fly (*Sarcophaga bullata*), the blow fly (*Phormia regina*), and four strains of the house fly (*Musca domestica*), and tentatively identified the metabolite as a C-5,6/C-9,10 diepoxide. This metabolite, in the presence of epoxide hydrolase, forms tetrahydrofuran diols of JH. Additional *in vitro* experiments with the house fly indicated that metabolism was dependent on NADPH, induced by 1% sodium phenobarbital in the diet, and inhibited by carbon monoxide and piperonyl butoxide as would be expected. JH I monooxygenase activity peaked late in larval development and in the adult stage. All the metabolites noted *in vitro* in the house fly were also detected when JH I was applied topically. Although this latter experiment provides evidence for an *in vivo* role for monooxygenase in the primary metabolism of JH, it is not clear what effect the topical application may have had on the results. JH is normally released directly into the hemolymph. Additional evidence for the oxidation of JH by insects was provided by Ajami and Riddiford (1973) and Hammock et al. (1977).

6.2.3. *Conjugation Reactions*

A number of *in vivo* experiments of JH conjugation reactions have been reported (Slade and Zibitt, 1972; White, 1972; Ajami and Riddiford, 1973; Slade and Wilkinson, 1974; Yu and Terriere, 1978). In the research just discussed with the house fly (Yu and Terriere, 1978), polar metabolites were identified that were partially cleaved by glucosidase, resulting in the detection of JH I diol and JH I acid, and by sulfatase, producing JH I diol. In these studies, 24% of the polar metabolites produced were resistant to glucosidase and sulfatase degradation. The major metabolite produced from similar experiments with the flesh fly was JH diol (Slade and Zibitt, 1972). As much as 68% of the conjugate was resistant to digestion in these latter experiments. The limited evidence available suggests that conjugation is a secondary route of metabolism. The importance of secondary metabolism in JH titer regulation has not been addressed. With the prospect that hemolymph JH acid can be converted by tissue to JH (Sparagana et al., 1985), secondary reactions could be more important in JH titer regulation than was formerly believed (Hammock, 1985). The same could be said for excretion. Erley et al. (1975) identified the Malpighian tubules of *Locusta migratoria* as the site of JH, JH acid, JH diol, and JH diol–acid excretion. No unmetabolized JH was found in the feces of the tobacco hornworm, *Manduca sexta* (Slade and Zibbitt, 1972). Secondary metabolism and excretion ensures that JH in the hemolymph and tissues is derived from *de novo* biosynthesis only.

6.2.4. *Primary Route of* In Vivo *JH Metabolism*

In summary, studies indicate that the possible primary routes of JH metabolism in insects so far studied are ester hydrolysis, epoxide hydration, and oxidation. These reactions can also occur secondarily along with sulfate and glucose conjugation. There is some evidence for the formation of tetrahydrofuran diols from the secondary metabolites JH diepoxide or JH diol, through the action of hydrolase or monooxygenase activity, respectively. No *in vivo* information is available on the direct interaction of JH with glutathione S-transferase. From the few *in vivo* studies that have been conducted, JH esterase appears to be the major route of primary metabolism (Slade and Zibitt, 1972; White, 1972; Ajami and Riddiford, 1973; Erley et al., 1975; Yu and Terriere, 1978). Note, however, that *in vivo* studies are often limited in scope and do not include kinetics or consider developmental effects, different substrates and routes of metabolism, differences between tissues, sequestration, excretion, and/or the effects of different routes and methods of introduction of radiotracers.

6.3. Role of JH Metabolism in Development

The appearance of hemolymph JH esterase activity at specific times during development in a number of insect species has provided correlative evidence for a possible functional role for JH metabolism in the regulation of JH titer. The major focus of this research has been on the last instar development of Lepidoptera, and it provides the strongest evidence that JH metabolism regulates metamorphosis. More recently, however, information has become available on JH metabolism in earlier instar larvae and adults of Lepidoptera, in insect embryos, and in adult ticks.

6.3.1. *Larval Growth*

In the tobacco hornworm (Fig. 6.2) (Gate I and II larvae; P. Jesudason, K. Venkatesh, and R. M. Roe, unpublished data) and the cabbage looper (G. Jones and Click, 1987), a single peak of hemolymph JH esterase activity was noted during each stadium from the second through the penultimate instar. This peak activity occurred at the end of development, and in *M. sexta* there was a successive increase in peak activity from the first through the fifth instar (Fig. 6.2). The exponential increase in JH esterase activity during larval development is correlated with an exponential increase in size or weight during this same period. Intermolt peaks have also been reported prior to the last instar in *Heliothis virescens*, *Hypantria cunea*, *Diploptera punctata*, and *Schistocerca vaga* (Weirich and Wren, 1976; Szibbo et al., 1982; G. Jones

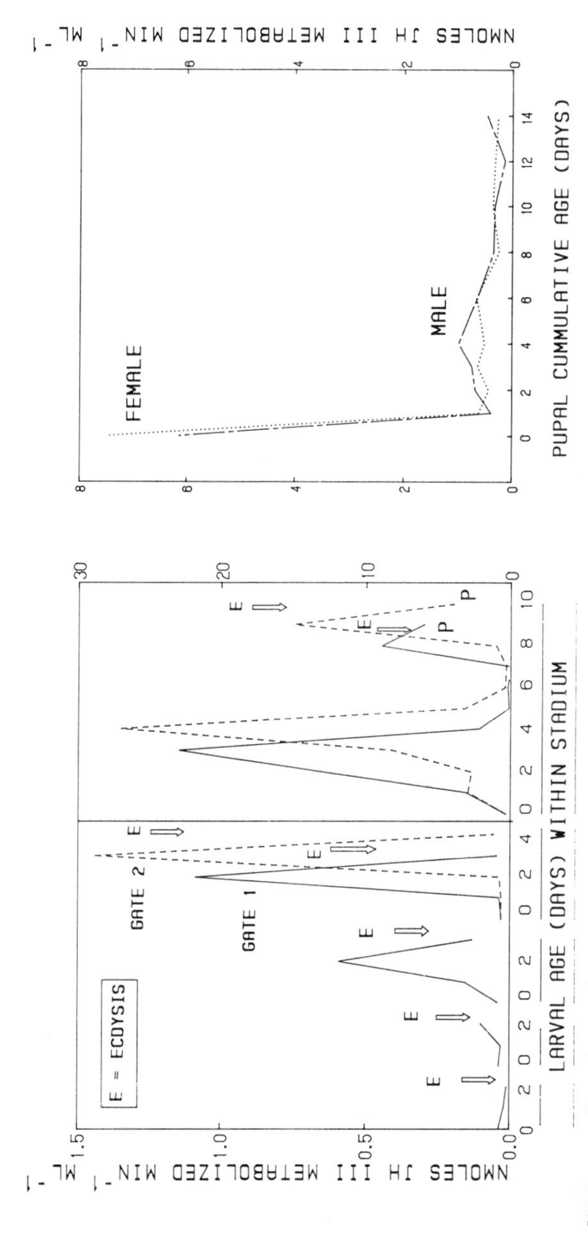

FIGURE 6.2. Daily changes in the JH esterase activity in gate 1 and gate 2 larvae from the first to the last instar and in female vs. male pupae of the tobacco hornworm, *Manduca sexta*. Key: JH = juvenile hormone; P = pupae.

and Click, 1987). No longer applicable is the original idea that the appearance of hemolymph JH esterase activity was something unique to the last instar of the Lepidoptera; in fact, JH esterase is also found in embryos and adults as will be discussed later (see Sections 6.3.4 and 6.3.5, below).

The function of these elevated levels of JH esterase activity during early larval development remains unknown. The general assumption is that elevated esterase activity occurs when low JH levels are required. The *in vivo* inhibition of this activity during peak periods with the JH esterase inhibitor, EPPAT (*O*-ethyl-*S*-phenylphosphoramidothioate), had no detectable effect on larval development of *Trichoplusia ni* (G. Jones and Click, 1987). These larvae were similarly unresponsive to juvenoid applications. It is possible that JH is not the endogenous substrate for the measured activity, but substrate and inhibitor specificity studies to be discussed later (Section 6.4) argue against this hypothesis. Peak levels of activity could be a result of changes in overall metabolic activity or hemolymph volume during this time (related to molting for example), but other hemolymph esterases like α-naphthyl acetate esterase do not follow a similar pattern as that of JH esterase in *M. sexta* (P. Jesudason, K. Venkatesh, and R. M. Roe, unpublished data). Furthermore, intermolt peaks in JH esterase activity are not correlated with changes in hemolymph protein (Jesudason, Venkatesh, and Roe, unpublished data).

6.3.2. *Metamorphosis*

The main focus of research on JH metabolism has been JH esterase activity in last-instar larval Lepidoptera. The insect models most often used are *Manduca sexta*, *Trichoplusia ni*, and *Galleria mellonella*. Two peaks of hemolymph JH esterase activity have been found during the last stadium of most Lepidoptera studied (Sparks et al., 1979b; Mane and Chippendale, 1981; Bean et al., 1982; D. Jones et al., 1982; Masner et al., 1983; Ożyhar et al., 1983; Roe et al., 1983; Fescemyer et al., 1986). An example is shown in Fig. 6.2 for the last instar of the tobacco hornworm. The first peak occurs just prior to wandering, and the second during the prepupal stage—with the level of activity between the two being variable from one insect species to another. The correspondence of maximum levels of activity with specific developmental markers occurs irrespective of gate (Sparks and Hammock, 1979b; D. Jones et al., 1981; Jesudason, Venkatesh, and Roe, unpublished data).

Wing et al. (1981) also found two peaks of tissue JH esterase activity in the fat body during the time of the first and second hemolymph peak, and in the midgut during the time of the first peak in the cabbage looper, *T. ni*. Maximum levels of R-20458 epoxide hydrolase activity were also noted in midgut tissue during the time of the prewandering peak and in fat body after the first peak. The significance of the relative timing of peak JH esterase activity and epoxide hydrolase activity in the fat body is unknown. Wiśniewski et al. (1986b) were unable to correlate total fat body metabolism of JH with hemolymph JH metabolism for the second peak in the greater wax moth, *Galleria mellonella*.

In the holometabola studied, the prewandering (midstadium) JH esterase peak of the hemolymph was found only in the Lepidoptera. The prepupal peak, however, also occurred in the mealworm larva *Tenebrio molitor* (Coleoptera), and in the fruit fly *Drosophila hydei* (Diptera) (Klages and Emmerich, 1979; Connat, 1983), where there is only a single peak of hemolymph JH esterase activity at the very end of last-stadium development. The orthopteran insects studied also have a single peak of activity in the last instar, but they differ from the Coleoptera and Diptera in that the peak occurs midway through the stadium. This was the case for the cockroach *Diploptera punctata* (Szibbo et al., 1982) and the house cricket *Acheta domesticus* (Woodring and Sparks, 1987). High levels of hemolymph JH esterase activity were also noted midway through penultimate development in the cockroach (Szibbo et al., 1982). It will be important in the future to understand the rationale of these different strategies for hemolymph JH degradation. Comparative studies of last-stadium JH degradation in different insect orders are severely lacking.

The functional role of high levels of tissue and hemolymph JH esterase, as would be expected, is to reduce JH levels at specific and critical periods in last-stadium larval development. Surprisingly, however, the majority of hemolymph JH turnover in the last stadium occurs well before the appearance of peak JH esterase levels. Figure 6.3 summarizes the sequence of events that occurs during the last stadium of the tobacco hornworm, *M. sexta* (Bhaskaran et al., 1986; Baker et al., 1987). The blocked areas in Fig. 6.3 represent periods when hormone or activity levels are elevated above background, and the arrows illustrate the amount relative to the peak levels observed during each period. The last stadium of many Lepidoptera can roughly be divided into halves accentuated by the onset of wandering behavior midway through development (Fig. 6.3). Peak levels of JH occur in the hemolymph early during the first half of development (Baker et al., 1987)

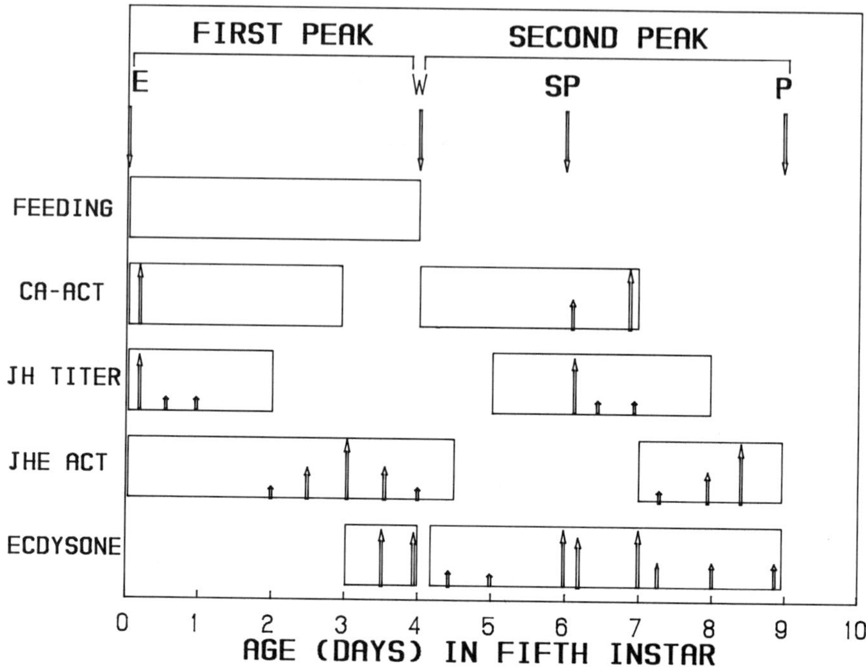

FIGURE 6.3. Changes in the rate of feeding, corpus allata activity, hemolymph JH titer, JH esterase activity, and ecdysone concentration during the last stadium of the tobacco hornworm, *Manduca sexta*. The instar is divided into two periods marked by the appearance of the first and second peak of hemolymph JH, JH esterase, and ecdysone. The blocked areas represent periods when hormone or activity levels are elevated above background, and the arrows illustrate the amount relative to the peak levels observed during each period. Key: E = ecdysis to the last instar; CA-ACT = corpus allata JH biosynthesis for the first half and JH acid biosynthesis for the second half of the last instar; JHE ACT = juvenile hormone esterase activity; P = pupa; SP = stationary phase; W = wandering. Derived from Bhaskaran et al. (1986) and Baker et al. (1987).

and are positively correlated with corpus allatum (CA) activity during this period (Bhaskaran et al., 1986; Granger et al., 1982). The CA on day 0 produces JH but later in development produces JH acid only. Interestingly, hemolymph JH titer drops to background levels well before the first JH esterase peak. The appearance of JH acid in the hemolymph prior to the first JH esterase peak suggests that the decline in hemolymph JH titer is at least partly a result of background JH esterase activity in the hemolymph and tissue. Also significant in this decline could be other routes of metabolism, the cessation of JH biosynthesis, tissue sequestration (Hammock et al., 1975; Nowock et al., 1976; Tobe et al., 1984, 1985), and excretion. The exact switchover point where the CA stops producing JH and starts producing JH acid has not been determined. Since peak JH esterase levels occur after JH levels have already declined to background, Baker et al. (1987) concluded that elevated esterase levels served mainly to purge the hemolymph of the last traces of JH. Elevated JH esterase levels may also be important in the elimination of JH resulting from any residual CA activity at this time and also at the time of the second JH esterase peak (to be discussed later). Studies by Hammock et al. (1975) and Nowock et al. (1976) have shown considerable fat body uptake and reflux of JH with hemolymph in *M. sexta*. These results suggest that peak levels of tissue JH esterase activity may also be an important factor in tissue and hemolymph JH metabolism. Based on the absence of hemolymph JH and JH acid during the time of the first JH esterase peak, the amount of JH of tissue origin that is metabolized in the hemolymph must be small, and there is very little release of tissue JH or JH acid into the hemolymph during this time.

The elimination of JH during the first half of last-instar development appears to be critical to the initiation of wandering behavior. Rountree and Bollenbacher (1986) showed that JH blocked the production of ecdysone during the first half of larval development by inhibiting the brain–retrocerebral complex in *M. sexta*. The normal drop in prewandering JH titer causes a gradual acquisition by prothoracic glands of biosynthetic competence, and the gated release of prothoracicotropic hormone (PTTH) (Watson et al., 1987). Dominick and Truman (1985) also found that it was the absence of JH in the presence of ecdysone that initiated wandering behavior. Note from Fig. 6.3 that the JH esterase peak precedes the first peak of ecdysone. Studies in the cabbage looper, *T. ni*, have demonstrated that the inhibition of JH esterase with specific and nonspecific inhibitors delays the appearance of the first ecdysone peak, increases the time to wandering and weight gain, delays pupation, and produces supernumerary molts in which the larvae do not survive (Sparks and Hammock, 1980; Hammock et al., 1984a; Prestwich et al., 1984; G. Jones, 1985; Newitt and Hammock, 1986). Sparks et al. (1983) also found that EPPAT treatment delayed the onset of wandering and increased the maximum weight attained in the last instar of the tobacco hornworm, *M. sexta*. These studies illustrate the critical nature of the first JH esterase peak.

At the beginning of the stationary phase during the second half of last-stadium development (Reinecke et al., 1980), a second peak of hemolymph JH appears in *M. sexta*, simultaneously with a second peak of ecdysone [Fig. 6.3 (Baker et al., 1987)]. The CA during this period produce the prohormone JH acid, which is converted to JH by imaginal disks (Sparagana et al., 1985; Bhaskaran et al., 1986). Studies with *M. sexta, Mamestra bassicae, Spodoptera littoralis,* and *T. ni* have shown that JH promotes prepupal ecdysone production and precocious development during this time (Cymborowski and Stolarz, 1979; Hiruma, 1980; Safranek et al., 1980; Gruetzmacher et al., 1984; Sparks, 1984; G. Jones et al., 1986). These studies are in contrast to other research with *T. ni* (Newitt and Hammock, 1986) and *Ephestia cautella* (Lazarovici et al., 1984) where there was no comparable JH effect. The appearance of the second JH peak appears to be important in *M. sexta, M. brassicae,* and *S. littoralis* in the suppression of adult characters during the pupal stage (Kiguchi and Riddiford, 1978; Cymborowski and Stolarz, 1979; Hiruma, 1980), and its absence in *T. ni* blocks pupation. Just prior to pupal ecdysis and after hemolymph JH titers have declined to low levels, a second prepupal peak of JH esterase activity occurs in *M. sexta* (Fig. 6.3). In this case, the ecdysone peak precedes the second JH esterase peak, but the ecdysone levels remain relatively high through pupation and during the time of maximum esterase activity. The role of prepupal JH esterase in last-stadium development has been studied principally in *T. ni*. G. Jones and Hammock (1985) have shown that the application of JH or esterase inhibitors during the time of the prepupal JH esterase peak disrupts larval–pupal ecdysis, promotes pupal–pupal molts, or produces pupal–adult intermediates. G. Jones (1983) found an increase in JH levels in larvae treated with JH esterase inhibitors, and G. Jones et al. (1986) found that *in vivo* esterase inhibition increased ecydsone levels. The latter authors hypothesized that elevated levels of JH as a result of the inhibition of the prepupal JH esterase peak elevated ecdysone levels and possibly blocked the release of eclosion hormone (Truman, 1981). Evidence in *T. ni*, at least, suggests that the appearance of the prepupal peak of JH esterase and the elimination of the last traces of JH from the larva just prior to pupal ecdysis can be critical for normal larval and pupal development. We have recently found that EPPAT treatment to eliminate the final JH esterase peak in last-stadium *M. sexta* likewise blocks ecdysis and produces larval–pupal intermediates (Venkatesh and Roe, unpublished data).

6.3.3. *Pupal Development*

JH esterase activity in the hemolymph of pupae is either declining from the prepupal peak in activity in *Tenebrio molitor* and *Drosophila hydei* (Klages and Emmerich, 1979; Connat, 1983) or is at low levels in *M. sexta* (Fig. 6.2) (P. Jesudason, K. Venkatesh, and R. M. Roe, unpublished data), *Galleria mellonella* (Hwang-Hsu et al., 1979), and *T. ni* (Wing et al., 1981). Interestingly, Wing et al. (1981) found high fat body JH esterase activity during the first half of pupal development of *T. ni* even though hemolymph JH esterase levels were very low. These fat body levels exceeded that found in the last instar. This suggests the possibility of a specific regulatory mechanism that in last-instar larvae permits the release of JH esterase activity from fat body but in pupae blocks this release. This assumes, of course, that the origin of the hemolymph JH esterase is the fat body and the JH esterase is the same in pupae and adults. These questions will be discussed later (Section 6.4.4). There is some evidence in *T. ni* of factors from the head that cause the release of JH esterase from larval tissues into the hemolymph but do not affect overall animal JH esterase activity (see Section 6.5.2).

6.3.4. *Reproduction*

There are several correlative studies between hemolymph JH esterase levels, JH titer, and egg development, suggesting a functional role for JH metabolism in female reproduction. More direct studies like the elimination of JH esterase with specific inhibitors have rarely been conducted, however. Surprisingly, considering the amount of research on the larval JH esterases of Lepidoptera, little is known about JH metabolism in adult moths and butterflies.

Lessman and Herman (1984) demonstrated JH esterase activity in the hemolymph of the adult monarch butterfly, *Danaus plexippus*, which indicated that JH esterase may be involved in reducing JH levels during autumn migration. It appears that elevated JH esterase activity resulted in a lowering of JH levels and a low level of reproductive development in migrating monarchs. The developmental changes in hemolymph JH esterase in moths *(T. ni)* were studied for the first time only recently by our laboratory (Venkatesh et al., 1987). JH esterase activity was high in newly emerged female adults and remained at a high level in virgin females (Fig. 6.4). In mated females, however, the JH esterase activity declined, reaching background levels by day 5 after emergence (Fig. 6.4). These changes were in contrast to that for hemolymph α-naphthyl acetate esterase activity, which was minimal a few days after eclosion in both mated and virgin females (data not shown). It is generally accepted that elevated JH levels are essential for vitellogenesis (Koeppe et al., 1985). Declining or low JH esterase levels, coupled with an increase in JH biosynthesis and JH titer, is likely responsible for egg development and a heightened ovipositional rate in mated females of *T. ni* (Fig. 6.4). High JH esterase levels in virgin moths, on the other hand, appear to be a contributing factor in the inhibition of egg development and oviposition (Fig. 6.4).

The most complete work conducted on JH titer regulation and its functional role in female reproduction in a hemimetabolous insect involved the cockroach *Diploptera punctata* (Rotin et al., 1982; Rotin and Tobe, 1983; Tobe et al., 1985). Tobe et al. (1985) reported that egg maturation, as measured by increases in oocyte length, ceased as JH biosynthesis declined, and JH esterase levels increased. Oocyte growth was directly correlated with high JH levels. It was also reported that as much as 30% of the whole-animal JH is found in the tissues of the cockroach. No studies are yet available of any life stage that effectively addresses the role of metabolism on the elimination of tissue JH.

In the hemolymph of adult *D. punctata* (Tobe et al., 1985) and the whole body of the mosquito *Aedes aegypti* (Shapiro et al., 1986), the majority of the JH titer decline occurred well prior to the appearance of peak hemolymph JH esterase activity. Recall (from Section 6.3.2) that the same finding was reported for larval *M. sexta* (Baker et al., 1987). The general model evolving from these studies for representatives from larval Lepidoptera and adult Diptera and Orthoptera is that background levels of JH esterase activity are adequate for the elimination of most of the animal JH and that peak JH esterase levels function primarily to scavenge residual hormone.

Shapiro et al. (1986) provided evidence in adult mosquitoes that the scavenging role of JH esterase may affect reproduction. The application of JH esterase inhibitors and/or methoprene at the time of peak hemolymph esterase activity reduced egg hatch. These results are inconclusive, however, in that they do not identify the mode of action of these treatments that could be explained as a direct effect on the egg (see Section 6.3.5, below). It appears in the mosquito that high JH levels, after adult emergence, prepares the mosquito behaviorally and

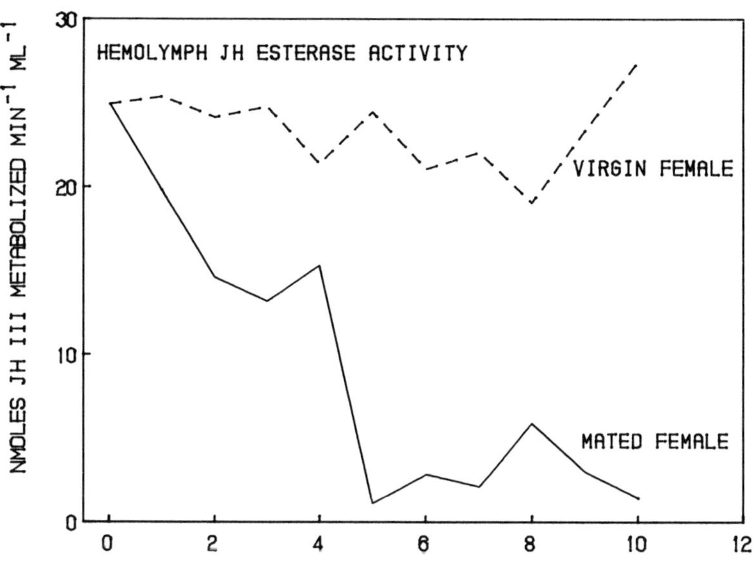

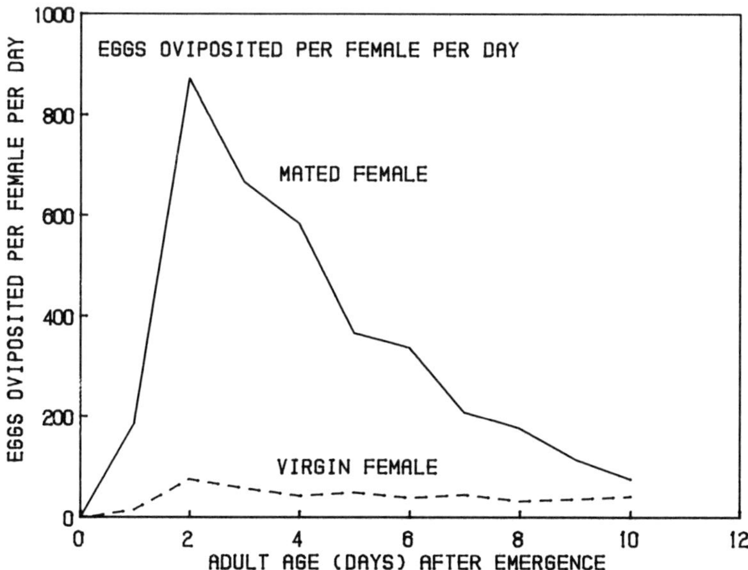

FIGURE 6.4. Oviposition rate and hemolymph juvenile hormone esterase activity in mated and virgin, female adults of the cabbage looper, *Trichoplusia ni*. Key: JH = juvenile hormone.

physiologically for a blood meal (Shapiro et al., 1986). Once a blood meal is taken, low levels of JH are required to stimulate vitellogenic egg development, likely by an increase in ecdysone levels and egg-development neurosecretory hormone. Shapiro et al. (1986) and our laboratory (C. S. Apperson and R. M. Roe, unpublished data) have shown that hemolymph JH esterase activity in *Aedes aegypti* and whole-body JH esterase and epoxide hydrolase activity in *Culex quinquefasciatus*, respectively, are induced by the acquisition of the blood meal. Additional developmental studies correlating low JH esterase levels to egg maturation have been reported in the house cricket *Acheta domesticus* and the Colorado potato beetle, *Leptinotarsa decemlineata* (Kramer and de Kort, 1976a; Renucci et al., 1984; Woodring and Sparks, 1987). JH metabolism was characterized in the adults of several more species (Erley et al., 1975; Weirich and Wren, 1976; Edwards and Rowlands, 1977, 1978; Peter et al., 1979).

Most JH metabolism studies have been in the Insecta, but there is evidence for the existence of JHs in other arthropods. Most recently, it was reported using LC (liquid chromatographic) and GC/MS (gas chromatographic/mass spectroscopic) identification that the mandibular organ of adult spider crabs *(Libinia emarginata)*, for example, produced methyl farnesoate and that the rate of synthesis varied with the stage of vitellogenesis (Laufer et al., 1987a). Less definitive, indirect evidence suggesting the existence of JH-like substances is also available in ticks (Solomon et al., 1982). Most noteworthy were studies in which oogenesis was disrupted by the anti-JH, precocene 2, in the soft tick *Ornithodoros pakeri* (Pound and Oliver, 1979). In these studies, JH III rescued the tick from precocene treatment and reinitiated egg laying.

Developmental studies on JH metabolism outside the Insecta are lacking, however. Recent studies were conducted to consider the possible existence of developmentally regulated JH esterases or epoxide hydrolases during adult feeding and reproduction in the American dog tick, *Dermacentor variabilis* (R. M. Roe, D. E. Sonenshine, C. S. Apperson, and K. Venkatesh, unpublished data). Both JH esterase and epoxide hydrolase activity was found in the hemolymph of blood-fed (virgin) females, with both enzymes preferring JH III over JH I as substrate. The hemolymph epoxide hydrolase activity noted could possibly be the result of fat body contamination. The same routes of JH metabolism were noted in whole-body homogenates (the gut removed), but substrate specificity for JH III was only noted for the epoxide hydrolase activity. Feeding by virgin females resulted in a 14-fold decrease in the JH III epoxide hydrolase activity per unit whole body wet weight, and mating in fed females, but not feeding alone, caused a small but significant decrease (30%) in the JH III esterase activity over that of unfed, virgin controls. The α-naphthyl acetate esterase activity followed the same pattern of change as that for the epoxide hydrolase activity during adult development. The changes noted for epoxide hydrolase activity and α-naphthyl acetate esterase expressed per milligram of protein also demonstrated the same pattern of change, but in the case of the JH esterase activity there was an increase in activity in fed female ticks as compared with that of unfeds. *In vitro* studies in our laboratories have demonstrated that synganglia or associated tissues of fed, mated females of the Rocky Mountain wood tick, *Dermacentor andersoni*, are capable of synthesizing JH using ^{3}H-methionine, radiolabeled substances that coelute with known insect juvenoids on TLC and HPLC.

6.3.5. *Embryogenesis*

Recent studies have also provided evidence supporting a functional role for JH metabolism during embryogenesis of the tobacco hornworm, *M. sexta*, and the house cricket *(A. domesticus)* (Roe et al., 1987a,b; Share et al., 1988). JH esterase activity was highest early in embryonic development in *M. sexta*, decreasing by twofold within 1 day after oviposition (Fig. 6.5). The decrease in JH esterase activity occurred prior to the egg–embryo transformation (before the blastoderm is formed). A similar decrease in JH esterase activity was also noted during embryogenesis in the house cricket, during the same relative period of embryonic development (Fig. 6.5). Like that in *M. sexta*, JH esterase activity was highest early in embryogenesis and decreased by threefold after dorsal closure.

These changes in JH esterase activity can also be inversely correlated with changes in JH titer. In the tobacco hornworm, JH 0 and JH I are virtually nonexistent in newly laid eggs, but titers peaked after 2–3 days of development (Bergot et al., 1980, 1981). This peak in JH titer is concurrent with CA development in the embryo (Dorn et al., 1987). In the cockroach *Nauphoeta cinerea*, JH III titers were undetectable before dorsal closure and then peaked during the latter half of embryogenesis (Brüning et al., 1985). In the locust *Locusta migratoria*, JH III is low or absent until the latter part of embryogenesis, when the CA are well developed (Pener et al., 1986; Temin et al., 1986).

It appears that the absence of JH early in embryogenesis is critical to normal development and hatching. Juvenoid treatment of adult

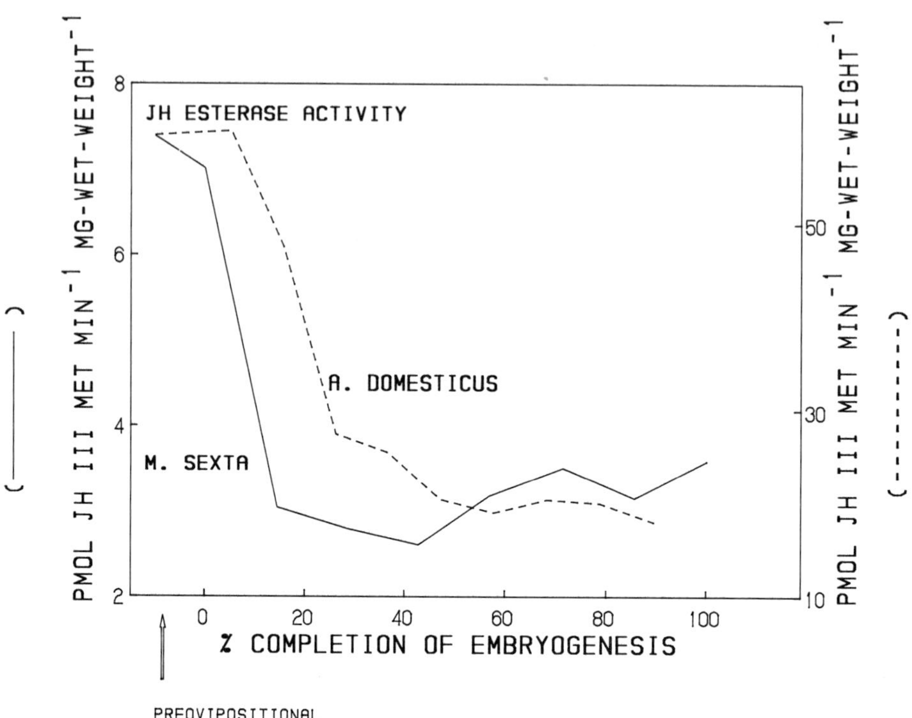

FIGURE 6.5. Juvenile hormone esterase activity in homogenates of eggs of the house cricket, *Acheta domesticus*, and the tobacco hornworm, *Manduca sexta*, during embryogenesis. The duration of embryogenesis was 9.5 days at 30°C for the house cricket and 3.5 days at 27°C for the tobacco hornworm. Key: JH = juvenile hormone.

females and newly oviposited eggs of *Hyalophora cecropia* and *Antheraea pernyi*, for example, blocked embryonic development shortly after germ band formation (Riddiford and Williams, 1967; Riddiford, 1970). These results were duplicated in a number of other insects (Hoffmann and Lagueux, 1985). The high JH esterase activity early in embryogenesis may then play a functional role in maintaining low JH titers early in embryogenesis, necessary for normal development.

Gilbert and Schneiderman (1961) showed that eggs from allatectomized females of *H. cecropia* contained no JH-active material, unlike that of normal females. This was the first evidence of the possible transfer of JH from females to developing oocytes. Temin et al. (1986) found

JH III in female adults and also in eggs of *L. migratoria* prior to the formation of the embryonic CA. The concentration in the adult was 50 times higher than that of the egg, thereby suggesting at least a passive transfer of maternal JH to the egg. The absence of JH in newly laid eggs of *M. sexta* (Bergot et al., 1981) may be explained in terms of the higher JH esterase activity seen during this period, and JH esterase may have a similar function in the house cricket.

Bürgin and Lanzrein (1988) did not consider early embryogenesis, but in their studies found that the JH metabolic activity increased in the embryonic gut, fat body and epidermis combined, and hemolymph of *N. cinerea* just before hatching. This was not found in *A. domesticus*, but there was a small increase in *M. sexta* at the end of development (Fig. 6.5), which did not prove to be statistically significant but was repeated with both JH I and JH III as substrate. The increase in JH metabolism in the cockroach just before hatching is correlated with a decrease in the *in vitro* biosynthesis of JH III and methyl farnesoate together (Bürgin and Lanzrein, 1988) and a decrease in their titer (Brüning et al., 1985).

6.3.6. *Diapause*

There is correlative evidence that JH esterase is involved in the regulation of JH titer in diapausing insects, but the mechanism differs between larvae and adults. Under short-day conditions, which are conducive to larval diapause in the southwestern corn borer, *Diatraea grandiosella*, and the European corn borer, *Ostrinia nubilalis*, hemolymph JH levels are high for an extended period at the beginning of the last instar (Yin and Chippendale, 1979; Bean et al., 1982), and it is believed that these elevated JH levels are responsible for or related to the last instar diapause of these species. JH esterase levels are relatively low at this time when JH levels are high (Mane and Chippendale, 1981; Bean et al., 1982). In *D. grandiosella*, elevated hemolymph JH esterase activity is also correlated with a decline in the JH titer associated with the termination of diapause and postdiapause pupation (Mane and Chippendale, 1981). In adults of the Colorado potato beetle, *Leptinotarsa decemlineata*, a low JH titer is necessary for the initiation of diapause (Kramer and de Kort, 1976a). In these insects, JH esterase levels remain high in diapausing beetles reared under short photophases presumably to ensure low JH levels needed to induce diapause. Likewise, JH esterase activity is reduced in beetles exiting diapause or in beetles reared under long-day conditions. It appears that the photoperiodic effect on the level of JH esterase activity in these beetles is mediated by a factor from the head (Kramer, 1978). Additional comparative studies are needed to more fully assess the significance of JH metabolism in the regulation of larval and adult diapause. In last-instar Lepidoptera, the role of metabolism remains to be examined in detail with respect to JH biosynthesis, JH titer, and the interaction of JH and ecdysone in the initiation, maintenance, and termination of diapause and in relation to the regulation of metamorphosis as we currently understand it (see Section 6.3.2, above).

6.4. Characterization of Ester Hydrolysis

Significant advances have been made since the last comprehensive review on JH metabolism (Hammock, 1985), in the areas of JH esterase purification, characterization of JH esterase from different life stages, and esterase kinetics and inhibition. It appears likely that these will also be areas of focus in the future to better understand the regulation of JH esterase biosynthesis at the level of the gene, to understand posttranscriptional events and JH esterase inactivation, and to delineate the critical interactions that occur between inhibitors and the esteratic active site.

6.4.1. *JH-Specific Esterase*

The term "juvenile hormone–specific esterase" was coined in 1975 to distinguish between what appeared to be two classes of hemolymph JH esterases in last instar *M. sexta*, separated as two peaks on gel filtration (Sanburg et al., 1975). The JH-specific esterases were capable of metabolizing JH and were not inhibited by O,O-diisopropyl phosphorofluoridate (DFP; see Fig. 6.6), whereas the general esterases were inhibited by DFP and also metabolized the substrate α-naphthyl acetate. It was believed in this model that the JH-specific esterase could metabolize JH whether it was free in solution or bound to specific JH-binding protein in the hemolymph, but nonspecific JH esterase could only metabolize unbound JH. Despite the attractiveness of this model and its wide acceptance, Sparks et al. (1983) and P. Jesu-

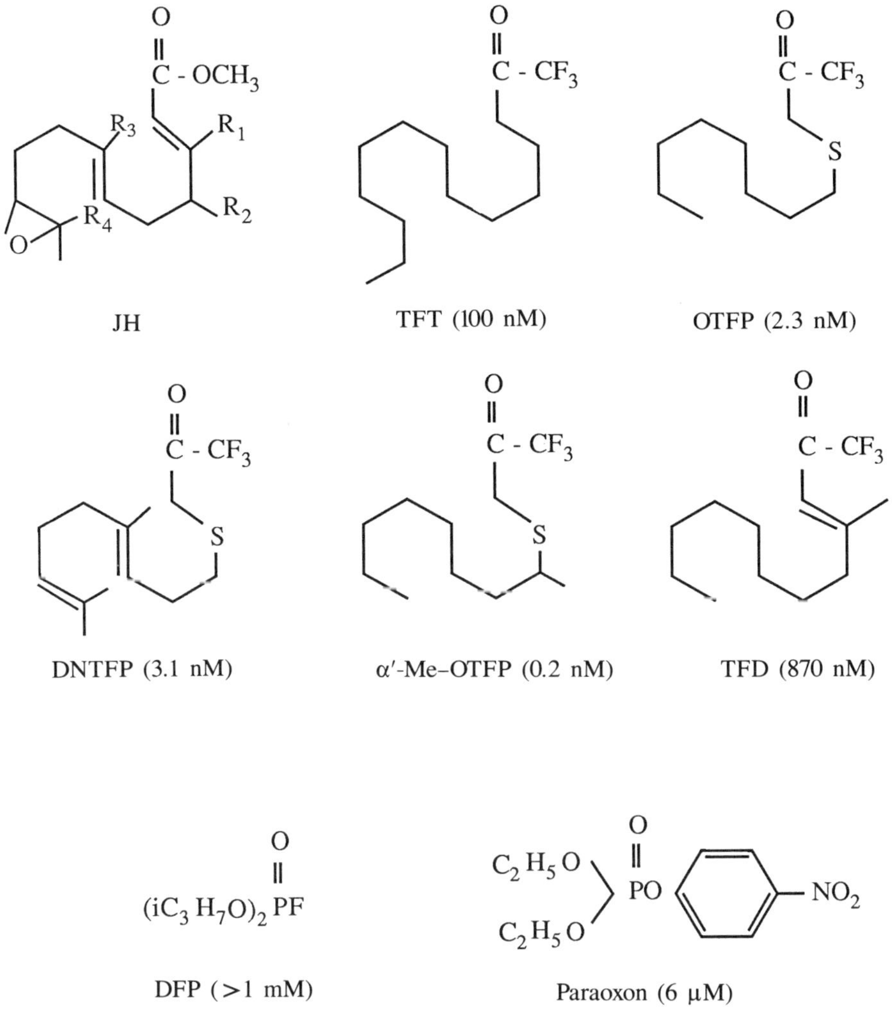

FIGURE 6.6. Juvenile hormone (JH 0, $R_1 = R_3 = R_4 = C_2H_5$, $R_2 = H$; iso-JH 0, $R_1 = R_2 = CH_3$, $R_3 = R_4 = C_2H_5$; JH I, $R_1 = CH_3$, $R_2 = H$, $R_3 = R_4 = C_2H_5$; JH II, $R_1 = R_3 = CH_3$, $R_2 = H$, $R_4 = C_2H_5$; and JH III, $R_1 = R_3 = R_4 = CH_3$, $R_2 = H$) and inhibitors of hemolymph juvenile esterase from the cabbage looper, *Trichoplusia ni*: TFT (1,1,1-trifluorotetradecan-2-one); OTFP (3-octylthio-1,1,1-trifluoro-2-propanone); DNTFP (3-[(E)-4,8-dimethyl-3,7-nonadienylthio]-1,1,1-trifluoro-2-propanone); α'Me–OTFP (3-(2-nonanethio)-1,1,1-trifluoropropane-2-one); TFD (1,1,1-trifluoro-3Z,4-methyl-dodecene-2-one); DFP (*O,O*-diisopropyl phosphorofluoridate); and paraoxon (*O,O*-diethyl-*O-p*-nitrophenyl phosphate). The concentration in paranthesis is the I_{50}, the inhibitor concentration necessary to inhibit 50% of the JH esterase activity. See the text for references.

dason, K. Venkatesh, and R. M. Roe (unpublished data), using the methods of Sanburg et al. (1975), found only a single peak of DFP-resistant JH-esterase activity. The α-naphthyl acetate was ineffective as a competitive inhibitor of this JH esterase activity. Furthermore, no DFP-susceptible JH esterase was found in the hemolymph of fourth-instar larvae, pupae, and adults of *M. sexta* (Jesudason, Venkatesh, and Roe, unpublished data) or in whole-egg homogenates (Share et al., 1988). Extensive developmental studies on the hemolymph of last-instar cabbage looper, *T. ni* (Sparks and Hammock, 1979a; Sparks et al., 1979b; Yuhas et al., 1983), and the sugarcane borer, *Diatraea saccharalis* (Roe et al., 1983), using gel filtration, isoelectric focusing, selective inhibition, thermal stability, and purification, have shown that there is no DFP-susceptible α-naphthyl acetate esterase that metabolizes JH in these species as well.

It is clear from inhibitor (see Section 6.4.2, below) and substrate-specificity studies (Peter, 1981; Yuhas et al., 1983; Abdel-Aal and Hammock, 1985c, 1988; D. Jones et al., 1986a; Hanzlik and Hammock, 1987; Abdel-Aal et al., 1988) that there are a number of critical recognition features that govern and also restrict the interactions of JH esterase with various potential substrates. For example, it was recently reported that the hemolymph JH esterase of *Trichoplusia ni*, *Heliothis zea*, and *Heliothis virescens* was able to distinguish between JH substrate molecules that differed only in the orientation of the epoxide moiety (Hanzlik and Hammock, 1987; Abdel-Aal et al., 1988). In *T. ni*, the 10*R*,11*S* enantiomer, which is believed to be the natural JH substrate, was metabolized at twice the rate of the 10*S*,11*R* enantiomer but had a twofold lower apparent K_m (Hanzlik and Hammock, 1987). There was no significant difference in the catalytic efficiency between the two enantiomers, however. There have also been a number of reports of differences in the rate of hydrolysis and the affinity of JH esterase between JH homologues. Abdel-Aal and Hammock (1985b), for example, reported a twofold greater V_{max} and K_m for JH I over that of JH III for the hemolymph JH esterase of *T. ni*. There are discrepancies between and within laboratories as to whether the K_m for JH esterase from this insect is in the 10^{-8} or $10^{-7}M$ range for JH I, JH II, and JH III (Sparks and Rose, 1983; Abdel-Aal and Hammock, 1985b, 1988; D. Jones et al., 1986a; Hanzlik and Hammock, 1987). Whatever the case, the enantiomeric selectivity, the apparent high affinity of JH esterase for JH, and structure activity studies with JH esterase inhibitors (see Section 6.4.2, below) supports the application of the term ''JH-specific esterase'' (Abdel-Aal et al., 1988). This conclusion has important functional significance in that it suggests that this esterase has the specific task of metabolizing only JH and not other substrates. This does not necessarily mean that the substrate specificity for JH has to be absolute, however, but only suggests that JH is the preferred substrate. Purified larval JH esterase from *T. ni* is also capable of metabolizing α-naphthyl acetate (Yuhas et al., 1983; Hanzlik and Hammock, 1987; Rudnicka and Jones, 1987) as well as several other non-JH substrates (Hanzlik and Hammock, 1987). The α-naphthyl acetate was metabolized at one-fifth the rate of JH III by hemolymph JH esterase from *T. ni*, showing that it was not the preferred substrate. Several studies have also indicated that this α-naphthyl acetate esterase activity of JH-specific esterase is a minor component of the total α-naphthyl acetate esterase activity in insect hemolymph and tissues (Sparks and Hammock, 1979a; Sparks et al., 1979b; Roe et al., 1983; Share et al., 1988). In contrast, Coudron et al. (1981) reported that purified JH esterase from larvae of *M. sexta* had no α-naphthyl acetate esterase activity. The question of JH esterase substrate specificity, although clearly defined above, should not be assumed but rather, should be characterized for each new system being investigated. This is often not considered in the literature.

The final point to be considered is the question of the interaction between the hemolymph JH-binding protein and JH esterase, a question which was addressed in the original definition by Sanburg et al. (1975) for JH-specific esterase (see above). Sanburg et al. (1975) and Hammock et al. (1975) found that hemolymph JH-binding protein provided a degree of protection for JH against JH esterases. It is probably safe to assume that JH bound to JH-binding protein is not susceptible to ester hydrolysis. Therefore, the rate of JH metabolism in insect hemolymph is a function of the association and dissociation rate constants for the binding protein, the catalytic efficiency for JH esterase (K_{cat}/K_m), and the concentration of these components in insect hemolymph. This of course assumes a simple compartment with no additional interactions. Abdel-Aal and Hammock (1988) have shown that in *T. ni* the rate-limiting step for this interaction is the release of JH from the binding protein. The half-life for the dissociation of JH II is approximately 7.5 min. The catalytic efficiency for JH esterase is such that, once JH is released into solution, the enzyme will turnover JH 150 times faster than the association rate with the binding protein. Abdel-Aal and Hammock (1988) concluded that any general esterase metabolism would be unlikely under these conditions since the catalytic efficiency of these esterases would be much lower than that of

JH-specific esterase. What needs to be factored into this model in the future is the interaction of insect tissues.

6.4.2. *JH Esterase Inhibition*

A number of potent inhibitors of JH esterase have been synthesized that have been useful in new assay techniques for insect JH epoxide hydrolase (Share and Roe, 1988), for the assessment of the biological role of JH metabolism in insect development (Sparks and Hammock, 1980; Hammock et al., 1984a; G. Jones and Hammock, 1985), for the purification of JH esterase (Abdel-Aal and Hammock, 1985a, 1986; Hammock et al., 1987; Hanzlik and Hammock, 1987), and in studies of enzyme–substrate interaction (Hammock et al., 1984a; Abdel-Aal and Hammock, 1985b; Linderman et al., 1987). Some of the first indications of the specificity of the JH-esterase active site came from the studies of Sanburg et al. (1975) when they reported that the JH-specific esterase of *M. sexta* was not inhibited by the esterase inhibitor DFP (Fig. 6.6). This stimulated a number of structure–activity and kinetic studies, and most recently physical studies to better understand the critical interactions that occur at the active site of JH esterase.

Hammock et al. (1982) were the first to report that aliphatic trifluoromethylketones, like that of TFT (Fig. 6.6), were potent inhibitors of JH esterase, and hypothesized that they formed tetrahedral adducts to the theoretical serine hydroxyl of JH esterase, thus mimicking the transition state in JH metabolism. It was interesting that the most effective inhibitor of those studied (TFT, $I_{50} = 100$ nM) closely mimicked the backbone of JH, as illustrated in Fig. 6.6. Surprisingly, the substitution of a sulfur atom for the C-4 methylene of TFT ultimately led to a second series of JH esterase inhibitors of even greater potency (Hammock et al., 1984a). OTFP, illustrated in Fig. 6.6, emerged as the most potent inhibitor of this series ($I_{50} = 2.3$ nM) and exhibited a molar refractivity identical to that observed for JH I. It is now well established from these and other studies in this laboratory that the backbone mimicry of JH is an important recognition feature in JH esterase inhibition and is also important in substrate interaction. Hammock et al. (1982) hypothesized that the increased potency of the OTFP series over that of TFT was the result of the sulfur mimicking the α,β unsaturation of JH. Attempts by Prestwich et al. (1984) to further increase the inhibitor potency of OTFP by mimicking the alkyl substitutions of JH at C-7 and C-11 did not lead to any improvements (see DNTFP, Fig. 6.6), but the addition of α- and α'-methyl substitutions by Linderman et al. (1987) produced the most potent *in vitro* JH esterase inhibitors available. The best of the series was α'-Me-OTFP, with approximately 10 times greater activity than that of OTFP (Fig. 6.6). The structure–activity studies conducted indicated that the carbonyl—but not the sulfur—interaction occurred in a sterically restricted environment and that a hydrophobic pocket near the presumed serine hydroxyl was important in the enzyme–inhibitor interaction. The α substitutions are likely mimicking the R_1 substituent of JH (Fig. 6.6). It now appears that the increased potency of the sulfur-containing trifluoromethylketones cannot be explained simply as mimicry of the α,β unsaturation of JH. The unsaturated trifluoromethylketone TFD (Fig. 6.6) was no better an inhibitor than was TFT despite TFD's obvious structural mimicry of JH and regardless of whether the Z or E isomer was considered (R. J. Linderman, L. Upchurch, M. Lonikar, K. Venkatesh, and R. M. Roe, unpublished data). High-field ^{19}F nuclear magnetic resonance (NMR) studies have shown that the polarized ketone of OTFP in water is almost exclusively hydrated, and in the active site of a common esterase, like that of electric eel acetylcholinesterase, is in a tetrahedral form (Linderman et al., 1988), supporting the hypothesis of Hammock et al. (1982) that these were transition state mimics. Most important in terms of their usefulness as affinity ligands, the trifluoromethylketones are reversible, competitive inhibitors of JH esterase (Hammock et al., 1982; Abdel-Aal et al., 1984; Hammock et al., 1984a; Prestwich et al., 1984; Abdel-Aal and Hammock, 1985b,c) and can be effectively dialyzed from the esteratic active site, regenerating JH esterase activity (Abdel-Aal et al., 1984).

Trifluoromethylketones are not only potent inhibitors of JH esterase but are also reasonably selective when their activity is compared with that of insect α-naphthyl acetate esterase and electric eel acetylcholinesterase (Hammock et al., 1982, 1984a). New inhibitors recently developed from this model by R. J. Linderman and R. M. Roe (unpublished data) have retained their JH esterase inhibitory potency but have absolute *in vivo* and *in vitro* selectivity. Trifluoromethylketone inhibitors are very effective and selective when applied topically and have been used successfully to demonstrate the function of JH metabolism in metamorphosis (Hammock et al., 1984a; Prestwich et al., 1984). Unfortunately, their *in vivo* activity is mostly short lived and less active, as compared with that of the organophosphate EPPAT (*O*-

ethyl-*S*-phenylphosphoramidothioate), even though EPPAT has moderate *in vitro* activity toward JH esterase. Until recently, structure–activity studies with organophosphates have been severely lacking. Sparks and Rose (1983) and Sparks et al. (1983) found that the inhibitory activity like that for trifluoromethylketones was dependent to some extent on the degree of JH mimicry. These researchers showed that stearic bulk near the phosphorous of paraoxon (Fig. 6.6) reduced inhibitory potency, presumably explaining the inactivity of DFP as a JH esterase inhibitor. Recent studies by R. J. Linderman, T. Tshering, K. Venkatesh, W. C. Dauterman, and R. M. Roe (unpublished data) have now demonstrated the importance of JH mimicry and have produced the most potent and selective *in vitro* and *in vivo* organophosphate inhibitors of JH esterase available. Organophosphates are irreversible inhibitors of JH esterase that follow Main kinetics (Abdel-Aal et al., 1984). Carbamate inhibitors of JH esterase have been reported (Mumby et al., 1979), and selective and persistent trifluoromethylketones are now also available that demonstrate no inhibition of α-naphthyl acetate esterase or acetylcholinesterase. (R. J. Linderman, M. Lonikar, K. Venkatesh, and R. M. Roe, unpublished data).

6.4.3. *JH Esterase Purification*

A major focus in JH esterase purification and subsequent characterization has been with the hemolymph of the cabbage looper, *T. ni*, and the tobacco hornworm, *M. sexta*. JH esterase has also been purified in *Galleria mellonella* (Rudnicka and Kochman, 1984), and in *Bombyx mori*, *Heliothis virescens*, and *H. zea* (Abdel-Aal and Hammock, 1986; Abdel-Aal et al., 1988). A classical purification scheme for *T. ni* has been available since 1981 (Yuhas et al., 1983) that utilized ammonium sulfate precipitation, anion exchange chromatography, gel filtration, and sucrose gradient ultracentrifugation to produce homogeneous, DFP-resistant JH esterase. Variations of the classical approach have been used by other researchers, the most noteworthy being that of Rudnicka and Jones (1987) and Wozniak and Jones (1987) for studies with *T. ni*, and Coudron et al. (1981) with *M. sexta*.

From structure–activity and kinetic studies with polarized trifluoromethylketones (Section 6.4.2, above), a number of potential affinity ligands were made available for the purification of JH esterase, the most useful being Sepharose $[S(CH_2)_4SCH_2COCF_3]$. This material, used with either column or batching techniques, had a high affinity for hemolymph and tissue JH esterase and moderate affinity for α-naphthyl acetate esterase; the latter can be prevented by pretreatment of insect samples with DFP. Treatment of JH esterase bound to this affinity matrix with salt, pH extremes, detergents, and urea were unsuccessful in causing its release, but exposure to OTFP concentrations greater than $10^{-5}M$ eluted the esterase in high purity (R. M. Roe and B. D. Hammock, unpublished data). The activity of the esterase can be regenerated by dialysis (Abdel-Aal et al., 1984) or ultrafiltration. The critical feature of this procedure is the proper preparation of the affinity gel (Abdel-Aal and Hammock, 1985a). JH esterase can be purified to homogeneity from crude hemolymph in one step with greater than 80% recovery in many cases, and has been used successfully on hemolymph and tissue JH esterase from *T. ni* (Hanzlik and Hammock, 1987) and on hemolymph from *M. sexta, B. mori, H. virescens*, and *H. zea* (Abdel-Aal and Hammock, 1985a, 1986; Abdel-Aal et al., 1988).

The recent availability of purified JH esterase and heightened interest in JH esterase structure has resulted in a significant amount of new information about this protein. Sparks and Hammock (1979a) were the first to separate two forms of JH esterase from the plasma of *T. ni* by high-resolution isoelectric focusing, but these studies were misinterpreted in subsequent research by a number of investigators. Roe et al. (1983), using isoelectric focusing and two different gel filtration techniques, found that there were two JH esterases present in the hemolymph of the sugarcane borer, *Diatraea saccharalis*. The existence of multiple forms of JH esterase has also been reported in other species (Kramer and de Kort, 1976b; Rudnicka and Kochman, 1984; Woodring and Sparks, 1987). D. Jones et al. (1986a,b) and Rudnicka and Jones (1987) have recently shown that *T. ni* JH esterase was heterogeneous with respect to charge; and even though the amino acid composition and the antigenic determinants were the same for the two major forms, there were differences between them in the fragmentation pattern when the proteins were subjected to cyanogen bromide degradation, suggesting possible differences in the amino acid sequence (Wozniak et al., 1987). It is interesting that purified protein from *T. ni* appeared as a single band on SDS–PAGE (sodium dodecyl sulfate–polyacrylamide gel electrophoresis) but consisted of multiple forms on narrow-range isoelectric focusing (Hanzlik and Hammock, 1987; Rudnicka and Jones, 1987). Even more intriguing, two kinetic forms with different apparent K_m values occurred in diluted crude hemolymph (D. Jones et al., 1986a; G. Jones and Click, 1987; Rudnicka and Jones, 1987), but only one form is present when the JH esterase is purified (Rudnicka and Jones, 1987) or the esterases separated by isoelectric focusing and analyzed alone (D. Jones et al., 1986a). Two esterase forms were also noted in inhibitor studies with crude *T. ni* hemolymph (Linderman et al., 1987). The molecular weight by SDS–PAGE reported for *T. ni* hemolymph JH esterase by Hanzlik and Hammock (1987) is 64 kD and by Rudnicka and Jones (1987) is 66 kD. Hanzlik and Hammock (1987) found that JH esterase was glycosylated and suggested that part of the differences between isoelectric forms could likely be explained in terms of the degree of glycosylation. Wozniak and Jones (1987) found some cross-reactivity with other species using anti-JH esterase of *T. ni*. The amino composition of the hemolymph JH esterase showed that aspartic and glutamic acid were in the highest amounts and cysteine and methionine in low amounts in *T. ni* (Wozniak and Jones, 1987), which is in general agreement with the results found for *M. sexta* [Table 6.1 (K. Venkatesh, Y. A. I. Abdel-Aal, M. E. Kelly, F. B. Armstrong, and R. M. Roe, unpublished data)]. The N-terminal sequence for *M. sexta* JH esterase has been determined: Arg-Ile-Pro-Ser-Thr-Glu-Glu-Val-Val-Val-Arg-Thr-Glu-Ser-Gly (Venkatesh, et al., unpublished data). Inhibitor studies with purified *M. sexta* JH esterase indicated two kinetic forms as in the native enzyme, suggesting the possibility of two active sites on the same protein (Abdel-Aal and Hammock, 1985a). These different kinetic forms in *M. sexta* and *T. ni* could be explained, as suggested by D. Jones et al. (1986a) as the result of aggregation, and it would be interesting to see the effect of mixing of different isoelectric forms or hemolymph fractions in the case of *T. ni*. With these questions in mind, it is likely that the protein chemistry of JH esterase will continue to be a focus of research at least in the near future.

6.4.4. *Distribution and Source of JH Esterase*

As illustrated earlier (Fig. 6.1), ester hydrolysis is the only route of metabolism of JH in insect hemolymph. Numerous studies have documented JH metabolism in insect tissues, such as whole-animal homogenates and homogenates of fat body, epidermis, body wall and tracheae, midgut, imaginal wing disks, corpora allata and cardiaca, testes, brain, subesophageal ganglion, and silk glands (Ajami and Riddiford, 1973; Terriere and Yu, 1973; Slade and Wilkinson, 1974; Hammock et al., 1975, 1977; Brown et al., 1977; Mane and Rembold, 1977; Yu and Terriere, 1978; Klages and Emmerich, 1979; Mitsui et al., 1979; Bisser and Emmerich, 1981; Mane and Chippendale, 1981; Wing et al.,

TABLE 6.1. Amino Acid Composition of the Juvenile Hormone Esterase Isolated from the Hemolymph of the Last-Instar Larvae of *Manduca sexta*[a]

Amino acid	Mol%	Residues/molecule
Aspartic acid	10.7	67
Glutamic acid	10.7	67
Serine	7.5	47
Glycine	8.9	56
Histidine	1.8	11
Arginine	5.6	35
Threonine	5.3	32
Alanine	12.2	74
Proline	7.0	42
Tyrosine	3.2	19
Valine	5.1	30
Methionine	1.4	9
Cysteine	ND	ND
Isoleucine	4.2	26
Leucine	7.4	47
Phenylalanine	5.1	30
Tryptophan	ND	ND
Lysine	4.0	21
Total	100.1	613

[a] ND = not determined.

1981; G. Jones, 1986; Renucci, 1986; Wiśniewski et al., 1986a,b, 1987; Hanzlik and Hammock, 1987; Roe et al., 1987a,b; Woodring and Sparks, 1987; Bürgin and Lanzrein, 1988; Share et al., 1988). The JH monooxygenase and epoxide hydrolase activity is microsomal in distribution, whereas the JH esterase activity is found in the cytosol. There have also been reports of soluble JH epoxide hydrolase (Wiśniewski et al., 1986b; M. R. Share and R. M. Roe, unpublished data). Despite considering several homogenization protocols and centrifugation procedures, soluble epoxide hydrolase always appeared in embryo homogenates of *M. sexta*, but it is not yet clear whether this soluble form is microsomal in origin (Share and Roe, unpublished data). The contribution of epoxide hydration to JH metabolism in tissue homogenates can vary from no role in eggs of *A. domesticus* (Roe et al., 1987a,b) to the major route of metabolism in integument homogenates of *G. mellonella* (Wiśniewski et al., 1986b). The same is true for tissue JH esterase. The distribution of JH monooxygenase and conjugation reactions in different tissues has not been given significant attention. Based on the correlation of tissue JH esterase levels to that of the hemolymph during last-stadium larval development of Lepidoptera, the detected *in vitro* release of esterase activity from the fat body held in temporary organ culture, the proportionally greater size of the fat body in comparison with other tissues, and the similarity of the fat body JH esterase to that of the hemolymph, it appears that the major source of the hemolymph JH esterase activity is the fat body (Whitmore et al., 1974; Hammock et al., 1975; Wing, 1981; Wing et al., 1981; G. Jones et al., 1987).

Comparative studies in the cabbage looper, *Trichoplusia ni*, indicated that the isoelectric point and inhibition kinetics were the same for JH esterase from fat body and hemolymph, but the hemolymph esterase was different from that of the midgut (Wing et al., 1981). Hanzlik and Hammock (1987) also found tissue JH esterase from whole-body homogenates of *T. ni* that was distinguishable from that of the hemolymph by SDS–PAGE as well as isoelectric focusing. Tissue JH esterases were also found in the embryo of the tobacco hornworm, *Manduca sexta*, that had the same and larger molecular weights to that of hemolymph JH esterase of last-instar larvae (Share et al., 1988). The same JH esterase was found in the tissue (excluding the gut) and hemolymph of the house cricket *Acheta domesticus* (Woodring and Sparks, 1987).

Recent studies have also examined the distribution of JH esterase in different life stages within the same species. The combined studies of Venkatesh et al. (1987) and G. Jones and Click (1987) have shown by isoelectric focusing, gel filtration, and substrate specificity that the hemolymph JH esterase is very similar or identical in the hemolymph of larvae from the second to the last instar and female adults of *T. ni*. Studies by Share et al. (1988) and P. Jesudason, K. Venkatesh, and R. M. Roe (unpublished data) have also shown, using the same methods, that the JH esterase of *M. sexta* embryos and that from the hemolymph of larvae from the second through the last instar and adults are similar. It is likely that the same gene is responsible for JH esterase biosynthesis in several tissues and at different life stages, and that it is the regulation of this single gene that changes in response to the stage of development (Venkatesh et al., 1987). It appears at least in the embryos of *M. sexta* and in the midgut of *T. ni* that there may also be a JH esterase peculiar to that life stage and tissue, respectively.

6.5. Regulation of JH Esterase

The developmental events that lead to the release of prothoracico-tropic hormone (PTTH) and the initiation of wandering behavior and pupation is regulated by a complex of events, including reduced JH biosynthesis and increased JH catabolism. The regulation of JH titer during various developmental events such as reproduction and diapause is also critical, and JH esterase appears to play a role in these processes. For example, in Lepidoptera, the appearance of JH esterase at precise times before wandering and pupation indicates that it is under some kind of neuroendocrine control. Progress in research made during the last 10 years suggests that JH esterase levels are regulated by JH and neuroendocrine and environmental factors (Fig. 6.7). However, the precise mechanism of JH esterase regulation by these various factors in determining the ultimate JH esterase levels in the hemolymph is far from clear (see also Hammock, 1985; Hammock et al., 1981, 1986). Therefore, investigation of the factor(s) that regulate JH degradative enzymes is an important area of research from a practical standpoint, particularly as it provides a rare opportunity to study the regulation of a protein, which appears to be under the variable control of neuroendocrine factors as well as the hormone JH.

6.5.1. *Role of JH*

The role of JH in the regulation of JH esterase activity was investigated in many insect species. Early studies by Whitmore et al. (1972, 1974) showed that the injection of JH into pupae of *Hyalophora gloveri* resulted in the appearance of carboxylesterases in the hemolymph capable of degrading JH. Increase in the JH esterase activity to topically applied or injected JH was also shown in the pupae of *G. mellonella* (Reddy et al., 1979), *T. ni* (Wing et al., 1981), and *Pieris rapae* (D. Jones et al., 1982) and in the adults of the cockroach *Diploptera punctata* (Rotin et al., 1982) and the Colorado potato bettle, *Leptinotarsa decemlineata* (Kramer, 1978).

The role of JH in the regulation of JH esterase activity was also documented in experiments involving ligation, allatectomy, and JH replacement. In newly emerged short-day adult beetles of *L. decemlineata*, allatectomy prevented the increase in the hemolymph JH esterase activity seen in normal insects and the application of JH to allatectomized beetles restored the JH esterase activity (Kramer, 1978). The JH esterase activity in the hemolymph of allatectomized females of *D. punctata* resembled that of virgin females, and topical application

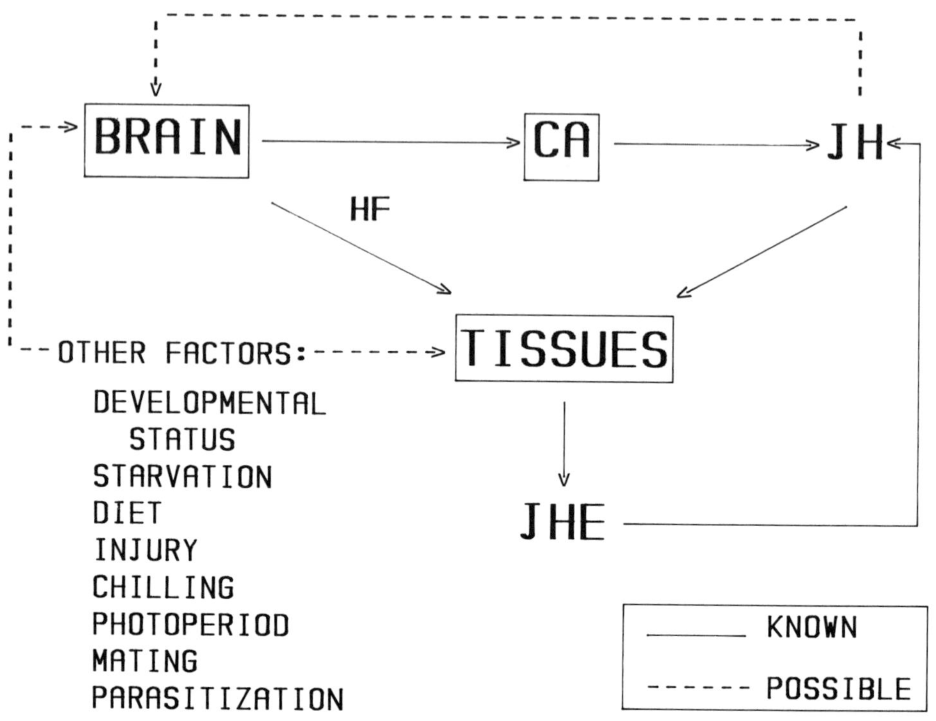

FIGURE 6.7. Factors that regulate the hemolymph juvenile hormone esterase in insects. Key: CA = corpora allata; HF = head factor(s); JH = juvenile hormone; JHE = juvenile hormone esterase.

of JH restored the JH esterase activity peak seen in normal mated females (Rotin et al., 1982). Allatectomy or ligation of wandering *T. ni* larvae (Sparks and Hammock, 1979b; G. Jones and Hammock, 1983) and allatectomy (Sparks et al., 1983) or ligation of postwandering *M. sexta* larvae (Venkatesh and Roe, 1988), prevented the appearance of the postwandering hemolymph JH esterase peak and the topical application of JH reversed the effect of ligation or allatectomy. It appears from the above studies that JH titer is influencing the JH esterase activity and that JH is directly acting on abdominal tissues, presumably the fat body (see Section 6.4.4, above), causing an increase in the hemolymph JH esterase levels and therefore its own hydrolysis at least at these specific stages. It is of interest to note that addition of JH caused a dose-dependent increase in the JH esterase released into the culture medium by the prepupal fat body of *T. ni* (Wing, 1981; G. Jones et al., 1987). In the pupae of *H. gloveri* (Whitmore et al., 1974) and adults of *L. decemlineata* (Kramer, 1978), injection of puromycin or actinomycin D prevented the increase in JH esterase activity in JH-treated insects, suggesting that JH is acting at the transcriptional level. By several enzyme characterization methods, the JH-induced JH esterase in prepupae of *T. ni* was shown to be similar to that of the naturally occurring enzyme (Sparks and Hammock, 1979a).

Extensive studies conducted in *T. ni* and *M. sexta* larvae have shown differences in the mechanism regulating the prewandering and postwandering peaks of hemolymph JH esterase activity. As discussed above, in both *T. ni* (Sparks and Hammock, 1979a,b; Sparks et al., 1979a; G. Jones and Hammock, 1983; Sparks, 1984; Venkatesh and Roe, unpublished data) and *M. sexta* (Venkatesh and Roe, 1988), and in other lepidopteran larvae, including *G. mellonella* (McCaleb and Kumaran, 1980) and *Ostrinia nubilalis* (Bean et al., 1983), JH increases the hemolymph JH esterase activity in both ligated and unligated postwandering and postfeeding diapausing larvae, suggesting a possible direct induction of JH esterase activity by JH. In prewandering larvae of *T. ni*, JH increased the hemolymph (Sparks and Hammock, 1979b; Venkatesh and Roe, unpublished data) and whole-body (Venkatesh and Roe, unpublished data) JH esterase activity in unligated larvae but not in ligated larvae. In the last-instar larvae of *M. sexta*, JH increased the hemolymph JH esterase activity in both unligated (Sparks et al., 1983; Venkatesh and Roe, 1988) and ligated prewandering larvae above respective controls, but the activity in ligated larvae never reached the levels observed in unligated larvae (Venkatesh and Roe, 1988). Ligation alone essentially eliminates the JH esterase activity from the hemolymph. Unlike postwandering larvae, the inability of prewandering ligated larvae to respond to JH provides indirect evidence for the presence of a head factor(s) other than JH involved in the regulation of JH esterase activity at this stage. However, compared to newly ligated prewandering larvae of *M. sexta* (0–2 days of age), JH treatment of larvae 24 h following ligation restored normal levels of JH esterase activity, indicating that the fat body of prewandering larvae was capable of synthesizing and/or releasing JH esterase into the hemolymph (Venkatesh and Roe, 1988). The same treatment had no effect in larvae that were starved for 24 h prior to the treatment, suggesting that the head in some way may be regulating the responsiveness of the abdomen to the topically applied JH. A similar effect of JH esterase activity was also seen in prewandering ligated *T. ni* larvae when the JH treatment was delayed for 24 h after ligation (Venkatesh and Roe, unpublished data). The turnover of inhibitory factor(s) in the bodies of ligated larvae or a reprogramming of fat body in the absence of head factor(s) may explain the increase in JH esterase when treatments were delayed for 24 h after ligation.

6.5.2. *Role of Head Factors*

The injection of prewandering brain extract into the abdomens of ligated *T. ni* larvae increased the hemolymph JH esterase activity above that of controls in prewandering and postwandering larvae (G. Jones et al., 1981). The increases by injection were only a fraction of those of unligated controls, but the transplantation of the intact brain–subesophageal complex from prewandering larvae into ligated larvae of the same age produced essentially normal levels of JH esterase activity. It was hypothesized that in *T. ni* the first prewandering peak of hemolymph JH esterase activity is induced by a factor from the head, whereas the second postwandering peak by JH itself is due possibly to a switchover in fat body responsiveness to JH between the first and second peak (Hammock, 1985). However, recent experiments with *T. ni* (Venkatesh and Roe, unpublished data) suggest that the regulation of JH esterase activity may be more complex than was formerly believed. In prewandering larvae of *T. ni*, Venkatesh and Roe (unpublished data) found that injection of head extract of 1-day-old larvae into the same-aged ligated larvae increased not only the JH esterase activity but also the α-naphthyl acetate esterase activity in the hemolymph by about threefold over that of the control. Surprisingly, the increase in the hemolymph JH esterase activity was not reflected in a similar increase in whole-body JH esterase levels. Therefore, it appears that the increase in the JH esterase activity in the hemolymph of head-injected *T. ni* larvae may be due to the release of the enzyme from tissues rather than *de novo* biosynthesis. In contrast to these findings, injection of head extracts into unligated prewandering *T. ni* larvae caused a decrease in the hemolymph JH esterase activity, compared with controls, whereas it increased the α-naphthyl acetate esterase activity. This decrease also occurred in whole-body preparations, suggesting that the head extract is directly acting on the tissues producing JH esterase. Head extract had no direct effect on JH esterase itself. In *T. ni*, both inhibitory and induction factor(s) may interact along with JH to regulate the prewandering peak of JH esterase activity. Evidence for an inhibitory factor in wandering larvae and prepupae was presented earlier by G. Jones et al. (1981). These authors found that the injection of brain extract from wandering larvae into ligated prewandering larvae and the transplantation of late last-instar brain and subesophageal ganglion into abdomens of ligated larvae of the same age reduced the hemolymph JH esterase activity as compared with the controls.

The idea of a head factor involved in the regulation of JH esterase was further supported by the work of McCaleb and Kumaran (1980) with *G. mellonella*. Brain and subesophageal ganglion from early-instar larvae transplanted into abdomens of head-ligated larvae of the same age increased the JH esterase activity to almost 50% that of unligated controls in 9 out of 23 brain-implanted larvae and to about 25% of unligated controls in 11 out of 22 subesophageal ganglion–implanted larvae. Transplants were made at a time in normal larvae when JH esterase activity was low but rising. Oźyhar et al. (1981) found later in development at the time of peak JH esterase activity that transplanted brains did not induce JH esterase but induced α-naphthyl acetate esterase activity, as has been noted in both *T. ni* and *M. sexta* for injections of brain extract (Venkatesh and Roe, 1988; Venkatesh and Roe, unpublished data). Ligated larvae during this period receiving both brains and exogenous JH in four out of nine cases had higher JH esterase than controls (Kumaran et al., 1981). In these studies it is unclear what effect the unreported results may have had on the interpretation of the data. Retnakaran and Joly (1976) reported that cauterization of A- and B-type neurosecretory cells of *Locusta migratoria* resulted in a 40% decrease in JH esterase activity, with a concomitant increase in general esterase activity.

A number of experiments conducted in *M. sexta* point to the possible existence of an inhibitory factor but not an induction factor (other than JH). This inhibitory factor is highly specific for JH esterase activity as compared with α-naphthyl acetate esterase activity (Venkatesh and Roe, 1988). Injection of head extracts into prewandering ligated larvae, unlike the studies discussed earlier in *T. ni*, had no effect on JH esterase activity, whereas it increased the α-naphthyl acetate esterase activity as before. In unligated larvae, injection of head or brain extract, but not subesophageal ganglion extract, could eliminate the normal increase in hemolymph JH esterase activity seen in prewandering larvae. This inhibitory effect was specific in action not affecting the α-naphthyl acetate esterase activity, was dose dependent, was present in head extract prepared from different-aged larvae, and acted at least in part independently of the head and thorax. Head extract injection also significantly decreased JH esterase activity in the whole body without affecting the α-naphthyl acetate esterase activity and did not directly affect the JH esterase enzyme itself. It therefore appears that, in *M. sexta*, JH increases the hemolymph JH esterase activity throughout the last stadium, but the response is possibly moderated by a factor from the brain, and the relative levels of JH and inhibitory factor may regulate the appearance of the wandering and postwandering JH esterase peak (Venkatesh and Roe, 1988). If the idea of a switchover described for *T. ni* is applicable to *M. sexta*, then the switchover would have to occur three times, since the insect is sensitive to JH at two separate points during last-stadium development. This explanation does not appear reasonable.

6.5.3. *Role of Environmental Factors*

In addition to JH and head factor(s), studies conducted in various insect species point to the existence of several additional factors that may influence JH esterase activity. Environmental input may affect the sites of JH esterase synthesis directly or act indirectly via the brain and CA to bring about a change in the JH esterase activity in the hemolymph (Fig. 6.7). These factors are briefly described below. The role of mating and the light:dark cycle on JH esterase activity was discussed earlier (in Sections 6.3.4 and 6.3.6, respectively).

Feeding

Starvation can drastically affect the hemolymph JH esterase activity in insects. In *M. sexta* (Cymborowski et al., 1982; Sparks et al., 1983; Venkatesh and Roe, 1988), *T. ni* (Venkatesh and Roe, unpublished data), and *G. mellonella* (Reddy et al., 1979), starvation of prewandering larvae blocked the appearance of the first peak of JH esterase activity in the hemolymph. In *M. sexta* larvae (Cymborowski et al., 1982), starvation leads to an increase in the JH titer. In starved *M. sexta* larvae, however, JH treatment failed to increase JH esterase activity and JH esterase levels remained low—even though normal larvae at this time are generally sensitive to JH. Factors responsible for increasing *in vivo* JH titers may also be regulating JH esterase levels, thus preventing esteratic degradation of the hormone in starved larvae.

Stress

In *G. mellonella*, chilling of larvae early in the last instar leads to a supernumerary molt, thus preventing metamorphosis (Reddy et al., 1979). This treatment also prevents the appearance of the first JH esterase peak. Similarly, injury of early last-instar larvae of *G. mel-*

lonella delayed metamorphosis and the appearance of JH esterase (Reddy et al., 1979). Injection of JH during the first 5 days of the last stadium caused a decrease in the JH esterase activity, compared with the controls in ligated and unligated larvae (Reddy et al., 1979; Kumaran et al., 1981). It appears then that environmental stress raises JH levels, which in turn directly inhibit JH esterase biosynthesis in the tissues.

Nutrition

Evidence also exists that the quality of an insect's diet can affect JH esterase levels. Studies have shown that feeding *M. sexta* larvae with a nonnutritive diet has the same effect on the hemolymph JH esterase activity as does starvation (Venkatesh and Roe, unpublished data). In *T. ni*, the increase in the hemolymph JH esterase activity was proportional to the amount of casein added to a minimal nutritional diet, suggesting that protein is an important factor influencing JH esterase levels (S. Levinson and B. D. Hammock, unpublished data).

Parasitism

In some insects, parasitization has an anti-JH effect leading to precocious metamorphosis. Parasitization of *T. ni* larvae by a *Chelonus* sp. caused larvae to form a prepupa one instar early. It was shown that in these parasitized larvae, changes occur that include the appearance of peak JH esterase levels in what would normally be the penultimate instar (Bühler et al., 1985; D. Jones 1985). These high JH esterase levels are a contributing factor to precocious development.

6.6.   Characterization of Epoxide Hydrolase

Research on insect JH epoxide hydrolase has been limited primarily to studies of the rate of hydration of the C-10,11 epoxide of JH in homogenates of various insect species or tissues, or when radiolabeled JH is introduced topically or by injection (Slade and Zibitt, 1972; White, 1972; Ajami and Riddiford, 1973; Slade and Wilkinson, 1974; Erley et al., 1975; Hammock et al., 1975; Slade et al., 1976; Edwards and Rowlands, 1978; Yu and Terriere, 1978; Mitsui et al., 1979; Wing et al., 1981; Renucci, 1986; Wiśniewski et al., 1986b; Share et al., 1988). JH epoxide hydrolase activity was measured during development in a small number of these studies (Slade et al., 1976; Yu and Terriere, 1978; Mitsui et al., 1979; Wing et al., 1981; Wiśniewski et al., 1986b; Share and Roe, 1988), some of which have already been discussed (Sections 6.3.2 and 6.3.4). These studies sometimes show some correlation of JH epoxide hydrolase activity in the tissues with changes in the JH esterase activity and in one case with JH monooxygenase activity, but this correlation is variable depending on the tissue being considered. Fat body, integument, and silk gland JH epoxide hydrolase in *G. mellonella* and fat body activity in *T. ni* appear to be low at the time of peak hemolymph JH esterase levels in the last instar but in the midgut of *T. ni* and in whole-body homogenates of young larvae and adults of the house fly are positively correlated with high JH esterase levels (Yu and Terriere, 1978; Wing et al., 1981; Wiśniewski et al., 1986b). The natural function of these changes in epoxide hydrolase activity is unknown, but Wing et al. (1981) hypothesized that the peak in fat body epoxide hydrolase activity following a peak in JH esterase activity might be important in preventing the synthesis of JH from JH acid. The significance of this latter point has been discussed earlier (Section 6.2.3).

During the embryogenesis of the tobacco hornworm, *M. sexta,* there were no changes in the JH epoxide hydrolase activity while there were significant changes in the JH esterase activity (Fig. 6.5). As would be expected, based on steric arguments, there was a significant difference in the rate of JH epoxide hydration between homologues, with JH III being metabolized 6.6 times faster than JH I. The reverse occurs for JH esterase, with JH I being metabolized roughly twice as fast as JH III. The interpretation of the importance of epoxide hydration in these experiments, as compared with ester hydrolysis can be drastically different depending on what substrate is being considered, but it was concluded in these studies considering the natural substrate and the relative rates of JH epoxide hydrolase and esterase activity, the *in vivo* information on JH metabolism, the significant changes that occurred in the esterase activity, and the correlation of these changes to developmental events that JH esterase had a functional role in embryogenesis. It is obvious, based on the significant differences between homologues, that caution is needed in the choice of the proper substrate, especially when studies are designed to compare different routes of metabolism. This point is usually not considered in the literature. With this idea in mind, and taking into account the recent reports of developmentally regulated JH esterase in adult mosquitoes (Shapiro et al., 1986; C. S. Apperson and R. M. Roe, unpublished data), the belief that JH epoxide hydrolase activity in the Diptera is more important than in the Lepidoptera (Wing et al., 1981) should be reexamined.

Characterization studies of insect JH epoxide hydrolase activity are almost completely lacking. Wiśniewski et al. (1986b) recently found that the soluble JH epoxide hydrolase of *G. mellonella* was greater than 1000 kD in molecular weight and concluded that the enzyme must aggregate to other proteins. Substrate specificity studies with purified epoxide hydrolase from the midgut microsomes of the southern armyworm, *Spodoptera eridania,* were conducted by Mullin and Wilkinson (1980), and the induction of activity by methoprene was reported in the house fly (Terriere and Yu, 1973). Effective inhibitors, always important in the assessment of biological function, are not yet available, although there appear now to be significant leads (Brooks, 1973; Slade et al., 1975; Hammock, 1985). Comprehensive studies analogous to those conducted with JH esterase are needed to clearly evaluate the function of epoxide hydrolase in JH liter regulation.

6.7. Prospectus

In a recent review, Hammock et al. (1986) emphasized the utility of using JH esterase as a potential strategy for the regulation of insect populations. It is clear that elevated JH esterase levels due to the injection of JH esterase or the elimination of JH esterase with powerful inhibitors can disrupt normal development. A number of significant leads are now available on the neuroendocrine regulation of JH esterase by the insect brain, on specific inhibitors of esterase activity, on the structure of JH esterase protein, and on the role of JH esterase in insect development. Not yet considered is the inactivation of JH esterase, which *in vitro* appears to be an extremely stable enzyme but *in vivo* is short lived. No doubt a major focus of future research will be to identify the structure and delineate the mechanism for the regulation of the JH esterase gene, especially considering the contemporary interest in this direction of research. In addition, the identity of the head factors that appear to regulate JH esterase biosynthesis will be critical to our understanding regulation. Not to be overshadowed by these more biochemical approaches, we should not forget that our understanding of JH metabolism is severely limited by an emphasis on specific life stages and insect models and that little is known outside the Insecta. Moreover, in our quest for understanding JH regulation at the molecular level, it is important that we not lose site of the physiology of the animal and its environment. Whole-animal modeling of JH metabolism that includes the consideration of different routes of metabolism and different tissues is a worthwhile goal for the future.

6.8.　Summary

JH titer in insects is regulated by a balance between the rate of JH biosynthesis and the rate of JH degradation. The primary route of *in vivo* metabolism of JH appears to be ester hydrolysis, with secondary metabolism occurring by epoxide hydration, oxidation, and conjugation. High hemolymph JH esterase activity is almost always associated with declining or low JH levels. A peak of JH esterase activity occurs prior to ecdysis in each larval instar, and a peak of activity occurs at wandering in the last instar of Lepidoptera. High metabolic activity in the last instar appears to have a scavenging role, removing the last traces of JH from the insect, which promotes ecdysone production in wandering larvae and regulates the development of the pupa just prior to the final larval molt. The timing of the appearance of this activity appears to be regulated by factors from the head and by JH itself. Mating in adult Lepidoptera eliminates JH esterase activity from the hemolymph, and it is believed that JH biosynthesis in the absence of JH esterase initiates vitellogenesis. High JH esterase activity also appears to be important in the inhibition of reproduction in migrating butterflies. Feeding in adult mosquitoes increases JH esterase activity, lowers JH levels, and increases ecdysone and the neurosecretory hormone responsible for egg maturation. Peak levels of JH esterase activity were noted prior to dorsal closure during embryogenesis in both hemimetabolous and holometabolous insects, suggesting that high activity during this time was important for the metabolism of maternal JH found in the egg and for normal development. High JH esterase activity also is correlated with the termination of larval diapause and the initiation of adult diapause. Several additional factors can affect JH esterase activity including photoperiod, nutrition, parasitism, and stress.

JH esterase is often referred to as JH-''specific'' esterase because of its specificity for JH. Studies indicate that ester hydrolysis of JH in insects is primarily restricted to this class of esterases and that general esterases or α-naphthyl acetate esterases are not a factor in JH metabolism. Although JH esterase demonstrates a high degree of substrate, inhibitor, JH homologue, and JH enantiomer selectivity, substrate specificity does not necessarily have to be absolute. Multiple, closely related isoelectric forms of JH esterase are distributed both in the hemolymph and tissue cytosol of insects. These similar forms are also noted in the egg, larva, and adult stages, but it appears that the mechanism that regulates their titer changes during development. The likely source of hemolymph JH esterase activity is the fat body. Potent trifluoromethylketone inhibitors of JH esterase have been synthesized, some of which have a high selectivity toward JH esterase as compared to α-naphthyl acetate esterase and acetylcholinesterase. NMR studies have shown that these inhibitors mimic the transition state in the metabolism of JH. Trifluoromethylketones have been used as an affinity ligand for the purification of JH esterase from hemolymph and tissues of several Lepidoptera. JH esterase in these studies is approximately 65 kD in molecular weight, and in some cases may be glycosylated or have two catalytic sites per protein. JH esterase exists as a monomer, and there appears to be different catalytic forms of the esterase in circulation in some insect species. The amino acid composition of the hemolymph indicates that aspartic and glutamic acid are in the highest amounts and cysteine and methionine in low amounts. The N-terminal sequence for *Manduca sexta* JH esterase begins with arginine, isoleucine, proline, and serine. Recent studies have shown that the synganglion or associated tissues of the tick are capable of synthesizing JH-like substances and that JH esterase metabolism is induced by feeding.

Acknowledgment

Our research reported in this chapter was supported by the USDA (85-CRCR-1-1661 and 87-CRCR-1-2417), the North Carolina Biotechnology Center (85-G-00701), the North Carolina State University Agricultural Foundation and the North Carolina Agricultural Research Service. We thank Janet Roe for her editorial work in the preparation of this manuscript.

References

Abdel-Aal, Y. A. I. and B. D. Hammock. 1985a. Apparent multiple catalytic sites involved in the ester hydrolysis of juvenile hormones by the hemolymph and by an affinity-purified esterase from *Manduca sexta* Johannson (Lepidoptera: Sphingidae). Arch. Biochem. Biophys. 243: 206–219.

Abdel-Aal, Y. A. I. and B. D. Hammock. 1985b. 3-Octylthio-1,1,1-trifluoro-2-propanone, a high affinity and slow binding inhibitor of juvenile hormone esterase from *Trichoplusia ni* (Hübner). Insect Biochem. 15: 111–122.

Abdel-Aal, Y. A. I. and B. D. Hammock. 1985c. Use of transition-state theory in the development of bioactive molecules. Pp. 135–160 *in* P. Hedin (ed.), *New Concepts and Trends in Pesticide Chemistry*, American Chemical Society Symposium Ser. No. 276. American Chemical Society, Washington, DC.

Abdel-Aal, Y. A. I. and B. D. Hammock. 1986. Transition state analogs as ligands for affinity purification of juvenile hormone esterase. Science (Wash., DC) 233: 1073–1076.

Abdel-Aal, Y. A. I. and B. D. Hammock. 1988. Kinetics of binding and hydrolysis of JH II in the hemolymph of *Trichoplusia ni* (Hübner). Insect Biochem. (in press).

Abdel-Aal, Y. A. I., R. M. Roe, and B. D. Hammock. 1984. Kinetic properties of the inhibition of juvenile hormone esterase by two trifluoromethylketones and *O*-ethyl,*S*-phenyl phosphoramidothioate. Pestic. Biochem. Physiol. 21: 232–241.

Abdel-Aal, Y. A. I., T. N. Hanzlik, B. D. Hammock, L. G. Harshman, and G. Prestwich. 1988. Juvenile hormone esterases in two heliothines: kinetic, biochemical and immunogenic characterization. Comp. Biochem. Physiol. (in press).

Ajami, A. M. and L. M. Riddiford. 1973. Comparative metabolism of the cecropia juvenile hormone. J. Insect Physiol. 19: 635–645.

Akamatsu, Y., P. E. Dunn, F. J. Kézdy, K. J. Kramer, J. H. Law, D. Reibstein, and L. L. Sanburg. 1975. Biochemical aspects of juvenile hormone action in insects. Pp. 123–149 *in* R. H. Meints and E. Davies (eds.), *Control Mechanisms in Development*. Plenum Press, New York.

Baker, F. C., L. W. Tsai, C. C. Reuter, and D. A. Schooley. 1987. *In vivo* fluctuation of JH, JH acid, and ecdysteroid titer, and JH esterase activity, during development of fifth stadium *Manduca sexta*. Insect Biochem. 17: 989–996.

Bean, D. W., S. D. Beck, and W. G. Goodman. 1982. Juvenile hormone esterases in diapause and nondiapause larvae of the European corn borer, *Ostrinia nubilalis*. J. Insect Physiol. 28: 485–492.

Bean, D. W., W. G. Goodman, and S. D. Beck. 1983. Regulation of juvenile hormone esterase activity in the European corn borer, *Ostrinia nubilalis*. J. Insect Physiol. 29: 877–883.

Bergot, B. J., G. C. Jamieson, M. A. Ratcliff, and D. A. Schooley. 1980. JH zero: new naturally occurring insect juvenile hormone from developing embryos of the tobacco hornworm. Science (Wash., DC) 210: 336–338.

Bergot, B. J., F. C. Baker, D. C. Cerf, G. Jamieson, and D. A. Schooley. 1981. Qualitative and quantitative aspects of juvenile hormone titers in developing embryos of several insect species: discovery of a new JH-like substance extracted from eggs of *Manduca sexta*. Pp. 33–45 *in* G. E. Pratt and G. T. Brooks (eds.), *Juvenile Hormone Biochemistry*. Elsevier/North-Holland, Publ., Amsterdam and New York.

Bhaskaran, G., S. P. Sparagana, P. Barrera, and K. H. Dahm. 1986. Changes in corpus allatum function during metamorphosis of the tobacco hornworm *Manduca sexta:* regulation at the terminal step in juvenile hormone biosynthesis. Arch. Insect Biochem. Physiol. 3: 321–338.

Bisser, A. and H. Emmerich. 1981. Hydrolysis of juvenile hormone and juvenile hormone analogues in the larval integument of *Drosophila hydei.* Pp. 199–212 *in* G. E. Pratt and G. T. Brooks (eds.), *Juvenile Hormone Biochemistry.* Elsevier/North-Holland Publ., Amsterdam and New York.

Brooks, G. T. 1973. Insect epoxide hydrase inhibition by juvenile hormone analogues and metabolic inhibitors. Nature (Lond.) 245: 382–384.

Brown, J. J., G. M. Chippendale, and S. Turunen. 1977. Larval esterases of the southwestern corn borer, *Diatraea grandiosella:* temporal changes and specificity. J. Insect Physiol. 23: 1255–1260.

Brüning, E., A. Saxer, and B. Lanzrein. 1985. Methyl farnesoate and juvenile hormone III in normal and precocene treated embryos of the ovoviviparous cockroach *Nauphoeta cinerea.* Int. J. Invertebr. Reprod. Dev. 8: 269–278.

Bühler, A., T. N. Hanzlik, and B. D. Hammock. 1985. Effects of parasitization of *Trichoplusia ni* by *Chelonus* sp. Physiol. Entomol. 10: 383–394.

Bürgin, C. and B. Lanzrein. 1988. Stage-dependent biosynthesis of methyl farnesoate and juvenile hormone III and metabolism of juvenile hormone III in embryos of the cockroach, *Nauphoeta cinerea.* Insect Biochem. 18: 3–9.

Connat, J. L. 1983. Juvenile hormone esterase activity during the last larval and pupal stages of *Tenebrio molitor.* J. Insect Physiol. 29: 515–521.

Coudron, T. A., P. E. Dunn, H. L. Seballos, R. E. Wharen, L. L. Sanburg, and J. H. Law. 1981. Preparation of homogeneous juvenile hormone specific esterase from the haemolymph of the tobacco hornworm, *Manduca sexta.* Insect Biochem. 11: 453–461.

Cymborowski, B. and G. Stolarz. 1979. The role of juvenile hormone during larval–pupal transformation of *Spodoptera littoralis:* switchover in the sensitivity of the prothoracic gland to juvenile hormone. J. Insect Physiol. 25: 939–942.

Cymborowski, B., M. Bogus, N. E. Beckage, C. M. Williams, and L. M. Riddiford. 1982. Juvenile hormone titres and metabolism during starvation-induced supernumerary larval moulting of the tobacco hornworm, *Manduca sexta* (L.). J. Insect Physiol. 28: 129–135.

de Kort, C. A. D. and N. A. Granger. 1981. Regulation of the juvenile hormone titer. Annu. Rev. Entomol. 26: 1–28.

Denlinger, D. L. 1985. Hormonal control of diapause. Pp. 354–412 *in* G. A. Kerkut and L. I. Gilbert (eds.), *Comprehensive Insect Physiology, Biochemistry and Pharmacology,* Vol. 8. Pergamon Press, Oxford and Elmsford, New York.

Dominick, O. S. and J. W. Truman. 1985. The physiology of wandering behaviour in *Manduca sexta.* II. The endocrine control of wandering behaviour. J. Exp. Biol. 117: 45–68.

Dorn, A., S. T. Bishoff, and L. I. Gilbert. 1987. An incremental analysis of the embryonic development of the tobacco hornworm, *Manduca sexta*. Int. J. Invertebr. Reprod. Dev. 11: 137–157.

Edwards, J. P. and D. G. Rowlands. 1977. Metabolism of a synthetic insect juvenile hormone (JH-I) during the development of *Tribolium castaneum* (Herbst) (Coleoptera: Tenebrionidae). Pestic. Biochem. Physiol. 7: 194–201.

Edwards, J. P. and D. G. Rowlands. 1978. Metabolism of a synthetic juvenile hormone (JH-I) in two strains of the grain weevil, *Sitophilus granarius*. Insect Biochem. 8: 23–28.

Erley, D., S. Southard, and H. Emmerich. 1975. Excretion of juvenile hormone and its metabolites in the locust, *Locusta migratoria*. J. Insect Physiol. 21: 61–70.

Fescemyer, H. W., R. L. Rose, T. C. Sparks, and A. M. Hammond. 1986. Juvenile hormone esterase activity in developmentally synchronous ultimate stadium larvae of the migrant insect, *Anticarsia gemmatalis*. J. Insect Physiol. 32: 1055–1063.

Feyereisen, R. 1985. Regulation of juvenile hormone titer: synthesis. Pp. 391–429 *in* G. A. Kerkut and L. I. Gilbert (eds.), *Comprehensive Insect Physiology, Biochemistry and Pharmacology*, Vol. 7. Pergamon Press, Oxford and Elmsford, New York.

Gilbert, L. I. and H. A. Schneiderman. 1961. The content of juvenile hormone and lipid in Lepidoptera: sexual differences and developmental changes. Gen. Comp. Endocrinol. 1: 453–472.

Goodman, W. G. and E. S. Chang. 1985. Juvenile hormone cellular and hemolymph binding proteins. Pp. 491–510 *in* G. A. Kerkut and L. I. Gilbert (eds.), *Comprehensive Insect Physiology, Biochemistry and Pharmacology*, Vol. 7. Pergamon Press, Oxford and Elmsford, New York.

Granger, N. A., S. M. Niemiec, L. I. Gilbert, and W. E. Bollenbacher. 1982. Juvenile hormone synthesis in vitro by larval and pupal corpora allata of *Manduca sexta*. Mol. Cell. Endocrinol. 28: 587–604.

Gruetzmacher, M. C., L. I. Gilbert, N. A. Granger, W. Goodman, and W. E. Bollenbacher. 1984. The effect of juvenile hormone on prothoracic gland function during the larval–pupal development of *Manduca sexta:* an *in situ* and *in vitro* analysis. J. Insect Physiol. 30: 331–340.

Hammock, B. D. 1981. Metabolism and environmental fate of juvenoids: a decade's research in perspective and an assessment of the future. Part 2. Pp. 841–852 *in* F. Sehnal, A. Zabża, J. J. Menn, and B. Cymborowski (eds.), *Regulation of Insect Development and Behavior*. Wrocław Technical University Press, Wrocław, Poland.

Hammock, B. D. 1985. Regulation of juvenile hormone titer: degradation. Pp. 431–472 *in* G. A. Kerkut and L. I. Gilbert (eds.), *Comprehensive Insect Physiology, Biochemistry and Pharmacology*, Vol. 7. Pergamon Press, Oxford and Elmsford, New York.

Hammock, B. D. and G. B. Quistad. 1976. The degradative metabolism of juvenoids by insects. Pp. 374–393 *in* L. I. Gilbert (ed.), *The Juvenile Hormones*. Plenum Press, New York.

Hammock, B. D. and G. B. Quistad. 1981. Metabolism and mode of action of juvenile hormone, juvenoids, and other insect growth regulators. Pp. 1–83 *in* D. H. Hutson and T. R. Roberts (eds.), *Progress in Pesticide Biochemistry*, Vol. 1. Wiley, New York.

Hammock, B. D. and R. M. Roe. 1985. Analysis of juvenile hormone esterase activity. Methods Enzymol. 111: 487–494.

Hammock, B. D. and T. C. Sparks. 1977. A rapid assay for insect juvenile hormone esterase activity. Anal. Biochem. 82: 573–579.

Hammock, B. [D.], J. Nowock, W. Goodman, V. Stamoudis, and L. I. Gilbert. 1975. The influence of hemolymph-binding protein of juvenile hormone stability and distribution in *Manduca sexta* fat body and imaginal discs in vitro. Mol. Cell. Endocrinol. 3: 167–184.

Hammock, B. D., S. M. Mumby, and P. W. Lee. 1977. Mechanisms of resistance to the juvenoid methoprene in the house fly *Musca domestica* L. Pest. Biochem. Physiol. 7: 261–272.

Hammock, B. D., D. Jones, G. Jones, M. Rudnicka, T. C. Sparks, and K. D. Wing. 1981. Regulation of juvenile hormone esterase in the cabbage looper, *Trichoplusia ni*. Pp. 219–235 *in* F. Sehnal, A. Zabża, J. J. Menn, and B. Cymborowski (eds.), *Regulation of Insect Development and Behavior*. Wrocław Technical University Press, Wrocław, Poland.

Hammock, B. D., K. D. Wing, J. McLaughlin, V. M. Lovell, and T. C. Sparks. 1982. Trifluoromethylketones as possible transition state analog inhibitors of juvenile hormone esterase. Pestic. Biochem. Physiol. 17: 76–88.

Hammock, B. D., Y. A. I. Abdel-Aal, C. A. Mullin, T. N. Hanzlik, and R. M. Roe. 1984a. Substituted thiotrifluoropropanones as potent selective inhibitors of juvenile hormone esterase. Pestic. Biochem. Physiol. 22: 209–223.

Hammock, B. D., Y. A. I. Abdel-Aal, T. Hanzlik, D. Jones, G. Jones, R. M. Roe, M. Rudnicka. T. C. Sparks, and K. D. Wing. 1984b. The role of juvenile hormone metabolism in the metamorphosis of selected Lepidoptera. Pp. 416–425 *in* J. Hoffmann and M. Porchet (eds.), *Biosynthesis, Metabolism and Mode of Action of Invertebrate Hormones*. Springer-Verlag, Berlin and New York.

Hammock, B. D., Y. A. I. Abdel-Aal, M. Ashour, A. Buhler, T. N. Hanzlik, R. Newitt, and T. C. Sparks. 1986. Paradigms for the discovery of new insect control agents. Pp. 53–72 *in* M. Sasa, S. Matsunaka, I. Yamomoto, and K. Ohsawa (eds.), *Human Welfare and Pest Control Chemicals: Approaches to Safe and Effective Control of Medical and Agricultural Pests*. Pesticide Science Society of Japan, Tokyo.

Hammock, B. D., Y. A. I. Abdel-Aal, T. N. Hanzlik, G. E. Croston, and R. M. Roe. 1987. Affinity purification and characteristics of juvenile hormone esterase from Lepidoptera. Pp. 315–328 *in* J. H. Law (ed.), *Molecular Entomology*. Liss, New York.

Hanzlik, T. N. and B. D. Hammock. 1987. Characterization of affinity-purified juvenile hormone esterase from *Trichoplusia ni*. J. Biol. Chem. 1987: 13584–13591.

Hiruma, K. 1980. Possible roles of juvenile hormone in the prepupal stage of

Mamestra brassicae. Gen. Comp. Endocrinol. 41: 392–399.

Hoffmann, J. A. and M. Lagueux. 1985. Endocrine aspects of embryonic development in insects. Pp. 435–460 *in* G. A. Kerkut and L. I. Gilbert (eds.), *Comprehensive Insect Physiology, Biochemistry and Pharmacology,* Vol. 1. Pergamon Press, Oxford and Elmsford, New York.

Hwang-Hsu, K., G. Reddy, A. K. Kumaran, W. E. Bollenbacher, and L. I. Gilbert. 1979. Correlations between juvenile hormone esterase activity, ecdysone titre and cellular reprogramming in *Galleria mellonella.* J. Insect Physiol. 25: 105–111.

Jones, D. 1985. The endocrine basis for developmentally stationary prepupae in larvae of *Trichoplusia ni* pseudoparasitized by *Chelonus insularis.* J. Comp. Physiol. 155: 235–240.

Jones, D., G. Jones, and B. D. Hammock. 1981. Growth parameters associated with endocrine events in larval *Trichoplusia ni* (Hübner) and timing of these events with developmental markers. J. Insect Physiol. 27: 779–788.

Jones, D., G. Jones, K. D. Wing, M. Rudnicka, and B. D. Hammock. 1982. Juvenile hormone esterases of Lepidoptera. I. Activity in the hemolymph during the last larval instar of 11 species. J. Comp. Physiol. 148: 1–10.

Jones, D., G. Jones, A. Click, M. Rudnicka, and S. Sreekrishna. 1986a. Multiple forms of juvenile hormone esterase active sites in the hemolymph of larvae of *Trichoplusia ni.* Comp. Biochem. Physiol. 85B: 773–781.

Jones, D., G. Jones, M. Rudnicka, A. J. Click, and S. C. Sreekrishna. 1986b. High resolution isoelectric focusing of juvenile hormone esterase from the hemolymph of *Trichoplusia ni* (Hübner). Experientia (Basel) 42: 45–47.

Jones, G. 1983. The regulation and role of juvenile hormone esterase during larval development of the cabbage looper, *Trichoplusia ni.* Ph.D. Thesis, University of California, Davis.

Jones, G. 1985. The role of juvenile hormone esterase in terminating larval feeding and initiating metamorphic development in *Trichoplusia ni.* Entomol. Exp. Appl. 39: 171–176.

Jones, G. 1986. Studies on the regulation, function and properties of juvenile hormone esterases of *Trichoplusia ni.* Pp. 315–318 *in* A. B. Bořkovec and D. B. Gelman (eds.), *Insect Neurochemistry and Neurophysiology.* Humana Press, Clifton, New Jersey.

Jones, G. and A. Click. 1987. Developmental regulation of juvenile hormone esterase in *Trichoplusia ni:* its multiple electrophoretic forms occur during each larval ecdysis. J. Insect Physiol. 33: 207–213.

Jones, G. and B. D. Hammock. 1983. Prepupal regulation of juvenile hormone esterase through direct induction by juvenile hormone. J. Insect Physiol. 29: 471–475.

Jones, G. and B. D. Hammock. 1985. Critical roles for juvenile hormone and its esterase in the prepupa of *Trichoplusia ni.* Arch. Insect Biochem. Physiol. 2: 397–404.

Jones, G., D. Jones, and S. Hiremath. 1987. An *in vitro* system for studying juvenile hormone induction of juvenile hormone esterase from the fat body of *Trichoplusia ni* (Hübner). Insect Biochem. 17: 897–904.

Jones, G., K. D. Wing, D. Jones, and B. D. Hammock. 1981. The source and action of head factors regulating juvenile hormone esterase in larvae of the cabbage looper, *Trichoplusia ni*. J. Insect Physiol. 27: 85–91.

Jones, G., A. Click, V. Reck-Malleczewen, M. J. Loeb, E. P. Brandt, and C. W. Woods. 1986. Haemolymph ecdysteroid titre in larvae of the cabbage looper *Trichoplusia ni*, and its modification by juvenile hormones. J. Insect Physiol. 32: 561–566.

Kiguchi, K. and L. M. Riddiford. 1978. A role of juvenile hormone in pupal development of the tobacco hornworm, *Manduca sexta*. J. Insect Physiol. 24: 673–680.

Klages, G. and H. Emmerich. 1979. Juvenile hormone metabolism and juvenile hormone esterase titer in hemolymph and peripheral tissues of *Drosophila hydei*. J. Comp. Physiol. 132: 319–325.

Koeppe, J. K., M. Fuchs, T. T. Chen, L.-M. Hunt, G. E. Kovalick, and T. Briers. 1985. The role of juvenile hormone in reproduction. Pp. 165–203 *in* G. A. Kerkut and L. I. Gilbert (eds.), *Comprehensive Insect Physiology, Biochemistry and Pharmacology*, Vol. 8. Pergamon Press, Oxford and Elmsford, New York.

Kramer, S. J. 1978. Regulation of the activity of JH-specific esterases in the Colorado potato beetle, *Leptinotarsa decemlineata*. J. Insect Physiol. 24: 743–747.

Kramer, S. J. and C. A. D. de Kort. 1976a. Age-dependent changes in juvenile hormone esterase and general carboxyesterase activity in the hemolymph of the Colorado potato beetle, *Leptinotarsa decemlineata*. Mol. Cell. Endocrinol. 4: 43–53.

Kramer, S. J. and C. A. D. de Kort. 1976b. Some properties of hemolymph esterases from *Leptinotarsa decemlineata*. Life Sci. 19: 211–218.

Kramer, S. J. and J. H. Law. 1980. Synthesis and transport of juvenile hormones in insects. Ac. Chem. Res. 13: 297–303.

Kumaran, A. K., M. M. Brook and S. Khipple. 1981. Juvenile hormone esterase activity in *Galleria*: effect of JH, ligation and brain implantation. Pp. 177–184 *in* G. E. Pratt and G. T. Brooks (eds.), *Juvenile Hormone Biochemistry*. Elsevier/North-Holland Publ., Amsterdam and New York.

Laufer, H., M. Landau, E. Homola, and D. W. Borst. 1987a. Methyl farnesoate: its site of synthesis and regulation of secretion in a juvenile crustacean. Insect Biochem. 17: 1129–1131.

Laufer, H., D. Borst, F. C. Baker, C. Carrasco, M. Sinkus, C. C. Reuter, L. W. Tsai, and D. A. Schooley. 1987b. The identification of a juvenile hormone-like compound in a crustacean. Science (Wash., DC) 235: 202–205.

Lazarovici, P., D. Shapira, V. Pisarev, and E. Shaaya. 1984. Ecdysteroid level and the change in sensitivity to a juvenile hormone analogue in *Ephestia cautella* larvae (Lepidoptera, Phycitidae). Arch. Insect Biochem. Physiol. 1: 409–415.

Lessman, C. A. and W. S. Herman. 1984. Juvenile hormone metabolism, binding and esterase activities in the haemolymph of the adult monarch butterfly (*Danaus p. plexippus* L.: Lepidoptera). Insect Biochem. 14: 445–452.

Linderman, R. J., J. Leazer, K. Venkatesh, and R. M. Roe. 1987. The inhibition of insect juvenile hormone esterase by trifluoromethylketones: steric parameters at the active site. Pestic. Biochem. Physiol. 29: 266–277.

Linderman, R. J., J. Leazer, R. M. Roe, K. Venkatesh, B. S. Selinsky, and R. E. London. 1988. ^{19}F-NMR spectral evidence that 3-octylthio-1,1,1-trifluoropropan-2-one, a potent inhibitor of insect juvenile hormone esterase, functions as a transition state analog inhibitor of acetylcholinesterase. Pestic. Biochem. Physiol. (in press).

Mane, S. D. and G. M. Chippendale. 1981. Hydrolysis of juvenile hormone in diapausing and non-diapausing larvae of the southwestern corn borer, *Diatraea grandiosella*. J. Comp. Physiol. 144: 205–214.

Mane, S. D. and H. Rembold. 1977. Developmental kinetics of juvenile hormone inactivation in queen and worker castes of the honey bee, *Apis mellifera*. Insect Biochem. 7: 463–467.

Masner, P., G. Hüsler, A. Pryde, and S. Dorn. 1983. Interactions between a non-terpenoid carbamate with juvenile hormone activity and esterases in the last larval instar of *Adoxophyes reticulana* Huebn. J. Insect Physiol. 29: 569–574.

McCaleb, D. C. and A. K. Kumaran. 1980. Control of juvenile hormone esterase activity in *Galleria mellonella* larvae. J. Insect Physiol. 26: 171–177.

Mitsui, T., L. M. Riddiford, and G. Bellamy. 1979. Metabolism of juvenile hormone by the epidermis of the tobacco hornworm, *Manduca sexta*. Insect Biochem. 9: 637–643.

Mullin, C. A. and C. F. Wilkinson. 1980. Insect epoxide hydrolase: properties of a purified enzyme from the southern armyworm *(Spodoptera eridania)*. Pestic. Biochem. Physiol. 14: 192–207.

Mumby, S. M. and B. D. Hammock. 1979. A partition assay for epoxide hydrases acting on insect juvenile hormone and an epoxide-containing juvenoid. Anal. Biochem. 92: 16–21.

Mumby, S. M., B. D. Hammock, T. C. Sparks, and K. Ota. 1979. Synthesis and bioassay of carbamate inhibitors of the juvenile hormone hydrolyzing esterases from the housefly, *Musca domestica*. J. Agric. Food Chem. 27: 763–765.

Newitt, R. A. and B. D. Hammock. 1986. Relationship between juvenile hormone and ecdysteroids in larval–pupal development of *Trichoplusia ni* (Lepidoptera: Noctuiidae). J. Insect Physiol. 32: 835–844.

Nowock, J., B. Hammock, and L. I. Gilbert. 1976. The binding protein as a modulator of juvenile hormone stability and uptake. Pp. 354–373 *in* L. I. Gilbert (ed.), *The Juvenile Hormones*. Plenum Press, New York.

Ożyhar, A., F. Sehnal, J. R. Wiśniewski, and M. Kochman. 1981. Regulation of juvenile hormone hydrolysis in the haemolymph. Pp. 193–198 *in* G.

E. Pratt and G. T. Brooks (eds.), *Juvenile Hormone Biochemistry*. Elsevier/ North-Holland Publ., Amsterdam and New York.

Ożyhar, A., J. R. Wiśniewski, F. Sehnal, and M. Kochman. 1983. Age dependent changes in the binding and hydrolysis of juvenile hormone in the haemolymph of last instar larvae of *Galleria mellonella*. Insect Biochem. 13: 435–441.

Pener, M. P. 1985. Hormonal effects on flight and migration. Pp. 491–550 *in* G. A. Kerkut and L. I. Gilbert (eds.), *Comprehensive Insect Physiology, Biochemistry and Pharmacology*, Vol. 8. Pergamon Press, Oxford and Elmsford, New York.

Pener, M. P., D. Dessberg, P. Lazarovici, C. C. Reuter, L. W. Tsai, and F. C. Baker. 1986. The effect of a synthetic precocene on juvenile hormone III titre in late *Locusta* eggs. J. Insect Physiol. 32: 853–857.

Peter, M. G. 1981. Stereochemistry of juvenile hormone hydrolysis in insect hemolymph. Pp. 237–244 *in* F. Sehnal, A. Zabża, J. J. Menn, and B. Cymborowski (eds.), *Regulation of Insect Development and Behaviour*. Wrocław Technical University Press, Wrocław, Poland.

Peter, M. G., S. Gunawan, G. Gellissen, and H. Emmerich. 1979. Differences in hydrolysis and binding of homologous juvenile hormones in *Locusta migratoria* hemolymph. Z. Naturforsch. 34C: 588–598.

Pound, M. and J. H. Oliver, Jr. 1979. Juvenile hormone: evidence of its role in the reproduction of ticks. Science (Wash., DC) 206: 355–357.

Pratt, G. E. 1975. Inhibition of juvenile hormone carboxyesterase of locust haemolymph by organophosphates *in vitro*. Insect Biochem. 5: 595–607.

Prestwich, G. D., W.-S. Eng, R. M. Roe, and B. D. Hammock. 1984. Synthesis and bioassay of isoprenoid 3-alkylthio-1,1,1-trifluoro-2-propanones: potent, selective inhibitors of juvenile hormone esterase. Arch. Biochem. Biophys. 228: 639–645.

Reddy, G., K. Hwang-Hsu, and A. K. Kumaran. 1979. Factors influencing juvenile hormone esterase activity in the wax moth, *Galleria mellonella*. J. Insect Physiol. 25: 65–71.

Reinecke, J. P., J. S. Buckner, and S. R. Grugel. 1980. Life cycle of laboratory-reared tobacco hornworms, *Manduca sexta*, a study of development and behavior, using time-lapse cinematography. Biol. Bull. (Woods Hole) 158: 129–140.

Renucci, M. 1986. Juvenile hormone degradation in nerve tissues and fat body of female *Acheta domesticus* (Insecta: Orthoptera). Comp. Biochem. Physiol. 84A: 101–106.

Renucci, M., N. Martin, and C. Strambi. 1984. Temporal variations of hemolymph esterase activity and juvenile hormone titers during ovocyte maturation in *Acheta domesticus* (Orthoptera). Gen. Comp. Endocrinol. 55: 480–487.

Retnakaran, A. and P. Joly. 1976. Neurosecretory control of juvenile hormone inactivation in *Locusta migratoria* L. Pp. 317–323 *in Actualitiés sur les hormones d'invertebrés*. Colloques Internationaux CNRS No. 251, Paris.

Riddiford, L. M. 1970. Effects of juvenile hormone on the programming of postembryonic development in eggs of the silkworm, *Hyalophora cecropia.* Dev. Biol. 22: 249–263.

Riddiford, L. M. and C. M. Williams. 1967. The effects of juvenile hormone analogues on the embryonic development of silkworms. Proc. Nat. Acad. Sci. USA 57: 595–601.

Roe, R. M., A. M. Hammond, Jr., and T. C. Sparks. 1983. Characterization of the plasma juvenile hormone esterase in synchronous last stadium female larvae of the sugar cane borer, *Diatraea saccharalis* (F.). Insect Biochem. 13: 163–170.

Roe, R. M., C. L. Crawford, C. W. Clifford, J. P. Woodring, T. C. Sparks, and B. D. Hammock. 1987a. Characterization of the juvenile hormone esterases during embryogenesis of the house cricket, *Acheta domesticus.* Int. J. Invertebr. Reprod. Dev. 12: 57–72.

Roe, R. M., C. L. Crawford, C. W. Clifford, J. P. Woodring, T. C. Sparks, and B. D. Hammock. 1987b. Role of juvenile hormone metabolism during embryogenesis of the house cricket, *Acheta domesticus.* Insect Biochem. 17: 1023–1026.

Rotin, D. and S. S. Tobe. 1983. The possible role of juvenile hormone esterase in the regulation of juvenile hormone titre in the female cockroach *Diploptera punctata.* Can. J. Biochem. Cell Biol. 61: 811–817.

Rotin, D., R. Feyereisen, J. Koener, and S. S. Tobe. 1982. Haemolymph juvenile hormone esterase activity during the reproductive cycle of the viviparous cockroach *Diploptera punctata.* Insect Biochem. 12: 263–268.

Rountree, D. B. and W. E. Bollenbacher. 1986. The release of the prothoracicotropic hormone in the tobacco hornworm, *Manduca sexta,* is controlled intrinsically by juvenile hormone. J. Exp. Biol. 120: 41–58.

Rudnicka, M. and D. Jones. 1987. Characterization of homogeneous juvenile hormone esterase from larvae of *Trichoplusia ni.* Insect Biochem. 17: 373–382.

Rudnicka, M. and M. Kochman. 1984. Purification of the juvenile hormone esterase from the haemolymph of the wax moth *Galleria mellonella* (Lepidoptera). Insect Biochem. 14: 189–198.

Safranek, L., B. Cymborowski, and C. M. Williams. 1980. Effects of juvenile hormone on ecdysone-dependent development in the tobacco hornworm, *Manduca sexta.* Biol. Bull. (Woods Hole) 158: 248–256.

Sanburg, L. L., K. J. Kramer, F. J. Kézdy and J. H. Law. 1975. Juvenile hormone-specific esterases in the haemolymph of the tobacco hornworm, *Manduca sexta.* J. Insect Physiol. 21: 873–887.

Shapiro, A. B., G. D. Wheelock, H. H. Hagedorn, F. C. Baker, L. W. Tsai, and D. A. Schooley. 1986. Juvenile hormone and juvenile hormone esterase in adult females of the mosquito *Aedes aegypti.* J. Insect Physiol. 32: 867–877.

Share, M. R. and R. M. Roe. 1988. A partition assay for the simultaneous determination of insect juvenile hormone esterase and epoxide hydrolase activity. Anal. Biochem. 169: 81–88.

Share, M. R., K. Venkatesh, P. Jesudason, and R. M. Roe. 1988. Juvenile hormone metabolism during embryogenesis in the tobacco hornworm, *Manduca sexta* (L.). Arch. Insect Biochem. Physiol. (in press).

Slade, M. and C. F. Wilkinson. 1974. Degradation and conjugation of cecropia juvenile hormone by the southern armyworm *(Prodenia eridania)*. Comp. Biochem. Physiol. 49B: 99–103.

Slade, M. and C. H. Zibitt. 1972. Metabolism of cecropia juvenile hormone in insects and in mammals. Pp. 155–176 *in* J. J. Menn and M. Beroza (eds.), *Insect Juvenile Hormones: Chemistry and Action.* Academic Press, Orlando, Florida.

Slade, M., G. T. Brooks, H. K. Hetnarski, and C. F. Wilkinson. 1975. Inhibition of the enzymatic hydration of the epoxide HEOM in insects. Pestic. Biochem. Physiol. 5: 35–46.

Slade, M., H. K. Hetnarski, and C. F. Wilkinson. 1976. Epoxide hydrase activity and its relationship to development in the southern armyworm, *Prodenia eridania.* J. Insect Physiol. 22: 619–622.

Solomon, K. R., C. K. A. Mango, and F. D. Obenchain. 1982. Endocrine mechanisms in ticks: effects of insect hormones and their mimics on development and reproduction. Pp. 399–438 *in* F. D. Obenchain and R. Galun (eds.), *Physiology of Ticks.* Pergamon Press, Oxford and Elmsford, New York.

Sparagana, S. P., G. Bhaskaran, and P. Barrera. 1985. Juvenile hormone acid methyltransferase activity in imaginal discs of *Manduca sexta* prepupae. Arch. Insect Biochem. Physiol. 2: 191–202.

Sparks, T. C. 1984. Effects of juvenile hormone I and the anti–juvenile hormone fluoromevalonolactone on development and juvenile hormone esterase activity in post-feeding last-stadium larvae of *Trichoplusia ni* (Hübner). J. Insect Physiol. 30: 225–234.

Sparks, T. C. and B. D. Hammock. 1979a. A comparison of the induced and naturally occurring juvenile hormone esterases from instar larvae of *Trichoplusia ni.* Insect Biochem. 9: 411–421.

Sparks, T. C. and B. D. Hammock. 1979b. Induction and regulation of juvenile hormone esterases during the last larval instar of the cabbage looper, *Trichoplusia ni.* J. Insect Physiol. 25: 551–560.

Sparks, T. C. and B. D. Hammock. 1980. Comparative inhibition of the juvenile hormone esterases from *Trichoplusia ni, Tenebrio molitor,* and *Musca domestica.* Pestic. Biochem. Physiol. 14: 290–302.

Sparks, T. C. and R. L. Rose. 1983. Inhibition and substrate specificity of the haemolymph juvenile hormone esterase of the cabbage looper, *Trichoplusia ni* (Hübner). Insect Biochem. 13: 633–640.

Sparks, T. C., K. D. Wing, and B. D. Hammock. 1979a. Effects of the anti-hormone–hormone mimic ETB on the induction of insect juvenile hormone esterase in *Trichoplusia ni.* Life Sci. 25: 445–450.

Sparks, T. C., W. S. Willis, H. H. Shorey, and B. D. Hammock. 1979b. Haemolymph juvenile hormone esterase activity in synchronous last instar larvae of the cabbage looper, *Trichoplusia ni*. J. Insect Physiol. 25: 125–132.

Sparks, T. C., B. D. Hammock, and L. M. Riddiford. 1983. The haemolymph juvenile hormone esterase of *Manduca sexta* (L.): inhibition and regulation. Insect Biochem. 13: 529–541.

Szibbo, C. M., D. Rotin, R. Feyereisen, and S. S. Tobe. 1982. Synthesis and degradation of C_{16} juvenile hormone (JH III) during the final two stadia of the cockroach, *Diploptera punctata*. Gen. Comp. Endocrinol. 48: 25–32.

Temin, G., M. Zander, and J.-P. Roussel. 1986. Physico-chemical (GC–MS) measurements of juvenile hormone III titres during embryogenesis of *Locusta migratoria*. Int. J. Invertebr. Reprod. Dev. 9: 105–112.

Terriere, L. C. and S. J. Yu. 1973. Insect juvenile hormones: induction of detoxifying enzymes in the housefly and detoxication by housefly enzymes. Pestic. Biochem. Physiol. 3: 96–107.

Tobe, S. S. and B. Stay. 1985. Structure and regulation of the corpus allatum. Adv. Insect Physiol. 18: 305–432.

Tobe, S. S., B. Stay, F. C. Baker, and D. A. Schooley. 1984. Regulation of juvenile hormone titre in the adult female cockroach *Diploptera punctata*. Pp. 397–406 *in* J. A. Hoffmann and M. Porchet (eds.), *Biosynthesis, Metabolism and Mode of Action of Invertebrate Hormones*. Springer-Verlag, Berlin and New York.

Tobe, S. S., R. P. Ruegg, B. A. Stay, F. C. Baker, C. A. Miller, and D. A. Schooley. 1985. Juvenile hormone titre and regulation in the cockroach *Diploptera punctata*. Experientia (Basel) 41: 1028–1034.

Truman, J. W. 1981. Interaction between ecdysterone, eclosion hormone, and bursicon titers in *Manduca sexta*. Am. Zool. 21: 655–661.

Venkatesh, K. and R. M. Roe. 1988. The role of juvenile hormone and brain factor(s) in the regulation of plasma juvenile hormone esterase activity during the last larval stadium of the tobacco hornworm, *Manduca sexta*. J. Insect Physiol. 34: 415–425.

Venkatesh, K., C. L. Crawford, and R. M. Roe. 1987. Characterization and the developmental role of plasma juvenile hormone esterase in the adult cabbage looper, *Trichoplusia ni*. Insect Biochem. 18: 53–61.

Watson, R. D., N. Agui, M. E. Haire, and W. E. Bollenbacher. 1987. Juvenile hormone coordinates the regulation of the hemolymph ecdysteroid titer during pupal commitment in *Manduca sexta*. Insect Biochem. 17: 955–959.

Weirich, G. and J. Wren. 1976. Juvenile-hormone esterase in insect development: a comparative study. Physiol. Zool. 49: 341–350.

White, A. F. 1972. Metabolism of the juvenile hormone analogue methyl farnesoate-10,11-epoxide in two insect species. Life Sci. 11: 201–210.

Whitmore, D., Jr., E. Whitmore, and L. I. Gilbert. 1972. Juvenile hormone induction of esterases: a mechanism for the regulation of juvenile hormone titer. Proc. Nat. Acad. Sci. USA 69: 1592–1595.

Whitmore, D., Jr., L. I. Gilbert, and P. I. Ittycheriah. 1974. The origin of hemolymph carboxylesterases "induced" by the insect juvenile hormone. Mol. Cell. Endocrinol. 1: 37–54.

Wing, K. D. 1981. Juvenile hormone esterases of *Trichoplusia ni* and other Lepidoptera: characteristics, site of synthesis, interaction with other proteins and inhibition. Ph.D. thesis. University of California, Riverside.

Wing, K. D., T. C. Sparks, V. M. Lovell, S. O. Levinson, and B. D. Hammock. 1981. The distribution of juvenile hormone esterase and its interrelationship with other proteins influencing juvenile hormone metabolism in the cabbage looper,*Trichoplusia ni.* Insect Biochem. 11: 473–485.

Wiśniewski, J. R., M. Muszyńska-Pytel, and M. Kochman. 1986a. Juvenile hormone degradation in brain and corpora cardiaca–corpora allata complex during the last larval instar of *Galleria mellonella* (Lepidoptera: Pyralidae). Experientia (Basel) 42: 167–169.

Wiśniewski, J. R., M. Rudnicka, and M. Kochman. 1986b. Tissue specific juvenile hormone degradation in *Galleria mellonella.* Insect Biochem. 16: 843–849.

Wiśniewski, J. R., M. Muszyńska-Pytel, K. Grzelak, and M. Kochman. 1987. Biosynthesis and degradation of juvenile hormone in corpora allata and imaginal wing discs of *Galleria mellonella* (L.). Insect Biochem. 17: 249–254.

Woodring, J. P. and T. C. Sparks. 1987. Juvenile hormone esterase activity in the plasma and body tissue during the larval and adult stages of the house cricket. Insect Biochem. 17: 751–758.

Wozniak, M. and D. Jones. 1987. Immunochemical characterization of juvenile hormone esterase from different species of Lepidoptera. Biochem. Biophys. Res. Commun. 144: 1281–1286.

Wozniak, M., G. Jones, S. Hiremath, and D. Jones. 1987. Biochemical and immunological properties of different electrophoretic forms of juvenile hormone esterase from *Trichoplusia ni* (Hübner). Biochem. Biophys. Acta 926: 26–39.

Yin, C.-M. and G. M. Chippendale. 1979. Diapause of the southwestern corn borer, *Diatraea grandiosella:* further evidence showing juvenile hormone to be the regulator. J. Insect Physiol. 25: 513–523.

Yu, S. J. and L. C. Terriere. 1978. Metabolism of juvenile hormone I by microsomal oxidase, esterase, and epoxide hydrase of *Musca domestica* and some comparisons with *Phormia regina* and *Sarcophaga bullata.* Pestic. Biochem. Physiol. 9: 237–246.

Yuhas, D. A., R. M. Roe, T. C. Sparks, and A. M. Hammond, Jr. 1983. Purification and kinetics of juvenile hormone esterase from the cabbage looper, *Trichoplusia ni* (Hübner). Insect Biochem. 13: 129–136.

Modes of Action of Juvenile Hormones at Cellular and Molecular Levels   7

A. KRISHNA KUMARAN

7.1. Introduction   183
    7.1.1. Chemistry of Juvenile Hormones   184
7.2. Effect of JH on Metamorphosis   185
    7.2.1. Morphology and Chemistry of the Cuticle   187
    7.2.2. Effect of JH on Imaginal Primordia   191
    7.2.3. JH and Cell Death   193
    7.2.4. Metamorphosis-Associated Changes in
       the Nervous System   194
    7.2.5. JH and the Fat Body   195
7.3. Other Morphogenetic Effects of JH   197
    7.3.1. Embryogenesis   197
    7.3.2. Caste Determination   199
    7.3.3. Phase Polymorphism   200
7.4. Effects on Reproduction   201
    7.4.1. Control of Vitellogenin Synthesis   202
    7.4.2. Vitellogenin Entry into the Oocyte   203
    7.4.3. Accessory Gland Function   205
7.5. Effect of JH on Life Cycle, Behavior,
  and Metabolism   206
    7.5.1. Diapause   206
    7.5.2. Effect on Flight and Migratory Behavior   210
    7.5.3. JH and Metabolism   210
7.6. Molecular Models of JH Action   210
    7.6.1. Nuclear Receptor   210
    7.6.2. Nature and Mechanisms of JH Receptor
       Activity   212
    7.6.3. Effect of JH on Membranes   213
7.7. Summary   217
Acknowledgments   218
References   218

7.1. Introduction

Juvenile hormone (JH) of insects is a unique morphogenetic hormone in that it inhibits expression of the phenotypic characteristics of the more advanced developmental stage of the insect. In the presence of JH, metamorphosis of a juvenile insect to the next developmental stage is blocked. From this simple observation, one can infer that JH blocks expression of the subsets of genes that are normally expressed during metamorphosis. Thus, JH acts as the typical genetic switch that seems to channelize the developmental potential of embryonic cells.

The existence of JH and its uniqueness among hormones was first surmised by Wigglesworth in 1934 during the course of his studies on the developmental physiology of *Rhodnius prolixus.* Following decapitation, fed *Rhodnius* nymphs underwent precocious metamorphosis. Parabiosis of nymphs at different stages of the life cycle revealed that the inhibition of metamorphosis was due to a blood-borne factor. Extirpation and implantation of the different tissues in the head of *Rhodnius* nymph identified the corpora allata (CA) as the source of the hormone that blocks metamorphosis (Wigglesworth, 1934, 1936). Soon thereafter, Bounhiol (1937) demonstrated that extirpation of CA from third-instar *Bombyx mori* caterpillars resulted in precocious metamorphosis, thereby showing that allata are the source of the metamorphosis-inhibiting hormone in lepidopterans as well.

Following these pioneering studies, several insect species belonging to diverse insect orders were studied, and in each case CA were shown to be the source of the JH (Schooley, 1977). Furthermore, in all hemimetabolous and holometabolous insects, JH was shown to block metamorphosis. Even in apterygote insects, which do not exhibit conspicuous metamorphic changes, CA produce compounds with juvenilizing effects (Watson, 1967). Thus CA, paired ectodermal ingrowths of the embryonic mandibular segment (Anderson, 1972), produce JH in all insects.

JH is a unique developmental hormone in another sense; it has many other functions. In addition to its ability to block metamorphosis, Wigglesworth (1936) observed that in *Rhodnius* JH is required for normal growth of the ovary and for the proper formation of spermatophores. This JH effect on reproduction was soon documented in cockroaches (Scharrer, 1946) and a variety of other insects but is by no means as universal as its morphogenetic effect. In its gonadotropic function, however, JH acts like other widely studied vertebrate hormones, namely, estradiol, progesterone, and testosterone. It selectively activates a set of genes necessary for successful reproduction. Thus, during larval life of an insect, JH blocks expression of the subset of genes that specify the imaginal phenotype, and during the adult reproductive stage it activates expression of the subset of genes that are necessary for reproduction.

In addition to its effects on metamorphosis and reproduction, which have been widely studied, JH affects many other aspects of insect physiology. Some of these effects are limited to a small group, whereas other effects may be seen in a wide variety of insects. For example, its effects on caste determination, phase polymorphism, and flight behavior may be limited to only a few groups of insects in which these phenomena are observed, whereas it affects embryonic development in a wide variety of insects. For an understanding of the mode of the action of JH at the cellular and molecular levels, all the diverse effects of JH on insect life have to be taken into consideration. Therefore, I shall describe in this chapter not only the effects of JH on metamorphosis, reproduction, and embryonic development but also its effects on other aspects of insect life.

7.1.1. *Chemistry of Juvenile Hormones*

The preparation in 1956 by Williams of a "golden oil" with the morphogenetic and gonadotropic effects of JH, from the abdomens of adult male *Hyalophora cecropia,* and the concomitant suggestion that JH could serve as a selective ecologically safe insecticide, heightened interest in the chemistry of the hormone and its chemical synthesis. In the next decade, the chemical structure of cecropia juvenile hormone was established to be a sesquiterpenoid derivative, methyl 10,11-epoxy-3-methyl-7,11-diethyl-2-*trans*,6-*trans*-tridecadienoate (Roeller et al., 1967), now designated as JH I. Soon thereafter, a second minor component of the golden oil, now designated as JH II, was identified as methyl 10,11-epoxy-3,7-dimethyl-11-ethyltricadienoate (Meyer et al., 1968). Two other juvenile hormones, JH III and JH 0, characterized as 3,7,11-trimethyl and 3,7,11-triethyl derivatives of methyl epoxytridecadienoate, respectively (Fig. 7.1), were later identified (Judy et al., 1973; Bergot et al., 1980). The relative concentrations of the different JH homologues vary in different groups of insects and also within the same species at different stages of the life cycle. Although these variations are very interesting and important from comparative and evolutionary points of view, from the perspective of their action at the cellular and molecular levels all the JH homologues may be expected to act in a similar manner. Hence, no special mention of the specific homologue acting at any specific developmental stage of the insect is included in these discussions. Similarly, the mechanisms of regulation of JH titers, though important for an overall understanding of the role of JH in regulation of insect life, are beyond the purview of this article.

The fact that JHs of all insects are chemically similar has an important bearing on the evolution of this hormone. As noted above, all the JH homologues identified thus far are derivatives of methyl-10,11-epoxytridecadienoate, suggesting that JH must have first evolved in the group ancestral to all insects. This view is further strengthened by the fact that the mandibular glands of the crab *Labinia marginata* and some related decapod crustaceans produce methyl farnesoate (Fig. 7.1), a compound with JH activity in insects (Laufer et al., 1987; Borst et al., 1987). Since the mandibular glands of crustaceans and the CA of all insects have identical embryological origins (Anderson, 1972), and because some insect embryos produce methyl farnesoate (Burgin and Lanzrein, 1988), it is reasonable to suggest that compounds with JH activity might be the invention of the mandibulate ancestors that gave rise to crustaceans, myriapods, and insects. This observation may also be relevant in an analysis of the primary role of JH in insect life during the early stages of insect evolution. Because it is not possible to collect data on this question, any discussion is largely conjectural and is not included in this chapter.

7.2. Effect of JH on Metamorphosis

JH derived its name from the fact that it blocks metamorphosis of *Rhodnius* nymphs into imagos. In silkworm larvae, JH blocks their metamorphosis into pupae. It also blocks development of pupae into adult insects. The magnitude of change in the shape of the insect body at metamorphosis is different in the three subclasses of insects. In Apterygota, the juvenile stage does not differ very much from the adult except in size. In *Thermobia domestica*, the first four instar nymphs have a smooth integument, whereas the next instar nymph has scales on its integument. This morphological change is under the control of CA. Allatal activity (as measured by bioassay) was low in the fourth-stage nymphs, suggesting that low JH titer permits this morphogenetic change (Watson, 1967). Morphological changes in juvenile *Lepisma* and *Ctenolepisma* were also shown to be under allatal control (see Yashika, 1960). Thus, morphological changes, regulated by JH in apterygotes, are not conspicuous. In hemimetabolous insects

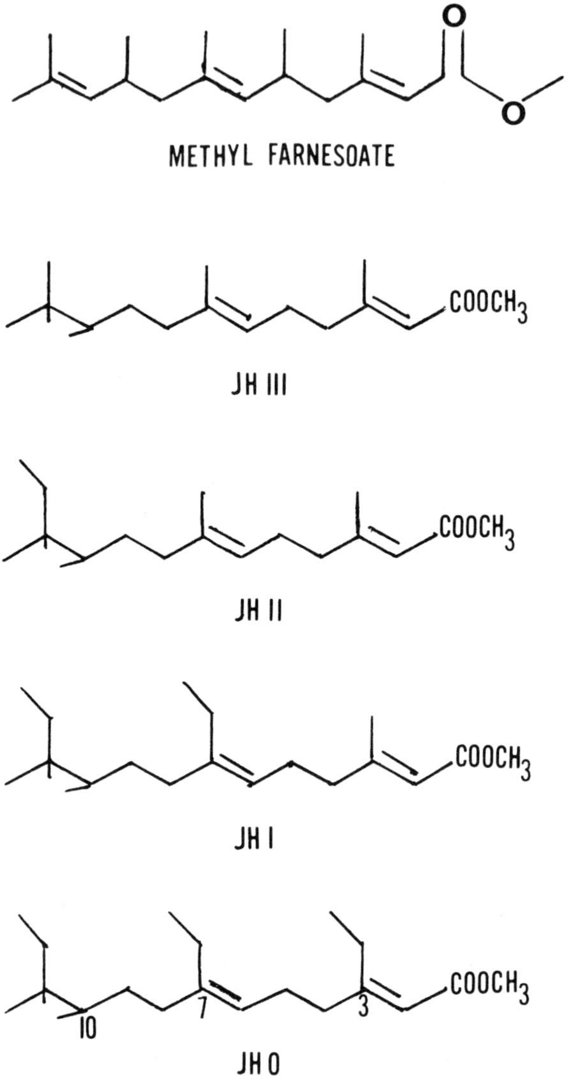

FIGURE 7.1. The molecular structure of the four juvenile hormone homologues and of methyl farnesoate, a compound secreted by the mandibular glands of crustaceans and by some insect embryos.

metamorphosis involves wing growth, heteromorphic growth of the segments of the legs and/or body segments. In addition, the insect develops musculature necessary for flight and acquires the ability to respond to hormonal and/or environmental cues involved in repro- duction. However, no major change in body shape occurs. JH blocks all these metamorphosis-associated changes (Wigglesworth, 1934).

In holometabolous insects, on the other hand, metamorphosis of a larva entails a gross change in the morphology of the insect. In a lepi- dopteran, for example, the wormlike, motile, feeding caterpillar with a soft pliable cuticle is changed into an immotile, nonfeeding stage with a thick, brittle cuticle and no obvious resemblance to a larva. This metamorphic change involves many organ systems. The epidermal cells that secrete a soft pliable cuticle in a larva synthesize a different type of cuticle at the metamorphic molt. The imaginal disk cells, which are represented in the larva as nests of cells with no specific phenotype, attain organ-specific shape and secrete a cuticle at metamorphosis. The pupa lacks certain organs that are present in the larva, e.g., the prolegs, cuticular linings of foregut and hindgut. Metamorphosis involves tissue-specific cell death in these tissues. The larva is longer than the pupa. Furthermore, the larva is active and motile, while the pupa is quiescent and immotile. Associated with these changes are changes in the nervous system and musculature. All these changes are blocked by the presence of JH. Similarly, during the course of development of an imago from a pupa several morpho- logical changes occur, all of which are blocked by JH. These develop- mental changes include deposition of an adult-type cuticle, reconstruc- struction of the fat body, development of flight muscles, degeneration of intersegmental muscles, and—more importantly—acquisition of the ability to reproduce. All these developmental changes are also blocked by JH. In this section, I shall analyze the cellular and molecular basis of the JH inhibition of metamorphosis.

7.2.1. *Morphology and Chemistry of the Cuticle*

The body of an insect, as that of all arthropods, is covered by a chitin–protein complex structure, the cuticle. The cuticle is deposited by a single-cell thick epidermal layer covering the entire body. Each cell secretes the piece of cuticle that is immediately overlying it. Thus, the body cuticle is a record of the synthetic potential of the epidermal cells that produced it. The cuticle in different parts of the body differs very much in thickness, chemical composition, and appearance. And the chemistry and morphology of the cuticle varies with its function.

The cuticle covering the tergites and sternites in a lepidopteran larva is moderately thick and slightly pliable. The cuticle of the intersegmental membrane is softer, thinner, and more pliable than that of tergites and sternites. The cuticle of the head capsule and mandibles, on the other hand, is hard, thick, and not pliable. The cuticles covering the bristles and the hooks of the prolegs are tanned and thus differ from the other three types of cuticle. Therefore, the integument of an insect larva is a mosaic of different types. Although the epidermal cells from each region of the insect body secrete distinctly different types of cuticles, all these are traditionally classified as the larval type cuticles. Secretion of the larval type cuticle requires the presence of JH. At each larval molt, in the presence of JH and under the influence of ecdysteroids, the epidermal cells secrete the larval type cuticle appropriate for each region of the larval body. The same epidermal cells secrete pupal-type cuticle if stimulated by ecdysteroids in the absence of JH. During larval–pupal metamorphosis, the endogenous JH titer declines; and at the succeeding molt, a pupal cuticle is formed.

The pupal integument is also a patchwork of areas serving different functions. The tergite and sternite cuticles are thicker and more tanned than that of the intersegmental membrane. The sternite cuticle, covered by the pupal wing and leg anlage, is also very thin. When pupal epidermal cells are exposed to JH and ecdysteroids, they again deposit a pupal cuticle. In the absence of JH, ecdysteroids stimulate pupal epidermal cells to secrete the adult type cuticle. Again, the adult-type cuticle, like the larval and pupal cuticles, differs in its structure and chemistry in different regions of the body. Despite these differences, all the types of cuticles in a moth are designated as adult cuticle, a reference to the developmental stage at which these cuticle types are formed. Similar differences in the pigmentation patterns and/or distribution of bristles and hairs of the nymphal and adult cuticles in *Rhodnius* and *Oncopeltus fasciatus* were described.

The effect of JH is cell autonomous, i.e., independent of the developmental fate of neighboring cells. If JH is topically applied to selected regions of the tergites, only the epidermal cells underlying this region of the cuticle fail to metamorphose. Application of low concentration of JH to abraded abdominal tergites in fifth-instar *Rhodnius* nymph resulted in patches of nymphal cuticle in a sea of adult cuticle at the ensuing molt (Wigglesworth, 1958). Application of JH to one wing lobe resulted in the production of an adult with one pupal wing. The *Galleria* bioassay, which involves inflicting a small injury on a pupal tergite and covering it with wax containing JH (Schneiderman et al., 1965), is based on the principle of cell-autonomous action of JH.

Similarly, the time of application of JH in the molt cycle has an effect on the magnitude of the morphogenetic effect of the hormone. Application of JH to day-3 to day-4 last-instar *Galleria mellonella* larvae leads to formation of perfect supernumerary larvae. On the other hand, applicaiton of JH to day-5 last-instar larvae leads to formation of larval-pupal intermediates. The pupal patches may be limited to everted wings, if applied early on day 5. Depending upon the age at the time of JH application, the pupal or larval cuticular patches predominate in the resulting chimeric larva (Krishnakumaran, 1972). Similar results were obtained with *Manduca sexta* and other insects. These observations show that different tissues and different regions of the body independently become insensitive to JH (Piepho, 1951). In addition, the same cells could make a composite cuticle, which combines the morphological features of two metamorphic stages, thereby suggesting that JH needs to be present continuously all through the period when the epidermal cell is secreting a cuticle in order to prevent metamorphosis (Willis et al., 1982). In *Pyrrhocoris apterus*, pigmentation, tanning, surface sculpturing, and microtrichiae of the cuticle may exhibit nymphal or adult characteristics, depending upon the time of JH administration in the last larval stadium. A cell can express some nymphal and some adult characteristics. Furthermore, the ability to synthesize larval cuticle is lost (conversely, ability to secrete imaginal cuticle is acquired) long before actual initiation of cuticle synthesis.

Because JH acts in a cell-autonomous manner, and because the type of cuticle synthesized by cells in different regions of the body differ, Wigglesworth (1966) suggested that two developmental programs exist in *Rhodnius* epidermal cells: one program specifies the larval phenotype, and the other specifies the adult phenotype. By analogy in holometabolous insects, three developmental programs—the larval, pupal, and adult programs—exist in their epidermal cells. In the presence of JH and ecdysone, larval epidermal cells express the larval program whereas the pupal epidermal cells express the pupal program. In the absence of JH, ecdysteroids activate the pupal programs in larval epidermal cells and the adult programs in the pupal cells (see Kumaran, 1976). All the same, the developmental program is cell specific. The developmental program activated in larval imaginal wing disk cells in the absence of JH leads to formation of a pupal wing cuticle, whereas the tergite and intersegmental epidermal cells synthesize pupal tergite and intersegmental cuticles, respectively.

By corollary, JH does not alter the developmental program of the chitogenous cells. The developmental program of each cell appears to be established during the embryonic development when the tissues in different regions of the embryo become determined in accordance with their position. Once the final developmental fate of the embryonic cells is fixed, this developmental program is normally inherited by all its descendents. According to this view, a chitogenous epidermal cell in a lepidopteran tergite possesses the developmental programs to secrete a larval, pupal, or adult tergite cuticle. Similarly, the cells of intersegmental membrane can secrete the intersegmental cuticle of the three developmental stages (see Wigglesworth, 1966; Kumaran, 1981; Willis, 1986).

Molecular analysis of the cuticular proteins from diverse regions of the body from *H. cecropia* larvae, pupae, and adults suggests that the subset of cuticular protein genes active in the epidermal cells that secrete sclerite, head capsule, and the intersegmental cuticles in larvae are different. Furthermore, the proteins in the intersegmental cuticle of a larva are more like those of the homologous region of a pupa or adult than those of the hard cuticles of the head capsule or the sclerite in a larva. Similarly, the hard cuticles of larval head capsule and of pupal and adult sclerites have a common subset of cuticular proteins, as defined by their electrophoretic mobility (Cox and Willis, 1985, 1987). These observations suggest that there is no stage-specific subset of cuticular genes (Willis, 1986). To put it another way, JH apparently blocks expression of different sets of cuticular protein genes in different regions of the larval or pupal body. These observations support the view that the particular anatomic region of the body that is involved plays a more important role in determining the protein composition of the cuticle than does any given developmental stage.

The question then revolves around the probable molecular mechanisms of JH control of metamorphosis. As noted above, Wigglesworth (1966) proposed the presence of larval, pupal, and adult sets of genes in the genome of a holometabolous insect. Furthermore, it was suggested that the stage-specific sets of genes are activated when the appropriate hormonal conditions are encountered. In the presence of ecdysteroids and JH, the larval set of genes is activated in a larval cell. In the absence of JH, ecdysteriods activate the pupal set of genes in a larva and the adult gene set in a pupa. However, there is an apparent diversity in the sets of cuticular protein genes expressed in different anatomical regions of a larva (see Fig. 7.2 in Section 7.2.2, below). And apparently the same set of cuticular protein genes may be expressed in a certain anatomical region in all three metamorphic stages. In light of these observations, the concept of developmental stage-specific sets of cuticular protein genes (by analogy other genes) is questioned.

A detailed analysis of the current data, however, does not repudiate the concept of the stage-specific gene sets. The insect epidermis is a patchwork of cells with different developmental potentialities (Wigglesworth, 1966). The type of cuticle secreted by any anatomic region of the insect body, is irreversibly determined early in embryonic development. As a consequence, the subset of cuticular protein (or other) genes expressed in the cells in each of these diverse anatomic regions is different. Furthermore, the subsets of cuticular protein genes that would be activated in the different regions of the body at the larval, pupal, and adult stages would have been preprogrammed during the process of embryonic determination. JH, by its presence or absence, allows selection of the appropriate gene sets for activation. If the developmental program in the cells in the intersegmental membrane specifies a common subset of cuticular protein genes in all three metamorphic stages, then JH cannot and does not block their expression. On the other hand, if the developmental program in the cells that secrete the sclerite in larva, pupa, and adult specifies three different subsets of genes, then JH blocks the appropriate subsets in larval and pupal cells. Thus, the presence of similar (or same) proteins in the intersegmental membranes of larvae, pupae, and adults, or the presence of a specific cuticular protein in a specific region of a larva and in a different region of a pupa or adult, does not repudiate the concept of stage-specific gene sets. These considerations support the concept that the type of response to JH in the diverse cell types is irreversibly fixed during embryonic determination (see Kumaran, 1981).

7.2.2. *Effect of JH on Imaginal Primordia*

Another group of tissues of epidermal origin that are profoundly affected by JH are the imaginal primordia. In all holometabolous insects, the wings, legs, antennae, and other adult appendages are represented in the larva as nests of rapidly dividing cells. Although these cells are potentially capable of depositing a cuticle, they do not synthesize a cuticle during larval life. At pupation, these cells undergo dramatic morphogenetic changes and then secrete a cuticle. The same cells again secrete an adult-type cuticle during the course of adult development. The genetic program used at each of the developmental stages is apparently affected by JH. In larval insects, the imaginal primordia are exposed to JH and ecdysteroids. During this phase, they do not deposit a cuticle (see Fig. 7.2). At pupation, the primordia are exposed to the molting hormone in the absence of JH. Under these conditions, pupal-type cuticle is made by these cells. During adult development, these cells are again exposed to ecdysteroids in

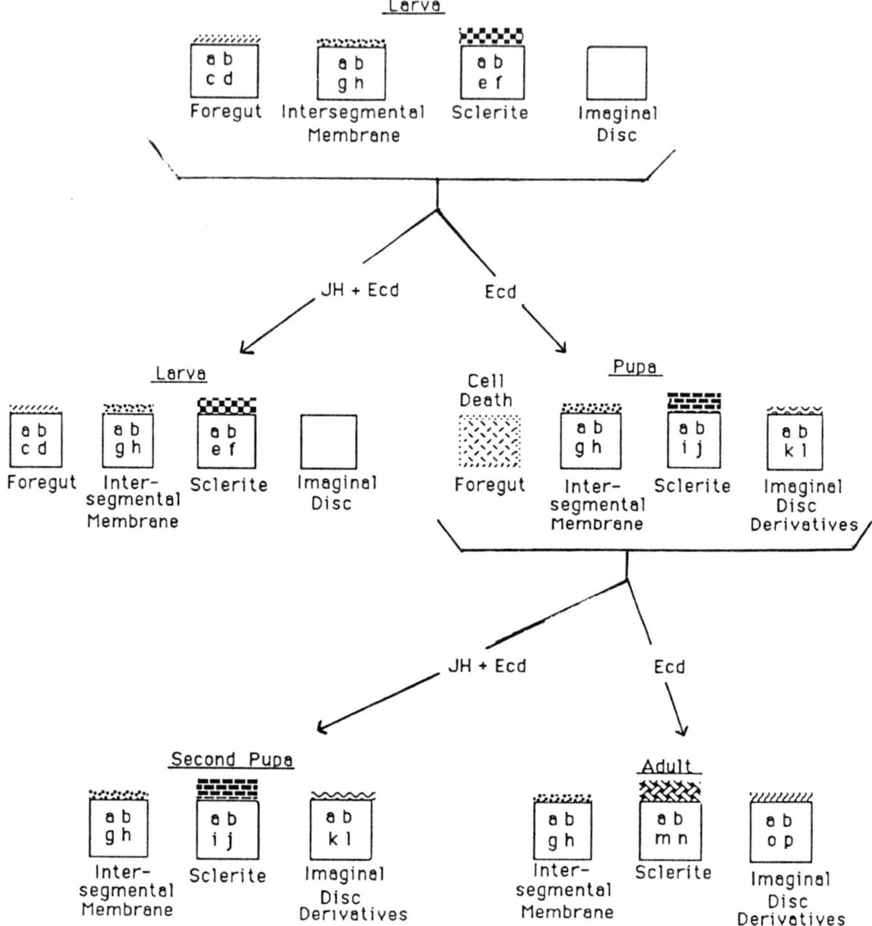

FIGURE 7.2. A diagrammatic sketch of the effect of JH on metamorphic changes in different larval and pupal tissues: a,b = hypothetical groups of cuticular protein genes expressed in all chitogenous cells; c–o = hypothetical groups of cuticular protein genes expressed in selected tissues at specific developmental stages; Ecd = ecdysteroid; JH = juvenile hormone.

the absence of JH and secrete adult type cuticle. However, if exposed to JH during this phase, they make a second pupal cuticle, i.e., express the gene sets blocked by JH during larval life. Furthermore, larval imaginal disks are competent to make cuticle whenever they are exposed to ecdysone in the absence of JH. Therefore, it is generally accepted that JH blocks expression of the genetic program that controls cuticulogenesis in larval imaginal disks. Once the imaginal primordia have expressed the cuticle-generating genetic program, JH cannot block this program. However, JH can still block a subprogram that specifies the adult phenotype of the resulting cuticle.

As in all chitogenous cells, JH effects are cell autonomous in the imaginal primordia. Furthermore, the type of cuticle made by the various imaginal primordia both at the pupal stage and in the adult has been programmed in the genome at the point of embryonic determination. JH merely acts as a switch. It blocks expression of the cuticulogenesis program in larval life and of the adult morphogenesis program in pupae. It is intriguing that, once the cuticulogenesis program has been operational, presence of JH cannot block production of cuticle by the imaginal disk–derived cells. However, the ability of imaginal disk cells to respond to JH is highly variable in different orders of holometabolous insects. Whereas in lepidopteran pupae a second pupal wing cuticle forms in the presence of JH, in dipterans it does not. The differences between the two orders of insects in their response to JH need to be explored at the molecular level.

7.2.3. *JH and Cell Death*

In addition to its effect on the choice of the genetic program that specifies the type of cuticle made by the chitogenous epithelia, JH also affects expression of the genetic program that results in programmed cell death. In holometabolous insect larvae, examples of this class of cells include the prolegs, silk glands, the tissues that produce larval mandibles, and the cells lining the foregut and hindgut. Some of these tissues secrete cuticle while others, e.g., silk glands, are not involved in cuticulogenesis during larval life. Last-instar *Galleria* larvae can be induced to undergo an extra larval molt either by treating them with exogenous JH or by exposing them to appropriate environmental cues. In such superlarvae, all these larval stage–specific tissues survive and function normally. Similar results were obtained with *Manduca sexta* and other lepidopterans. When superlarvae are allowed to undergo normal metamorphosis, however, these tissues die during the metamorphic molt. *In vitro* analysis of chitogenous cells from the prolegs in last-instar *Manduca* larva showed that programmed death of these cells was initiated only in the absence of JH and that JH can block their death (Fain and Riddiford, 1977). Thus, JH blocks the genetic program that leads to programmed cell death in certain cells in lepidopteran larvae. Activation of the program that culminates in cell death, like the program that leads to secretion of pupal cuticle in other larval cells, is under the control of ecdysteroids. However, cell death is not an all-or-none phenomenon in some cells. Depending on the time of JH application, the extent of loss of ability to survive and deposit a typical cuticle varies. The proventricular cells in *Galleria* larvae may survive and produce a typical cuticle or merely survive with no cuticulogenic property or a whole range of features in between (Cymborowski and Sehnal, 1980). In this respect, JH blockage of cell death resembles JH blockage of secretion of adult cuticle. Several steps are involved in expression of the metamorphic phenotype and each step requires the presence of JH. In addition, JH blockage of cell death also appears to be cell autonomous, as indicated by the fact that only a limited number of crochet-producing chitogenous cells on the prolegs survive when JH is applied either late in the instar or in subthreshold levels.

7.2.4. *Metamorphosis-Associated Changes in the Nervous System*

Cell death is by no means limited to contiguous cells of an organ. Within the larval brain, many individual neurons selectively die during the prepupal period in *Pieris* (Sánchez y Sánchez, 1925) and in *Ephestia* (Schrader, 1938). These cells apparently degenerate in their original location within the brain without the aid of phagocytosis. Simultaneously with the selective cell death, some neuroblasts in the brain proliferate rapidly. The newly produced neurons help construct adult neural centers. Similar changes involving cell death and neurogenesis occur in the ventral ganglia. All these changes in the neuronal populations lead to new neuronal connections and behavior patterns.

For example, a last-instar lepidopteran larva exhibits wandering behavior after completion of the feeding phase. In some species, these larvae wander out of the food, spin a cocoon, and pupate within the cocoon. The spinning behavior involved in cocoon formation is very different from the spinning behavior involved in preparing a molting pad that occurs at each larval molt. Although no direct role of JH has been adduced either in neuronal cell death or in the behavioral changes associated with metamorphosis, their temporal correlation suggests a functional association. In *Galleria*, application of exogenous JH that leads to formation of a perfect supernumerary larva suppresses spinning behavior. Similar behavioral changes can be observed in the larvae of other holometabolous insects. Thus, the change in behavior associated with metamorphosis may be universal among insects. The role of selective larval brain neuronal death and/or formation of pupal/adult brain centers in regulating insect behavior can provide a suitable model whereby we can study neural integration in moderately complex brains and the role of hormones in control of behavior.

7.2.5. *JH and the Fat Body*

The fat body cells in insects have diverse functions. In addition to their function as a nutrient store, fat body cells are the source of many hemolymph proteins, some of which play an important role in development and reproduction. The intermediary metabolism involved in conversion of carbohydrates into fats and vice versa, also takes place in part in the fat body. Some of these activities are regulated by the neurosecretory hormones released via the corpora cardiaca (CC) and are not discussed here. On the other hand, synthesis of specific proteins in the fat body cells, a tissue of mesodermal origin, is under the direct control of JH. For example, synthesis of JH esterase, a specific JH-degrading enzyme with an important developmental role, appears to be selectively controlled by JH in lepidopterans. In the late last instar of *Trichoplusia ni* (Sparks and Hammock, 1979) and in pupae of *Galleria* (Reddy et al., 1979) and of *Manduca* (Whitmore et al., 1972), exogenous JH stimulates production of JH esterase. Normally JH esterase production occurs during the last larval stadium when the endogenous JH titers decline (Vince and Gilbert, 1977; Hwang-Hsu et al., 1979; Sparks and Hammock, 1979). Furthermore, prevention of metamorphosis by induction of an extra larval molt suppresses JH esterase production. Thus, JH may repress JH esterase production in early last-instar larvae, whereas later in the stadium JH activates production of this enzyme.

In addition to this subtle change in the response of the fat body cells to JH during larval–pupal metamorphosis, other major changes occur in the fat body cells during this period. For example, fat body cells accumulate protein granules and fat globules just before pupation. At least in *Calpodes ethlius,* larval mitochondria disintegrate and are enclosed in autophagic vacuoles. Following loss of bulk of mitochondria, some mitochondria proliferate rapidly and the normal number for each cell is reestablished during the prepupal period (Larson, 1970). The role of JH, if any, in these changes in the mitochondrial populations has not been fully analyzed.

A more widely studied phenomenon is the production of vitellogenin. Fat body is the source of the yolk precursor in all insects. Furthermore, vitellogenin production in many insects is under the positive control of JH. This aspect of the role of JH will be discussed in the next section (7.3). It is, however, important to note that larval fat body cells do not synthesize vitellogenin in response to JH. Only after metamorphosis do the fat body cells largely become competent to respond to JH in production of vitellogenin. In *Drosophila melanogaster* and other higher dipterans, the adult fat body arises from histoblasts set aside during embryonic development. In these insects, larval fat body cells die (< 72 h) after adult emergence. The adult fat body is thus composed of cells that have an embryonic history different from that of the larval fat body. Therefore, the difference in the ability to respond to JH between larval and adult fat body cells can be explained on the basis of differences in their embryological lineage. This is not, however, a general feature. In hemimetabolous insects and also in some of the holometabolous insects, the larval fat body cells are generally believed to survive metamorphosis and reconstitute into the pupal and later the adult fat body. *Leptinotarsa decemlineata* serves as an example: vitellogenin is produced only by the fat body cells of adult insects in response to JH, but not by the larval fat body cells (DeWilde and Stegwee, 1958); it is obvious that metamorphosis of fat body cells changes their response to JH.

Regulation of larval hemolymph protein (LHP) genes in *Galleria mellonella* is interesting in this context. In this insect, larval fat body cells produce four hemolymph proteins. These four LHPs fall into two classes. One class, represented by LHP74, 76, and 81 are expressed both in the penultimate- and the last-instar larvae, whereas the other, LHP82 (more recently designated as LHP84), is expressed only in the last larval stadium. Interestingly, transcripts of two of these genes are not detectable in pupae older than 2 days or adults (Ray et al., 1987). Thus, metamorphosis of fat body involves repression of larval stage–specific genes, in addition to rendering the adult stage–specific genes, such as the vitellogenin genes, accessible for activation. Since this change in the accessibility of different gene sets for transcription occurs following metamorphosis, i.e., in the absence of JH, it is reasonable to suggest that JH may have an effect on the process. This phenomenon is all the more interesting because the last instar–specific LHP gene (LHP82) is normally repressed in earlier instars. Its activation in the last larval stadium and its suppression in the penultimate larval stadium appear to be mediated by JH (Memmel and Kumaran, 1988).

7.3. Other Morphogenetic Effects of JH

7.3.1. *Embryogenesis*

In addition to its effects during postembryonic life on a variety of tissues (chitogenous epithelia, imaginal disks and the nervous system, which are ectodermal in origin, and fat body cells, which are of mesodermal origin; see Section 7.2), JH affects embryonic development. Slama and Williams (1966) observed that a JH analogue, called the "paper factor," blocked embryonic development in *Pyrrhocoris apterus* eggs. This effect was related to the morphogenetic activity of the paper factor, these authors suggested, because this chemical had neither morphogenetic effect nor embryotoxic effect in *Oncopeltus* and *Rhodnius*, two other hemipterans. Microscopic examination of the arrested *Pyrrhocoris* embryos suggested that the paper factor affected development of cuticle, hairs, setae, compound eyes, and the appendages. In a few eggs, the embryo did not develop beyond the blastoderm stage.

Later studies on the time of onset of JH synthesis during embryonic life and on the effects of exogenous JH (or JH analogue) and/or antiallatal compounds on embryogenesis in a variety of insects confirmed the view that JH plays an important developmental role during embryogenesis. In embryos of *Nauphoeta cinerea*, a cockroach, JH titers begin to increase at 21 days of development. They peak at day 25, thereafter decline briefly, and then increase to a second peak around days 32–34 of embryonic development (Lanzrein et al., 1984). In addition to JH, methyl farnesoate is also synthesized in a stage-specific manner (Burgin and Lanzrein, 1988). The embryo hatches into a first instar nymph around day 40. Similar observations were made on the embryos of *Hyalophora cecropia* and *Manduca sexta* (Bergot et al., 1981) and in *Locusta* (Lageux et al., 1979). JH titers were low or undetectable in the early part of embryonic life but were measurable just before deposition of embryonic and larval cuticles.

Histological study of the embryonic development of CA and of the brain neurosecretory cells in *Oncopeltus fasciatus* (Dorn, 1975, 1984) show that the allata are distinguishable just before the stage when dorsal closure occurs. The embryonic allata appear to be active just before the embryo deposits a larval cuticle.

Studies of the effect of exogenous JH on embryos of diverse ages show that the presence of this hormone in early stages of embryogeneis—i.e., cleavage, blastoderm formation, and gastrulation—is deleterious to normal development in *Pyrrhocoris* (Riddiford, 1972), *H. cecropia* (Riddiford and Williams, 1967), and *Schistocerca gregaria* (Novak, 1969). The development was blocked at the blastoderm stage or the germ band stage. JH esterase activity was high in preovipositional and day-1-oviposited eggs of *Acheta domesticus* but was low during blastokinesis and dorsal closure (Roe et al., 1987), thereby suggesting that JH may be deleterious in the early stages of embryogenesis.

Similarly, physiological JH titers are essential for normal development beyond the germ band stage. Application of precocene, an allatotoxin, to early embryos of *Oncopeltus* led to defects in dorsal closure (Dorn, 1984; Bowers et al., 1976). In *Manduca sexta*, application of fluoromevalonate ($100\mu g/g$) to 0- to 24-h-old eggs blocked production of JH and caused a decline in endogenous JH in the embryo, and the resulting embryos exhibited precocious pupal cuticle or melanizaion at hatching (Bergot et al., 1981).

The cellular basis of JH action in early embryos is not known. Failure of the JH-treated embryos to develop beyond the blastoderm or germ band stage appears to be directly related to JH's hormonal effect rather than the result of a pharmacologic effect. Defects in dorsal closure, embryokinesis, and shortened appendages may be the consequences of inhibition of cell division and/or cell orientation. DNA (deoxyribonucleic acid) synthesis was curtailed in JH-treated *Acheta domesticus* embryos (Rao and Krishnakumaran, 1974). However, it is not known whether reduced DNA synthesis is the primary or secondary effect of JH. Thus, the cellular basis of the effects of exogenous JH on early (preblastoderm to predorsal closure) embryos needs a detailed study with modern methods.

The effects of JH deprivation on the latter stages of embryogenesis can be explained on the basis of JH effects on metamorphosis. Embryos deprived of JH by chemical allatectomy generally show signs of precocious metamorphosis. In these cases, the ''larval set'' of genes, which would direct production of a larval type cuticle for the first-instar larva that emerges from the egg, appears not to have been activated or was activated simultaneously with the "pupal set" of genes (Bergot et al., 1981).

7.3.2. Caste Determination

Social insects, such as bees, wasps, and termites, exhibit polymorphism. In these insects, different morphotypes occur simultaneously in a colony and each morphotype has a specific role. JH has been shown to play an important role in caste determination and thus in elaboration of the different morphotypes in honey bees (Wirtz, 1973) and bumble bees (Roseler, 1976). Fourth- and fifth-instar larvae of *Apis mellifera* fed with queen substance have higher JH titers than those destined to develop into worker bees (Rembold, 1987; see also Chapter 9 in Part 3). Moreover, application of exogenous JH to honey bee larvae promotes queen development. In particular, fourth- and fifth-instar larvae are most sensitive to JH-mediated queen production (Dietz et al., 1979). It is interesting to note that the JH-sensitive period is different in different species. In the stingless bee *Scaptotrigona* the cocoon-spinning stage of the last larval stadium is the most sensitive, whereas in the bumble bee *Bombus hyponorum* the prepupal period is the most sensitive stage (Campos et al., 1973). In *Scaptotrigona* the rate of JH synthesis in queen CA was 30–80% higher than that of workers just in the cocoon-spinning stage (Hartfelder, 1987). The major differences between worker and queen bees are in the body size and functional gonads. Queens are usually larger and have functional ovaries. Thus, the presence of high JH titers at critical stages in the last larval stadium or prepupal period somehow alters the ability of the gonad to respond to reproductive stimuli after metamorphosis.

The role of JH in caste determination in lower isopterans, e.g., *Kalotermes flavicollis*, is more complex. Nymphs can develop as pseudergates (false workers), soldiers, or sexual winged imagos. In pseudergates, JH titers are low and they undergo stationary molts without growth or differentiation. Implantation of active allata or application of JH to pseudergate nymphs leads to their development into soldiers. Pseudergates can also develop under appropriate stimuli into reproductives. During the course of this development, JH titers are low at the beginning of the stadium and increase in the latter part of the stadium. Thus, the developmental stage within the stadium at which JH is present affects the morphotype of the resulting organism (Nijhout and Wheeler, 1982).

JH has been implicated in soldier production in other genuses of termites, namely *Macrotermes* and *Nasutermes* (Lüscher, 1976). However the mechanism of action of JH in differentiation of caste morphotypes is not known. Obviously, JH acts as a true genetic switch. Hypothetically, JH may activate the set of genes that specify the caste phenotype. Identification of caste specific molecular markers and their regulation by JH will help elucidate the cellular/molecular basis of JH action with respect to caste determination.

7.3.3. *Phase Polymorphism*

Polymorphic forms occur not only in social insects but also in other insect groups. In locusts, aphids, and armyworms, polymorphism is associated with seasonal and/or environmental factors. The different polymorphs exhibit behavioral differences in addition to morphological differences. In locusts, armyworms, and cutworms, for example, they may be solitary or gregarious. In aphids, the major difference between the morphotypes is alate or apterous and parthenogenetic or sexual individuals. JH was shown to affect solitary and gregarious forms of locusts, armyworms (*Leucania separata*) and cutworms (*Spodoptera litura*). However, in aphids the role of JH in polymorphism is indicated but is not unequivocal (Hardie and Lees, 1985; Hales and Mittler, 1981).

Solitary locusts are characterized by green colorization, persistent prothoracic glands, large ovaries and short wings, whereas gregarious locusts are pinkish brown or yellow in color, have small ovaries and large wings, and have an altered ratio of wing length to femur length. Solitary locust nymphs have higher JH titers than gregarious forms, and application of JH or implantation of active allata into gregarious nymphs results in their development into solitary forms. In nature, crowded conditions induce the gregarious morphotype. The cellular and molecular mechanisms of JH action on the morphology and behavior of solitary and gregarious locusts are largely unknown. Since development of the adult nervous sytem in all insects appears to be under JH control, presence of high or low JH titers at critical development stages may affect brain development, and hence behavior of the resulting adult. Similarly, the size of wing and relative length of femur are different in the solitary and gregarious forms. Since the size of an organ is believed to be genetically determined in all animals, the presence of two locust morphotypes implies existence of two genetic programs that determine wing size and femur length. The molecular mechanisms of JH selection of the appropriate developmental program, when elucidated, should shed light on growth control in general.

7.4. Effects on Reproduction

In addition to its morphogenetic effect, JH possesses a gonadotropic effect. (The term *gonadotropic* is used here in the broad sense of affecting reproduction.) In adult insects of most species CA, which were functionally inactive just before metamorphosis, regain activity after metamorphosis (Tobe and Stay, 1985). In the adults of many species JH regulates reproduction both in males and females. In males, its regulatory function is mainly confined to the accessory glands. In females, JH affects egg production at different levels. JH regulates vitellogenin production and/or uptake in many insects. In some insects, JH also regulates the function of female accessory reproductive organs. The chemical nature of the allatal hormone produced in juvenile and adult insects is identical. Furthermore, all JH analogues that possess morphogenetic activity also exhibit gonadotropic effect. Moreover, the two activities of JH analogues run parallel, i.e., if a compound has high morphogenetic activity, it also has high gonadotropic activity. Thus, it is reasonable to assume that the same molecule has two divergent functions.

Another enigmatic feature of JH is that the two apparently divergent features are exhibited at different stages of ontogeny. This led to a discussion of its primary role in the course of evolution. Because JH regulates oogenesis and egg production even in apterygotes, which do not show overt morphogenetic changes during ontogeny, it was suggested that the primary role of JH is its gonadotropic function (Novak, 1956). According to this view, JH was utilized in primitive insects only as a gonadotropic agent. This hormone was secondarily adapted to serve a morphogenetic role in the course of evolution of pterygotes. It was argued that prolongation of the juvenile stage, during which all growth occurs, has survival value to insects. An insect with a longer period of feeding and growth would be larger than the one with a short juvenile stage. The larger insect not only can fly greater distances but can also be more fecund than a smaller one. Because of these advantages, it was argued that those individuals in which JH was secondarily commandeered to prolong juvenile life survived in larger numbers. The present-day insects are descendents of these insects.

Similar arguments can be adduced in favor of the view that the primary role of JH, evolutionarily speaking, was its morphogenetic role. For example, the CA in *Thermobia*, a thysanuran apterygote, produce JH in the juvenile stages, cease production during the fourth instar, i.e., before molting into a reproductive, and resume cyclical production in adults (Watson, 1967). It is, however, not known whether the allatal physiology of modern thysanurans represents a primitive condition or a secondarily derived state. Therefore, any discussion of the evolutionary history of the allatal role in insect developmental physiology can only be conjectural. In this context, knowledge of the physiological role of methyl farnesoate produced by the mandibular glands of decapod crustaceans (Laufer et al., 1987) will be valuable in surmising the evolutionary history of allatal function.

Whatever its evolutionary history might be, the dual role of JH raises another intriguing problem. JH is present in high titers all through the juvenile life of an insect. However, gonads, accessory reproductive organs and other somatic tissues that are targets of its gonadotropic actions in adult life are affected by JH differently during the larval stage. Some tissues, such as the fat body cells, which synthesize vitellogenins in response to JH in adult insects, are cytologically indistinguishable between the larval and adult stages, except for their synthetic potential. The molecular changes that these cells undergo during metamorphosis are completely unknown; yet such changes are undoubtedly responsible for their differential response to JH between the larval and adult stages. Thus, the gonadotropic role of JH raises many interesting questions. For a detailed account of the role of JH in reproductive physiology of insects, excellent reviews are available and may be consulted (Telfer, 1965; Wyatt, 1972; Hagedorn and Kunkel, 1979; Koeppe et al., 1985). In this section, we shall discuss the gonadotropic effects of JH with a view to assess its mode of action at the cellular and molecular levels.

7.4.1. *Control of Vitellogenin Synthesis*

Among its functions as a regulator of reproduction, the role of JH in control of vitellogenin synthesis is one of the best documented. Wigglesworth (1936) was the first to note that the presence of active CA was necessary for yolk deposition and completion of oogenesis in *Rhodnius prolixus*. Since then, the positive role of JH in vitellogenesis has been documented in a number of insects (see Doane, 1973, for a review of the early literature). In these early studies, it was shown that extirpation of the CA or destruction of the neurosecretory cells of the brain prevented vitellogenesis in a variety of insects. Only in the late 1960s was the role of JH in specific stimulation of yolk protein synthesis demonstrated. Yolk in insects, as in most coelomate oviparous animals, is synthesized outside the ovary. In insects, yolk was shown to be produced by the fat body (Telfer, 1954). In studies on *Leucophaea maderae*, Brookes (1969) and Engelmann (1969) showed by immunological methods that JH stimulates selective synthesis of vitellogenins by the fat body. Furthermore, these studies indicated that JH may act directly on the fat body because vitellogenin production increases in isolated abdomens treated with a JH analogue.

That JH acts directly on the fat body to stimulate yolk protein synthesis was unequivocally demonstrated by *in vitro* studies (Wyatt et al., 1976). In these studies, fat body from allatectomized locusts was placed in tissue culture and exposed to JH. Under these conditions, JH stimulated production of vitellogenin. Furthermore, measurements of rates of total and vitellogenin-specific mRNA (ribonucleic acid) synthesis showed that JH stimulated total RNA synthesis and specifically activated vitellogenin genes in *Locusta migratoria* fat body (Wyatt et al., 1987).

The effects of JH on vitellogenin synthesis in fat body are not universal. Its effect has been documented in *Leucophaea*, a cockroach (Engelmann, 1984), and in *Triatoma* (Regis, 1979), among others. On the other hand, JH has no effect on yolk protein synthesis in *Hyalophora cecropia* and *Bombyx mori*, the short-lived lepidopterans. Similarly, in *Apis mellifera*, an insect that lays eggs more or less continuously all through adult life, JH has no effect on vitellogenin synthesis (Ramamurthy and Engels, 1977). In *Drosophila* and other dipterans that produce eggs more or less continuously, the role of JH in regulation of vitellogenin synthesis is not definitive (Kelly et al., 1987). Ecdysteroids seem to intervene in regulation of vitellogenin synthesis in these insects. Lastly, in *Carausius morosus*, the orthopteran stick insect, JH appears to have no specific role in regulation of vitellogenin synthesis (Giorgi and Macchi, 1980). Thus, no definitive evolutionary trend in JH regulation of vitellogenin synthesis by fat body is obvious, thus calling into question the evolutionary primary role of JH in reproduction.

7.4.2. *Vitellogenin Entry into the Oocyte*

Besides its stimulatory effect on vitellogenin synthesis by the fat body, JH also has a direct effect on the gonad. JH exerts its effect on the entry of vitellogenins and other yolk components into the oocytes.

This effect of JH is mediated by its action on the follicle cells. In a study of *Periplaneta americana*, vitellogenesis was blocked by allatectomy. However, when vitellogenins were injected into hemocoel, vitellogenesis did not occur in allatectomized cockroaches. Only on application of JH was vitellogenesis reestablished, thereby showing that JH has a dual role in regulating vitellogenin production and its entry into oocytes (Bell, 1969). In other experiments, the female sterile phenotype of homozygous *apterous*[4] mutants of *Drosophila* was corrected by application of JH (Gavin and Williamson, 1976). These mutants produce vitellogenins, but their entry into the oocyte is blocked. Application of JH analogues to homozygous *ap*[4] females stimulated formation of microvilli and pinocytotic vesicles in their oocytes within 12 h after topical application of the analog. Furthermore, the JH analogue was effective when applied to ovarioles *in vitro*, showing that JH acts directly on the ovary to stimulate pinocytotic activity (Tedesco et al., 1981).

In *Rhodnius*, extensive analysis of the effects of JH on follicle cells in stimulation of vitellogenin uptake by the oocytes, showed that JH affects the follicle cell membrane (Pratt and Davey, 1972; Abu-Hakima and Davey, 1979; Illenchuk and Davey, 1982, 1987). Following application of JH, the follicle cells undergo rapid reversible shrinkage resulting in formation of large intercellular spaces, a condition termed *patency*. Application of colchicine, which depolymerizes microtubules, major cytoskeletal components, suppresses this effect of JH. Similarly ouabain, a drug that blocks the membrane Na^+/K^+ ion pump, also suppresses the effect of JH on follicle cell patency. These observations suggest that JH may regulate vitellogenin entry into oocytes by its effect on the Na^+/K^+ ATPase (adenosine triphosphatase) in the follicle cell membrane (Illenchuk and Davey, 1987).

In addition to entry of vitellogenins into oocytes, JH was shown to affect synthesis and/or uptake of lipids (Gilbert, 1967) and carbohydrates (Ramamurthy, 1968) in *Leucophaea* and *Panorpa*, respectively. However, the mode of JH action on these aspects of metabolism is not known.

7.4.3. *Accessory Gland Function*

As noted in the section on the role of JH in metamorphosis, JH blocks growth and differentiation of the gonadal ducts and associated accessory glands in larval stages. In the imago, JH was shown to affect the function of the male and female accessory reproductive organs. Wigglesworth (1936) was the first to show that secretory activity of the male accessory gland in *Rhodnius prolixus* require normal allatal activity. Since these pioneering observations, accessory reproductive gland activity was shown to be dependent on CA in *Melanoplus, Calliphora, Periplaneta, Locusta, Schistocerca, Gomphocerus,* and many other insects (see Raabe, 1986, for a review). In addition to accessory reproductive glands, JH also affects secretions by the reproductive ducts and associated parts. For example, in *Danaus* (Herman, 1982), JH stimulated secretory activity in male ejaculatory ducts and female bursa copulatrix. Recent studies on the male accessory glands in *Drosophila* suggest that JH may act at the membrane level (Yamamoto et al., 1988). Physiological doses of JH III were shown to stimulate RNA and protein synthesis in accessory glands from unmated males within 1 h following application of hormone *in vitro*. Furthermore, phorbol esters, which activate proitein kinase C, mimic the effect of JH, and *Drosophila* mutants deficient in protein kinase C did not respond to JH, thereby suggesting that JH may act at the level of the membrane by activating protein kinase C. Calcium (Ca^{2+}) ions are also essential for JH stimulation of protein synthesis in these glands.

An interesting function of the male accessory glands in saturniid and other lepidopterans was observed. In *Hyalophora cecropia* and related lepidopterans, adult males store a large supply of lipids with JH activity. The JH stores are interestingly located in the male accessory gland, and some of the hormone is transferred during mating to the female bursa copulatrix (Shirk et al., 1983). It is also interesting to note that the final steps in JH biosynthesis, viz. methyl esterification of the triterpenoid acid, occurs in the male accessory gland. The function of this store of JH is as yet unresolved.

Accessory reproductive glands of some female insects are also selective targets of JH action. The collaterial glands of cockroaches have been studied in great detail in this respect. The paired collaterial glands in *Periplaneta* are asymmetrical: the left gland is large and produces a protein and the glucoside of protocatechuic acid; the smaller right gland produces a glucosidase. It is interesting to note that JH regulates the secretory activity of the left collaterial gland but not that of the right collaterial gland (Bodenstein and Sprague, 1959; Willis and Brunet, 1966). The rate of RNA and protein synthesis in the left collaterial gland shows a cyclical variation corresponding to the cyclical allatal activity, suggesting that JH stimulates macromolecular syntheses in these cells (Zalokar, 1968).

The left gland in *Periplaneta* was shown to synthesize and secrete 11 glycine-rich proteins called *oothecins*. All the oothecins are encoded by members of a multigene family that have some similarity to members of the chorion gene family (Pau, 1987). Allatectomy and hormone replacement studies have shown that JH is responsible for induction of oothecin synthesis and that JH specifically activates oothecin gene transcription, resulting in increased oothecin messenger RNA (mRNAs) (Pau et al., 1986). In *Glossina austeni* also CA were shown to stimulate production of milk by the milk glands (Ejezie and Davey, 1976), but the molecular mechanism in activation of milk production remains to be explored. In *Diploptera punctata*, an ovoviviparous cockroach, however, JH inhibits milk production in the brood pouch (Stay and Lin, 1981). Thus, the effect of JH on the accessory reproductive glands of diverse insects is varied. It is stimulatory in some cases and inhibitory in others. JH acts at the level of the plasma membrane in some instances, whereas in others it apparently exerts a transcriptional control.

7.5. Effect of JH on Life Cycle, Behavior, and Metabolism

In addition to its morphogenetic and gonadotropic effects, JH affects several aspects of life cycle, behavior, and metabolism. Some of these effects result from the direct action of JH on its target tissues, whereas other effects are the result of JH action on some other neuroendocrine organ. In this section a brief account of the effect of JH on diapause, flight behavior, and specific synthesis and/or release of materials in selected target tissues is given.

7.5.1. *Diapause*

Many insects exhibit during their life cycle a period of developmental arrest called diapause. During the period of diapause, the insect is quiescent and thus can survive long periods of harsh environmental conditions, such as high or low temperatures with very low levels of consumption of O_2 and of stored nutriment. Diapause occurs at different developmental stages in different insect species. In *Bombyx mori*, diapause, mediated by a neurosecretory hormone, the diapause hormone, occurs during embryonic life. In *Hyalophora cecropia*, a pupal diapause is caused by cessation of secretion of prothoracicotropic hormone (PTTH). We are, however, here concerned only with JH-mediated diapause. Larval diapause in some lepidopteran larvae and adult diapause in some beetles are shown to be mediated by JH.

The larvae of various species of Lepidoptera, Diptera, Odonata, Hymenoptera, Coleoptera, and Orthoptera enter diapause (see Chippendale, 1977). In some of these insects, e.g., *Chilo suppressalis* and *Diatrea grandiosella* JH was shown to be the primary regulator of diapause (Yin and Chippendale, 1973). In diapausing larvae, a low level of JH persists and thus blocks activation of ecdysial glands. Although the precise target of JH has not been identified, release of PTTH by the neurosecretory cells in the brain is implicated, as are the prothoracic gland cells themselves. In the presence of low levels of JH, prothoracic gland cells may become less sensitive to PTTH and thus fail to produce normal levels of ecdysteroids for continued development. Alternatively, low levels of JH might block release of PTTH, thereby depriving the prothoracic gland cells of their normal developmental cue.

Although the mechanism of JH action is unknown, that JH is responsible for diapause in *Diatraea* is well documented. Allatectomy of larvae terminates diapause prematurely. Application of JH or JH analogues to allatectomized larvae maintains them in diapause state (Yin and Chippendale, 1979). JH-induced delay in pupation by cessation of development is also observed in other insects under experimental conditions. Application of JH or a JH analogue to last-instar larvae of *Bombyx mori* (Akai et al., 1973), *Galleria mellonella* (Reddy and Krishnakumaran, 1973) and *Cerura vinula* (Hintze-Podufal, 1975) delays metamorphosis and produces dauer larvae. Similarly, larval quiescence and/or delayed pupation were induced by feeding a JH analogue to *Trogoderma* larvae (Nair, 1974) and by topical application of JH to cockroaches (Masner et al., 1974). Thus, the effect of JH on PTTH release and/or on the sensitivity of prothoracic gland cells to PTTH may be more widespread among insects. A molecular analysis of the mode of JH action on the neurosecretory cells of brain and on the prothoracic glands will provide new insights to the mechanisms of JH action.

Diapause in the adult Colorado potato beetle, *Leptinotarsa decemlineata*, was also shown to be mediated via JH (DeWilde and Stegwee, 1958). When the insects are reared under short photoperiod, the adults enter diapause. They are characterized by low basal metabolism, atrophied flight muscles, and small previtellogenic ovaries. Induction of diapause was correlated with low endogenous JH titer, which very soon after adult emergence becomes undetectable (DeWilde et al., 1968). Low JH titers are also correlated with blockage of vitellogenin synthesis and with reduced numbers of mitochondria in flight muscles. Furthermore, the mitochondria were small in size.

Application of exogenous JH to diapausing beetles leads to biogenesis of mitochondria and initiation of vitellogenin synthesis (Bartelink and DeKort, 1975). Cytochromes aa_3, b, and c, which were virtually undetectable in mitochondria of diapause beetle flight muscles, reappear on JH application. However, this effect was observed only when JH was applied to whole insect. Application of JH to isolated mitochondria had no effect on cytochrome synthesis and/or mitochondrial protein synthesis. Since some of the component polypeptides of the cytochromes are now known to be encoded for by nuclear genes, this observation is not surprising. All the same, it is important to determine whether JH acts directly on the flight muscle cells. It is conceivable that JH may affect neurons in the thoracic ganglia or another neuroendocrine organ, which in turn stimulate growth of flight muscles.

7.5.2.   *Effect on Flight and Migratory Behavior*

Young adults of many insect species exhibit a flight behavior that aids dispersal of the species. Most animals have evolved mechanisms that ensure efficient dispersal of the species. In aquatic invertebrates that have a sedentary or sessile adult life, the larval stages are adapted for dispersal. In insects, the adult with its wings is more suited for dispersal from the site of oviposition and larval feeding. JH has been implicated in the migratory behavior and/or active flight of preovipositing adult insects.

In the large milkweed bug, *Oncopeltus fasciatus*, for example, newly emerged adults exhibit a migratory behavior, normally induced by short days coupled with low temperatures (Rankin, 1974). Migratory behavior can be assayed in the laboratory by measuring an insect's sustained flight for 30 min or longer under tethered test conditions. Using this feature as a measure of migratory behavior, Rankin showed that application of JH or implantation of CA into young adult males increased flight duration. Even in females, migratory behavior was enhanced by JH application very early in adult life. In older females, however, application of JH stimulated oogenesis. Analysis of JH titers at different stages in adult females showed a correlation between low JH titers and migratory behavior and between high JH titers and oogenesis. Absence of migratory behavior in older *Oncopeltus* was attributed to blockage of migratory behavior by oogenesis (Rankin and Riddiford, 1978). Chemical allatectomy by precocene treatment of 10- to 11-day-old or 28-day-old or older *Oncopeltus* adult females confirmed that low JH titers induce migratory behavior and high JH titers initiate oogenesis.

The role of JH in induction of migratory behavior of many other insects has been examined. In *Schistocerca* and *Locusta*, allatectomy appears to reduce flight performance in young adults. The lady beetle *Hippodamia convergens* exhibits a migratory behavior at two different stages in its adult life. The migratory behavior in a young adult appears to be regulated by JH, because precocene treatment decreases the percentage of individuals that exhibit sustained flight for 30 min or longer. Furthermore, application of exogenous JH restores their ability for sustained flight (Rankin and Rankin, 1980). The role of JH, if any, in monarch butterflies, which exhibit a classical case of migratory behavior, is as yet unknown (see Pener, 1985, for a review).

In some insects, such as ants and termites, prenuptial flight is followed by loss of wings and the loss of the ability to fly. The role of JH, if any, in these events is not clear, although JH-mediated dealation of the fire ant *Solenopsis invicta* (Kearny et al., 1977) and of the termite *Kalotermes* (Lebrunn, 1969) was reported. Similarly, JH was implicated in flight muscle degeneration in *Acheta domesticus* (Chudakova and Guttmann, 1978), *Dysdercus* (Davis, 1975), and *Ips confusus* (Unnithan and Nair, 1977). In all these insects, the presence of JH caused flight muscle degeneration. This situation is in contrast to the condition in *Leptinotarsa*, referred to earlier in the subsection (7.5.1) on diapause. In this beetle, lack of JH causes flight muscle degeneration and its presence stimulates muscle regeneration.

The cellular and molecular mechanisms underlying JH-induced flight behavior, muscle degeneration and muscle regeneration, are not known. In the course of migratory or long duration flights, utilization of sugars and/or lipids increases. However, it is not known whether nutrient reserves are mobilized by JH directly or as the result of the primary action of another hormone. Since neurosecretory products released from the CC were shown to affect carbohydrate and lipid metabolism, it is unlikely that JH acts directly on mobilization of these nutrient reserves. Theoretically, JH may affect flight behavior and muscle regeneration or degeneration by acting on the locomotor centers of the nervous system either directly or indirectly through some other neuroendocrine agent. Identification of the primary target of JH in its modification of flight behavior and elucidation of the molecular mechanisms underlying JH action would provide valuable information regarding insect behavior that might well be useful in developing novel, environmentally safer insect pest control methods.

7.5.3. *JH and Metabolism*

In addition to all the diverse effects of JH described in the preceding pages, many reports describe a variety of other physiological features regulated by JH. For example, JH was shown to block its own synthesis by negative feedback control, to stimulate JH esterase synthesis by fat body at certain stages of the life cycle, and to inhibit JH esterase production at other stages of life. JH was also shown to increase the rate of O_2 consumption and RNA synthesis in adult female *Nauophoeta*, although some of these effects are coupled to its vitellogenic effects. Similarly JH affects migration of pigment granules in the epidermal cells of *Manduca* larva. In the absence of JH or in the presence of low JH titers, insecticyanin distribution as well as synthesis are altered to render the larva dark in color. We have already referred to the effect of JH on release of PTTH and sensitivity of ecdysial glands to PTTH.

7.6. Molecular Models of JH Action

A brief summary of the JH-mediated effects in diverse tissues of various insects given in the preceding discussion shows that the effects of JH are varied. The target cells respond to JH in diverse ways. Broadly speaking, JH affects morphogenesis, reproduction, behavior, and metabolism. The morphogenetic effects of JH involve blocking developmental changes, and thus, by inference, JH blocks expression of the genes that specify the morphotype of the succeeding stage. On the other hand, JH effects on reproduction include, among others, stimulation of vitellogenin and oothecin synthesis. These activities require selective activation of gene expression. Thus, JH apparently represses gene expression in exerting its morphogenetic effect and appears to activate gene expression in exerting control on reproduction. In addition, JH also alters membrane properties in causing patency changes in egg follicle cells or protein kinase C activity in male accessory reproductive organs. Therefore, various molecular mechanisms may be underlying the diverse effects of JH.

7.6.1. *Nuclear Receptor*

In theory, JH may activate or repress specific target genes or gene sets by binding in association with a specific receptor protein to the relevant regulatory sequences. Because JH is lipid soluble, like the sex steroid and other vertebrate steroid hormones, it was suggested that JH may exhibit molecular mechanisms of hormonal regulation of gene expression similar to the vertebrate steroid hormone model (Gronemeyer et al. 1987; Yamamoto, 1986; Compton et al., 1983). According to this model, JH would bind to a receptor protein and the two together would bind to the regulatory sequence(s) of each of the target genes. Based on these considerations, serious efforts were made to identify, isolate, and characterize a JH-specific intracellular receptor. The fat body in *Leucophaea* and *Locusta* produces vitellogenin in response to JH and was the subject of intense investigation in these efforts to identify a JH receptor (Engelmann et al., 1987; Roberts and Jeffries, 1986; Koeppe et al., 1987).

The search for JH-specific receptors was hindered by the lack of JH labeled with radioisotopes to a high specific activity. However, this problem was largely solved by synthesis of a tritiated JH analogue, hydroprene, at a specific activity of 60 Ci/mmol (Prestwich et al., 1987). These workers also synthesized other JH analogues labeled with ^{125}I for potential use in receptor studies. A second major impediment to rapid progress of these studies is the difficulty in distinguishing between true intracellular JH receptors and the JH-binding proteins that circulate in the hemolymph. The latter are ubiquitous and have high affinity for JH. Recently, Koeppe et al. (1987) showed that a JH-binding protein is synthesized in the fat body, secreted into the hemolymph, and later sequestered into the growing oocytes of the adult female cockroach. This JH-binding protein has a molecular weight of 275 kD (although a modified derivative may be 225 kD) and binds JH with a high affinity 2×10^{-8} M. Despite the difficulty in distinguishing between this JH-carrier protein and the intracellular JH-specific receptor(s), some progress has been made in this direction. Adult fat body cells in the cockroach *Leucophaea* were the source of a cytosolic and a nuclear JH-receptor protein (Engelmann et al., 1987). Both the cytosolic and the nuclear JH receptors have nearly identical sedimentation rates (6.5–6.6 S in a sucrose gradient) and specific affinity properties. Their affinity is approximately 2.5×10^{-9} M, and both receptors have greater affinity for the native enantiomer (10*R*) over the synthetic analogue (10*S*). On the basis of these observations, it has been suggested that these proteins may be the intracellular mediators of JH effects on fat body cells.

Using labeled JH compounds with high specific activity, similar putative JH receptors were identified in the larval epidermis in *Manduca sexta* (Riddiford et al., 1987) and *Galleria mellonella* (Wiśniewski et al., 1988). The JH binding has an affinity to the receptor in *Manduca* is 3.8×10^{-9} to 5.8×10^{-10} M. Furthermore, the JH-binding property of the epidermis declines during the last larval stadium as the larva nears metamorphosis. This loss in the JH-binding property corresponds to the loss in the ability of the larva to respond to exogenous JH. Taken together, these data suggest that a JH-specific receptor protein may be involved in intracellular transduction of JH effects even in larval insects. However, we do not know as yet whether the JH receptors in the larval epidermis and in the adult female fat body are the same molecules.

7.6.2. *Nature and Mechanisms of JH Receptor Activity*

Even if we tentatively accept the existence of a JH receptor we cannot explain the diverse aspects of JH activity based on the currently accepted models of steroid receptor activity without making other assumptions. Steroid receptors have DNA binding and hormone binding domains in addition to a third domain whose function is not known. The amino acid sequence in the DNA-binding domain of the receptor recognizes a specific nucleotide sequence in the regulatory element of the target genes and thus activates transcription of the gene (Gronemeyer et al., 1987). JH receptor (JHR) by analogy could hypothetically bind to the specific regulatory sequences of its target genes. Once bound, the receptor-hormone complex should be able to transduce the signal to the cell-specific response. However, the response to JH may be to activate transcription of selected genes, such as the vitellogenin and oothecin genes, or to selectively repress those genes which specify cuticular features of a succeeding developmental stage. Thus, JHR should act as a positive regulator of genes in exerting its reproductive control and as a negative regulator of genes in exerting its morphogenetic function. From our current knowledge of hormone receptor, this is not possible. All known hormone receptors have only one DNA-binding domain, which recognizes a specific nucleotide sequence (see Gronemeyer et al., 1987). Therefore, additional assumptions have to be made to explain the apparent dual role of the JHR as a negative regulator at one stage and as a positive regulator at another stage or in a different tissue.

Several explanations for this dual role of the postulated JHR complex are plausible. Conceivably there are two types of JH receptors, one inhibitory and the other stimulatory. Both types of receptors might share the same hormone-binding domain but bear different DNA-binding domains. The "stimulatory" receptor would have a DNA-binding domain that interacts with a positive regulatory element of the target gene, whereas that of the "inhibitory" receptor would bind to a negative regulatory element. As a corollary of this model, the stimulatory and inhibitory JH receptors would be synthesized only at the appropriate developmental stages in a tissue-specific manner. A second explanation for the dual effects of JH is that JHR acts alone as a positive regulator but interacts with another trans-acting factor as a negative regulator, or vice versa. According to this model, JHR binds to the regulatory domain of the target gene to serve as a positive gene regulator. On the other hand, to serve as a repressor of gene activity, JHR complex would interact with another tissue-specific nuclear protein before it could function as a regulator of gene expression. The apparent negative regulatory role of JH occurs mostly in its function as a morphogenetic agent. Since morphogenetic activity of JH is manifest only in the presence of ecdysteroid, it is conceivable that JHR interacts with an ecdysteroid-modulated trans-acting factor to bring about its properties as a negative regulator of genes. Thirdly, it is also possible that JH receptors have two DNA-binding domains, one of which binds to a negative regulatory sequence and the other to a positive regulatory genetic element. Were this the case, we would have to further assume that the choice of the appropriate DNA-binding domain is under yet another modulator. The characteristics of the putative JHR are summarized in Table 7.1. Thus, a molecular model involving a JHR in transducing JH signals into appropriate cell response has been oversimplified, given the current state of our knowledge, and has many limitations. Because of the great diversity in the JH-mediated developmental and physiological events, a detailed knowledge of the molecular mechanisms of JH action should inevitably lead to a better understanding of gene regulation in general.

Even if we assume that a JH–receptor complex is involved in exerting a negative regulatory role in exerting its morphogenetic effect, this model poses other problems. According to it, JHR, as a negative regulator of gene function, should bind to a specific regulatory genetic element of the target genes that are shut off during larval molts. However, as noted earlier in Section 7.2.1, the cuticular proteins exhibit body region specificity (as determined by their electrophoretic and immunological properties). In other words, the cuticular protein genes expressed in different regions of the body vary, depending on the physical and physiological properties of the cuticle. By corollary, the groups of cuticular protein genes blocked by JH in different regions of

---

TABLE 7.1. Predicted Properties of the Putative JH-Receptor Complex and Theoretical Models to Explain Its Functions

---

1. Repressor of metamorphosis-associated genes.

2. Activator of the genes involved in oogenesis and/or accessory reproductive gland function (may also serve as an activator of certain genes functional during a juvenile developmental stage).

3. The repressor and activator functions may be modulated (a) by two different molecules; (b) by virtue of its bearing two different DNA-binding domains; (c) by serving alone as an activator, and in concert with stage- and/or tissue-specific nuclear factors as a repressor or vice versa.

---

the body differ from one another. That is, a group of cuticular protein genes repressed by JH in one region of the larval body may not be repressed in another region of the body. This conclusion implies that the negative regulatory function of JHR must be further modified by additional tissue-specific trans-acting factors. There is a tacit assumption in all these theoretical considerations that two proteins with identical electrophoretic (size as well as charge) and immunological properties are encoded by the same gene. The reality may be more complex, as indicated by the formation of a second pupal cuticle in the presence of JH in lepidopteran and coleopteran pupae. In the last-instar larvae, pupal stage–specific cuticular genes are repressed by JH. But in a pupa, this set of genes is expressed only in the presence of JH. Thus, the same set of genes that is blocked by JH during larval life is expressed in pupa in the presence of JH. If these assumptions are valid, it is reasonable to assume further that stage and/or tissue-specific cofactors are essential for JH-mediated gene repression. Identification and characterization of specific JH-modulated genes should help develop testable molecular models to explain JH action.

Despite all the gaps in our knowledge regarding the molecular mechanisms of JH regulation of metamorphosis, the simplest theoretical model to explain JH repression of metamorphosis-associated genes should include a receptor or other trans-acting factor(s) that selectively block transcription. This assertion is supported by the absence of any evidence for a second messenger or other similar molecule in exerting its morphogenetic effects. In other words, it is argued that JH acts directly, and individually, on the genes in repressing their expression.

Furthermore, this model can explain all the observed morphogenetic effects of JH. For example, JH blocks expression of all cuticular protein genes in larval imaginal disks, whereas in their descendents in a pupa JH blocks expression of only those genes that specify the adult phenotype. Similarly, JH blocks expression of the genes that lead to programmed cell death in the crochet epidermal cells (on the proleg), and the cells of the foregut and hindgut in the larval insect. In the general chitogenous epithelia, different subsets of cuticular protein genes are blocked in different regions of the body. In addition, JH also blocks expression of some last instar–specific genes (e.g., genes coding for some larval hemolymph proteins, JH esterase, etc.) in the larval fat body during earlier instars. In conclusion, the groups of genes repressed by JH are tissue specific. Therefore, JH should be considered as a genetic switch whose selectivity appears to be dependent on a stage- and tissue-specific factor. The simplest model to explain this phenomenon is that the affinity of the JHR complex to the negative regulatory element of a gene can be modified by tissue-specific trans-acting factors. In larval imaginal disk cells the JHR, together with the tissue-specific factor, blocks expression of all cuticular protein genes that are normally activated by ecdysteroids. In pupae JHR does not block *all* cuticular protein genes but blocks only those cuticular genes that define adult phenotype. Therefore, during metamorphosis, the imaginal disk cells synthesize a factor that alters their response to JH. Similarly, in chitogenous epithelia, which make different types of cuticles at larval, pupal, and adult stages, the tissue-specific factor(s) at each stage directs JHR binding to the appropriate sets of genes. It remains an open question whether this factor directly interacts with the JH-receptor complex or affects the configuration of the JHR-binding regulatory element of the JH responsive gene so as to render the latter accessible or inaccessible to JHR.

This model can also explain the other morphogenetic effects of JH. In the larval cells destined to die at metamorphosis, JHR complex binds to the genes whose expression leads to cell death. In the absence of JH, these genes are expressed and as a consequence the cells die.

Knowledge of the organization of the JH target genes is essential for testing this hypothesis. Isolation of the regulatory sequences and identification of the putative trans-acting regulatory factors should throw light on the mode of action of this very interesting hormone. Knowledge of molecular mechanisms of the morphogenetic activity of JH should provide insights into regulation of gene expression not hitherto revealed by the other model systems currently under study.

One final consideration: if JHR modulation of specific gene expression is dependent on the tissue-specific factor, where is the need for JH? JH serves to coordinate metamorphosis of the individual organ systems and to prevent asynchrony in the developmental program of different tissues in the larva.

7.6.3. *Effect of JH on Membranes*

Not all effects of JH can be explained on the basis of its effect on selective gene activation or gene repression. Some JH effects are observed with a very short latent period. For example, changes in patency in vitellogenic follicles occurs within minutes after application of JH *in vitro* (see Section 7.4.2, above). This change in intercellular spaces is preceded by changes in ion transport by the Na^+/K^+ ion pump. Baumann (1969), in a study on the salivary glands in *Galleria mellonella* and artificial membranes, showed that JH analogues alter ionic conductances, possibly by interacting with membrane phospholipids. Kroeger (1967), in a summary of the studies on the probable effect of insect hormones on ion transport across membranes, hypothesized that hormone-induced puffing in *Chironomus thummi* salivary glands could be a consequence of the effect of the hormones on intranuclear ratio of Mg^{2+}, K^+, and Na^+ ion concentrations. According to this hypothesis, ecdysteroids and JH affect relative intranuclear cation concentrations of Na^+ and K^+, probably by affecting cell and nuclear membrane permeability. The altered ion ratios are believed to be responsible for differential gene expression. However, definitive evidence for this theory is not available.

The fact that JH can affect ion transport, at least in some insect cells, is now firmly acknowledged. Perhaps there exists in insects a mode of hormone action analogous to that of some prostaglandins which act on cell surface receptors to bring about long-term changes in cell function. Observations on the role of Ca^{2+} and protein kinase C in the JH-stimulated increase in protein synthesis in the male reproductive accessory gland suggest that JH may act by a membrane protein-mediated effect (see Section 7.4.3), at least in this tissue. Hence, a detailed analysis of the effect of JH on the plasma membrane and the nuclear membrane is important for a complete understanding of the molecular basis of JH action.

7.7. Summary

The juvenile hormone (JH) of insects is a unique developmental hormone. Its effects are very diverse. It blocks metamorphosis of a juvenile insect to a more advanced developmental stage. In blocking metamorphosis, JH alters the normal pattern of gene expression that would have occurred in response to ecdysteroids in the absence of JH. JH represses the sets of genes that specify the phenotype of the more advanced developmental stage. By corollary, JH also permits continued expression of the sets of genes that specify the juvenile phenotype.

JH effects are cell autonomous in the sense that the response of a cell is not dependent on the neighboring cells. The sets of cuticular genes that are repressed by JH, as well as the sets expressed in the presence of JH, are variable in the different regions of the body. Similarly, the sets of genes regulated by JH in nonchitogenous cells are different from those that are regulated in chitogenous cells. Thus, the sets of genes regulated by JH have been determined, probably irreversibly, during embryonic life when the cell lineages become fixed. JH modulates only their overt expression at the appropriate stage in the ontogeny of the insect. An additional feature of JH action is that, as a morphogenetic agent, the effects of JH are manifest only when JH acts in concert with the molting hormone. Another aspect of the morphogenetic effect of JH is that it is not an all-or-none phenomenon. In model systems amenable for dissection of the successive steps in expression of the phenotype of the advanced developmental stage, the need for continuous presence of JH during the entire period of expression has been documented. Therefore, JH is not operating at the level of a master genetic switch that, when activated, successively activates the component genes without additional regulatory input. JH is involved in regulating genes directly and individually. At the present time, we have no knowledge of the molecular basis of JH action in regulating these genetic events.

In addition to its effect on morphogenesis, JH also affects reproduction in some insects. The different aspects of reproduction affected by JH are highly variable in different insect species. JH stimulates vitellogenin synthesis by the fat body in some insects. In some others JH controls vitellogenin uptake, accessory reproductive gland function, or merely the reproductive behavior. The molecular mechanisms underlying control of reproduction appear to be diverse. JH effects on vitellogenin synthesis in *Locusta* have been shown to involve activation of vitellogenin gene transcription. JH-stimulated vitellogenin sequestration by oocytes in *Rhodnius* is mediated by its effect on ion pumps in the plasma membranes. Similarly, JH stimulates synthesis of protein in the male accessory glands in *Drosophila* by altering the function of protein kinase C in a Ca^{2+}-dependent interaction. Lastly, JH stimulates oothecin synthesis in the left collaterial gland of *Periplaneta* by activating oothecin gene transcription. All these diverse modes of function are definitely attributable to JH. It is not known, however, whether there is a common unifying molecular event that can explain these apparently diverse functions.

The effects of JH on behavior, polymorphism (including caste determination), and general metabolism in diverse groups of insects have not been studied at the cellular level. But the data in the literature certainly point to the fact that JH, as a hormonal agent, influences one or more of these facets of insect life in many insects. An understanding of the molecular mechanisms of JH action in its diverse physiological and developmental roles will provide not only a better understanding of the intricacies of gene regulation, in general, but also help us devise novel methods to control insect pests. In this chapter some molecular models proposed to explain JH regulation of gene expression are discussed. More information on individual JH-modulated genes from chitogenous and nonchitogenous cells is essential to test the validity of these theoretical models. In the absence of this information, it is not possible to understand JH action at the cellular and molecular levels.

Acknowledgments

I thank Drs. Larry J. Heilmann, Patrick M. Trewitt, and James B. Courtright for helpful discussions, Ms. Barbara DeNoyer for secretarial help, and Michael Koch for graphics. The research work from the author's laboratory was supported by the National Science Foundation. This chapter is dedicated to Prof. Howard A. Schneiderman, who introduced the author to the study of JH.

References

Abu-Hakima, R. and K. G. Davey. 1979. A possible relationship between ouabain-sensitive (Na^+/K^+)-dependent ATPase and the effect of juvenile hormone on the follicle cells in *Rhodnius prolixus*. Isect Biochem. 9:195–198.

Akai, H., K. Kiguchi, and K. Mori. 1973. The influence of juvenile hormone on the growth and metamorphosis of *Bombyx mori* larvae. Bull. Sericult. Exp. Stn. 25: 287–305.

Anderson, D. T. 1972. The development of hemimetabolous insects. Pp. 95–163 *in* S. J. Counce and C. H. Waddington (eds.), *Developmental Systems: Insects.* Academic Press, Orlando, Florida.

Bartelink, A. K. M. and C. A. D. DeKort. 1975. Effects of diapause induction and juvenile hormone administration on mitochondrial formation in the flight muscles of the Colorado beetle. Proc. K. Ned. Akad. Wet. 78C: 1–11.

Baumann, G. 1969. Juvenile hormone: effect on biomolecular lipid membranes. Nature (Lond.) 223: 316–317.

Bell, W. J. 1969. Dual role of juvenile hormone in the control of yolk formation in *Periplaneta americana.* J. Insect Physiol. 15: 1279–1290.

Bergot, B. J., G. C. Jamieson, M. A. Ratcliff, and D. A. Schooley. 1980. JH zero: new naturally occurring insect juvenile hormone from developing embryos of the tobacco hornworm. Science (Wash., DC) 210: 336–338.

Bergot, B. J., F. C. Baker, D. C. Cerf, G. Jamieson, and D. A. Schooley. 1981. Qualitative and quantitative aspects of juvenile hormone titers in developing embryos of several insect species: discovery of a new JH like substance extracted from eggs of *Manduca sexta.* Pp. 33–45 *in* G. E. Pratt and G. T. Brooks (eds.), *Juvenile Hormone Biochemistry.* Elsevier/North-Holland Publ., Amsterdam and New York.

Bodenstein, D. and I. B. Sprague. 1959. The developmental capacities of the accessory sex glands in *Periplaneta americana.* J. Exp. Zool. 142: 177–202.

Borst, D. W., H. Laufer, M. Landau, E. S. Chang, W. A. Hertz, F. C. Baker, and D. A. Schooley. 1987. Methyl farnesoate and its role in crustacean reproduction and development. Insect Biochem. 17: 1123–1127.

Bounhiol, J. J. 1937. Metamorphose prématurée par ablation des corpora allata chez le jeune ver à soie. C. R. Acad. Sci. Paris 205: 175–179.

Bowers, W. S., T. Ohta, J. S. Cleere and P. A. Marsella. 1976. Discovery of insect anti-juvenile hormones in plants. Science (Wash., DC) 193: 542–547.

Brookes, V. J. 1969. The induction of yolk protein synthesis in the fat body of *Leucophaea maderae* by an analog of the juvenile hormone. Dev. Biol. 20: 459–471.

Burgin, C. and B. Lanzrein. 1988. Stage dependent biosynthesis of methyl farnesoate and juvenile hormone III and metabolism of juvenile hormone III in embryos of the cockroach, *Nauphoeta cinera.* Insect Biochem. 18: 3–9.

Campos, L. A., Deo, F. M. Velthuis-Klupell, and H. H. W. Velthuis. 1975. Juvenile hormone and caste determination in a stingless bee. Naturwissenschaften. 62: 98–99.

Chippendale, G. M. 1977. Hormonal regulation of larval diapause. Annu. Rev. Entomol. 22: 121–138.

Chudakova, I. and E. Guttmann. 1978. Developmental changes of succinate dehydrogenase, ATPase and acid phosphatase activity in the flight muscles of normal and allatectomized adult cricket, *Acheta domestica* (Orthoptera). Zool. Jahrb. Physiol. 82: 1–15.

Compton, J. G., W. T. Schrader, and B. W. O'Malley. 1983. DNA sequence preference of the progesterone receptor. Proc. Nat. Acad. Sci. USA 80: 16–20.

Cox, D. L. and J. H. Willis. 1985. The cuticular proteins of *Hyalophora cecropia* from different anatomical regions and metamorphic stages. Insect Biochem. 15: 349–362.

Cox, D. L. and J. H. Willis. 1987. Analysis of the cuticular proteins of *Hyalophera cecropia* with two-dimensional electrophoresis. Insect Biochem. 17: 457–468.

Cymborowski, B. and F. Sehnal. 1980. Graded inhibition of cell disintegration by juvenile hormone. Cell Differ. 9: 105–115.

Davis, N. T. 1975. Hormonal control of flight muscle histolysis in *Dysdercus fulvoniger*. Ann. Entomol. Soc. Am. 68: 710–714.

DeWilde, J. and D. Stegwee. 1958. Two major effects of the corpus allatum in the adult Colorado beetle (*Leptinotarsa decimlineata* Say). Arch. Neerl. Zool. 13: 277–289.

DeWilde, J., G. B. Staal, C. A. D. Dekort, A. DeLoof, and G. Board. 1968. Juvenile hormone titer in the hemolymph as a function of the photoperiodic treatment in the adult Colorado beetle (*Leptinotarsa decimlineata* Say). Proc. K. Ned. Akad. 71C: 321–326.

Dietz, A., H. R. Hermann, and M. S. Blum. 1979. The role of exogenous JH I, JH III and anti-JH (precocene II) on queen induction of 4.5-day-old worker honeybee larvae. J. Insect Physiol. 25: 503–512.

Doane, W. W. 1973. Role of hormones in insect development. Pp. 291–497 *in* S. J. Counce and C. H. Waddington (eds.), *Developmental Systems: Insects.* Academic Press, Orlando, Florida.

Dorn, A. 1975. Struktur und funktion des embryonalen Corpus allatum von *Oncopeltus fasciatus* Dallas (Insecta: Heteroptera). Verh. Dtsch. Zool. Ges. 67: 85–89.

Dorn, A. 1984. Neurosecretion in the insect embryo. Proc. Arthropod Embryol. Soc. Jpn. pp. 1–16.

Ejezie, G. C. and K. G. Davey. 1976. Some effects of allatectomy in the female tsetse, *Glossina austeni*. J. Insect Physiol. 22: 1743–1749.

Engelmann, F. 1984. Regulation of vitellogenesis in insects: the pleiotropic role of juvenile hormone. Pp. 444–453 *in* J. A. Hoffmann and M. Porchet (eds.), *Biosynthesis, Metabolism and Mode of Action of Invertebrate Hormones.* Springer-Verlag, Berlin and New York.

Englemann, F. 1969. Female specific protein: biosynthesis controlled by corpus allatum in *Leucophaea maderae*. Science (Wash., DC) 165: 407–409.

Englemann, F., J. Mala and S. Tobe. 1987. Cytosolic and nuclear receptors for juvenile hormone in fat bodies of *Leucophaea maderae*. Insect Biochem. 17: 1045–1052.

Fain, M. J. and L. M. Riddiford. 1977. Requirements for molting of the crochet epidermis of the tobacco hornworm larva *in vivo* and *in vitro*. Wilhelm Roux' Arch. Dev. Biol. 181: 255–307.

Gavin, J.A. and J. H. Williamson. 1976. Juvenile hormone induced vitellogenesis in *apterous*[4], a nonvitellogenic mutant in *Drosophila melanogaster*. J. Insect Physiol. 22: 1737–1742.

Gilbert, L. I. 1967. Changes in lipid content during the reproductive cycle of *Leucophaea maderae* and effects of the juvenile hormone on lipid metabolism *in vitro*. Comp. Biochem. Physiol. 21: 237–257.

Giorgi, F. and F. Macchi. 1980. Vitellogenesis in the stick insect, *Carausius morosus*. I. Specific protein synthesis during ovarian development. J. Cell Sci. 46: 1–16.

Gronemeyer, H., B. Tucotte, C. Quinn-Stricker, M. T. Bocquel, M. E. Meyer, Z. Krozowski, J. M. Jeltsch, T. Leroze, J. M. Garnier, and P. Chambon. 1987. The chicken progesterone receptor: sequence, expression and functional analysis. EMBO (Eur. Mol. Biol. Organ.) J. 6: 3985–3994.

Hagedorn, H. H. and J. G. Kunkel. 1979. Vitellogenin and vitellin in insects. Annu. Rev. Entomol. 24: 475–505.

Hales, D. F. and T. E. Mittler. 1981. Precocious metamorphosis of the aphid *Myzus persicae* induced by the precocene analogue 6-methoxy-7-ethoxy-2,2-dimethylchromene. J. Insect. Physiol. 27: 333–337.

Hardie, J. and A. Lees. 1985. Endocrine control of polymorphism and polyphenism. Pp. 441–490 *in* G. A. Kerkut and L. I. Gilbert (eds.), *Comprehensive Insect Physiology, Biochemistry and Pharmacology*, Vol. 8. Pergamon Press, Oxford and Elmsford, New York.

Hartfelder, K. H. 1987. Rates of juvenile hormone synthesis control caste differentiation in the stingless bee, *Scaptotrigona postica dipilis*. Wilhelm Roux' Arch. Dev. Biol. 196: 552–526.

Herman, W. S. 1982. Endocrine regulation of the bursa copulatrix and receptacle glands of *Danaus plexippus* L. (Lepidotera: Danaidae). Experientia 38: 631–632.

Hintze-Podufal, C. 1975. Über die morphogenetische Wirkungsweise von Juvenilehormon und analogen Substanzen und die adult Entwicklung von Schwarmern and Spinnern. Z. Angew. Entomol. 77: 286–291.

Hwang-Hsu, K., G. Reddy, A. K. Kumaran, W. E. Bollenbacher, and L. I. Gilbert. 1979. Correlations between juvenile hormone esterase activity, ecdysone titre and cellular reprogamming in *Galleria mellonella*. J. Insect Physiol. 25: 105–111.

Illenchuk, T. T. and K. G. Davey. 1982. Some properties of Na/K ATPase in the follicle cells of *Rhodnius prolixus* (Stal). Insect Biochem. 12: 675–679.

Illenchuk, T. T. and K. G. Davey. 1987. Effects of various compounds on Na/K ATPase activity, JH I binding capacity and patency response in follicles of *Rhodnius prolixus*. Insect Biochem. 17: 1085–1088.

Judy, K. J., D. A. Schooley, L. L. Dunham, M. S. Hall, B. J. Bergot, and J. B. Siddall. 1973. Isolation, structure and absolute configuration of a new insect juvenile hormone from *Manduca sexta*. Proc. Nat. Acad. Sci. USA 70: 1509–1513.

Kearney, G. P., P. M. Toom, and G. J. Bloomquist. 1977. Induction of dealation in virgin female *Solenopsis invicta* with juvenile hormones. Ann. Entomol. Soc. Am. 70: 699–701.

Kelly, T. J., T. S. Adams, M. B. Schwartz, M. J. Birnbaum, E. Rubenstein, and R. Imberski. 1987. Juvenile hormone and ovarian maturation in the diptera: a review of recent results. Insect Biochem. 17: 1089–1093.

Koeppe, J. K., M. Fuchs, T. T. Chen, L. M. Hunt, G. E. Kovalick, and T. Briers. 1985. The role of juvenile hormone in reproduction. Pp. 165–205 *in* G. A. Kekut and L. I. Gilbert (eds.), *Comprehensive Insect Physiology, Biochemistry and Pharmacology*, Vol. 8. Pergamon Press, Oxford and Elmsford, New York.

Koeppe, J. K., R. C. Rayne, E. A. Whitsel, and M. E. Pooler. 1987. Synthesis and secretion of juvenile hormone binding protein by fat body from the cockroach, *Leucophaea maderae* protein A immunoassay. Insect Biochem. 17: 1027–1032.

Krishnakumaran, A. 1972. Injury induced molting in *Galleria mellonella*. Biol. Bull. (Woods Hole) 142: 281–292.

Kroeger, H. 1967. Hormones, ion balances and gene activity in dipteran chromosomes. Mem. Endocrinol. Soc. 15: 55–66.

Kumaran, A. K. 1976. Relationship between DNA synthesis, juvenile hormone and metamorphosis in *Galleria* larvae. Pp. 184–197 *in* L. I. Gilbert (ed.), *Insect Juvenile Hormones*. Plenum Press, New York.

Kumaran, A. K. 1981. Reprogramming in insect epidermal cells: role of hormones. Pp. 461–472 *in* F. Sehnal, A. Zabza, J. J. Menn, and B. Cymborowski (eds.), *Regulation of Insect Development and Behavior*. Wrocław Technical University Press, Wrocław, Poland.

Lageaux, M., C. Hetru, F. Goltzene, E. Kappler, and J. A. Hoffmann. 1979. Ecdysone titer and metabolism in relation to cuticulogenesis in embryos of *Locusta migratoria*. J. Insect Physiol. 25: 709–723.

Lanzrein, B., H. Imboden, C. Burgin, E. Bruning, and H. Gfeller. 1984. On titers, origin and functions of juvenile hormone III, methyl farnesoate and ecdysteroids in embryonic development of ovoviviparous cockroach, *Nauphoeta cinerae*. Pp. 454–465 *in* J. A. Hoffmann and M. Porchet (eds.), *Biosynthesis, Metabolism and Mode of Action of Invertebrate Hormones*. Springer-Verlag, Berlin and New York.

Larson, W. 1970. Genesis of mitochondria in the fat body of an insect. J. Cell Biol. 47: 373–383.

Laufer, H., M. Landau, E. Homolo, and D. W. Borst. 1987. Methyl farnesoate, its site of synthesis and regulation of secretion in a juvenile crustacean. Insect Biochem. 17: 1129–1131.

Lebrun, D. 1969. Corps allates et instinct génésique de *Calotermes flavicollis* Fabr. nécessié la présence dans l'organisme d'un taux élevé d'hormone juvénile. C. R. Acad. Sci. Paris 269D: 632–634.

Lees, A. D. 1966. Control of polymorphism in aphids. Adv. Insect Physiol. 3: 207–207.

Lüscher, M. 1976. Evidence for an endocrine control of caste determination in higher termites. Pp. 91–103 *in* M. Lüscher (ed.), *Phase and Caste Determination in Insects: Endocrine Aspects*. Pergamon Press, Oxford and Elmsford, New York.

Masner, P. W. Hangartner, and M. Suchy. 1975. Reduced titers of ecdysone following juvenile hormone treatment in the German cockroach, *Blatella germanica*. J. Insect Physiol. 21: 1755–1762.

Memmel, N. A. and A. K. Kumaran. 1988. Role of ecdysteroids and juvenile hormone in regulation of larval haemolymph protein gene expression in *Galleria mellonella*. J. Insect Physiol. 34:585–591.

Meyer, A. S., H. A. Schneiderman, E. Hanzmann, and J. H. Ko. 1968. The two juvenile hormones from the cecropia silkmoth. Proc. Nat. Acad. Sci. USA 60: 853–860.

Nair, K. S. S. 1974. Studies on the diapause of *Trogoderma granarum*: Effects of juvenile hormone analogues on growth and metamorphosis. J. Insect Physiol. 20: 231–244.

Nijhout, H. F. and D. E. Wheeler. 1982. Juvenile hormone and the physiological basis of insect polymorphisms. Q. Rev. Biol. 57: 109–132.

Novak, V. J. A. 1956. The gradient factor theory: a general conception of metamorphosis of insects. Ann. Soc. Zool. Fr. 11: 335–337.

Novak, V. J. A. 1969. Morphogenetic analysis of the effects of juvenile hormone analogues and other morphogenetically active substances on embryos of *Schistocerca gregaria* (Forskal). J. Embryol. Exp. Morphol. 21: 1–21.

Pau, R. N. 1987. Characterization of juvenile hormone-regulated cockroach oothecin genes. Insect Biochem. 17: 1075–1078.

Pau, R. N., R. J. Weaver, and K. Edwards-Jones. 1986. Regulation of cockroach oothecin synthesis by juvenile hormone. Arch. Insect Biochem. Physiol. (Suppl.) Pp. 59–73.

Pener, M. P. 1985. Hormonal effects on flight and migration. Pp. 491–550 *in* G. A. Kerkut and L. I. Gilbert (eds.), *Comprehensive Insect Physiology, Biochemistry and Pharmacology*. Pergamon Press, Oxford and Elmsford, New York.

Piepho, H. 1951. Über die Lenkung der Insekten metamorphose durch hormone. Verh. Dtsch Zool. Ges. 1951: 62–75.

Pratt, G. E. and K. G. Davey. 1972. The corpus allatum and oogenesis in *Rhodnius prolixus*: the effects of allatectomy. J. Exp. Biol. 56: 201–214.

Prestwich, G. D., W. S. Eng, M. R. Boehm, and C. Wawrzenczyk. 1987. High specific activity tritium and iodine labeled JH homologs and analogs for receptor binding studies. Insect Biochem. 17: 1033–1037.

Raabe, M. 1986. Insect reproduction: regulation of successive steps. Adv. Insect Physiol. 19: 29–154.

Ramamurthy, P. S. 1968. Origin and distribution of glycogen during vitellogenesis of the scorpion fly, *Panorpa communis*. J. Insect Physiol. 14: 1325–1330.

Ramamurthy, P. S. and W. Engels. 1977. Allatektomie-und Juvenilehormonwirkungen und Synthese und Einlagerung von Vitellogenin bei der Bienenkönigin *(Apis mellifica)*. Zool. Jahrb. Physiol. 81: 165–176.

Rankin, M. A. 1974. The hormonal control of flight in the milkweed bug. *Oncopeltus fasciatus*. Pp. 317–328 *in* L. Barton-Browne (ed.), *Experimental*

Analysis of Insect Behavior. Springer-Verlag, Berlin and New York.

Rankin, M. A. and S. Rankin. 1980. Some factors affecting presumed migratory flight activity of the convergent lady beetle, *Hippodamia convergens* (Coccinellides: Coleoptera). Biol. Bull. (Woods Hole) 158: 356–359.

Rankin, M. A. and L. M. Riddiford. 1978. Significance of haemolymph juvenile hormone titer changes in timing of migration and reproduction in adult *Oncopeltus fasciatus.* J. Insect Physiol. 24: 31–38.

Rao, K. D. P. and A. Krishnakumaran. 1974. Effect of juvenile hormone on DNA synthesis in *Acheta* embryos. Wilhelm Roux' Arch. Dev. Biol. 174: 276–285.

Ray, A., N. A. Memmel, and A. K. Kumaran. 1987. Developmental regulation of the larval hemolymph protein genes in *Galleria mellonella.* Wilhelm Roux' Arch. Dev. Biol. 196: 414–420.

Reddy, G. and A. Krishnakumaran. 1973. Effect of diet on response to juvenile hormone in *Galleria mellonella* larvae. Experientia 29: 621–622.

Reddy, G., K. Hwang-Hsu, and A. K. Kumaran. 1979. Factors influencing juvenile hormone esterase activity in the waxmoth, *Galleria mellonella.* J. Insect Physiol. 25: 65–71.

Regis, L. 1979. The role of blood meal in egg-laying periodicity and fecundity in *Triatoma infestans.* Int. J. Invertebr. Reprod. 1: 187–195.

Rembold, H. 1987. Caste specific modulation of juvenile hormone titers in *Apis mellifera.* Insect Biochem. 17: 1003–1006.

Riddiford, L. M. 1972. Juvenile hormone and insect embryonic development: Its potential role as an ovicide, Pp. 95–111 *in* J. J. Menn and M. Beroza (eds.), *Insect Juvenile Hormone: Chemistry and Action.* Academic Press, Orlando, Florida.

Riddiford, L. M. and C. M. Williams. 1967. The effects of juvenile hormone analogues on the embryonic development of silkworms. Proc. Nat. Acad. Sci. USA 57: 595–601.

Riddiford, L. M., E. O. Osir, C. M. Fittinghoff, and J. M. Green. 1987. Juvenile hormone analog binding in *Manduca* epidermis. Insect Biochem. 17: 1039–1043.

Roberts, P. E. and L. S. Jefferies. 1986. Grasshopper as a model for the analysis of juvenile hormone delivery to chromatin acceptor sites. Arch. Insect Biochem. Physiol. (Suppl.) pp. 7–23.

Roe, R. M., C. L. Crawford, C. W. Clifford, J. P. Woodring, T. C. Sparks, and B. D. Hammock. 1987. Role of juvenile hormone metabolism during embryogenesis of the house cricket, *Acheta domesticus.* Insect Biochem. 17: 1023–1026.

Roeller, H., K. H. Dahm, C. C. Sweeley, and B. M. Trost. 1967. The structure of the juvenile hormone. Angew. Chem. Int. Engl. 6: 179–180.

Roseler, P. F. 1976. Juvenile hormone and queen rearing in bumble bees. Pp. 55–61 *in* M. Lüscher (ed.), *Phase and Caste Determinaton in Insects: Endocrine Aspects.* Pergamon Press, Oxford and Elmsford, New York.

Sánchez y Sánchez, D. 1925. L'histogenèse dan le centres nerveaux des insectes pendant les métamorphoses. Trab. Lab. Invest. Biol. Univ. Madr. 23: 29–52.

Scharrer, B. 1946. The relationship between corpora allata and reproductive organs in adult *Leocophaea maderae* (Orthoptera). Endocrinology 38: 46–55.

Schneiderman, H. A., A. Krishnakumaran, V. G. Kulkarni, and L. Friedman. 1965. Juvenile hormone activity of structurally unrelated compounds. J. Insect Physiol. 11: 1641–1647.

Schooley, D. A. 1977. Analysis of the naturally occurring juvenile hormones: their isolation, identification and titer determination at physiological levels. Pp. 241–287 *in* R. B. Turner (ed.), *Analytical Biochemistry of Insects.* Elsevier, Amsterdam and New York.

Schrader, K. 1938. Untersuchungen über die Normalentwicklung des Gehirns und Gehirntransplantation bei der Mehlmotte *Ephestia kuhniella* Zeller nebst einigen Bemerkungen über das Corpus allatum. Biol. Zentralbl. 58: 52–90.

Shirk, P., G. Bhaskaran, and H. Roller. 1983. Developmental physiology of corpora allata and accessory glands in the *Cecropia* silkmoth. J. Exp. Zool. 227: 69–79.

Slama, K. and C. M. Williams. 1966. "Paper factor" as an inhibitor of the embryonic development of the European bug *Pyrrhocoris apterus.* Nature (Lond.) 210: 329–330.

Sparks, T. C. and B. D. Hammock. 1979. Induction and regulation of juvenile hormone esterases during the last larval instar of the cabbage looper, *Trichoplusia ni.* J. Insect Physiol. 25: 551–560.

Stay, B. and H. L. Lin. 1981. The inhibition of milk synthesis by juvenile hormone in the viviparous cockroach, *Diploptera punctata.* J. Insect Physiol. 27: 551–557.

Tedesco, J., J. B. Courtright, and A. K. Kumaran. 1981. Ultrastructural changes induced by juvenile hormone analogues in oocyte membranes of *apterous*[4] *Drosophila melanogaster.* J. Insect Physiol. 27: 895–902.

Telfer, W. H. 1954. Immunological studies of insect metamorphosis. II. The role of a sex-limited blood protein in egg formation by the *Cecropia* silkworm. J. Gen. Physiol. 37: 539–558.

Telfer, W. H. 1965. The mechanism and control of yolk formation. Annu. Rev. Entomol. 10: 161–182.

Tobe, S. S. and B. Stay. 1985. Structure and regulation of the corpus allatum. Adv. Insect Physiol. 18: 305–432.

Unnithan, G. C. and K. K. Nair. 1977. Ultrastructure of juvenile hormone induced degenerating flight muscles in a bark beetle, *Ips paraconfusus.* Cell Tissue Res. 155: 481–490.

Vince, R. K. and L. I. Gilbert. 1977. Juvenile hormone esterase activity in precisely timed last instar larvae and pharate pupae of *Manduca sexta.* Insect Biochem. 7: 115–120.

Watson, J. A. L. 1967. The growth and activity of the corpora allata in the larval firebrat, *Thermobia domestica* (Packard) (Thysanura: Lepismatidae). Biol. Bull. (Woods Hole) 132: 277–291.

Whitmore, D., E. Whitmore, and L. I. Gilbert. 1972. Juvenile hormone induction of esterases: a mechanism for the regulation of juvenile hormone titer. Proc. Nat. Acad. Sci. USA 69: 1592–1595.

Wigglesworth, V. B. 1934. The physiology of ecdysis in *Rhodnius prolixus*. II. Factors controlling moulting and metamorphosis. Quart. J. Microsc. Sci. 77: 191–222.

Wigglesworth, V. B. 1936. The function of the corpus allatum in the growth and reproduction of *Rhodnius prolixus* (Hemiptera). Q. J. Microsc. Sci. 79: 91–121.

Wigglesworth, V. B. 1958. Some methods for assaying extracts of the juvenile hormone in insects. J. Insect Physiol. 2: 73–84.

Wigglesworth, V. B. 1966. Hormonal regulation of differentiation in insects. Pp. 180–209 *in* W. Beerman (ed.), *Cell Differentiation and Morphogenesis*. North-Holland Publ., Amsterdam and New York.

Williams, C. M. 1956. The juvenile hormone of insects. Nature (Lond.) 178: 212–213.

Willis, J. H. 1986. The paradigm of stage-specific gene sets in insect metamorphosis: time for revision. Arch. Insect Biochem. Physiol. (Suppl.) pp. 47–57.

Willis, J. H. and P. C. J. Brunet. 1966. The hormonal control of collaterial gland secretion. J. Exp. Biol. 44: 363–378.

Willis, J. H., R. Rezaur, and F. Sehnal. 1982. Juvenoids cause some insects to form composite cuticles. J. Embryol. Exp. Morphol. 71: 25–40.

Wirtz, P. 1973. Differentiation in honeybee larva. Meded. Landbouwhogesch. Wageningen. 73/75: 1–66.

Wiśniewski, J. R., C. Wawrzenczyk, G. D. Prestwich, and M. Kochman. 1988. Juvenile hormone binding proteins from the epidermis of *Galleria mellonella*. Insect Biochem. 18: 29–36.

Wyatt, G. R. 1972. Insect hormones. Pp. 385–490 *in* G. R. Litwack (ed.), *Biochemical Actions of Hormones*. Academic Press, Orlando, Florida.

Wyatt, G. R., T. T. Chen, and P. Couble. 1976. Juvenile hormone-induced vitellogenin synthesis in locust fat body *in vitro*. Pp. 195–202 *in* E. Kurstak and K. Marmmorosch (eds.), *Invertebrate Tissue Culture: Applications to Biology, Medicine and Agriculture*. Academic Press, Orlando, Florida.

Wyatt, G. R., K. E. Cook, H. Firko, and T. Dhadialla. 1987. Juvenile hormone action on locust fat body. Insect Biochem. 17: 1071–1073.

Yamamoto, K., A. Chadarevian, and M. Pelligrini. 1988. Juvenile hormone actions mediated in male accessory glands of *Drosophila* by calcium and kinase C. *Science (Wash., DC)* 239: 916–919.

Yamamoto, K. R. 1986. Hormone-dependent transcriptional enhancement and its implication for mechanisms of multifactor gene regulation. Pp. 131–148 *in* L. Bogorad (ed.), *Molecular Developmental Biology*. Liss, New York.

Yashica, K. 1960. Studies on the neurosecretory system in Apterygota. I. Histological observations on the corpus allatum and neurosecretory cells in *Ctenolepisma*. Mem. Coll. Sci. Kyoto Univ. B 27: 1–7.

Yin, C.-M. and G. M. Chippendale. 1973. Juvenile hormone regulation of the larval diapause of the southwestern corn borer, *Diatraea grandiosella*. J.

Insect Physiol. 19: 2403–2420.

Yin, C.-M. and G. M. Chippendale. 1979. Diapause of the southwestern corn borer, *Diatraea grandiosella:* further evidence showing juvenile hormone to be the regulator. J. Insect Physiol. 25: 513–523.

Zalokar, M. 1968. Effect of corpora allata on protein and RNA synthesis in collateral glands of *Blatella germanica.* J. Insect Physiol. 14: 1177–1184.

Evidence for Ecdysteroids as Molting Hormones in Chelicerata, Crustacea, and Myriapoda

8

T. C. JEGLA

8.1. Introduction	231
8.2. Extraction, Chemical Identification, and Other Considerations	232
8.3. Evidence for Molting Hormone in Specific Taxonomic Groups	235
8.3.1. Aquatic Chelicerata	235
8.3.1.1. Merostomata	235
8.3.1.2. Pycnogonida	237
8.3.2. Terrestrial Chelicerata	237
8.3.2.1. Arachnida	237
8.3.2.1.1. Scorpiones	238
8.3.2.1.2. Araneae	239
8.3.2.1.3. Opiliones	241
8.3.2.1.4. Acarina	241
8.3.3. Crustacea	245
8.3.3.1. Branchiopoda	245
8.3.3.2. Maxillopoda	246
8.3.3.3. Malacostraca	248
8.3.3.3.1. Decapoda	248
8.3.3.3.2. Isopoda	258
8.3.3.3.3. Amphipoda	259
8.3.4. Myriapoda	260
8.3.4.1. Chilopoda	260
8.3.4.2. Symphyla	261
8.4. Summary	261
References	262

8.1. Introduction

Molting hormones, the ecdysteroids, were first isolated in a noninsect arthropod, at least in crude preparations, in 1955 by Karlson (see Karlson, 1956). He obtained an extract from the shrimp *Crangon vulgaris* by fractionating the animals according to the methods for isolation of insect prothoracic gland hormone. The extract provoked a positive pupariation response in the fly (*Calliphora*) larval bioassay, thus showing molting hormone activity. This observation suggested to Karlson that "ecdysone, or a related hormone, is present in all [a]rthropods," a notion that has been repeated many times since his discovery. As I will show in this review, we cannot yet state that the notion is indeed correct, but the probability that it is true is now very high. Molting hormones (MHs) have been identified in at least 8 major classes (of 13 in the classification system presented by Barnes, 1987, as modified for crustaceans by Bowman and Abele, 1982), 13 orders, and all of the major forms (chelicerates, mandibulates, myriapods).

MHs are members of a group of similar compounds, the ecdysteroids, which are synthesized from cholesterol or closely related sterols. By classical definition, an MH is a compound that is secreted by a gland, distributed through the hemolymph, and absorbed by the target cells, the epidermal cells, which secrete a new shell or cuticle. By general acceptance, a molting response has occurred when apolysis or erosion of the original cuticle and/or deposition of a new cuticle is observed. In the general arthropod system, ecdysone (EC) or a precursor is secreted by the molting gland and is converted to 20-hydroxyecdysone (20-OHE) by the target cell. The hydroxylation of EC to form 20-OHE is a common reaction that occurs in all arthropods tested (LaFont and Koolman, 1984). It is widely assumed that 20-OHE is the actual compound that stimulates the molting response in most species of arthropods studied; exceptions at this date may be the several species of larval hemipteran insects in which makisterone A is the major hemolymph ecdysteroid while EC and 20-OHE are only minor constituents (Kelly et al., 1984).

EC and 20-OHE, originally known as α- and β-ecdysone, were the first ecdysteroids discovered, but they are only two of many ecdysteroids that provoke a molting response in arthropods, either through interconversion or mimicry. These two compounds have commonly been assumed to play a role in control of molting in a particular species after injection and subsequent observation of a molting response and/or after extraction and identification by bioassay, radioimmunoassay, various chromatographic techniques, or physical methods (e.g., mass spectrometry).

8.2. Extraction, Chemical Identification, and Other Considerations

Horn's group (see Hampshire and Horn, 1966) in Australia was the first to extract a noninsect arthropod, the crustacean spiny lobster *Jasus lalandei,* on a mass scale and to obtain a pure compound in large enough quantity for chemical identification. Crustecdysone, as it was first named, is now known as 20-OHE, the same substance Karlson (1956) termed β-ecdysone. From 1000 kg of spiny lobster they obtained 2.3 mg of the pure hormone (Horn et al., 1968). The extraction procedures began by homogenizing lobster waste from commercial operations in a large amount of aqueous ethanol. The liquid extract was reduced to an aqueous concentrate under vacuum, about one-tenth of its original volume. This concentrate was extracted with a hexane:2-propanol (1:3) system to remove less-polar materials (e.g., neutral lipids) in the hexane layer and again reduced to an aqueous concentrate by low-temperature distillation. A countercurrent extraction with *n*-butanol separated the active material from inactive polar constituents. From this point to final purification, ecdysteroids were concentrated enough to be detected by bioassay, and active extracts were subsequently identified by use of the *Calliphora* test (see Karlson, 1956, for details of the fly bioassay). Degree of purification was indicated by results of the bioassays.

The active extract was subjected to countercurrent extraction in a chloroform : methanol : water (1:2:1) system to remove polar lipids. The methanol layers were retained and evaporated, and the original mass had been reduced by $14,085\times$. It was treated with aqueous potassium bicarbonate and extracted countercurrently in a chloroform : ethanol : water system to remove acidic impurities. The active material was retained in the chloroform layer, which was evaporated and then subjected to reversed-phase column chromatography, eluting the active material with an *n*-butanol : water solvent system. This separated 20-OHE from other active ecdysteroids (EC and 2-deoxy-20-OHE). Final purification of 20-OHE was achieved by column chromatography on CM-Sephadex (acid form) and silicic acid (acidic form of silica gel).

Few other studies on noninsect arthropods have involved such massive-scale extractions, although some of the procedures have been used in whole-animal extractions of many species, e.g., *Limulus polyphemus* larvae (Jegla and Costlow, 1979a). Most studies since the original work of Horn and associates have incorporated such techniques as analytical thin-layer chromatography (TLC) and high-performance liquid chromatography (HPLC). These techniques allow separation of microgram-to-nanogram quantities of MHs, and therefore much less starting material is required. The paper by LaFont et al. (1980; see also Chapter 13 by Lafont and Beydon herein) and reviews by Morgan and Poole (1976) and Morgan and Wilson (1980) present detailed considerations and discussions of ecdysteroid extraction and identification from arthropod materials. Further, development of a radioimmunoassay (RIA) for ecdysteroids by Borst and O'Connor (1972) gradually displaced the insect bioassays and has allowed detection of picogram quantities. The horseshoe crab *Limulus* larva bioassay developed by Jegla (1972; Jegla and Costlow, 1979b) can also detect picogram quantities, but it too has given way to RIAs, which rapidly became accessible to most laboratories working on arthropod MHs.

Whole animals were used as starting material for most of the studies covered in this review largely because of the small size of the species investigated. However, evidence for MHs in noninsect arthropods has also accumulated from studies that have started with hemolymph or organ culture fluid. Both of these fluids can be tested directly in an RIA for ecdysteroid content, thus circumventing the time-consuming steps of extraction from whole animals. But, since the majority of arthropods are probably too small for hemolymph sampling (a minimum of about 10 μl are required for RIA testing) or for culturing of individual organs, whole-animal extraction may be the most practical method for isolating and identifying specific ecdysteroids.

Since the early 1980s, the isolation, purification, and identification of EC and 20-OHE from arthropod hemolymph, tissues, and whole bodies has been considerably streamlined with the use of disposable SEP-PAK reversed-phase cartridges (see LaFont et al., 1982; Watson and Spaziani, 1982). Less than a gram of starting material gives satisfactory results. Watson and Spaziani used a one-step methanol extraction of crustacean hemolymph and tissues (Table 8.1). The extract was applied to the cartridge packing in water, polar materials were washed off with 20% aqueous methanol, and the ecdysteroids were eluted with methanol. After evaporation the material was dissolved in chloroform : 95% ethanol, and ecdysteroids were separated and quantified by HPLC (Watson and Spaziani, 1982). Although they used normal-phase HPLC, reversed-phase HPLC is also excellent for separating ecdysteroids. However, for accurate identification of an unknown ecdysteroid, both normal-phase and reversed-phase systems should be used (LaFont et al., 1980). Positive identification may also

TABLE 8.1.   Simplified Scheme of Extraction and Purification of Ecdysteroids from Crustacean Tissues

1.   Homogenize material in methanol.
2.   Centrifuge; reextract the pellet with methanol.
3.   Combine supernatants; evaporate.
4.   Dissolve sediment in water; apply to primed SEP-PAK cartridge.
5.   Flush cartridge with 10 ml rinse water.
6.   Flush with 10 ml 20% aqueous methanol.
7.   Elute ecdysteroids with 10 ml methanol.
8.   Collect and dry methanol fraction.
9.   Dissolve in solvent for HPLC; filter.
10. Inject sample onto HPLC column.
11. If $\leq$low–nanogram range, evaporate, dissolve in water, and quantify in an RIA.

SOURCE: Modified from Watson and Spaziani (1982).

involve mass spectrometry (MS), mass fragmentography (MF), or nuclear magnetic resonance spectrometry (NMR), and LaFont et al. (1980) have suggested that a coupled HPLC/MS system may be the ideal analytical tool for quantification and identification of specific ecdysteroids. If the amount of ecdysteroids in a sample is below HPLC detection (picogram or low-nanogram range), ecdysteroid-containing fractions could be detected and quantified by use of an RIA.

Injection of pure ecdysteroids into noninsect arthropods has been accomplished in a number of the larger species and has provided evidence for the action of specific compounds in molt stimulation. Although many pure ecdysteroids have been synthesized or isolated from animal and plant material, relatively few of these have been injected into more than a few species to assess their action. A structure–function study based on dose–response relationships of a wide variety of ecdysteroids and comparable to some similar studies on insects has been performed on *Limulus* larvae (Jegla and Costlow, 1979b), and in a few species tritiated EC or 20-OHE has been injected for study of the metabolites produced. Connat and Diehl (1986) investigated the metabolism of these two compounds in soft and hard ticks, two species of scorpions, five species of spiders, one centipede species, and species of crustaceans belonging to three different orders.

Their conclusion—that metabolism of these two major ecdysteroids to apolar conjugates with long-chain fatty acids is very common in arthropods—is yet further evidence for a common system linking the chemical control of molting by ecdysteroids in all arthropod species.

8.3. Evidence for Molting Hormone in Specific Taxonomic Groups

8.3.1. *Aquatic Chelicerata*

8.3.1.1. MEROSTOMATA

Evidence from injection. Krishnakumaran and Schneiderman (1970) injected batches of small (3–10 g) and large (35–50 g) horseshoe crabs (*Limulus polyphemus*) with large doses of 20-OHE (40 μg/g) and observed apolysis and new cuticle formation, but no spontaneous molting. The same basic result was observed by Herman (1972) after injection of five pure ecdysteroids into juvenile (20–75 g) *Limulus,* apparently during intermolt, at doses of 5–50 μg/g. Daily injections of 20-OHE from 0.01–5.00 μg/g for 20 days also stimulated a molting response (apolysis in starved animals; apolysis plus new cuticle in fed animals). Extensive experiments on injection of plant and animal ecdysteroids and synthetic analogues in *Limulus* larvae (Jegla, 1972, 1982; Jegla and Costlow, 1970, 1979b; Jegla et al., 1972) have clarified responses to these compounds. There is a programmed sequence of events during the molt cycle. Exogenous ecdysteroids, injected anytime at doses of 4–40 μg/g, cause abnormal development; although larvae may deposit new cuticle, they are unable to shed it and eventually die. However, larvae, injected after mid-intermolt, respond to doses in the low-picogram to 2-ng range (4–400 ng/g) with an accelerated but perfectly normal molt (Jegla, 1982). Responses to EC and 20-OHE are very similar, but at least three jumps in sensitivity occur (Fig. 8.1) and these are positively correlated with specific events in the molt cycle (peak epidermal cell size, apolysis and initial stages of new cuticle formation, and onset of new cuticle deposition). The time of new cuticle deposition is when the larvae are most sensitive to injected ecdysteroids. The accelerated yet normal molt in responses to EC, 20-OHE, and similar ecdysteroids, in contrast to the lack of or reduced molting response to compounds with specific molecular modifications, strongly implicate the ecdysteroids as the normal MHs in *Limulus*.

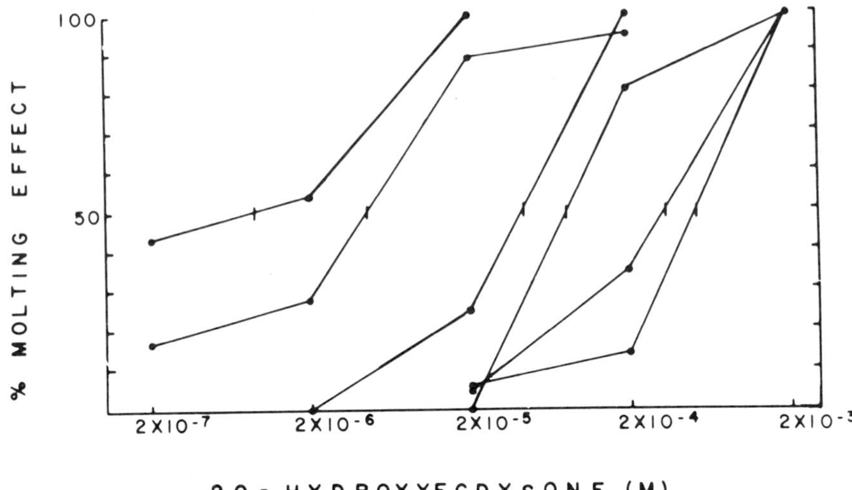

FIGURE 8.1. Differential sensitivity of first-instar *Limulus* to 20-hydroxyecdysone injected at different times (2, 4, 6, 8, 10, 12 days—lines from right to left, respectively) after hatching during a hatch to molt period of 20 days. Injected doses given in molarity (*M*) range from 4 ng to 40 μg/g. Data from about 400 animals. (From Jegla, 1982.)

Evidence from extraction. The principal free ecdysteroid in whole *Limulus* larvae appears to be 20-OHE (Jegla, 1974; Jegla and Costlow, 1979a). Extraction of 200 g of first-stage larvae was based on the method of Karlson and Shaaya (1964), and the ecdysteroids were purified by use of HPLC and TLC systems. The HPLC system separated ecdysteroids in a polarity range from slightly more polar than 20-OHE to slightly less polar than ponasterone A (PNA). All collected fractions were tested for ecdysteroids by an RIA, and only that fraction co-eluting with 20-OHE showed any significant activity above background. The same result was obtained when extracts were purified in three different TLC systems. Quantification of extracted and purified ecdysteroid by RIA and the *Limulus* bioassay gave nearly identical results. The amount of free ecdysteroid as estimated by both RIA and bioassay was three times higher in early premolt than mid-intermolt. Extraction of hemolymph taken from juvenile *Limulus* collected during summer months, partial purification in several TLC systems, and quantification by the *Limulus* bioassay indicated the presence of ecdysteroids at a level of 9 ng/ml, although specific ecdysteroids were not determined (Winget and Herman, 1976). Hemolymph concentration of free ecdysteroids in juvenile horseshoe crabs is the same concentration found in whole intermolt larvae by Jegla and Cost-low (1979a).

8.3.1.2. PYCNOGONIDA

No reports on molt-promoting effects of exogenous ecdysteroids have been published for the sea spiders, but ecdysteroids were first extracted from these animals by Behrens (1981). Two ecdysteroids that were active in the housefly (*Musca*) bioassay were isolated from eggs, juveniles, and adults of *Pycnogonum litorale* (Behrens and Bückmann, 1983). One was identified as 20-OHE and the other more abundant unknown compound was similar but not identical to EC in its behavior in an HPLC and GLC (gas–liquid chromatography) system. As in many other species of arthropods, adult pycnogonids do not molt yet they contain ecdysteroid compounds. Eight pure ecdysteroids were extracted from adult *P. litorale* by the methods of LaFont et al. (1982), isolated by successive use of reversed-phase and normal-phase HPLC systems, and identified by a combination of NMR and MS (Bückmann et al., 1986). The molecular weights and structural formulas were determined for all eight compounds. EC was not detected, and 20-OHE comprised only 1.2% of the total ecdysteroids identified. All of the seven other compounds lacked a free 20-OH group. It was absent in two and esterified with acetic acid or glycolic acid in the other five compounds. Presumably, these are inactive as molt-inducing compounds (Bückmann et al., 1986).

8.3.2. *Terrestrial Chelicerata*

8.3.2.1. ARACHNIDA

Evidence for ecdysteroids as MHs in arachnids has been obtained from four different types (orders): scorpions, spiders, harvestmen, and ticks. The most extensive information is available for ticks from many published works. In addition to their induction of molting, many other effects have been ascribed to the ecdysteroids. In at least one species of ticks, exogenous ecdysteroids terminate larval diapause (Wright, 1969), induce molting in adults (supermolting) that normally do not molt (Mango et al., 1976), inhibit oogenesis (see Diehl et al., 1986), increase sex-attractant pheremone production (Dees et al., 1984a), increase the number of DNA-synthesizing germ cells in unfed males (Dumser and Oliver, 1981), and induce degeneration in salivary glands and cause high mortality under certain conditions (Diehl et al., 1986). Since ticks are pests and disease vectors, the latter effect has been the basis for some studies on the use of ecdysteroids for control of ticks (Solomon et al., 1982).

8.3.2.1.1. Scorpiones

El Bakary et al. (1987) presented strong evidence that ecdysteroids are the MHs in scorpions. These investigators extracted for ecdysteroids in eggs, embryos, three immature instars, and adult hemolymph of the scorpion *Leiurus quinquestriatus* and tested the extracts with an RIA for ecdysteroids. They extracted hemolymph and homogenates of eggs and immature instars with methanol, evaporated the supernatant, and dissolved the sediment in a phosphate buffer for direct measurement in the RIA or in a methanol : water system for separation of ecdysteroids by HPLC. One-milliliter HPLC fractions were collected, evaporated, dissolved in phosphate buffer, and tested by RIA. During the first three instars, RIA-reactive material that had the same retention time as 20-OHE in their HPLC system represented a major ecdysteroid component, whereas RIA-reactive material with the retention time of EC represented a minor component. Together, the ecdysteroids from the EC- and 20-OHE-containing fractions constituted a majority of the ecdysteroids detected by RIA in most of their extracts from the early instars. Although adult scorpions do not molt, the hemolymph contained RIA-detectable ecdysteroids at a mean concentration of 1.51 ng/g 20-OHE equivalents (El Bakary et al., 1987).

Adult scorpions can also metabolize EC and 20-OHE to polar and apolar products (Connat and Diehl, 1986). They injected the tritiated hormones into two species of scorpions, *Androctonus australis* and *Buthus occitanus*, and after 88 h extracted whole bodies for ecdysteroids with methanol. Extracts were purified by HPLC, and eluted fractions were monitored for radioactivity by liquid scintillation counting. Both species converted the hormones into polar and apolar metabolites. Apolar metabolites contained about 70% of the injected radioactivity in *B. occitanus* and 40–50% in *A. australis*. Treatment of these products with hog liver esterase liberated the original free hormone and polar compounds. Apolar conjugates of arthropod MHs may represent a storage form of the hormones for future embryos or hormone inactivation products. In the adult scorpion, only a small amount of the injected labeled ecdysone was converted to a labeled compound that comigrated in their HPLC system with authentic 20-OHE (Connat and Diehl, 1986).

8.3.2.1.2. Araneae

Evidence from injection. Spiders (*Araneus*) and tarantulas (*Dugesiella*) respond to exogenous ecdysteroids (Krishnakumaran and Schneiderman, 1968, 1970). Spiders injected with about 130 μg/g 20-OHE spun a molting pad instead of the usual webs, underwent apolysis, and synthesized a new cuticle during the 29-day experimental period (62% vs. 13% in saline-injected controls), but died before shedding the new cuticle. All controls that molted did so normally and survived. Tarantulas injected with about 20 μg/g of 20-OHE, and 2 months later with a second dose of 50–100 μg/g, were also stimulated to molt, one of four normally. The others died before shedding, and synthesized a new cuticle that lacked the spines and bristles of a normal cuticle.

The most extensive and insightful work has been performed by Bonaric on the spider *Pisaura mirabilis.* Injections of 20-OHE ''activates the molting mechanism,'' but the end result depends on concentration of the hormone and time of injection (Bonaric, 1976; see also Chapter 3 by Bonaric and Juberthie herein). The response was abnormal when a high dose of 100 μg/g was injected into seventh- to ninth-instar nymphs at the end of intermolt or beginning of premolt. They either synthesized a new cuticle without normal shedding of the old one or died at the beginning of new cuticle formation. Injections of 10 μg/g at the beginning of the intermolt period provoked a slightly accelerated but normal molt, whereas doses of 50–200 μg/g decreased the postexperimental time for a definite molting response by more than 50% compared to controls injected only with solvent (Bonaric, 1977a). However, the percentage of normal ecdyses decreased from 90% at doses of 50 μg/g to 7.7% at 200 μg/g. A dose of 50 μg/g injected at any time during the intermolt period of seventh- to ninth-instar nymphs (from about days 4–16 in an average molt–molt cycle of 23 days) decreased the postinjection time to a molt in *P. mirabilis* (Bonaric, 1977b). It is clear from this work that 20-OHE can stimulate the normal molting phenomena in spiders (Fig. 8.2).

Evidence from extraction. Free edysteroids have been extracted from the spider *P. mirabilis* (Bonaric and DeReggi, 1977; Bonaric, 1980). Methanol homogenates of whole eighth-instar nymphs were centrifuged and chromatographed on silica gel TLC. Sections of the gels

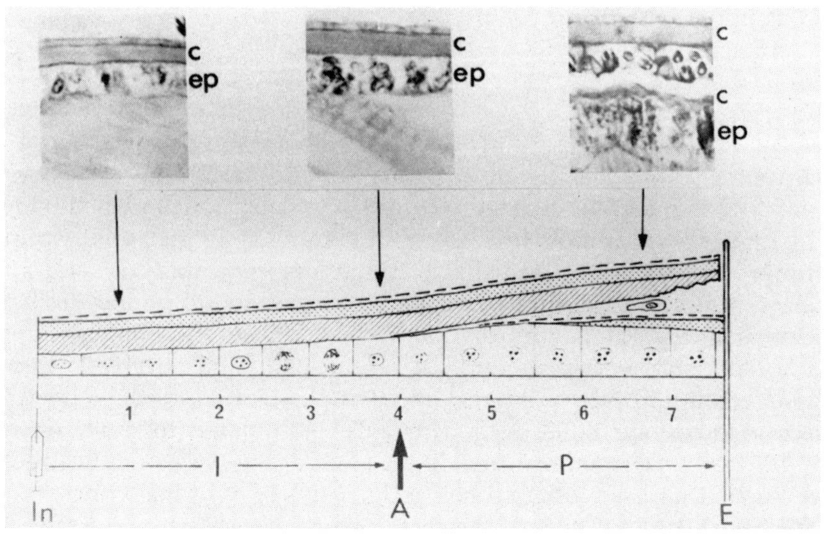

FIGURE 8.2. Normal molting responses provoked by an effective dose of 20-hydroxyecdysone injected into immature spiders *Pisaura mirabilis*. Times of injection (In), apolysis (A), and ecdysis (E) are shown, along with the duration (in days) of the intermolt (I) and premolt (P) periods and the times when photomicrographs of prepared tissues were taken. Epidermis (Ep) and cuticles (C) can be seen in the drawings and photographs. (Modified from Bonaric, 1980.)

co-chromatographing with pure standards were eluted with methanol, sonicated, and tested by an ecdysteroid RIA. The only RIA-reactive material they detected by these methods co-chromatographed with 20-OHE. Tests of methanolic extracts of whole eighth-instar nymphs homogenized at different times during the molt cycle by RIA, after evaporation and suspension of the sediment in a citrate buffer showed a pattern (Bonaric, 1980) that, in general features, has been demonstrated in many other species of arthropods.

A mean value of 75 ng/g was measured at the time of molting to the eighth instar. Although elevated somewhat during the 4-day postmolt, free ecdysteroid levels plateaued at a mean of 48 ng/g during the 12-day intermolt period. With the onset of premolt, ecdysteroid titers increased rapidly to a peak of 372 ng/g within 2 days, then decreased rapidly to low values of 25–50 ng/g shortly before molting to the next instar, about 22 days after the previous molt (Bonaric, 1980).

8.3.2.1.3.   Opiliones

Ecdysteroids were detected by RIA in organ culture extracts after incubation of femoral leg segments (containing the large oenocytes), ovaries, testes, and opisthosomal tergites of the harvestman *Opilio ravennae* (Romer and Gnatzy, 1981). All of these structures apparently synthesized and secreted EC and 20-OHE [identified by GC (gas chromatographic) detection of the silylated derivatives (see Gnatzy and Romer, 1980)] into the culture medium. This is the first report of evidence for secretion of molting hormones by specific tissues or organs in arachnids. Organs from *O. ravennae* cultured in early autumn synthesized large combined amounts of EC and 20-OHE (a high of 179 ng in 16 h by femoral oenocytes) compared to the same organ from animals in winter dormancy (1.5–3.2 ng in 16 h). Some similar results were also obtained from three other species of harvestmen by Romer and Gnatzy (1981).

8.3.2.1.4.   Acarina

More work on evidence for ecdysteroids as MHs has been performed on ticks than in any other group of arachnids. The evidence is extensive for stimulation of molting by ecdysteroids and for the occurrence of these compounds in tick extracts.

Evidence from feeding, injections, and topical applications. Although adult ticks are not known to molt in nature, adults of the soft tick *Ornithodoros porcinus* (Family Argasidae) were induced to supermolt (many of them apparently ecdysed normally) one or more times after feeding on blood that contained 1–5 µg/ml of 20-OHE or PNA (Mango et al., 1976). Females were more responsive than males, and at least 70% of the females molted at 20-OHE concentrations of 2–5 µg/ml. Interestingly, more than 50% of the females fed doses of 4 or 5 µg/ml molted a second time after the one feeding. Considerable mortality was produced by the two compounds but was greatest after PNA ingestion. Campbell and Oliver (1984) induced supernumerary molting in *O. parkeri* females at ingested doses of 20-OHE at 6 and 9 µg/ml of blood, although only one of six was a completely normal molt. When third-instar nymphs were fed hormone at 1.5–6.0 µg/ml of blood there was no significant difference from controls in the number of nymphs that molted or in the time from treatment to molting, but the 20-OHE treatment stimulated a second molt in some of the experimental nymphs. However, the same compound did significantly shorten time to the next molt in third and fourth, but not second, instars of *O. porcinus* (Mango, 1978). The time for 50% of a batch of 50 animals to molt was reduced 33% by 20-OHE doses of 4 and 8 μg/ml of blood in the four instar. In *O. moubata* 20 μg/ml of blood of 20-OHE was necessary to provoke supernumary molts, but only 15–20 ng of an EC analogue, 22,25-dideoxyecdysone, was sufficient for a 100% effect (Connat et al., 1983). Ingested ecdysteroids with a 22-OH group were rapidly converted to apolar esterase-labile metabolites, but this EC analogue was not; hence this difference could account for the much greater sensitivity of the ticks to 22,25-dideoxyecdysone (Connat et al., 1986).

Topical application of 20-OHE has been shown to stimulate molting in immature ticks in at least one study. In the hard tick (Family Ixodidae) *Hyalomma dromedarii*, doses of 10–20 μg significantly shortened the time to the next molt in all groups of nymphs tested and doses of 1 and 5 μg were effective in feeding nymphs (Khalil et al., 1984).

Injected 20-OHE at doses of 1–100 ng per tick stimulated apolysis and new cuticle formation in adult females of *O. parkeri* in a dose-dependent manner; a dose–response relationship was obscured at higher doses, because of the high mortality and small number of animals used (Pound et al., 1984). Nymphs of the hard ticks *H. dromedarii* and *Dermacentor variabilis* did not show an accelerated molt after injection of 10 ng of 20-OHE (Dees et al., 1984b), but it is clear from evidence presented above or available elsewhere that ecdysteroids stimulate a molting response in hard as well as soft ticks.

Evidence from extraction and metabolism of radioactive ecdysteroids. Conclusive evidence on the presence of EC and 20-OHE in fed nymphs of the hard tick *Amblyomma hebraeum* was presented by Delbecque et al. (1978) and in fed fifth-instar nymphs of the soft tick *O. moubata* by Germond et al. (1982). The two families of ticks, Ixodidae (hard) and Argasidae (soft), are biologically distinct in terms of feeding behavior and number and length of immature stages required to reach sexual maturity, but it is clear from these two studies that the MH system is very similar in the two groups.

Similar methods were used for extraction of whole bodies of *A. hebraeum* and hemolymph and whole bodies of *O. moubata*, although large numbers were used for the former species (over 1000) whereas only a few individuals were used for the latter. Whole ticks were homogenized in aqueous methanol, and hemolymph in methanol. After centrifugation, pellets were reextracted and the combined supernatants were kept overnight in a deep-freezer, then centrifuged to remove precipitated lipids. The extracts were further purified by adsorption chromatography in a column of silicic acid, and the ecdysteroids were eluted with methanol.

Eluates of *A. hebraeum* were concentrated and further purified by silica gel TLC. After delipidation with diisopropyl ether, ecdysteroids were separated in a chloroform:methanol system and eluted from the gel in the bands comigrating with EC and 20-OHE. Extracts of *O. moubata* were delipified on the silicic acid column by chloroform and isoamyl acetate treatment, and the ecdysteroid-containing fractions (assayed by RIA) were further purified by reversed-phase HPLC. More than 80% of the RIA activity (85.6% in hemolymph) was found in the fractions co-eluting with EC and 20-OHE. The latter was by far the most abundant of the two MHs, comprising about 86% of the combined RIA activity in both hemolymph and whole body extracts of *O. moubata* (Germond et al., 1982). The TLC and HPLC fractions were silylated with *N*-trimethylsilylimidazole (TMIS), and the TMIS derivatives were extracted with hexane. The *A. hebraeum* derivatives were further purified by TLC (Delbecque et al., 1978). Derivatized ecdysteroids in the samples were separated by gas chromatography with an electron-capture detector (GC/ECD) and subjected to mass spectrometry (GC/MS) or mass fragmentography (GC/MF). Proof was obtained for EC and 20-OHE and in the TLC or HPLC fractions that comigrated with authentic standards. Derivatized ecdysteroids had the same retention times on the GC column as the fully silylated standards. The GC peak from *A. hebraeum* extract that co-chromatographed with silylated authentic 20-OHE showed a mass spectrum identical to the spectrum of the derivatized authentic compound. Further proof came from identification of characteristic ions of derivatized EC and 20-OHE in *A. hebraeum* extracts and of derivatized 20-OHE in *O. moubata* extracts. These evidences strongly support the conclusions by Delbecque et al. (1978) that EC and 20-OHE occur in nymphs of *A. hebraeum* and of Germond et al. (1982) that the major MH of *O. moubata* is 20-OHE.

MHs have been identified in several other species of ticks, and the following is only a selective account. Ellis and Obenchain (1984) found RIA-reactive ecdysteroids in replete nymphs of *A. variegatum* after methanol extraction of whole bodies. Dees et al. (1984a) found ecdysteroids in embryos, larvae, nymphs, and adults of *D. variabilis* after extensive extraction with methanol and methanol:benzene systems, purification by reversed-phase HPLC, and detection of eluted fractions by RIA. Material co-eluting with 20-OHE was absent in unfed larvae and nymphs but present throughout embryonic development in engorged larvae and nymphs and in unfed adults. It occurred in highest concentration during the latter two stages. The same analytical methods were used to detect 20-OHE in unfed and fed adults (male and female) of *H. dromedarii* (Dees et al., 1985). EC was the major free ecdysteroid extracted from adult females of *Boophilus microplus* just prior to engorgement, while 20-OHE occurred at much lower levels (Wigglesworth et al., 1985). They extracted whole ticks with a methanol : water system, followed by a methanol and then methanol : dichloromethane system. After apolar materials were removed with hexane, the extract was defatted with chloroform on a silicic acid column, and free ecdysteroids were eluted with 30% methanol in chloroform. Further purification was done on a reversed-phase SEP-PAK cartridge; then ecdysteroids were separated by HPLC and quantified by RIA. Eggs of this species also contain free ecdysteroids, although in comparatively low quantities. Most of the ecdysteroid content of eggs occurs as apolar metabolites of EC (Wigglesworth et al., 1985).

A blood meal activates the MH system in immature ticks, and molting occurs a set time later, as determined by a given set of environmental conditions. Ecdysteroid levels change during the molt cycle, and there is a pattern of ecdysteroids in hemolymph and whole-body extracts that correlates well with specific molt-related events in the cycle. The cycle is 9–10 days in last-stage nymphs (fifth) of *O. moubata* cultured at 27°C, with 30–40% relative humidity, and in darkness (Germond et al., 1982). RIA-reactive ecdysteroids extracted from hemolymph were at basal values of about 20 pg/μl 20-OHE equivalents during the first 3 days, rose gradually for 1½ days, then rapidly, and reached a peak (nearly 500 pg/μl 20-OHE equivalents) on about day 5. Most cell division in the epidermal layer occurred during days 2–4; apolysis occurred on day 4 when ecdysteroid levels rose rapidly; and epidermal cell volume reached maximal size when peak levels were first measured. Hemolymph levels remained high during days 5 and 6, the time of new epicuticle deposition; then digestion of nymphal cuticle began. Ecdysteroid levels began to decline, and deposition of adult procuticle commenced. On day 8, shortly before ecdysis, low values of 10–30 pg/μl 20-OHE equivalents were reached. Levels of ecdysteroids in whole-nymph extracts followed essentially the same pattern (Germond et al., 1982). Although the last immature stage has a greater duration in the hard tick *A. hebraeum* (31–34 days under similar culture conditions), Diehl et al. (1982) found a pattern of ecdysteroid levels remarkably similar to the pattern in the soft tick *O. moubata*. Correlations of specific portions of the pattern to specific molt-related events are nearly identical in the two species. The pattern of RIA-reactive ecdysteroids in whole nymphal extracts of the hard tick *A. variegatum* (Ellis and Obenchain, 1984) is also remarkably similar to the pattern found in the soft tick.

Pathways for synthesis of MHs from cholesterol and metabolism to primarily apolar products occur in ticks. After injection of radioactive cholesterol into nymphs of *Hyalomma dromedarii*, radioactive 20-OHE was detected in HPLC-purified extracts of female and male ticks, although conversion occurred at a very low level (Sonenshine et al., 1985). Further work confirmed the conversion of cholesterol to ecdysteroids, particularly 20-OHE in this species (Sonenshine et al., 1986). The metabolism of labeled EC to other ecdysteroids and principally apolar ecdysteroid conjugates has been studied primarily in adults but also occurs in nymphs. Metabolism of labeled EC has been observed in soft ticks [e.g., *O. moubata* nymphs (Bouvier et al., 1982) and adult females (Connat et al., 1986); *O. parkeri* adult females (Connat and Diehl, 1986)], as well as in hard ticks [e.g., *A. hebraeum* nymphs and adult females (Connat and Diehl, 1986); *B. microplus* adult females (Wigglesworth et al., 1985); *H. dromedarii* adult females (Sonenshine et al., 1986)]. Although, taken together, all of the experiments on ticks discussed in this section provide stong evidence for the role of ecdysteroids as MHs in these animals, experiments on identification of specific ecdysteroid cell receptors and induction of molt-specific genes by ecdysteroids have yet to be performed.

8.3.3. *Crustacea*

8.3.3.1. BRANCHIOPODA

Apparently the effects of exogenous MHs have not been tested in these animals, but K.-D. Spindler and Radi (unpublished results) detected EC and 20-OHE by RIA in dormant cysts and nauplii of the brine shrimp *Artemia salina* after extraction and purification by GC and HPLC (Spindler et al., 1980). Walgraeve et al. (1986) found RIA-reactive ecdysteroids in whole-body extracts and hemolymph of adult female *Artemia* (a Great Salt Lake strain). Animals at specific times during their vitellogenic cycle (and at the same time during their molt cycle) were homogenized in methanol : acetone, and the extracts were left at 4°C overnight and centrifuged. Supernatants were evaporated, and the residue was partitioned between *n*-hexane and water. The aqueous layer was injected onto a reversed-phase SEP-PAK cartridge, and ecdysteroids were eluted with methanol after a water rinse. After evaporation, the residue was suspended in a phosphate-buffered saline and quantified by RIA. Lowest levels of ecdysteroids (EC equivalents) occurred about a day after a molt; later, levels were elevated to a peak 3–4 days before the next molt and then declined rapidly to near-minimal levels that were measured shortly before molting (Walgraeve et al., 1986).

In a later paper (Walgraeve et al., 1988), an attempt was made to correlate concentrations of HPLC-purified EC, 20-OHE, and highly polar immunoreactive substances from whole-body extracts of both male and female *Artemia* at specific molt cycle stages. The staging was based on development of the setae. Apparently no significant differences of EC and 20-OHE concentrations were observed, although small molt-cycle-related fluctuations occurred. In females, 20-OHE changes seemed to correlate more with the reproductive cycle than the molt cycle, although these two physiological cycles are tightly linked. Determination of hemolymph ecdysteroid concentrations at different molt cycle stages and evidence for induction of molting by exogenous ecdysteroids are needed to decide whether these compounds serve as molting hormones in branchiopod crustaceans.

8.3.3.2. MAXILLOPODA

Evidence from injection, immersion, and in vitro *culture.* Injections of doses of 20-OHE as low as 50 ng/g stimulated molting in adult barnacles of the species *Semibalanus balanoides* (Tighe-Ford and Vaile, 1972; Tighe-Ford, 1977). This species has six free-swimming nauplii stages, followed by the cypris stage, which settles on a substrate and metamorphoses to the adult. Immersion of first-day cyprids in sea water solutions of 20-OHE from 300 ng to 250 μg/ml induced new cuticle formation and precocious, although abnormal, metamorphosis (Cheung, 1974). Successively larger doses caused correspondingly faster responses. Doses of 8 μg/ml or larger affected 80–100% of the experimental animals, while no premature metamorphosis was observed in any of the controls during the experimental period. Freeman and Costlow (1983) discovered that premolt in the *Balanus amphitrite* cyprid is biphasic, and apolysis does not occur in the dorsal carapace until the cypris attaches to a substrate. It is this dorsal area that is stimulated by solutions of 20-OHE; dorsal carapace apolysis, epidermal morphogenesis, and thoracic adductor muscle degeneration occurred after immersion in test solutions. Isolated appendages of *B. eburneus* (Cheung and Nigrelli, 1974) and mantle tissue of *B. amphitrite* (Freeman and Costlow, 1979) cultured in solutions of 20-OHE showed molting responses to the hormone. Inner mantle tissue explants taken from adult barnacles during intermolt already responded to the treatments by accelerated apolysis 8 h after immersion. Apolysis in nearly 100% of the experimentals occurred within 24 h at doses of 5–20 μg/ml, compared with less than 20% of the controls (Fig. 8.3). Normal enlargement of the epidermal cells occurred in these experiments; and under light microscopic observation, the cultured tissues that responded to hormone treatment appeared the same as tissues observed from natural premolt barnacles (Freeman and Costlow, 1979).

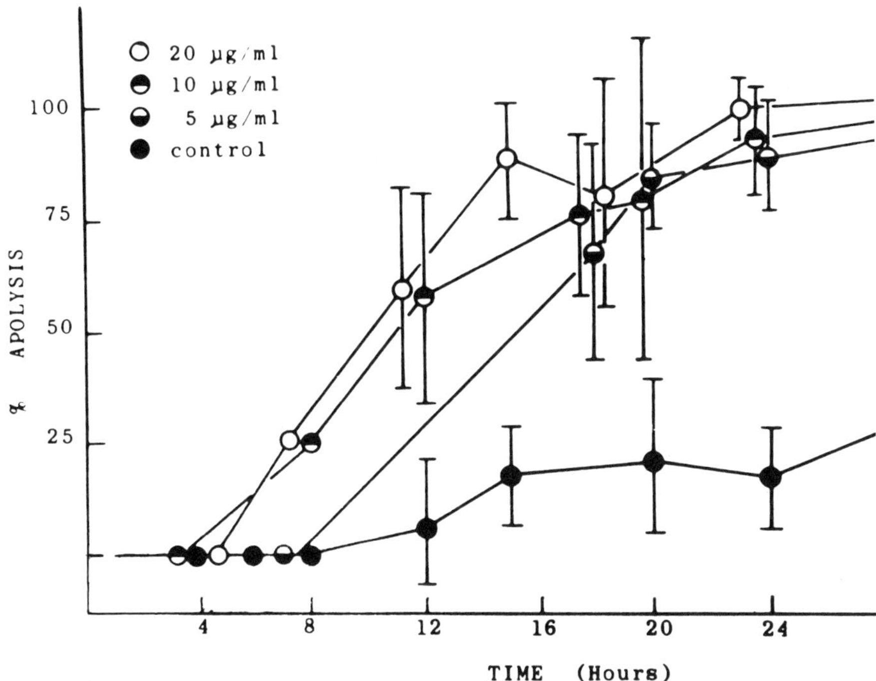

FIGURE 8.3 Effects of 20-hydroxyecdysone on apolysis in isolated mantle tissue explants (taken during intermolt) of the barnacle *Balanus amphitrite*. Each point means of 1–10 explants and SD (standard deviation; for 5–10 explants). (Modified from Freeman and Costlow, 1979.)

Evidence from extraction. Ecdysone (for the first time in a crustacean) and 20-OHE were extracted from adult *S. balanoides* by Bebbington and Morgan (1977) and Bebbington et al. (1977). They used basically the techniques that Galbraith et al. (1968) used on spiny lobsters to extract ecdysteroids from barnacles in mass. After initial solvent extractions of 1100 kg of barnacles, extracts were chromatographed on a silica gel column and 20-OHE-containing fractions were treated successively on two columns of hydrophobic celite and eluted by *n*-butanol:cyclohexane systems. Each column fraction during the purification scheme was treated with TMSI, and the TMSI ecdysteroid derivatives were detected and quantities estimated by GC. The identity of EC and 20-OHE was confirmed by comparing retention times of specific TMSI derivatives of the extracted compounds to the same derivatives of the authentic compounds (Bebbington and Morgan, 1977). The dominant compound was 20-OHE (40 ng/g vs. 230 pg/g wet weight of live tissue).

8.3.3.3. MALACOSTRACA

There are more than 42,000 described species of crustaceans, and about 75% of them are malacostracans (Barnes, 1987). Most of the larger crustaceans belong to the class Malacostraca, and hence most of the work on MHs in crustaceans has been done on this group. Most of the species used come from three subgroups (orders): decapods, isopods, and amphipods. The literature is large and somewhat complicated, and for ease of interpretation and presentation will be discussed separately for each of these orders. Various aspects of the evidence for ecdysteroids as molting hormones in crustaceans has been reviewed by Willig (1974), Vernet (1976), Vernet et al. (1978), Kleinholz and Keller (1979), Spindler et al. (1980), and Skinner (1985).

8.3.3.3.1. Decapoda

This group includes the familiar shrimps, crabs, crayfish, and lobsters. Evidence for the occurrence of ecdysteroids in decapods and their use as MHs in these animals will be reviewed selectively from several different types of investigations: effects of exogenous ecdysteroids; extraction of ecdysteroids; measurement of titer changes during the molt cycle; metabolic conversion mechanisms and products; Y-organ biosynthesis of ecdysteroids; uptake of ecdysteroids by specific tissues; and isolation of ecdysteroid receptors. Crustacean Y-organs, paired epidermal derivatives, are generally believed to be the sole, or at least primary, source of MHs (see Skinner, 1985). Proof for their role as MH-producing glands was provided for the crabs *Carcinus maenas* (by Echalier, 1959) and *Sesarma reticulatum* (by Passano and Jyssum, 1963) and for the crayfish *Orconectes limosus* (by Willig and Keller, 1976). Stimulation of precocious premolt events and molting by bilateral eyestalk removal has been observed in numerous species of crustaceans since the original observation in fiddler crabs by Zeleny (1905). When eyestalks were removed from Y-organ-less crabs, premolt was blocked in *Sesarma;* but when Y-organs were implanted, the animals (both *Sesarma* and *Carcinus*) entered premolt and duration of premolt was considerably shortened compared with the controls (Passano and Jyssum, 1963). Blockage of normal premolt activities by bilateral Y-organ ablation has been observed in many species. Although it has been commonly stated that only one cell type exists in the Y-organ (Skinner, 1985), some cells presumably are involved in synthesizing the cuticular covering, and it is not known whether two major physiological types exist—those synthesizing cuticle and those synthesizing hormone (Spindler et al., 1980). Buchholz and Adelung (1980) found two major cell types in the crab *Hemigrapsus nudus* and demonstrated important characteristics of mammalian steroid-secreting cells, namely, numerous mitochondria of tubular type and abundant smooth endoplasmic reticulum. In the crayfish *Astacus astacus*, the steroid-secreting ultrastructure of the Y-organ cells was most conspicuous during premolt, and intermolt cells contained only a narrow zone of cytoplasm around the nuclei (Birkenbeil and Gersch, 1979).

Evidence from exogenous ecdysteroid application. Ecdysteroids are known to accelerate the onset of premolt in more than two dozen species of decapod crustaceans (see also Vernet, 1976, and Skinner, 1985, for a more complete listing). Unlike insects and arachnids, many crustaceans continue to molt as adults, and in many of the studies adult animals were used. Shrimps respond to injected 20-OHE. Doses in the range of 5–500 ng per animal accelerated the onset of premolt in a dose–response relationship in *Palaemonetes pugio* (Freeman and Bartell, 1976). A majority (60%) of shrimp injected with a dose of 50 ng per animal showed an accelerated successful molt. Clawed lobsters (Infraorder Astacidea) also respond to injected ecdysteroids. Fourth-stage juvenile *Homarus americanus* responded to a dose of 6 μg 20-OHE per animal with an accelerated premolt and to 2 to 4 μg EC per animal with precocious premolt in most of the animals, and successful molting in about 30% of them (Rao et al., 1973). Accelerated premolt (apolysis and new cuticle formation) was also observed in adult male lobster after injected 20-OHE (Gilgan and Burns, 1976; Gilgan and Farquharson, 1977), although the animals died in late premolt. Under certain experimental regimens, injections of EC (0.5–2.0 μg/g) followed by 20-OHE (1 μg/g) or vice versa at different times produced a few accelerated but normal molts in adult males (Gilgan et al., 1977).

Apolysis, formation of gastroliths (calcium carbonate concretions), and new cuticle formation were stimulated by various pure ecdysteroids in crayfish *Procambarus* sp. (Krishnakumaran and Schneiderman, 1968, 1970). Injected doses of 20-OHE from 3 to 20 μg/g stimulated precocious but unsuccessful molts in all experimental animals. Digestion of the old cuticle occurred (most pronounced with the lowest dose) and the epidermal cells increased in size, although the new cuticle they deposited was abnormal in all except those receiving 3 μg/g. Based on gastrolith size after hormone treatment of crayfish *Orconectes virilis*, McWhinnie et al. (1972) suggested that the hormone dose and physiological state of the animal are both important in determining progress from premolt initiation to molting in ecdysteroid-injected animals. After all, eyestalk ablation stimulates premolt intiation at all times of the year in this species of crayfish. In a careful study of injection of about 2 μg/g of 20-OHE in immature males of *O. obscurus* during all four seasons, precocious molting was induced in all the experimental animals of summer, autumn, and winter, although only the summer animals molted successfully without dying (Warner and Stevenson, 1972). During the spring experiment, half the animals showed a normal but unaccelerated molt, while the other half died during ecdysis of an accelerated molt. Dose and time of injection during the molt cycle is probably important in all species of arthropods, if one is to simulate normal molting by exogenous ecdysteroids, as demonstrated here in crayfish and in horseshoe crab larvae by Jegla and Costlow (1979b).

Spiny lobsters (Infraorder Palinura), the crustacean type from which pure ecdysteroids were first extracted (Hampshire and Horn, 1966), also respond to injected 20-OHE (Dall and Barclay, 1977). Doses of 2 μg/g injected into early juvenile *Panulirus longipes* after about 70% of the intermolt had elapsed maximally shortened premolt and time to ecdysis, all of which were viable and apparently normal in all respects.

More than half a dozen species of brachyuran crabs have been tested and shown to respond to exogenous ecdysteroids. Two injections of 10 $\mu g/g$ 20-OHE some 24 h apart in *Carcinus maenas* induced an accelerated molt in 66% of the crabs, more than half being successful and normal (Bazin, 1977). Both inokosterone and 20-OHE at 33 $\mu g/g$ (in two or five separate smaller doses) induced precocious premolt and ecdysis, although none survived the molt, in the fiddler crab *Uca pugilator* (Rao, 1978). EC in a similar dose range is also an effective molt stimulator in this species (Rao et al., 1972). When these compounds were injected into crabs that had regenerating limb buds in the basal limb growth period, the exogenous ecdysteroids induced accelerated basal limb growth and promoted premolt limb bud growth. Thus, based on yet another criterion (regenerating limb buds), injected ecdysteroids were demonstrated to stimulate a normal premolt event in crustaceans. These compounds also stimulate molt-related events in brachyuran crab larvae. Immersion of various zoeal stages of *Rithropanopeus harrisii* in 1- and 5-$\mu g/g$ solutions of 20-OHE shortened the time to apolysis in a dose-dependent manner (McConaugha and Costlow, 1981), although only a few individuals actually survived the subsequent molt. McConaugha and Costlow showed with a radioactive tracer compound that the larvae do absorb ecdysteroids from an aqueous solution. A direct effect of 20-OHE on a target tissue, epidermis, was demonstrated in cultured dorsal spine explants of *R. harrisii* zoea (Freeman and Costlow, 1984). Accelerated apolysis in fourth-instar larval explants, in a dose-dependent manner, was observed with hormone concentrations of 0.1–10.0 $\mu g/g$.

Evidence from extraction in a variety of decapod species. Ecdysteroids have been extracted in many species, the majority for determination of titer changes during the molt cycle. The few studies concerned with absolute chemical identification will be reviewed here. Isolation and purification of 20-OHE from 1000 kg of the spiny lobster *Jasus lalandei* by Horn et al. (1968) was described above in Section 8.2.; its specific identity, different from that of EC, was determined on the bases of its migration in silica gel TLC, polarity in solvent partition systems, mass spectrum, infrared (IR) and ultraviolet (UV) spectra, and NMR (Hampshire and Horn, 1966; Horn, 1971). The same compound was extracted from silkworms, and it's structure confirmed (Hocks and Weichert, 1966). A second ecdysteroid, 2-deoxy-20-OHE, was isolated in large enough quantity (200 μg) from 3000 kg of *J. lalandei* for chemical characterization by the same methods used for identifying 20-OHE in this species (Galbraith et al., 1968).

Evidence for the chemical identity of 20-OHE in extracts of clawed lobster (*H. americanus*) was obtained by Gagosian et al. (1974) and

Gagosian and Bourbonniere (1976). After extensive organic solvent and water extraction, the extracts were purified by chromatography on three separate silicic acid columns. Material that eluted with retention times of authentic 20-OHE was treated with TMSI and the penta- and hexa-TMSI derivatives were identified by GC. Final proof for these derivatives was obtained from GC/MS. Mass spectra of TMSI derivatives from lobster extract were identical to spectra obtained from corresponding derivatives of authentic 20-OHE (Gagosian and Bourbonniere, 1976). Three ecdysteroids (20-OHE, inokosterone and makisterone A or its stereoisomer) were extracted and unequivocally identified in female brachyuran crabs of the species *Callinectes sapidus* (Faux et al., 1969). They used the same methods that were used to identify ecdysteroids in *Jasus*. Chang et al. (1976) identified 20-OHE and EC (ratio 8–12:1) in hemolymph extracts of *Pachygrapsus crassipes*; after they removed serum proteins, hemolymph was extracted with organic solvent and water systems and extracts were purified successively by TLC and two different reversed-phase HPLC columns. The ecdysteroids were treated with TMSI and the derivatives identified by GC. They were further characterized by their competitive binding behavior in two different RIA systems. Basically, the same procedures were used to identify EC as the only ecdysteroid secreted by the Y-organ of *P. crassipes* (Chang and O'Connor, 1977). But, in this paper, they provided additional proof for specific identity by showing that the mass spectrum of the extracted compound was identical to the spectrum for authentic EC. Large amounts of EC, 20-OHE, and PNA were identified in mature ovaries of *Carcinus maenas* by Lachaise et al. (1981), and this group also identified several deoxyecdysteroids that are presumably MH precursors in the ovary of this crab. Identifications were made by comparison of the mass spectrum of derivatized extracted compounds that were analyzed on a coupled GC/MS system to spectra of authentic ecdysteroids. PNA was also identified in developing embryos of the blue crab *Callinectes sapidus* and in hemolymph of the land crab *Gecarcinus lateralis* by McCarthy (1979), who based his identifications on UV spectral analysis, competitive binding in two RIA systems, and mass spectral analysis.

Ecdysteroid concentration during the molt cycle. These studies generally were not done with extensive purification and rigorous chemical identification as their object. Rather, they were done to determine if the presumed MHs show concentration fluctuations that correlate with phases or stages of the molt cycle (intermolt, premolt, postmolt, ecdysis). The same general pattern has been observed repeatedly in many species. In the shrimp *Palaemon serratus*, ecdysteroids purified by HPLC and detected by RIA from whole-animal extracts and a specific organ (integument) are relatively low in concentration during intermolt, but rise at first slowly, then rapidly during premolt to peak levels near the end of this period, and decline to low levels around the time of ecdysis (Baldaia et al., 1984). Although EC and 20-OHE apparently occurred during all molt cycle stages, 20-OHE was the predominant form, particularly during late premolt. Willig and Keller (1973) correlated rising titers of ecdysteroids in whole-body extracts with specific premolt events in the crayfish *Orconectes limosus*. Ecdysteroid levels are very low during intermolt in this species, rising slowly during early premolt when calcium is removed from the cuticle and deposited as gastroliths in the lining of the stomach wall. The rate of gastrolith growth closely parallels the rate of rise of ecdysteroid concentrations. New cuticle formation occurs during the period of premolt, when ecdysteroid levels are increasing steeply. This same pattern of concentration change and correlation with the same premolt events was observed for hemolymph ecdysteroids (Fig. 8.4) of this

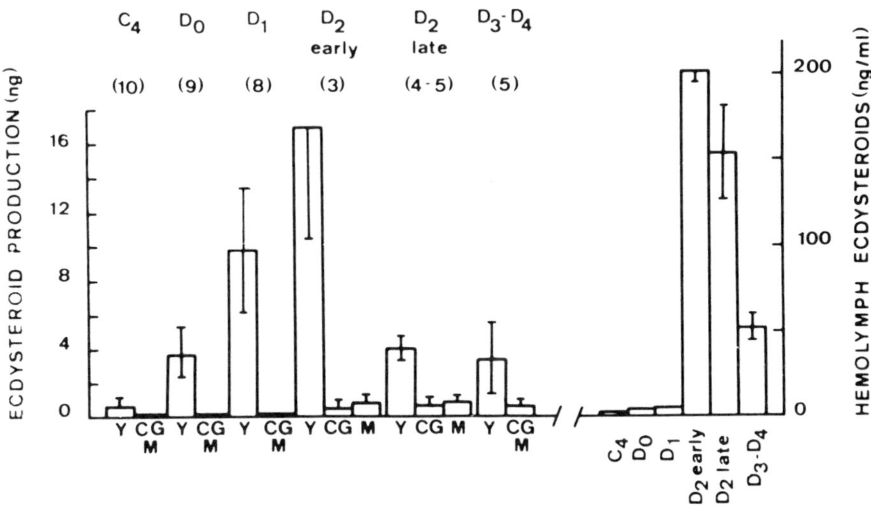

FIGURE 8.4. Hemolymph ecdysteroid concentrations (right-hand ordinate) and production of ecdysteroids *in vitro* (left-hand ordinate) by a pair of Y-organs (Y), a cephalic gland (CG), and a piece of muscle (M) as a function of premolt stage (D_0–D_4). Intermolt values are given for comparison. Values are means and SEM. Number of determinations ( ) are the same for the left and right part of the figure. (From Jegla et al., 1983.)

species (Jegla et al., 1983). A similar curve of titer changes was found in hemolymph ecdysteroids of a population of the crayfish *O. sanborni*, but there is a fair amount of variation around mean values (Stevenson et al., 1979). They suggested that individual crayfish may produce hormone intermittently and therefore control its own preparation for ecydsis separately from other individuals in the same population. Those individuals that had relatively high concentrations of hemolymph ecdysteroids at a specific point during premolt proceeded to the next stage of premolt more rapidly than did individuals with relatively low concentrations. During the first three larval molt periods of the lobster *H. americanus*, when intervening intermolt periods are very short or absent, a peak of ecdysteroid concentration was reached halfway through the period, and a low at the time of molting (Chang and Bruce, 1981). All the RIA-detectable ecdysteroid during peak concentrations co-chromatographed with 20-OHE.

Brachyuran crabs show the same pattern of molt cycle ecdysteroid levels as species already discussed. For example, *Carcinus maenas* (Andrieux et al., 1976), *Gecarcinus lateralis* (McCarthy and Skinner, 1977), *Callinectes sapidus* (Soumoff and Skinner, 1983), and *Uca pugilator* (Hopkins, 1983, 1986). Interestingly, the pattern of circulating ecdysteroids in *C. maenas* is reflected in a specific tissue, the Y-organ, although its concentrations were much lower (Andrieux et al., 1976). A terminal molt occurs in *Callinectes* and ecdysteroid concentrations in hemolymph of adult females remains low, but animals accumulate ecdysteroids in the ovary during vitellogenesis (Soumoff and Skinner, 1983). Ecdysteroids occur in the ovary of other species of arthropods, where they may regulate vitellogenesis or become incorporated in the egg for regulation of embryogenesis or embryonic molting. In *G. lateralis*, hemolymph ecdysteroid concentrations were closely correlated with premolt limb regeneration (McCarthy and Skinner, 1977). Initial (basal) regeneration occurred when hemolymph ecdysteroids were low, and rapid (premolt growth) size increase of the regenerate occurred during the linear increase of serum MHs. Although both EC and 20-OHE occurred in the hemolymph, 20-OHE was the major circulating ecdysteroid in *Gecarcinus*.

Ecdysteroid syntheses by the Y-organ and their correlation with hemolymph ecdysteroid levels and with stage of the molt cycle. Y-organ removal and replacement experiments demonstrated that it is necessary for molting to occur in the crabs *C. maenas* (Echalier, 1959) and *S. reticulatum* (Passano and Jyssum, 1963) and in the crayfish *O. limosus* (Burghause,

1975). In the crayfish, onset of premolt (gastrolith formation) was inhibited in most of the Y-organ-less animals (Keller and Willig, 1976), but there was no assurance that the Y-organ was completely removed in those that initiated premolt. Premolt was initiated, however, in 100% of Y-organ-less animals who were injected with sequential doses of 50 ng/g 20-OHE. It is therefore clear that the basis of Y-organ control of molting is its ability to secrete MH during critical stages of the molt cycle.

Secretion of ecdysteroids by Y-organs in isolated culture systems has been demonstrated for several decapod crustaceans. Ecdysone (identified by MS) is the only ecdysteroid secreted *in vitro* by Y-organs of *Pachygrapsus* and *Cancer antennarius* (Chang and O'Connor, 1977). Although the organs did not survive for more than a few days in their culture systems, one *Cancer* Y-organ secreted as much as 100 ng of EC during a 24-h period. *In vitro* secretion of RIA-detectable ecdysteroids by Y-organs was observed in *G. lateralis* (Soumoff and Skinner, 1980) and *O. limosus* (Keller and Schmid, 1979; Jegla et al., 1983). The Y-organ secretes large amounts of ecdysteroid but stores a correspondingly small amount; hence such substances were not detected in these organs (Karlson and Skinner, 1960) before the invention of an ecdysteroid RIA. Only 3% of the amount secreted *in vitro* during a 20-h incubation period was measured in Y-organs (Keller and Schmid, 1979). In the crayfish *O. immunis*, I have found that organ culture media of low pH inhibited the synthesis of Y-organ ecdysteroid, yet in both experimental animals and controls the organs contained only about 6% of the amount measured in the medium after a 2-h culture period (Table 8.2). Control Y-organs accumulated some ecdysteroid during the incubation, while ecdysteroid in experimental organs usually decreased, compared with their content at the beginning of the experiments.

Eyestalk ablation in many species of decapods induces premature premolt and a normal—although usually accelerated—pattern of hemolymph ecdysteroid concentrations. However, hemolymph ecdysteroids remained low in eyestalk-less Y-organectomized crabs *Pachygrapsus* (Chang et al., 1976) and *Eriocheir* (De Leersnyder et al., 1981) and the crayfish *O. limosus* (Jegla et al., 1983). No values higher than 8 ng/ml of hemolymph were recorded in Y-organ-less crayfish during the 40-day experiment, while sham-operated animals showed the normal pattern of premolt ecdysteroid concentrations and all molted in about 25 days (Fig. 8.5). Amounts of ecdysteroid secreted *in vitro* correlate closely with hemolymph concentrations in *Pachygrapsus* (Chang and

TABLE 8.2.　Effect of pH on *In Vitro* RIA-Reactive Ecdysteroid Content of Y-Organs of Early Premolt Crayfish *Orconectes immunis* and Organ Culture Medium (M199) After a 2-h Incubation Period

		Y-organ				Culture medium	
N	Initial	Control	pH	Exper.	pH	Control	Exper.
10	0	160	7.1	130	6.0	2.8 ± 0.8	1.8 ± 0.3
11	—	341	7.1	132	5.5	3.8 ± 0.4	2.3 ± 0.4
12	625	1,042	7.1	44	4.5	16.0 ± 2.7	1.0 ± 0.7

NOTE: One Y-organ of an animal served as experimental, the other as control. After incubation all Y-organs of a group were pooled and homogenized in distilled water. Mean values of duplicate samples in picograms per Y-organ are listed. Different control values merely represent interbatch variation. Values for hemolymph are means in nanograms ± SEM (standard error mean).

O'Connor, 1978) and *Orconectes* (Fig. 8.4). Y-organ secretion is low during intermolt, rises to maximal values in premolt during the time of new cuticle synthesis, and declines rapidly before ecdysis occurs.

Crayfish Y-organs synthesize EC from cholesterol (Willig and Keller, 1976). Labeled EC was extracted from Y-organs of animals previously injected with radiolabeled cholesterol. Y-organs of the crab *Hemigrapsus nudus* also took up radiolabeled cholesterol *in vivo* and synthesized radiolabeled ecdysteroids (Spaziani and Kater, 1973). In *Cancer antennarius*, activated Y-organs (eyestalks removed for 48 h) secreted labeled EC into organ culture medium after they had been incubated *in vivo* the previous 12 h with tritium-labeled cholesterol (Watson and Spaziani, 1985). Since EC appears to a secretory product of some decapod crustacean Y-organs and 20-OHE is the major ecdysteroid in the hemolymph, hydroxylation of EC must occur in tissues other than the Y-organ. High hydroxylase activity was demonstrated in the testis of *Pachygrapsus* and *Cancer* (Chang and O'Connor, 1978) and in the hepatopancreas of *Gecarcinus* (Soumoff and Skinner, 1982). The uptake of radiolabeled EC by cultured integument was demonstrated in the crayfish *Astacus leptodactylus* by Daig and Spindler (1980, 1983). An enrichment of about fourfold occurred compared with the culture medium. King and Siddall (1969) showed that crustaceans can convert EC to 20-OHE: "actively molting" *Crangon* converted 75% of the EC in 12 h; intermolt *Uca*, about 25%. Conversion to 20-OHE also occurs in *Carcinus* (Lachaise and Feyereisen, 1976), *Gecarcinus* (McCarthy, 1982), and *Orconectes* (Kuppert et al., 1978a). In *Carcinus*, the rate of hydroxylation is clearly a function of time during the molt cycle; it is highest during premolt stages compared to postmolt and intermolt (Lachaise et al., 1976). In addition, crayfish converted the MHs to polar and apolar metabolites in an organ-specific pattern. As in ticks, the MHs in crustaceans are also metabolized to esterase-labile apolar compounds (Connat and Diehl, 1986). These conjugates may represent inactivation products or a storage form of MHs.

A final proof for the role of ecdysteroids as MHs in crustaceans comes from the demonstration of ecdysteroid cell receptors. Hepatopancreas of premolt crayfish (*O. virilis*) have two cytosolic proteins that bind tritium-labeled EC (Gorell et al., 1972). Both cytoplasmic (Kuppert et al., 1978b) and nuclear receptors (Spindler-Barth et al., 1981) have been isolated from crayfish *O. limosus* integumental cells.

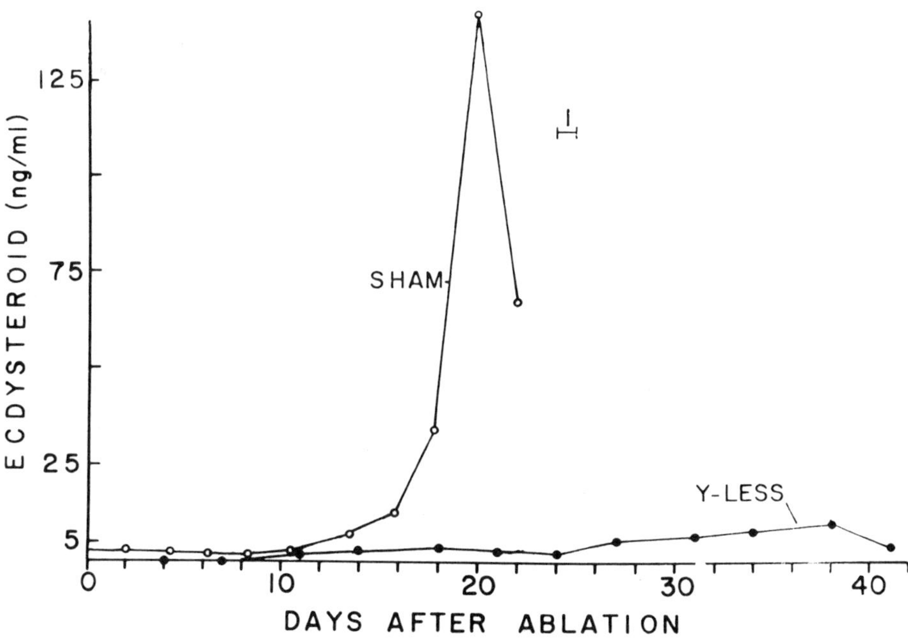

FIGURE 8.5. Hemolymph ecdysteroid concentrations in eyestalk-less, Y-organectomized crayfish *Orconectes limosus* and sham-operated controls. Points represent the mean value of 8-14 and 5 crayfish, respectively. I = time of molting for controls. No experimentals molted. (Modified from Jegla et al., 1983.)

Both protein molecules show several similar physical characteristics (Londershausen et al., 1982).

8.3.3.3.2. Isopoda

Evidence for ecdysteroids as MHs in several different morphological types of isopod crustaceans is convincing, although far fewer studies have been performed on these animals than on decapods. Large doses of 20-OHE initiated and accelerated premolt in *Armadillidium vulgare*, whether injected or applied topically or by dipping (Krishnakumaran and Schneiderman, 1969). Topical application of small doses of 20-OHE (about 150 ng/g) daily shortened intermolt and stimulated a precocious but normal ecdysis in this species (Madhavan, 1981). Madhavan suggested that high doses (which typically accelerate molting without normal exuviation in crustaceans) prematurely break muscle connections to the old cuticle, which is an alternate explanation to the one that invokes inhibition of an ''exuviation factor'' release by high ecdysteroid levels (see Skinner, 1985, for further discussion). Other species stimulated to molt by exogenous ecdysteroids include *Asellus aquaticus* (Balesdent, 1971), *Proasellus cavaticus* (Maissiat and Graf, 1973), *Idotea balthica* (Reidenbach, 1971), *Ligia oceanica* (Maissiat and Legrand, 1970), *Porcellio dilatatus* (Noulin and Maissiat, 1974), and *Sphaeroma serratum* (Charmantier and Trilles, 1976).

Molting is regulated by the Y-organ in isopods. Successful removal of this organ prevented molting in *Porcellio* (Maissiat and Legrand, 1970) and in *Sphaeroma* females and juvenile males (Charmantier and Trilles, 1977). Individual *Sphaeroma*, locked in intermolt by Y-organ removal, demonstrated premolt conditions after receiving Y-organ implants or injections of $5\,\mu$g 20-OHE per animal, while those receiving muscle implants remained in intermolt. The ecdysteroid 20-OHE was extracted from *S. serratum* by methods already described and identified from GC profiles of the silylated derivatives (Charmantier et al., 1976). The Y-organ degenerates in adult males of this species, and 20-OHE is no longer detectable in whole-animal extracts after degeneration is complete (Charmantier et al., 1977). Ecdysteroid concentrations in hemolymph of *Ligia* (Girard and Maissiat, 1983) and *Helleria brevicornis*, as well as whole-animal extracts of the latter species (Hoarau and Hirn, 1978), show a profile during the molt cycle that is essentially identical to the general decapod crustacean condition. The amount of MH is very low during intermolt, begins to rise during early premolt, reaches a maximal peak during late premolt, and declines to low levels before molting. In a study by Connat and Diehl (1986), some 18 h after injection of tritium-labeled EC, 34% of the label was recovered in apolar metabolites, both esterase-sensitive and esterase-resistant forms, in porcellionid isopods; thus they share common ecdysteroid-processing pathways with other arthropods.

8.3.3.3.3. Amphipoda

A molt-stimulating effect of 20-OHE has been demonstrated in several species of amphipod crustaceans: *Orchestia gammarella* (Blanchet, 1972; Blanchet and Charniaux-Cotton, 1971); *O. cavimana (Graf, 1972a); Niphargus virei* (Graf, 1972b); and *Gammarus pulex* and *G. fossarum* (Ducruet, 1975). Implanted crystals, injected solutions, or immersion in solutions of 20-OHE were effective in accelerating premolt and new cuticle synthesis when applied during intermolt, and normal ecdyses were also produced by immersion treatments in *Niphargus*. Cauterization of the Y-organ in *O. gammarella* during postmolt permanently locked the animals into an intermolt condition (Blanchet, 1974). However, when Blanchet implanted a crystal of 20-OHE into animals about 2 weeks after Y-organectomy, the animals initiated premolt within 3 days. RIA-detectable ecdysteroids have been demonstrated in whole-body extracts, after purification in organic solvent : water systems and TLC in *O. gammerella* (Blanchet et al., 1976, 1979), and in hemolymph and whole-body extracts after reversed-phase HPLC purification in *O. cavimana* (Graf and Delbecque, 1987). Both species show a typical crustacean profile of ecdysteroid concentrations during the molt cycle: low levels during intermolt, a maximal peak in late premolt, and a decrease to low levels before molting. The last-cited study is particularly insightful, since as many as 17 separate stages could be determined precisely during the 46-day molt cycle. Titers increased gradually but progressively during most of the intermolt, and peaked very sharply about 10 h before new epicuticle deposition. Concentrations increased from 320 to the peak value of 810 pg 20-OHE equivalents in 24 h, and decreased to about half that level equally rapidly (Graf and Delbecque, 1987). About 95% of the radiolabeled EC injected into *O. cavimana* during the ecdysteroid peak, a few days before and a few days after, was converted within 48 h to labeled apolar metabolites (but no 20-OHE), which liberated free EC and very small amounts of polar compounds after esterase hydrolysis (Connat and Diehl, 1986).

8.3.4. *Myriapoda*

The collective name Myriapoda is used here to describe the uniramian
arthropods having a head and a main body with many leg-bearing
segments. MHs have been isolated from centipedes and symphylans.
They have not been isolated from the Pauropoda. There are no published data on MHs in Diplopoda, although Bonaric (1980) refers to an
unpublished observation of R. Joly that ecdysteroids do occur in milli-
pedes.

8.3.4.1. CHILOPODA

Ecdysteroids are known to stimulate apolysis and new cuticle synthesis in adults and third-instar larvae of *Lithobius forficatus*. Injected
EC (0.9 μg/animal) stimulated cuticle formation and exuviation in more
than half of the surviving adults (Joly, 1964). Lower doses also cause a
significant increase in molting over controls (Descamps, 1977). Doses
of synthetic EC in the same range stimulated about 40% of the starved
third-instar larvae injected, some of these to a complete molt (Scheffel,
1969). Dipping in a solution of phytoecdysones–*Taxus* extract (Pollak
and Scheffel, 1973) and 20-OHE at 500 μg/ml (Scheffel and Wilke,
1974; Pollak, 1977) was also effective on starved larvae, although
exuviations were rare. But actinomycin D treatment 15 min before 20-
OHE application stimulated molting in 70% of the larvae. Apolysis was
also induced in cultured antennal explants of larvae by EC and 20-
OHE (Scheffel, 1983, 1986).

Apparently an organ ("lymph cord tissue") occurs in *Lithobius* that
is comparable to the crustacean Y-organ, and it secretes ecdysteroid
into an organ culture medium, as does the ovary (Leubert et al., 1982).
Ecdysteroids were extracted from adult animals with various organic
solvent systems and purified on a silica gel column and in several TLC
systems, with authentic EC and 20-OHE used as markers (Leubert et
al., 1979). Isolated materials were highly active in the *Musca* bioassay,
and the purified EC- and 20-OHE-containing fractions were quantified
by RIA. Ecdysteroid concentrations (RIA-reactive) in the hemolymph
vary during the molt cycle of adult *Lithobius forficatus* and are basically
similar to the general arthropod pattern: lowest values during intermolt; highest during premolt (Joly et al., 1979). A second peak
occurred during postmolt, although this is not consistently found in
other species of arthropods. Adults also metabolize injected tritiated
EC to similar apolar conjugates found in other species (Connat and

Diehl, 1986). The presumed ecdysial gland of centipedes is controlled by the cerebral gland (see Joly and Descamps, 1988; also Chapter 9 by Lachaise herein). Hemolymph ecdysteroid titers elevate prematurely after surgical removal of this organ (Joly et al., 1979).

8.3.4.2. SYMPHYLA

Ecdysteroids occur in the one species, *Hanseniella ivorensis,* so far studied. Antennalectomy, as in other myriapods, stimulates precocious molting in this species. RIA-reactive ecdysteroids from methanol extracts of whole bodies after antennalectomy showed the typical arthropod pattern—elevation from low levels to highest values in premolt, then decline to low levels at the time of molting (Juberthie-Jupeau et al., 1979).

8.4. Summary

Ecdysteroids occur in all the major types of noninsect arthropods—the chelicerates, crustaceans, and myriapods. As a whole, these compounds were shown to function as molting hormones (MHs), as they do in insects. Evidence for the occurrence of ecdysteroids comes from 8 classes and 13 orders of noninsect arthropods. Specific compounds have been unequivocally identified by physical chemical methods in many of these groups.

In crustaceans a major source of the molting hormones is the so-called ecdysial gland, or Y-organ. The Y-organ apparently synthesizes and secretes ecdysone (EC) into the hemolymph; EC is taken up by peripheral organs, particularly testes, integument, and hepatopancreas, and hydroxylated at the C-20 to form what most investigators consider the active MH in these animals, 20-hydroxyecdysone (20-OHE). One of its targets, epidermal cells of the integument, binds the hormone to specific intracellular protein receptors and is induced to secrete enzymes that erode the old cuticle. Under 20-OHE stimulation the epidermal cells secrete a new cuticle, in part before a molt, and the remainder after a molt. The rate of Y-organ secretion of ecdysteroid and the hydroxylation of EC by peripheral organs, as well as concentrations of 20-OHE in the hemolymph, are a function of time during the molt cycle—highest during premolt stages, lowest during intermolt. Exposure to exogenous ecdysteroids induces premolt events, including formation of new cuticle, and often molting was demonstrated, in numerous studies. EC is synthesized from cholesterol in some crustacean Y-organs and along with 20-OHE is metabolized to polar and apolar products in an organ-specific pattern. There is a widespread metabolism of EC to apolar conjugated forms that may represent inactivation products or storage forms of MH.

Ecdysteroids have been extracted from all three classes of chelicerates and function as MHs in these animals, although the tissue or organ of origin of EC is unknown. *Limulus* larvae, injected after mid-intermolt, respond to a variety of ecdysteroids in the picogram range with an accelerated but perfectly normal molt. EC is synthesized from cholesterol in ticks and converted, at least in part, to 20-OHE in horseshoe crabs and ticks. Apolar ecdysteroid conjugates are major conversion products in ticks. The epidermis is a major target organ of the ecdysteroids in chelicerates, but no intracellular receptors for these compounds have been isolated.

Comparatively little information about molting and its hormonal control is known for myriapods. It is clear, however, that ecdysteroids function as MHs in these animals. Various ecdysteroids stimulate a molting response. Hemolymph ecdysteroid concentrations exhibit the typical arthropod pattern of lowest values during intermolt, highest during premolt. In centipedes, EC is metabolized to similar apolar conjugates formed in other groups of arthropods.

References

Andrieux, M., P. Porcheron, J. Berreur-Bonnenfant, and F. Dray. 1976. Détermination du taux d'ecdysone au cours du cycle d'intermue chez le crabe *Carcinus maenas:* comparaison entre individus sains et parasités par *Sacculina carcini.* C. R. Acad. Sci. Paris 283: 1429–1432.

Baldaia, L., P. Porcheron, J. Coimbra, and P. Cassier. 1984. Ecdysteroids in the shrimp *Palaemon serratus:* relations with molt cycle. Gen. Comp. Endocrinol. 55: 437–443.

Balesdent, M.-L. 1971. Action d'une ecdysone de synthèse sur la mue et sur la sexualité du crustacé isopode femelle *Asellus aquaticus* L. C. R. Acad. Sci. Paris 273: 1972–1974.

Barnes, R. D. 1987. *Invertebrate Zoology,* 5th ed. Saunders, Philadelphia.

Bazin, F. 1977. Effets d'injections d'ecdystérone sur la mue et la régénération chez le crabe *Carcinus maenas* (L.). C. R. Acad. Sci. Paris 284: 765–768.

Bebbington, P. M. and E. D. Morgan. 1977. Detection and identification of moulting hormone (ecdysones) in the barnacle *Balanus balanoides.* Comp. Biochem. Physiol. 56B: 77–79.

Bebbington, P. M., E. D. Morgan, and C. F. Poole. 1977. Detection and identification of moulting hormone (ecdysones) in the barnacle *Balanus balanoides*. Pp. 367–381 *in* D. J. Faulkner and W. H. Fenical (eds.), *Marine Natural Products Chemistry*. Plenum Press, New York.

Behrens, W. 1981. Morphologische und hormonphysiologische Untersuchungen an einem Pantopoden, *Pycnogonum litorale* Ström. Dissertation, University of Ulm, Germany.

Behrens, W. and D. Bückmann. 1983. Ecdysteroids in the pycnogonid *Pycnogonum litorale* (Ström) (Arthropoda, Pantopoda). Gen. Comp. Endocrinol. 51: 8–14.

Birkenbeil, H. and M. Gersch. 1979. Ultrastructure of the Y-organ of *Astacus astacus* (L.) (Crustacea) in relation to the moult cycle. Cell Tissue Res. 196: 519–524.

Blanchet, M. F. 1972. Effets sur la mue et sur la vitellogenèse de la β-ecdysone introduite aux étapes A et D^2 du cycle d'intermue chez *Orchestia gammarella* Pallas (crustacé amphipode): comparaison avec les effets de la β et de l'α-ecdysone aux autres étapes de l'intermue. C. R. Acad. Sci. Paris 274: 3015–3018.

Blanchet, M. F. 1974. Etude du contrôle hormonal du cycle d'intermue et de l'exuviation chez *Orchestia gammarella* par microcautérisation des organes Y suivie d'introduction d'ecdystérone. C. R. Acad. Sci. Paris 278: 509–512.

Blanchet, M. F. and H. Charniaux-Cotton. 1971. Contrôle du déclenchement et de la durée de la période D du cycle d'intermue par l'ecdystérone, chez le crustacé amphipode *Orchestia gammarella* (Pallas): interaction avec la vitellogenèse. C. R. Acad. Sci. Paris 272: 307–310.

Blanchet, M. F., P. Porcheron, and F. Dray. 1976. Etude des variations du taux des ecdysones au cours du cycle d'intermue chez le mâle d'*Orchestia gammarella* Pallas (crustacé amphipode) par dosage radioimmunologique. C. R. Acad. Sci. Paris 283: 651–654.

Blanchet, M. F., P. Porcheron, and F. Dray. 1979. Variations du taux des ecdystéroides au cours des cycles de mue et de vitellogenèse chez le crustacé amphipode, *Orchestia gammarella*. Int. J. Invertebr. Reprod. 1: 133–139.

Bonaric, J. C. 1976. Effects of ecdysterone on the molting mechanisms and duration of the intermolt period in *Pisaura mirabilis* Cl. Gen. Comp. Endocrinol. 30: 267–272.

Bonaric, J. C. 1977a. Action de l'ecdystérone sur la réduction de durée de la période d'intermue chez *Pisaura mirabilis* Cl. (Araneae/ Pisauridae): détermination de l'effet optimal par inoculation de concentrations croissantes. Arch. Zool. Exp. Gén. 118: 43–51.

Bonaric, J. C. 1977b. Action de l'ecdystérone sur le cycle de mue de l'araignée *Pisaura mirabilis* Cl.: variation des réponses en fonction de la date d'intervention au cours de l'intermue nymphale. Arch. Zool. Exp. Gén. 118: 409–421.

Bonaric, J. C. 1980. Contribution à l'étude de la biologie du développement chez l'araignée *Pisaura mirabilis* (Clerk, 1758): approche physiologique des phénomènes de mue et de diapause hivernale. Doctoral thesis, University of Montpellier, France.

Bonaric, J. C. and M. DeReggi. 1977. Changes in ecdysone levels in the spider *Pisaura mirabilis* nymphs (Araneae, Pisauridae). Experientia (Basel) 33: 1664–1665.

Borst, D. W. and J. D. O'Connor. 1972. Arthropod molting hormone: radioimmune assay. Science (Wash., DC) 178: 418–419.

Bouvier, J., P. A. Diehl, and M. Morici. 1982. Ecdysone metabolism in the tick *Ornithodoros moubata* (Argasidae, Ixodoidea). Rev. Suisse Zool. 89: 967–976.

Bowman, T. E. and L. G. Abele. 1982. Classification of the recent crustacea. Pp. 1–27 *in* L. G. Abele (ed.), *The Biology of Crustacea*, Vol. I. Academic Press, Orlando, Florida.

Buchholz, C. and D. Adelung. 1980. The ultrastructural basis of steroid production in the Y-organ and the mandibular organ of the crabs *Hemigrapsus nudus* (Dana) and *Carcinus maenas* L. Cell Tissue Res. 206: 83–94.

Bückmann, D., G. Starnecker, K.-H. Tomaschko, E. Wilhelm, R. Lafont, and J.-P. Girault. 1986. Isolation and identification of major ecdysteroids from the pycnogonid *Pycnogonum litorale* (Ström) (Arthropoda, Pantopoda). J. Comp. Physiol. B156: 759–765.

Burghause, F. 1975. Das Y-Organ von *Orconectes limosus* (Malacostraca, Astacura). Z. Morphol. Tiere 80: 41–57.

Campbell, J. D. and J. H. Oliver, Jr. 1984. Membrane feeding and developmental effects of ingested β-ecdysone on *Ornithodoros parkeri* (Acari: Argasidae). Pp. 393–399 *in* D. A. Griffiths and C. E. Bowman (eds.), *Acarology VI*, Vol. 1. Ellis Horwood, Chichester, England.

Chang, E. S. and M. J. Bruce. 1981. Ecdysteroid titers of larval lobsters. Comp. Biochem. Physiol. 70A: 239–241.

Chang, E. S. and J. D. O'Connor. 1977. Secretion of α-ecdysone by crab Y-organs *in vitro*. Proc. Nat. Acad. Sci. USA 74: 615–618.

Chang, E. S. and J. D. O'Connor. 1978. *In vitro* secretion and hydroxylation of α-ecdysone as a function of the crustacean molt cycle. Gen. Comp. Endocrinol. 36: 151–160.

Chang, E. S., B. A. Sage, and J. D. O'Connor. 1976. The qualitative and quantitative determinations of ecdysones in tissues of the crab, *Pachygrapsus crassipes*, following molt induction. Gen. Comp. Endocrinol. 30: 21–33.

Charmantier, G. and J. P. Trilles. 1976. Ecdysterone, premue et exuviation chez *Sphaeroma serratum* (Fabricius, 1787) (Crustacea, Isopoda, Flabellifera). Gen. Comp. Endocrinol. 28: 249–254.

Charmantier, G. and J. P. Trilles. 1977. Contrôle endocrine des phénomènes de la mue par les organes Y chez *Sphaeroma serratum* (Fabricius, 1787) (Crustacea, Isopoda, Flabellifera). C. R. Acad. Sci. Paris 285: 905–908.

Charmantier, G., M. Ollé and J. P. Trilles. 1976. Aspects du dosage de l'ecdystérone chez *Sphaeroma serratum* (Crustacea, Isopoda, Flabellifera) et premiers résultats. C. R. Acad. Sci. Paris 283: 1329–1331.

Charmantier, G., M. Ollé, and J. P. Trilles. 1977. Evolution du taux d'ecdystérone, dégéenérescence des organes Y et sénescence chez les mâles pubères de *Sphaeroma serratum* (Fabricius, 1787) (Crustacea, Isopoda, Flabellifera). C. R. Acad. Sci. Paris 285: 1487–1489.

Cheung, P. J. 1974. The effect of ecdysterone on cyprids of *Balanus eburneus* Gould. J. Exp. Mar. Biol. Ecol. 15: 223–229.

Cheung, P. J. and R. F. Nigrelli. 1974. Ecdysterone-induced molting in isolated penis and appendages of barnacle, *Balanus eburneus* Gould, kept in sterilized sea water. Am. Zool. 14: 1266. (abstract).

Connat, J. L. and P. A. Diehl. 1986. Probable occurrence of ecdysteroid fatty acid esters in different classes of arthropods. Insect Biochem. 16: 91–97.

Connat, J. L., P. A. Diehl, and M. J. Thompson. 1986. Possible inactivation of ingested ecdysteroids by conjugation with long-chain fatty acids in the female tick *Ornithodoros moubata* (Acarina: Argasidae). Arch. Insect Biochem. Physiol. 3: 235–252.

Connat, J. L., P. A. Diehl, N. Dumont, S. Carminati, and M. J. Thompson. 1983. Effects of exogenous ecdysteroids on the female tick *Ornithodoros moubata*: induction of supermolting and influence on oogenesis. Z. Angew. Entomol. 96: 520–530.

Daig, K. and K.-D. Spindler. 1980. *In vitro* uptake and retention of molting hormones in the integument of the crayfish, *Astacus leptodactylus*. Pp. 273–278 *in* E. Kurstak, K. Maramorosch, and A. Dubendorfer (eds.), *Invertebrate Systems In Vitro*. Elsevier/North-Holland Publ., Amsterdam and New York.

Daig, K. and K.-D. Spindler. 1983. Uptake and retention of moulting hormones by the integument of crayfishes *in vitro*. I. Influence of temperature, hormone concentration and hormone structure. Mol. Cell. Endocrinol. 31: 93–104.

Dall, W. and M. C. Barclay. 1977. Induction of viable ecdysis in the western rock lobster by 20-hydroxyecdysone. Gen. Comp. Endocrinol. 31: 323–334.

Dees, W. H., D. E. Sonenshine, and E. Breidling. 1984a. Ecdysteroids in the american dog tick, *Dermacentor variabilis* (Acari: Ixodidae), during different periods of tick development. J. Med. Entomol. 21: 514–523.

Dees, W. H., D. E. Sonenshine, and E. Breidling. 1984b. Ecdysteroids in *Hyalomma dromedarii* and *Dermacentor variabilis* and their effects on sex pheromone activity. Pp. 406–413 *in* D. A. Griffiths and C. E. Bowman (eds.), *Acarology VI*, Vol. 1. Ellis Horwood, Chicester, England.

Dees, W. H., D. E. Sonnenshine, and E. Breidling. 1985. Ecdysteroids in the camel tick, *Hyalomma dromedarii* (Acari: Ixodidae), and comparison with sex pheromone activity. J. Med. Entomol. 22: 22–27.

Delbecque, J. P., P. A. Diehl, and J. D. O'Connor. 1978. Presence of ecdysone and ecdysterone in the tick *Amblyomma hebraeum* Koch. Experientia (Basel) 34: 1379–1380.

De Leersnyder, M., A. Dhainaut, and P. Porcheron. 1981. Influence de l'ablation des organes y sur l'ovogénèse du crabe *Eriocheir sinensis* dans les conditions naturelles et après épédonculation. Gen. Comp. Endocrinol. 43: 157–169.

Descamps, M. 1977. Influence de la croissance somatique sur le cycle spermatogénétique de *Lithobius forficatus* L. (Myriapode Chilopode). Gen. Comp. Endocrinol. 33: 412–422.

Diehl, P. A., J. E. Germond, and M. Morici. 1982. Correlations between ecdysteroid titers and integument structure in nymphs of the tick, *Ambylomma hebraeum* Koch (Acarina: Ixodidae). Rev. Suisse Zool. 89: 859–868.

Diehl, P. A., J.-L. Connat, and E. Dotson. 1986. Chemistry, function and metabolism of tick ecdysteroids. Pp. 165–193 *in* J. P. Sauer and J. A. Hair (eds.), *Morphology, Physiology, and Behavioral Biology of Ticks.* Ellis Horwood, Chichester, England.

Ducruet, J. 1975. Action de l'ecdystérone sur la mue de crustacés amphipodes femelles: *Gammarus pulex* (L.) et *G. fossarum* Koch. Crustaceana 28: 86–88.

Dumser, J. B. and J. H. Oliver. 1981. Kinetics of spermatogenesis, cell cycle analysis, and testis development in nymphs of the tick *Dermacentor variabilis.* J. Insect Physiol. 27: 743–753.

Echalier, G. 1959. L'organe Y et le déterminisme de la croissance et de la mue chez *Carcinus maenas* (L.), crustacé décapode. Ann. Sci. Nat. Zool. [Ser. 12] 1: 1–59.

El Bakary, Z., P. Porcheron, M. Morinière, and S. Fuzeau-Braesch. 1987. Mise en évidence et dosage d'ecdystéroïdes chez le scorpion *Leiurus quinquestriatus.* C. R. Acad. Sci. Paris 304: 453–456.

Ellis, B. J. and F. D. Obenchain. 1984. *In vivo* and *in vitro* production of ecdysteroids by nymphal *Amblyomma variegatum* ticks. Pp. 400–404 *in* D. A. Griffiths and C. E. Bowman (eds.), *Acarology VI,* Vol. I. Ellis Horwood, Chichester, England.

Faux, A., D. H. S. Horn, E. J. Middleton, H. M. Fales, and M. E. Lowe. 1969. Moulting hormones of a crab during ecdysis. J. Chem. Soc., Chem. Commun. Pp. 175–176.

Freeman, J. A. and C. K. Bartell. 1976. Some effects of the molt-inhibiting hormone and 20-hydroxyecdysone upon molting in the grass shrimp, *Palaemonetes pugio.* Gen. Comp. Endocrinol. 28: 131–142.

Freeman, J. A. and J. D. Costlow. 1979. Hormonal control of apolysis in barnacle mantle tissue epidermis, *in vitro.* J. Exp. Zool. 210: 333–346.

Freeman, J. A. and J. D. Costlow. 1983. The cyprid molt cycle and its hormonal control in the barnacle *Balanus amphitrite.* J. Crustacean Biol. 3: 173–182.

Freeman, J. A. and J. D. Costlow. 1984. Endocrine control of apolysis in *Rhithropanopeus harrisii* larvae. J. Crustacean Biol. 4: 1–6.

Gagosian, R. B. and R. A. Bourbonniere. 1976. Isolation and purification of the molting hormones from the American lobster (*Homarus americanus*). Comp. Biochem. Physiol. 53B: 155–161.

Gagosian, R. B., R. A. Bourbonniere, W. B. Smith, E. F. Couch, C. Blanton, and W. Novak. 1974. Lobster molting hormones: isolation and biosynthesis of ecdysterone. Experientia (Basel) 30: 723–724.

Galbraith, M. N., D. H. S. Horn, and E. J. Middleton. 1968. Structure of deoxycrustecdysone, a second crustacean moulting hormone. J. Chem. Soc., Chem. Commun. Pp. 83–85.

Germond, J.-E., P. A. Diehl, and M. Morici. 1982. Correlations between integument structure and ecdysteroid titers in fifth-stage nymphs of the tick, *Ornithodoros moubata* (Murray, 1877; *sensu* Walton, 1962). Gen. Comp. Endocrinol. 46: 255–266.

Gilgan, M. W. and B. G. Burns. 1976. The successful induction of molting in the adult male lobster (*Homarus americanus*) with a slow-release form of ecdysterone. Steroids 27: 571–580.

Gilgan, M. W. and T. E. Farquharson. 1977. A change in the sensitivity of adult male lobsters (*Homarus americanus*) to ecdysterone on changing from intermolt to active premolt development. Comp. Biochem. Physiol. 58A: 29–32.

Gilgan, M. W., T. E. Farquharson, and B. G. Burns. 1977. The effect of α-ecdysone, ecdysterone and inokosterone treatment, separately or in combinations, on premolt development and molting in adult male lobsters (*Homarus americanus*). Comp. Biochem. Physiol. 56A: 43–49.

Girard, P. and R. Maissiat. 1983. Variations du taux des ecdystéroïdes hémolymphatiques chez le mâle de *Ligia oceanica* (L.), (Crustacea, Isopoda, Oniscoidea) en fonction due cycle de mue et des modifications structurales de l'organe Y. Can. J. Zool. 61: 534–538.

Gnatzy, W. and F. Romer. 1980. Morphogenesis of mechanoreceptor and epidermal cells of crickets during the last instar, and its relation to molting-hormone level. Cell Tissue Res. 213: 369–391.

Gorell, T. A., L. I. Gilbert, and J. B. Siddall. 1972. Studies on hormone recognition by arthropod target tissues. Am. Zool. 12: 347–356.

Graf, M. F. 1972a. Action de l'ecdystérone sur la mue, la cuticle et le métabolisme due calcium chez *Orchestia cavimana* Heller (Crustacé, Amphipode, Talitridé). C. R. Acad. Sci. Paris 274: 1731–1734.

Graf, M. F. 1972b. Etude comparative de l'action de l'ecdystérone chez le gammaridé hypogé *Niphargus* et le talitridé épigé *Orchestia* (Crustacés, Amphipodes). C. R. Acad. Sci. Paris 275: 2045–2048.

Graf, F. and J. P. Delbecque. 1987. Ecdysteroid titers during the molt cycle of *Orchestia cavimana* (Crustacea, Amphipoda). Gen. Comp. Endocrinol. 65: 23–33.

Hampshire, F. and D. H. S. Horn. 1966. Structure of crustecdysone, a crustacean moulting hormone. J. Chem. Soc., Chem. Commun. Pp. 37–38.

Herman, W. S. 1972. Molt initiation in response to phytoecdysones and low doses of animal ecdysones in the chelicerate arthropod, *Limulus polyphemus.* Gen. Comp. Endocrinol. 18: 301–305.

Hoarau, F. and M. Hirn. 1978. Evolution du taux des ecdystéroïdes au cours du cycle de mue chez *Helleria brevicornis* Ebner (Isopode terrestre). C. R. Acad. Sci. Paris 286: 1443–1446.

Hocks, P. and R. Weichert. 1966. 20-Hydroxy ecdyson isoliert aus Insekten. Tetrahedron Lett. 26: 2989–2993.

Hopkins, P. 1983. Patterns of serum ecdysteroids during induced and uninduced proecdysis in the fiddler crab, *Uca pugilator.* Gen. Comp. Endocrinol. 52: 350–356.

Hopkins, P. 1986. Ecdysteroid titers and Y-organ activity during late anecdysis and proecdysis in the fiddler crab, *Uca pugilator.* Gen. Comp. Endocrinol. 63: 362–373.

Horn, D. H. S. 1971. The ecdysones. Pp. 333–459 *in* M. Jacobson and D. G. Crosby (eds.), *Naturally Occurring Insecticides.* Dekker, New York.

Horn, D. H. S., S. Fabbri, F. Hampshire, and M. E. Lowe. 1968. Isolation of crustecdysone (20*R*-hydroxyecdysone) from a crayfish (*Jasus lalandei* H. Milne-Edwards). Biochem. J. 109: 399–406.

Jegla, T. C. 1972. Development and molting physiology of horseshoe crab larvae. Am. Zool. 12: 724 (abstract).

Jegla, T. C. 1974. Ecdysone activity in *Limulus polyphemus.* Am. Zool. 14: 1288 (abstract).

Jegla, T. C. 1982. A review of the molting physiology of the trilobite larva of *Limulus.* Pp. 83–101 *in* J. Bonaventura, C. Bonaventura, and S. Tesh (eds.), *Physiology and Biology of Horseshoe Crabs.* Liss, New York.

Jegla, T. C. and J. D. Costlow, Jr. 1970. Induction of molting in horseshoe crab larvae by polyhydroxy steroids. Gen. Comp. Endocrinol. 14: 295–302.

Jegla, T. C. and J. D. Costlow. 1979a. Ecdysteroids in *Limulus* larvae. Experientia (Basel) 35: 554–555.

Jegla, T. C. and J. D. Costlow. 1979b. The *Limulus* bioassay for ecdysteroids. Biol. Bull. (Woods Hole) 156: 103–114.

Jegla, T. C., J. D. Costlow, and J. Alspaugh. 1972. Effects of ecdysones and some synthetic analogs on horseshoe crab larvae. Gen. Comp. Endocrinol. 19: 159–166.

Jegla, T. C., C. Ruland, G. Kegel, and R. Keller. 1983. The role of the Y-organ and cephalic gland in ecdysteroid production and the control of molting in the crayfish, *Orconectes limosus.* J. Comp. Physiol. 152: 91–95.

Joly, R. 1964. Action de l'ecdysone sur le cycle de mue de *Lithobius forficatus* L. (Myriapode, Chilopode). C. R. Acad. Sci. Paris 158: 548–550.

Joly, R. and M. Descamps. 1988. Endocrinology of myriapods. Pp. 429–449 *in* H. Laufer and R. G. H. Downer (eds.), *Endocrinology of Selected Invertebrate Types.* Liss, New York.

Joly, R., P. Porcheron, and F. Dray. 1979. Etude des variations du taux d'ecdystéroïdes au cours du cycle de mue dans l'hemolymphe de *Lithobius forficatus* L. (Myriapode, Chiopode), par dosage radioimmunologique. C. R. Acad. Sci. Paris 288: 243–246.

Juberthie-Jupeau, L., A. Strambi, M. de Reggi, and C. Juberthie. 1979. The presence of ecdysteroids and the variations of their level during the first adult stage of the myriapod *Hanseniella ivorensis* Juberthie-Jupeau and Kehe (Symphyla). Experientia (Basel) 35: 1406–1408.

Karlson, P. 1956. Biochemical studies on insect hormones. Pp. 227–266 *in* R. S. Harris, G. F. Marrian, and K. V. Thimann (eds.), *Vitamins and Hormones.* Academic Press, Orlando, Florida.

Karlson, P. and E. Shaaya. 1964. Der Ecdysontiter während der Insektenentwicklung. I. Eine Methode zur Bestimmung des Ecdysongehalts. J. Insects Physiol. 10: 797–804.

Karlson, P. and D. M. Skinner. 1960. Attempted extraction of crustacean moulting hormone from isolated Y-organs. Nature 185: 543–544.

Keller, R. and E. Schmid. 1979. *In vitro* secretion of ecdysteroids by Y-organs and lack of secretion by mandibular organs of the crayfish following molt induction. J. Comp. Physiol. 130: 347–353.

Keller, R. and A. Willig. 1976. Experimental evidence of the molt controlling function of the Y-organ of a macruran decapod, *Orconectes limosus.* J. Comp. Physiol. 108: 271–278.

Kelly, T. J., J. R. Aldrich, C. W. Woods, and A. J. Borkovec. 1984. Makisterone-A: its distribution and physiological role as the molting hormone of true bugs. Experientia (Basel) *40:* 996–997.

Khalil, G. M., A. A. A. Shaarawy, S. E. Sonenshine, and S. M. Gad. 1984. β-Ecdysone effects on the camel tick, *Hyalomma dromedarii* (Acari: Ixodidae). J. Med. Entomol. 21: 188–193.

King, D. S. and J. B. Siddall. 1969. Conversion of α-ecdysone to β-ecdysone by crustaceans and insects. Nature (Lond.) 221: 955–956.

Kleinholz, L. H. and R. Keller. 1979. Endocrine regulation in crustacea. Pp. 160–213 *in* E. J. W. Barrington (ed.), *Hormones and Evolution.* Academic Press, Orlando, Florida.

Krishnakumaran, A. and H. A. Schneiderman. 1968. Chemical control of moulting in arthropods. Nature (Lond.) 220: 601–603.

Krishnakumaran, A. and H. A. Schneiderman. 1969. Induction of molting in crustacea by an insect molting hormone. Gen. Comp. Endocrinol. 12: 515–518.

Krishnakumaran, A. and H. A. Schneiderman. 1970. Control of molting in mandibulate and chelicerate arthropods by ecdysones. Biol. Bull. (Woods Hole) 139: 520–538.

Kuppert, P., M. Büchler, and K.-D. Spindler. 1978a. Distribution and transport of molting hormones in the crayfish, *Orconectes limosus.* Z. Naturforsch. 33: 437–441.

Kuppert, P. M., S. Wilhelm, and K.-D. Spindler. 1978b. Demonstration of cytoplasmic receptors for the molting hormones in crayfish. J. Comp. Physiol. 128: 95–100.

Lachaise, F. and R. Feyereisen. 1976. Métabolisme de l'ecdysone par divers organes de *Carcinus maenas* L. incubés *in vitro.* C. R. Acad. Sci. Paris 283: 1445–1448.

Lachaise, F., M. Lagueux, R. Feyereisen, and J. A. Hoffmann. 1976. Metabolisme de l'ecdysone au cours du développement de *Carcinus maenas* (Brachyura, Decapoda). C. R. Acad. Sci. Paris 283: 943–946.

Lachaise, F., M. Goudeau, C. Hetru, C. Kappler, and J. A. Hoffmann. 1981. Ecdysteroids and ovarian development in the shore crab, *Carcinus maenas*. Hoppe-Seyler's Z. Physiol. Chem. 362: 521–529.

Lafont, R. and J. Koolman. 1984. Ecdysone metabolism. Pp. 196–225 *in* J. Hoffmann and M. Porchet (eds.), *Biosynthesis, Metabolism and Mode of Action of Invertebrate Hormones*. Springer-Verlag Berlin and New York.

Lafont, R., G. Sommé-Martin, B. Mauchamp, B. F. Maume, and J.-P. Delbecque. 1980. Analysis of ecdysteroids by high-performance liquid chromatography and coupled gas-liquid chromatography–mass spectrometry. Pp. 45–67 *in* J. A. Hoffmann (ed.), *Progress In Ecdysone Research*. Elsevier/North-Holland Publ., Amsterdam and New York.

Lafont, R., J.-L. Pennetier, M. Andrianjafintrimo, J. Claret, J.-F. Modde, and C. Blais. 1982. Sample processing for high-performance liquid chromatography of ecdysteroids. J. Chromatogr. 236: 137–149.

Leubert, F., H. Eibisch, and H. Scheffel. 1979. Isolierung und Charakterisierung des Häutungshormons von *Lithobius forficatus* (L.) (Chilopoda). Zool. Jahrb. Abt. Allg. Zool. Physiol. Tiere 83: 334–339.

Leubert, F., H. Eibisch, H. Kroschwitz, and H. Scheffel. 1982. Ecdysteroid-Biosynthese durch das Ovar von *Lithobius forficatus* (L.) (Chilopoda). Zool. Jahrb. Abt. Allg. Zool. Physiol. Tiere 86: 465–476.

Londershausen, M., P. Kuppert, and K.-D. Spindler. 1982. Ecdysteroid receptors: a comparison of cytoplasmic and nuclear receptors from crayfish hypodermis. Hoppe-Seyler's Z. Physiol. Chem. 363: 797–802.

Madhavan, K. 1981. Induction of precocious molting in the male and ovigerous female *Armadillidium vulgare* by topical application of 20-hydroxyecdysone. Gen. Comp. Physiol. 44: 28–36.

Maissiat, J. and F. Graf. 1973. Action de l'ecdystérone sur l'apolysis et l'ecdysis de divers crustacés isopodes. J. Insect Physiol. 19: 1265–1276.

Maissiat, J. and J.-J. Legrand. 1970. Contribution à l'étude expérimentale du contrôle hormonal du cycle de mue chez les oniscoïdes *Porcellio dilatatus* et *Ligia oceanica*. C. R. Soc. Biol. Poitiers 164: 359–362.

Mango, C. K. A. 1978. Effects of beta-ecdysone and ponasterone A on nymphs of the soft tick *Ornithodoros moubata*. Pp. 36–37 *in* J. K. H. Wilde (ed.), *Tick-Borne Diseases and Their Vectors*. Edinburgh University Press, Edinburgh.

Mango, C., T. R. Odhiambo, and R. Galun. 1976. Ecdysone and the super tick. Nature (Lond.) 260: 318–319.

McCarthy, J. F. 1979. Ponasterone A: a new ecdysteroid from the embryos and serum of brachyuran crustaceans. Steroids 34: 799–806.

McCarthy, J. F. 1982. Ecdysone metabolism in premolt land crabs (*Gecarcinus lateralis*). Gen. Comp. Endocrinol. 47: 323–332.

McCarthy, J. F. and D. M. Skinner. 1977. Proecdysial changes in serum ecdysone titers, gastrolith formation, and limb regeneration following molt induction by limb autotomy and/or eyestalk removal in the land crab, *Gecarcinus lateralis*. Gen. Comp. Endocrinol. 33: 278–292.

McConaugha, J. R. and J. D. Costlow. 1981. Ecdysone regulation of larval crustacean molting. Comp. Biochem. Physiol. 68A: 91–93.

McWhinnie, M. A., R. J. Kirchenberg, R. J. Urbanski, and J. E. Schwarz. 1972. Crustecdysone-mediated changes in crayfish. Am. Zool. 12: 357–372.

Morgan, E. D. and C. F. Poole. 1976. The extraction and determination of ecdysones in arthropods. Adv. Insect. Physiol. 12: 17–62.

Morgan, E. D. and I. D. Wilson. 1980. Progress in the analysis of ecdysteroids. Pp. 29–43 *in* J. A. Hoffmann (ed.), *Progress in Ecdysone Research*. Elsevier/North-Holland Publ., Amsterdam and New York.

Noulin, G. and J. Maissiat. 1974. Etude du rôle de l'organe Y et de l'effet de l'ecdystérone dans la régénération d'un appendice chez l'oniscoïde *Porcellio dilatatus*. J. Insect Physiol. 20: 1963–1974.

Passano, L. M. and S. Jyssum. 1963. The role of the Y-organ in crab proecdysis and limb regeneration. Comp. Biochem. Physiol. 9: 195–213.

Pollak, W. 1977. Untersuchungen zur quantitativen Wirkung von Ecdysteron auf die postembryonale Morphogenese des Chilopoden *Lithobius forficatus* (L.). Zool. Jahrb. Abt. Allg. Zool. Physiol. Tiere 81: 383–394.

Pollak, W. and H. Scheffel. 1973. Häutungsauslösung bei Chilopoden-Larven durch Phytecdysone der Eibe (*Taxus baccata* L.). Zool. Anz. 191: 86–92.

Pound, J. M., J. H. Oliver, Jr., and R. H. Andrews. 1984. Induction of apolysis and cuticle formation in female *Ornithodoros parkeri* (Acari: Argasidae) by hemocoelic injections of beta-ecdysone. J. Med. Entomol. 21: 612–614.

Rao, K. R. 1978. Effects of ecdysterone, inokosterone and eyestalk ablation on limb regeneration in the fiddler crab, *Uca pugilator*. J. Exp. Zool. 203: 257–270.

Rao, K. R., M. Fingerman, and C. Hays. 1972. Comparison of the abilities of α-ecdysone and 20-hydroxyecdysone to induce precocious proecdysis and ecdysis in the fiddler crab, *Uca pugilator*. Z. Vgl. Physiol. 76: 270–284.

Rao, K. R., S. W. Fingerman, and M. Fingerman. 1973. Effects of exogenous ecdysones on the molt cycles of fourth and fifth stage American lobsters, *Homarus americanus*. Comp. Biochem. Physiol. 44A: 1105–1120.

Reidenbach, J.-M. 1971. Action d'une ecdysone de synthèse sur la mue et le fonctionnement ovarien chez le crustacé isopode *Idotea balthica* (Pallas). C. R. Acad. Sci. Paris 273: 1614–1617.

Romer, F. and W. Gnatzy. 1981. Arachnid oenocytes: ecdysone synthesis in the legs of harvestmen (Opilionidae). Cell Tissue Res. 216: 449–453.

Scheffel, H. 1969. Untersuchungen über die hormonale Regulation von Häutung und Anamorphose von *Lithobius forficatus* (L.) (Myriapoda, Chilopoda). Zool. Jahrb. Abt. Allg. Zool. Physiol. Tiere 74: 436–505.

Scheffel, H. 1983. *In-vitro*-Untersuchungen über die Häutungsauslösende Wirkung von Ecdyson und 20-Hydroxyecdyson bei Chilopoden. Zool. Jahrb. Abt. Allg. Zool. Physiol. Tiere 87: 425–438.

Scheffel, H. 1986. Weitere Untersuchungen über die Ecdysteroid-Sensitivität der larvalen Epidermis von *Lithobius forficatus* (L.) (Chilopoda). Zool. Jahrb. Abt. Allg. Zool. Physiol. Tiere 90: 77–83.

Scheffel, H. and C. Wilke. 1974. Fördernder Einfluss von Actinomycin D auf die Häutungsauslösung durch exogenes Ecdysteron bei Chilopoden-Larven. Zool. Jahrb. Abt. Allg. Zool. Physiol. Tiere. 78: 33–39.

Skinner, D. M. 1985. Molting and regeneration. Pp. 44–146 *in* D. E. Bliss and L. H. Mantel (eds.), *The Biology of Crustacea*, Vol. 9. Academic Press, Orlando, Florida.

Solomon, K. R., C. K. A. Mango, and F. D. Obenchain. 1982. Endocrine mechanisms in ticks: effects of insect hormones and their mimics on development and reproduction. Pp. 399–438 *in* F. D. Obenchain and R. Galun (eds.), *Physiology of Ticks*. Pergamon Press, Oxford and Elmsford, New York.

Sonenshine, D. E., P. J. Homsher, M. Beveridge, and W. H. Dees. 1985. Occurrence of ecdysteroids in specific body organs of the camel tick, *Hyalomma dromedarii*, and the American dog tick, *Dermacentor variabilis* (Acari: Ixodidae), with notes on their synthesis from cholesterol. J. Med. Entomol. 22: 303–311.

Sonenshine, D. E., L. M. Boland, M. Beveridge, and B. J. Upchurch. 1986. Metabolism of ecdysone and 20-hydroxyecdysone in the camel tick, *Hyalomma dromedarii* (Acari: Ixodidae). J. Med. Entomol. 23: 630–650.

Soumoff, C. and D. M. Skinner. 1980. Effect of Y-organectomy and autotomy on circulating ecdysteroid levels and regeneration in the crab *Gecarcinus*. Am. Zool. 20: 933 (abstract).

Soumoff, C. and D. L. Skinner. 1982. *In vitro* assay of hydroxylase activity in the crab *Gecarcinus lateralis*. Am. Zool. 22: 938 (abstract).

Soumoff, C. and D. M. Skinner. 1983. Ecdysteroid titers during the molt cycle of the blue crab resemble those of other Crustacea. Biol. Bull. (Woods Hole) 165: 321–329.

Spaziani, E. and S. B. Kater. 1973. Uptake and turnover of cholesterol-^{14}C in Y-organs of the crab *Hemigrapsus* as a function of the molt cycle. Gen. Comp. Endocrinol. 20: 534–549.

Spindler, K.-D., R. Keller, and J. D. O'Connor. 1980. The role of ecdysteroids in the crustacean molting cycle. Pp. 247–280 *in* J. A. Hoffmann (ed.), *Progress in Ecdysone Research*. Elsevier/North-Holland Publ., Amsterdam and New York.

Spindler-Barth, M., U. Bassemir, P. Kuppert, and K.-D. Spindler. 1981. Isolation of nuclei from crayfish tissues and demonstration of nuclear ecdysteroid receptors. Z. Naturforsch. 36C: 326–332.

Stevenson, J. R., P. W. Armstrong, E. S. Chang, and J. D. O'Connor. 1979. Ecdysone titers during the molt cycle of the crayfish *Orconectes sanborni*. Gen. Comp. Endocrinol. 39: 20–25.

Tighe-Ford, D. J. 1977. Hormonal aspects of barnacle antifouling research. Pp. 383–402 *in* D. J. Faulkner and W. H. Fenical (eds.), *Marine Natural Products Chemistry*. Plenum Press, New York.

Tighe-Ford, D. J. and D. C. Vaile. 1972. The action of crustecdysone on the cirripede *Balanus balanoides* (L.). J. Exp. Mar. Biol. Ecol. 9: 19–28.

Vernet, G. 1976. Données actuelles sur le déterminisme de la mue chez les crustacés. Ann. Biol. 15: 155–188.

Vernet, G., C. Bressac, and J. P. Trilles. 1978. Quelques données récentes sur l'organe Y (glande de mue) des crustacés décapodes. Arch. Zool. Exp. Gén. 119: 201–225.

Walgraeve, H. R., G. R. Criel, P. Sorgeloos, and A. P. DeLeenheer. 1988. Determination of ecdysteroids during the moult cycle of adult *Artemia*. J. Insect Physiol. 34: 597–602.

Walgraeve, H., E. van Beek, G. Criel, K. van Brussel, and A. de Leenheer. 1986. Comparison of three separation systems for the radioimmunoassay of ecdysteroids in *Artemia* (Crustacea: Branchiopoda). Insect Biochem. 16: 41–44.

Watson, R. D. and E. Spaziani. 1982. Rapid isolation of ecdysteroids from crustacean tissues and culture media using SEP-PAK C_{18} cartridges. J. Liquid Chromatogr. 5: 525–535.

Watson, R. D. and E. Spaziani. 1985. Biosynthesis of ecdysteroids from cholesterol by crab Y-organs, and eyestalk suppression of cholesterol uptake and secretory activity, *in vitro*. Gen. Comp. Endocrinol. 59: 140–148.

Warner, A. C. and J. R. Stevenson. 1972. The influence of ecdysones and eyestalk removal on the molt cycle of the crayfish *Orconectes obscurus*. Gen. Comp. Endocrinol. 18: 454–462.

Wigglesworth, K. M., D. Lewis, and H. H. Rees. 1985. Ecdysteroid titre and metabolism to novel apolar derivatives in adult female *Boophilus microplus* (Ixodidae). Arch. Insect Biochem. Physiol. 2: 39–54.

Willig, A. 1974. Die Rolle der Ecdysone im Häutungszyklus der Crustaceen. Fortschr. Zool. 22: 55–74.

Willig, A. and R. Keller. 1973. Molting hormone content, cuticle growth and gastrolith growth in the molt cycle of the crayfish *Orconectes limosus*. J. Comp. Physiol. 86: 377–388.

Willig, A. and R. Keller. 1976. Biosynthesis of α- and β-ecdysone by the crayfish *Orconectes limosus in vivo* and by its Y-organs *in vitro*. Experientia (Basel) 32: 936–937.

Winget, R. R. and W. S. Herman. 1976. Occurrence of ecdysone in the blood of the chelicerate arthropod, *Limulus polyphemus*. Experientia (Basel) 32: 1345–1346.

Wright, J. E. 1969. Hormonal termination of larval diapause in *Dermacentor albipictus*. Science (Wash., DC) 163: 390–391.

Zeleny, C. 1905. Compensatory regulation. J. Exp. Zool. 2: 1–102.

Synthesis, Metabolism, and Effects on Molting of Ecdysteroids in Crustacea, Chelicerata, and Myriapoda

9

F. LACHAISE

9.1. Introduction	277
9.2. Various Ecdysteroids	278
9.3. Synthesis	282
9.3.1. Crustaceans	282
9.3.2. Chelicerates	284
9.3.3. Myriapods	285
9.4. Metabolism	285
9.4.1. Crustaceans	286
9.4.2. Chelicerates	287
9.5. Effect on Molting	288
9.5.1. Endogenous Titer and Molting	288
9.5.1.1. Molt Cycle	288
9.5.1.2. Titer Variations During the Molting Cycle	292
9.5.2. Effects of Organ Removal on Endogenous Titer	293
9.5.2.1. Removal of Y-Organs	294
9.5.2.2. Removal of Eyestalks	294
9.5.3. Effects of Exogenous Ecdysteroid on Endogenous Titer	295
9.5.3.1. Method of Application and Nature of Ecdysteroids	296
9.5.3.2. Breeding Conditions	298
9.5.3.3. Temporal Anecdysis	299
9.5.3.4. Terminal Anecdysis	301
9.5.3.5. Successively Molting Animals	301

9.6.   Effects on Target Tissues   302
    9.6.1.   Morphological Modifications of the
    Epidermis   302
        9.6.1.1.   *In Vivo* Experiments   302
        9.6.1.2.   *In Vitro* Experiments   303
    9.6.2.   Biochemical Changes in the Epidermis   303
        9.6.2.1.   Proteins and Nucleic Acids   303
        9.6.2.2.   Chitin   304
        9.6.2.3.   Protein Kinase   304
    9.6.3.   Other Target Tissues and Calcification   305
        9.6.3.1.   Hepatopancreas   305
        9.6.3.2.   Neurosecretory Cells and Y-Organs   305
        9.6.3.3.   Other Tissues   306
        9.6.3.4.   Calcification of Crustacean Cuticle   306
9.7.   Effect on Regeneration   307
    9.7.1.   Regeneration and Molt   307
    9.7.2.   Regeneration and Ecdysteroids   308
9.8.   Summary   309
Acknowledgments   310
References   310

9.1. Introduction

There is great molecular diversity of hormones in arthropods. These are either like those in vertebrates (e.g., neuropeptides or steroid hormones), or more specific to arthropod groups, such as juvenile hormone (JH) and JH-like compounds. Owing to their dual function in molt and regeneration, they are called morphogenetic hormones. In arthropods, this is achieved by the molting (ecdysteroids) and juvenile hormones.

Investigations on molting hormones, or ecdysteroids, have been done in all major arthropods groups (insects, crustaceans, myriapods, and chelicerates (Fig. 9.1). In 1954, Butenandt and Karlson isolated from the silk worm a 27-carbon steroid hormone, formed by a tetracyclic nucleus and a side chain: ecdysone (α-ecdysone, or E), or the molting hormone of insects. Later, Hampshire and Horn (1966) found in decapod crustaceans 20-hydroxyecdysone (20E), formerly called

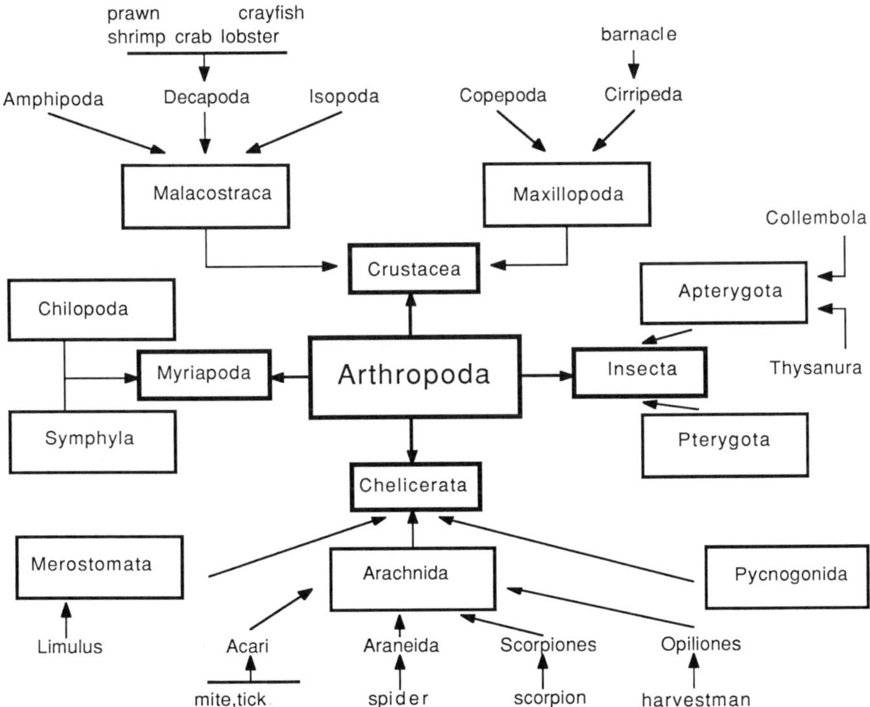

FIGURE 9.1. Simplified classification of arthropoda.

crustecdysone, ecdysterone, or β-ecdysone. Since then, a number of molecules, more or less closely related to ecdysone and causing molting, have been detected in these animals and grouped under the general term of ecdysteroids (Goodwin et al., 1978). Ecdysteroids were successively separated in chelicerates by Winget and Herman (1976) in *Limulus* larvae, by Bonaric and DeReggi (1977) in spider, and by Delbecque et al. (1978) in ticks. They were subsequently identified in myriapods (Leubert et al., 1979) and more recently in pycnogonids (Behrens and Bückmann, 1983; Bückmann et al., 1986).

Synthesis and degradation of these hormones result in maintaining a given level of hormone concentration in the hemolymph (or in tissues). Variation in the concentration of these hormones reflects their cyclic activity at the level of the target tissue. In all arthropods, an increase in the ecdysteroid concentration occurs just before molt.

This chapter will deal first with the identification, synthesis, and metabolism of ecdysteroids (Sections 9.2, 9.3, and 9.4, respectively). Then, Sections 9.4–9.7 will be devoted to the effects of these hormones on the morphogenetic changes in arthropods, such as molt and regeneration.

9.2. Various Ecdysteroids

The first studies of ecdysteroids in arthropods were performed with hemolymph, tissues, or whole-animal extracts subjected to either bioassay (Adelung, 1971; Carlisle and Connick, 1973; Willig and Keller, 1973; Winget and Herman, 1976) or radioimmunoassay [for crustaceans, see Spindler et al. (1980); for chilopod myriapods, Leubert et al. (1979) and Joly et al., (1979); for diplopod myriapods, Juberthie-Jupeau et al., (1979); for chelicerates, see respectively Bonaric and DeReggi (1977) for spiders, Delbecque et al. (1978) for ticks, and El Bakary et al. (1987) for scorpions; for pycnogonids, Behrens and Bückman (1983)]. These authors have demonstrated that shortly before molt there is an increase in hemolymph ecdysteroids. In a second type of study, the ecdysteroids present were identified, ecdysone and 20E (Fig. 9.2) being the most frequently characterized hormones (Spindler et al., 1980, and Skinner, 1985, for crustaceans; Solomon et al., 1982, and Diehl et al., 1986, for ticks; Jegla, 1982, for *Limulus*; Bonaric, 1986, for spiders). However, other ecdysteroids have been separated in crustaceans: 2-deoxy-20-hydroxyecdysone (Galbraith et al., 1968), makisterone A and inokosterone A (Faux et al., 1969), and more recently ponasterone A (PoA) (Fig. 9.2) (McCarthy,

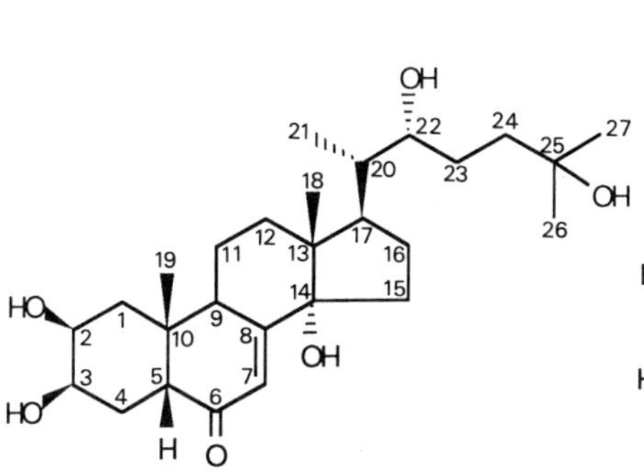

ECDYSONE

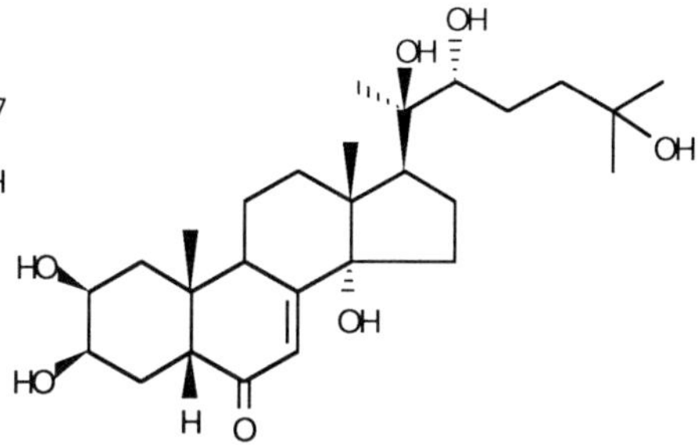

20-HYDROXYECDYSONE

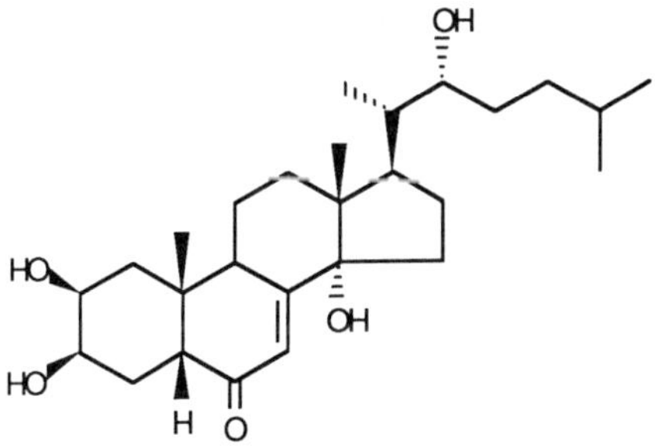

25-DEOXYECDYSONE

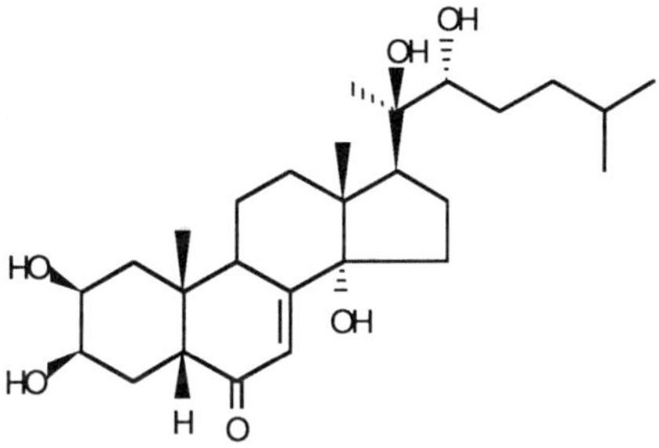

PONASTERONE A

FIGURE 9.2. Major hemolymph ecdysteroids found in arthropods other than insects and their precursors.

1979; Lachaise et al., 1981). PoA, discovered earlier in the conifer *Podocarpus narkaii* (Nakanishi et al., 1966), was previously unknown in animals. In addition to 20E, PoA was shown to be quantitatively important in the hemolymph of crab just prior to molting (McCarthy, 1979, 1982; Lachaise and Lafont, 1984); it is found in the ovary during ovarian maturation (Lachaise et al., 1981), and in eggs in the course of embryonic development (McCarthy and Skinner, 1979a; Lachaise and Hoffmann, 1982).

PoA apparently exists in a number of decapod crustaceans (Table 9.1), e.g., spider crab *(Maia squinado)*, shrimp *(Palaemon serratus)*, and lobster *(Homarus gammarus)*. It was also found in other crustaceans, such as the amphipod *Orchestia gammarella* and the isopod *Ligia oceanica*. Interestingly, PoA has been found in Thysanura and Collembola (Insecta: Apterygota) but is absent in ticks and pterygote insects (Lachaise, unpublished data; see Table 9.1). As will be seen later,

TABLE 9.1. Presence of Ponasterone A in Various Arthropods

			Ecdysone (ng/g)	20-hydroxyecdysone (ng/g)
Ticks	*Rhipicephalus appendiculatus*	vitellogenic females	traces	trace
		nymphs after engorgement	traces	10
		eggs	60	32
	Ligia oceanica		/	40
	Sphaeroma serratum			
	Orchestia gammarella		/	80
	Palaemon serratus		/	20
	Acanthonyx lunulatus		/	traces
		hemolymph	/	200

Crustaceans	*Maia squinado*	immature ovaries	traces	100
		ovaries in vitellogenesis	/	280
		newly laid eggs	/	410
	Carcinus maenas	ovaries	/	40
		eggs	/	traces
Apterygotan Insects	Collembola: *Allacna fusca*	premolt animals	traces	100
		premolt animals	10	150
	Thysanura: *Thermobia domestica*	eggs	/	/
		females in vitellogenesis	/	/
Pterygotan Insects	*Carausius morosus*	eggs	traces	50
	Tenebrio molitor	pupae	150	250
		nymphs	/	230
	Pieris brassicae	eggs	125	180
		caterpillars	/	200
	Drosophila melanogaster	pupae	/	120

studies of ecdysteroid synthesis and metabolism have revealed a number of other ecdysteroids (precursors or metabolites) likely to exist as endogenous hormones in the animals studied.

9.3. Synthesis

A number of factors affect changes in hormone concentration, including synthesis. Considering the presence of ecdysteroids in arthropods, two questions can be raised: Are these ecdysteroids synthesized by the animal? If so, which biochemical pathway is followed considering that arthropods do not synthesize cholesterol, the classical precursor of C_{27} ecdysteroids (Zandee, 1964, 1966, 1967; Maroun and Kamal, 1976), and where does this synthesis occur?

9.3.1. *Crustaceans*

Injection of cholesterol into decapod crustaceans results in its conversion into ecdysteroids: E and 20E (Spaziani and Kater, 1973; Gagosian et al., 1974; Willig and Keller, 1976) (see Fig. 9.3). In the crab *Carcinus maenas*, cholesterol is converted into E, 20E, and PoA (Lachaise, unpublished results). Which tissue or gland is involved in this synthesis? Echalier (1954, 1955) discovered in *C. maenas* the existence of a pair of ventral epidermal glands, the Y-organs, which had a major role in the molting of this animal: removal of the Y-organs resulted in preventing molting, whereas implantation restored it. As a consequence, Y-organs or molting gland were thought to be involved in the synthesis of ecdysteroids or molting hormones (see Chapter 5 by Spaziani in Part 2). However, other organs or tissues also seem to synthesize or release ecdysteroids: cephalic glands in crayfishes (Gersh et al., 1979) and ovaries and eggs of crabs (Lachaise et al., 1981; Lachaise and Hoffmann, 1982). Synthesis of ecdysone by the Y-organs has been demonstrated by using as precursor either (1) labeled cholesterol or (2) ketodiol (2,22,25-trideoxyecdysone), or (3) by incubation of Y-organs and detection of ecdysteroids by radioimmunoassay.

(1) The Y-organs of the crayfish incubated *in vitro* are able to convert tritiated cholesterol into ecdysone, yet the conversion obtained remains very weak: 0.015% of cholesterol is transformed into ecdysone (Willig and Keller, 1976). Some 12 h after injection of cholesterol, Y-organs of crabs incubated *in vitro* synthesized labeled ecdysone within 24 h at incubation (Watson and Spaziani, 1985b).

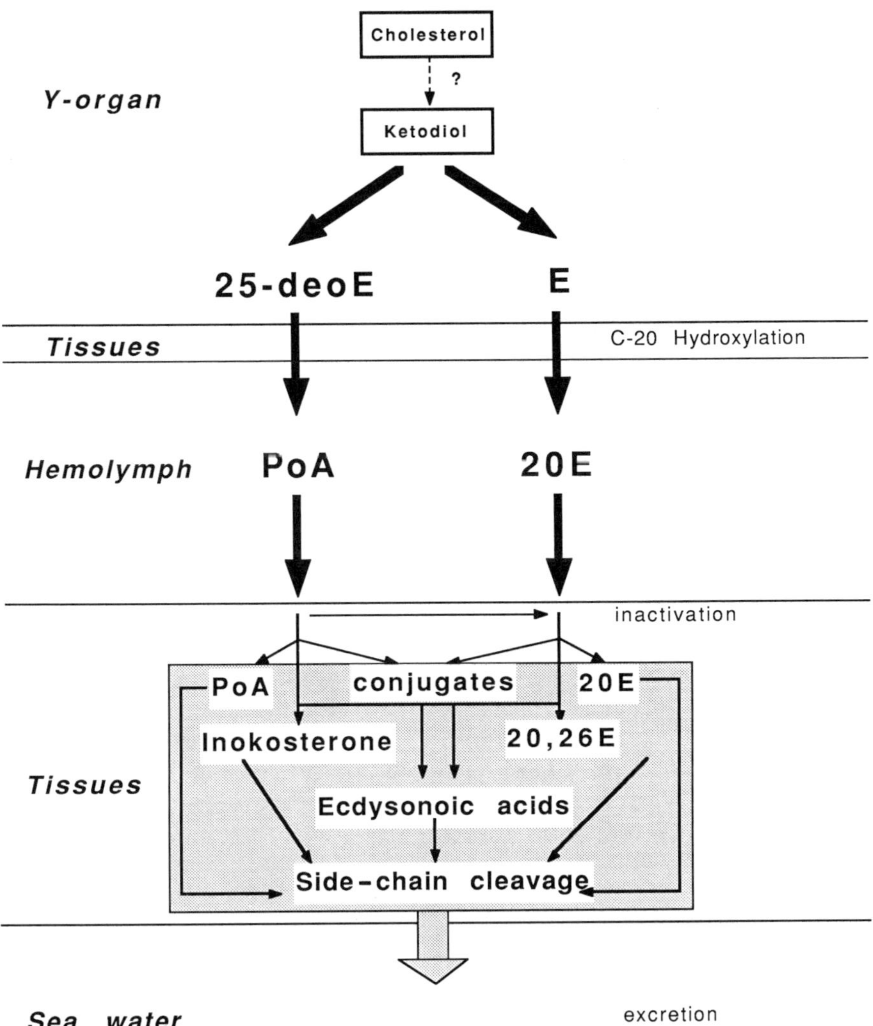

FIGURE 9.3. Ecdysteroid synthesis and metabolism in crustaceans: E = ecdysone; 20E = 20-hydroxyecdysone; PoA = ponasterone A; 25-deoE = 25-deoxyecdysone; 20,26E = 20,26-dihydroxyecdysone.

(2) Haag et al. (1985) synthesized a molecule with very high specific activity (107 Ci/mmol), the ketodiol that is probably involved in ecdysteroid synthesis (Hoffmann et al., 1980; Lachaise et al., 1981). *In vitro* this molecule is converted by Y-organs of the crab into three compounds: 22,25-dideoxyecdysone, E, and 25-deoxyecdysone (25-deoE) (Lachaise et al., 1986); 25-deoE and E alone are released into the medium (Lachaise et al., 1988), suggesting that in the intact animal they may be hydroxylated at the C-20 position by peripheral tissues, thereby producing PoA and 20E, respectively (Koolman, 1982; Lachaise et al., 1988). Moreover, Y-organs have shown a variable competence to convert ketodiol *in vitro:* Y-organs of crabs in anecdysis (females during vitellogenesis, or overwintering crabs) fail to convert ketodiol. In the successive molt period, that is, in spring and summer, the Y-organs of crabs convert ketodiol mainly into E if the animals are at the C_4 stage, but into 25-deoE if the animals are at the D_1 stage (Lachaise et al., 1986). *In vitro,* other tissues of crab are capable of converting ketodiol essentially into 2,22-dideoxyecdysone and 22,25-dideoxyecdysone (Lachaise et al., 1988). If injected, ketodiol is converted into a suite of unidentified products (Lachaise, unpublished results).

(3) During incubation, Y-organs of the crab *Pachygrapsus crassipes* synthesize ecdysteroids detectable by radioimmunoassay (Bollenbacher, 1974; Chang and O'Connor, 1977). Similarly, Y-organs of the crayfish secrete ecdysone (Keller and Schmidt, 1979). Thus, like the molting gland of insects, the Y-organs of crustaceans seem to synthesize ecdysone. Furthermore, the Y-organ synthesizes other cholesterol-derived products that have not yet been identified (Bollenbacher, 1974; Watson and Spaziani, 1985b). Ecdysteroid biosynthesis in crustaceans is regulated by a neuropeptide hormone (see Section 9.5.2.2).

9.3.2. *Chelicerates*

In the harvestman, certain cells, the oenocytes, seem capable of producing or releasing both E and 20E (Romer and Gnatzy, 1981). In ticks, ecdysteroid synthesis has not yet been elucidated either in terms of providing evidence of a precursor of ecdysone or with regard to the localization of one or more organs of synthesis. The data concerning possible precursors in ticks differ from those related to insects or crustaceans. Cholesterol, tested on argasids throughout their molting cycle and on ixodid females, is not converted into E or 20E (Connat, 1987).

Ketodiol, which is a possible ecdysteroid precursor in insects and crustaceans, is only weakly converted (0.5%) into E or 20E (Connat, 1987). Similarly, 22,25-dideoxyecdysone, transformed neither by the prothoracic glands of the insect *Locusta migratoria* (Meister et al., 1985) nor by the Y-organs of crabs (Lachaise et al., 1986), seems to be converted by females of the tick *Boophilus microplus* (Connat, 1987). Evidence for this comes from the finding of apolar conjugates (AP2), demonstrating that the compound was at least hydroxylated at the C-22 position (see Section 9.4.2, below). Another classical intermediate compound in the synthesis of ecdysteroids in insects, 2-deoxyecdysone, is mainly converted to AP1 and AP2 when injected into 5-day-old nymphs of *Ornithodoros moubata*. These apolar products liberate only 2-deoxyecdysone by esterase hydrolysis, indicating that this compound is unchanged prior to conjugation. Ecdysteroid synthesis in ticks may pass through intermediates other than those in insects and crustaceans (Diehl et al., 1986). According to Cox's (1960) experiments, the control of ecdysteroid biosynthesis in ticks might involve the synganglia.

9.3.3. *Myriapods*

In myriapods, as in both crustaceans and insects, there exists a molting gland that apparently secretes ecdysteroids (Scheffel, 1969; Seifert and Rosenberg, 1974) (see Chapter 6 by Seifert in Part 2). This synthesis seems to be regulated by an hormone originating from neurosecretory brain cells (Joly, 1966a,b).

9.4. Metabolism

Mechanisms affecting hormone inactivation or excretion are also of prime importance in the control of its action. Metabolism of ecdysteroids has been mostly studied in decapod crustaceans and ticks. It has been analyzed either by injection or ingestion of labeled compounds or by incubation of the organs in the presence of labeled hormones followed by chromatographic analysis of the metabolites formed.

9.4.1. *Crustaceans*

In crustaceans, the major ecdysteroids studied have been E, 20E, and PoA. A number of studies have provided evidence of the *in vivo* conversion of E into 20E (King and Siddall, 1969; Gorell et al., 1972; Kuppert et al., 1978; McCarthy, 1980, 1982; McCarthy and Skinner, 1979b). However, in the amphipod *Orchestia cavimana*, although the major ecdysteroid seems to be 20E, labeled E, when injected, is not converted into 20E after injection (Connat and Diehl, 1986; Graf and Delbecque, 1987). Similar experiments have been performed *in vitro* (Gorell et al., 1972; Chang et al., 1976; Lachaise and Feyereisen, 1976; Chang and O'Connor, 1978; Daig and Spindler, 1980). Most tissues other than Y-organs convert E into 20E *in vitro* (Lachaise and Feyereisen, 1976). Ecdysone metabolism in decapods proceeds via the usual inactivation pathways known in insects, i.e., conjugation and 26-hydroxylation, leading to the formation of polar products (Lafont and Koolman, 1984). In the crayfish, Gorell et al. (1972) and Kuppert et al. (1978) have detected unidentified apolar metabolites of ecdysone. Metabolism of 20E and PoA were studied concomitantly in *Carcinus maenas* in relation to the molt cycle (Lachaise and Lafont, 1984). In this crab, PoA is weakly converted into 20E both *in vitro* and *in vivo*. In contrast to 20E, PoA is little excreted into seawater as free ecdysteroid (possibly due to its poor solubility).

PoA and 20E are inactivated and released into seawater as ecdysonoic acids, via 26-hydroxylation, and as conjugates. As in insects or in ticks, these inactivation pathways may be more or less important, depending on the molting stage. Both hormones, 20E and PoA, can also be bound, in a noncovalent way, to polar products resistant to enzymatic hydrolysis by *Helix pomatia* juice. Such unidentified compounds have been found in the crab *Gecarcinus lateralis* after labeled ecdysone injection (McCarthy and Skinner, 1979b; McCarthy, 1980). The major inactivation pathway might be the side-chain cleavage between the C-20 and C-22 positions. This side-chain scission was proved to occur in insects (Hikino et al., 1975), but it only represents a minor inactivation pathway in this group of arthropods. In contrast, in crabs injected during spring or summer, this inactivation pathway accounts for up to 90% of the compounds formed. This cleavage can lead to a new class of endogenous poststerone-derived ecdysteroids (Lachaise and Lafont, 1984).

In the crab *Carcinus maenas*, the amphipod *Orchestia cavimana*, and the isopod *Porcellio dilalatus*, ecdysone-derived apolar compounds have been detected; the percentage conversion of ecdysone into these apolar compounds is weak in *Carcinus* (8.2%) compared with that obtained in the two other terrestrial crustaceans and in both chelicerates and insects, where it may represent up to 95%. The qualitative and quantitative variations in the metabolites formed or excreted suggest that the ecdysteroid ratio is regulated by both metabolism and synthesis.

9.4.2. *Chelicerates*

Like insects and crustaceans, the arachnids, and more especially the ticks, convert injected tritiated E into 20E, irrespective of the developmental stage (Bouvier et al., 1982; Connat et al., 1984, 1987) (see Fig. 9.4). In the 5-day-old nymph of *Ornithodoros moubata*, 20E is converted into 20,26E (Bouvier et al., 1982). More generally, in a number of ticks, E and 20E are converted into polar products whose nature has not always been identified (Connat et al., 1985, 1987). Some of these products correspond to ecdysonoic acids (Crosby et al., 1986). *In vitro*, the ovaries of the tick *Amblyomma habraeum* are able to produce a 3-epimer of ecdysone and probably also 3-dehydroecdysone (Connat et al., 1986b, 1987). Nevertheless, this inactivation pathway does not seem to be a major one in ticks. By contrast, formation of apolar con-

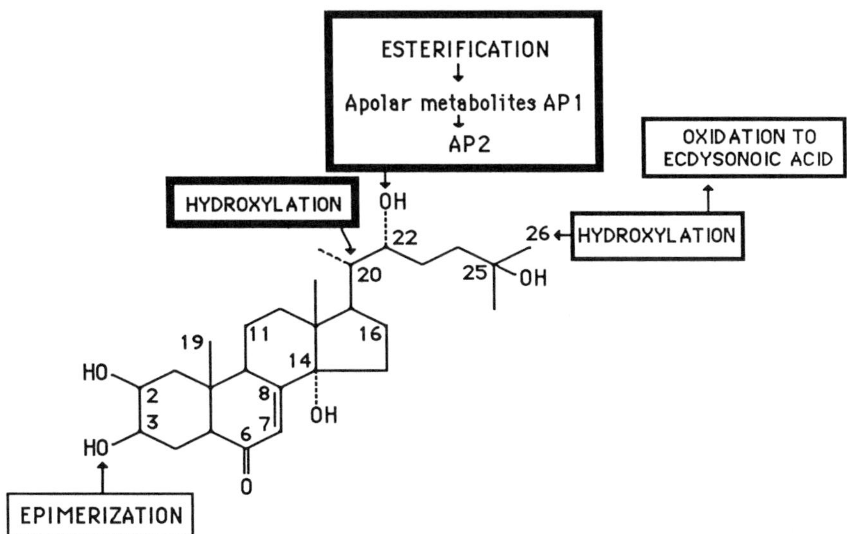

FIGURE 9.4. Ecdysteroid metabolism in chelicerates.

jugates is an important ecdysteroid inactivation pathway in these animals. These conjugates have been obtained from E and 20E after either injection (Bouvier et al., 1982; Connat et al., 1984, 1987), ingestion (Diehl et al., 1985), or incubation of tissues (Connat et al., 1984, 1986b). These apolar conjugates comprise two groups of compounds, called AP1 for the less apolar ones, and AP2 for the other.

The AP2 compounds are ecdysteroids esterified at C_{22} with palmitic, stearic, oleic, or linoleic acids (Diehl et al., 1985); they are further converted into AP1 (Connat, 1987). Such an esterification process can be attributed to detoxification of ingested hormones, as in the tick *Ornithodoros moubata* and the spider *Tegenaria ferruginea* (Connat et al., 1986a; Connat, 1987), but with endogenous hormone inactivation as well (Connat, 1987). Conjugation of fatty acids to ecdysteroids appears to be a general inactivation pathway of these hormones in arthropods (Connat and Diehl, 1986). However, although it is actually a major one in terrestrial arthropods (ticks, spiders, insects, terrestrial crustaceans), it seems to be noticeably less important in marine arthropods, where the principal inactivation pathway appears to be side-chain cleavage.

9.5. Effect on Molting

Ecdysteroids intervene in the postembryonic development of arthropods and, more especially, in their morphogenesis, that is, molting and regeneration. Their role in the molting has been studied either by quantitative analysis in relation to molt achievement, by assessment of their action on target tissues, or through modifications of the endogenous titer. The titer of endogenous ecdysteroids has been modified either by removing the organs involved in molt control or by the addition of exogenous hormones.

9.5.1. *Endogenous Titer and Molting*

9.5.1.1. MOLT CYCLE

In arthropods, the egg gives rise to a nymph or a larva, whose habit may greatly differ from that of the adult. These nymphs or larvae undergo one or several consecutive molts or a true metamorphosis; the last larval molt gives rise to an adult-looking juvenile that is unable to reproduce. It will pass through one or several additional molts, during which its reproductive organs will develop up to the adult size. Although this adult-looking juvenile has been called a nymph in chelicerates (except the Limulidae), it is not actually analogous to the true nymph in insects. In crustaceans, myriapods, and *Limulus* (Jegla, 1982), the juvenile is called a larva until the adult stage (Fig. 9.5). Postembryonic development of larvae, nymphs, and certain adults is accompanied by molting cycles, including an ecdysis (or molt) period during which the animal emerges from the exuviae and a between-molt period that separates two consecutive ecdyses.

The between-molt period can be subdivided into *postmolt*, or metecdysis (just after the molt); *intermolt* per se, or anecdysis (the animal has accomplished its molt but has not yet entered the following one); and *premolt*, or proecdysis (the animal secretes a new cuticle under the old one). There are two ways in which molts proceed in arthropods. There may either be larval molts followed by adult molts (e.g., most crustaceans, myriapods, and apterygote insects) or larval molts

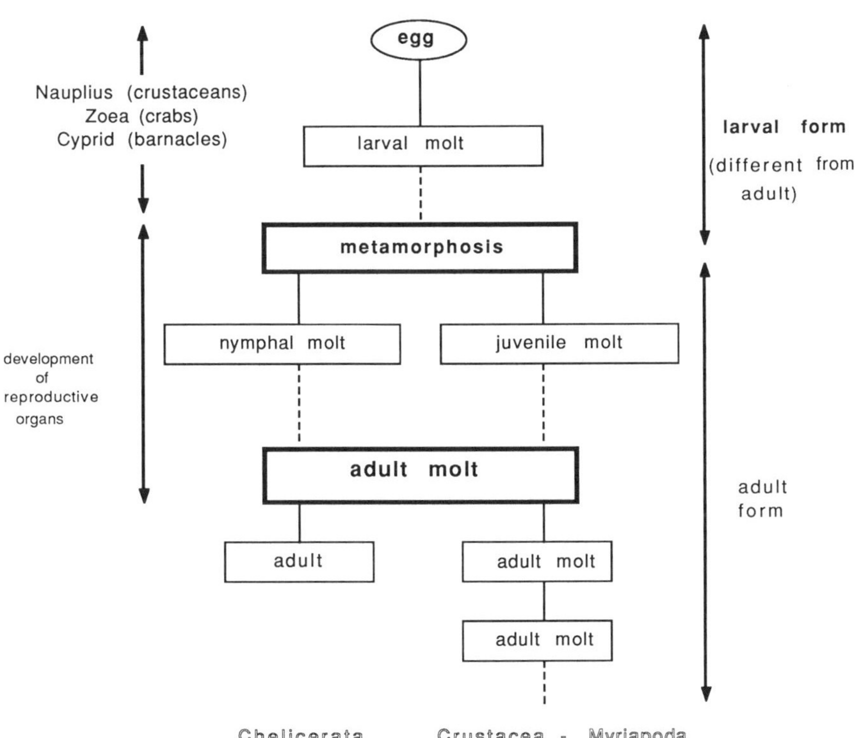

FIGURE 9.5. Development scheme of chelicerata, crustacea, and myriapoda.

exclusively (e.g., most chelicerates and pterygote insects). In animals exhibiting adult molts, the frequency of molting decreases with age (myriapods: Joly, 1966a,b; crustaceans: Skinner, 1985). However, whether the molt occurs in the larva or adult, the span of time between two molts has also been divided into more precise periods designated A, B, C, and D (Drach, 1939) (see Fig. 9.6). This partitioning of time, initially established in decapod crustaceans, enables one to always know the actual physiological state of the animal with respect to the molt. This appears to be more accurate than dating in days, say, from the molt (Stevenson, 1985).

Postmolt (metecdysis) is characterized by gradual cuticle rigidity. Just after the molt, the carapace of the starved animal is soft (A, B, C_1, C_2, and C_3 periods according to Drach). Generally, the duration of this period is constant for a given category of animal.

Intermolt per se (anecdysis) corresponds to a period that may be either nonexistent (e.g., in juvenile crabs during spring or summer time) or exceedingly prolonged (e.g., it may last several months in adult female crabs during vitellogenesis and during the embryonic development of eggs). Throughout this lengthy intermolt, the animals normally feed and no phenomena relevant to molt preparation are visible (C_4 period according to Drach).

Premolt (proecdysis) is characterized by an interruption of feeding and by secretion of a new cuticle under the old one (D period according to Drach). This D period has been subdivided further on the basis of microscopic observations concerning the tip of a transparent appendix like the ''gill epipodite'' in the crab (Drach and Tchernigovtzeff, 1967). Epidermis retraction (D_0 stage) or apolysis (Jenkin and Hinton, 1966) corresponds to the first manifestation of entry into premolt. The D_1', D_1'', and D_1''' stages correspond to a gradual invagination of the epidermis of the new setae, and in the D_2 stage enough cuticle has been secreted to be visible. In the course of the D_3 stage, the line of dehiscence will break under fingernail pressure. During the D_4 stage, the cuticle has split open along the line of dehiscence as the animal expands by absorption of water. At this stage, beyond a critical period that is species specific, nothing will prevent the occurrence of molting.

Attempts to define similar landmarks in chelicerates and myriapods have failed to reveal such a transparent appendix that would permit observation of the steps of new seta formation and new cuticle secretion. Most often, the stages have been determined, as in insects, by using finite time units from the molt. The number of days necessary for apolysis to be established was defined from histological analysis of teguments at various times after the molt.

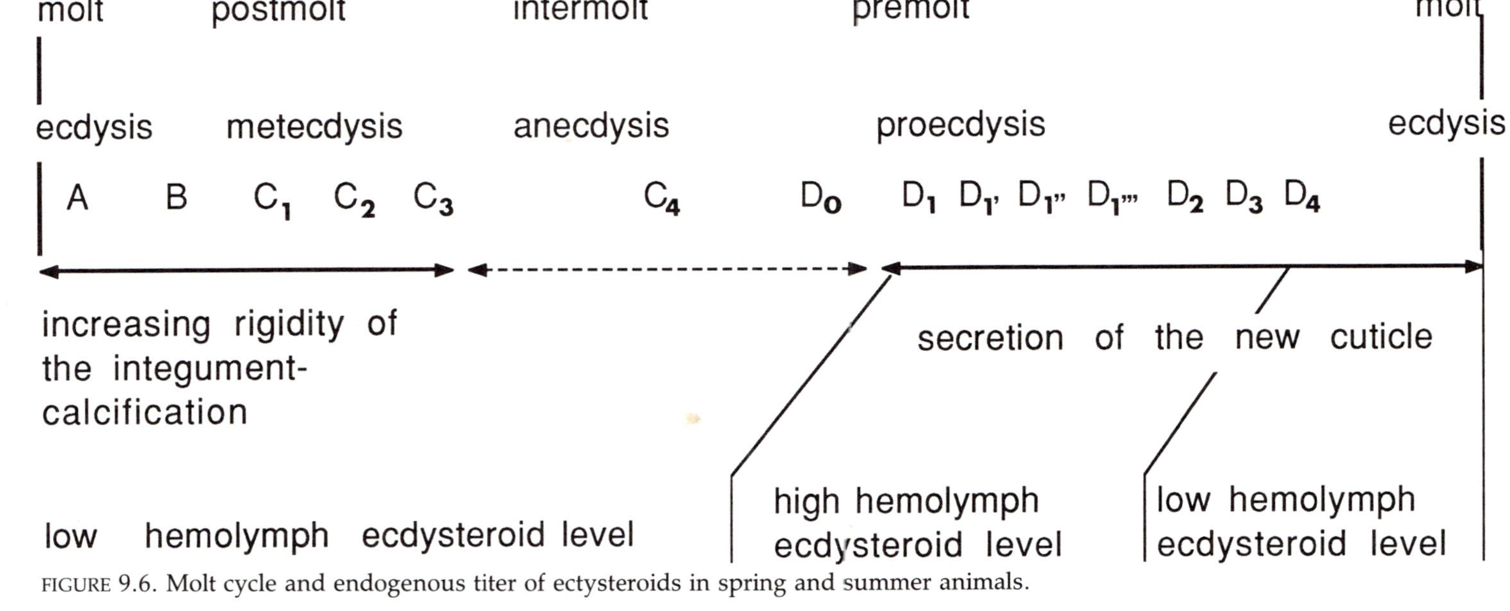

FIGURE 9.6. Molt cycle and endogenous titer of ectysteroids in spring and summer animals.

9.5.1.2. TITER VARIATIONS DURING THE MOLTING CYCLE

Molting is preceded by a peak in the concentration of hemolymphatic ecdysteroids in all arthropods, whether they be chelicerates, myriapods, or crustaceans (see Fig. 9.6). The reduction of circulating ecdysteroids, which occurs during late premolt, is due to a marked reduction in Y-organ activity, possibly induced by ecdysteroid feedback upon Y-organs itself (Hopkins, 1986) or eyestalk ganglia (Mattson and Spaziani, 1986). The maximal ecdysteroid titers vary depending on animals, the methods of extraction, and the tissue used (Spindler et al., 1980; Skinner, 1985). It should be noted that the increase in ecdysteroid concentration in tissues at the very time of the hemolymph peak may be due to the presence of hemolymph itself, if the relevant tissue is well irrigated and voluminous. After being synthesized in a given organ, the hormones are released into the hemolymph as free ecdysteroids or linked to carrier proteins. The cells of the target organ(s) take up the free hormone from the circulatory system and react to the steroid by a specific response. Metabolism occurs either by inactivation of the active hormone, whose metabolites can be found in the hemolymph and excreted, or by activation of a prohormone (Koolman, 1982). The action of hormones on the molt can be evaluated by studying the ecdysteroid levels in the hemolymph. There are two ecdysteroids, E and 20E, common to crustaceans, chelicerates, and myriapods, and present at the very time of the peak hormone concentration coinciding with the molt.

In the crab, PoA and 20E are the major ecdysteroids present at the time of the increase in the hemolymph ecdysteroid concentration (McCarthy, 1979; Lachaise and Lafont, 1984). PoA is quantitatively important in spring wild-caught animals that are undergoing successive molts (Lachaise et al., 1988). Inokosterone, also found in crustaceans (Faux et al, 1969), is a metabolite of PoA (Lachaise and Lafont, 1984). Its presence during late premolt in the crab *Callinectes sapidus* might be due to endogenous PoA degradation. PoA is present in crustaceans and apterygote insects, and absent in pterygote insects and ticks studied (Table 9.1). Considering that the two former groups of organism continue to molt as adults whereas the latter two do not, it is assumed that PoA plays a role in the preparation of adults molts. From the evidence that the activity of PoA may be 50 times greater than that of 20E (Bergamasco and Horn, 1980), the concern then becomes with the capacity of PoA to play a role in molting either in synergy with 20E or by itself.

9.5.2. *Effects of Organ Removal*
on Endogenous Titer

The titer of endogenous ecdysteroids can be modified in crustaceans
by means of Y-organ, eyestalk, or sinus gland removal, intensive limb
autotomy, or by means of exogenous ecdysteroid supply (see Fig. 9.7).

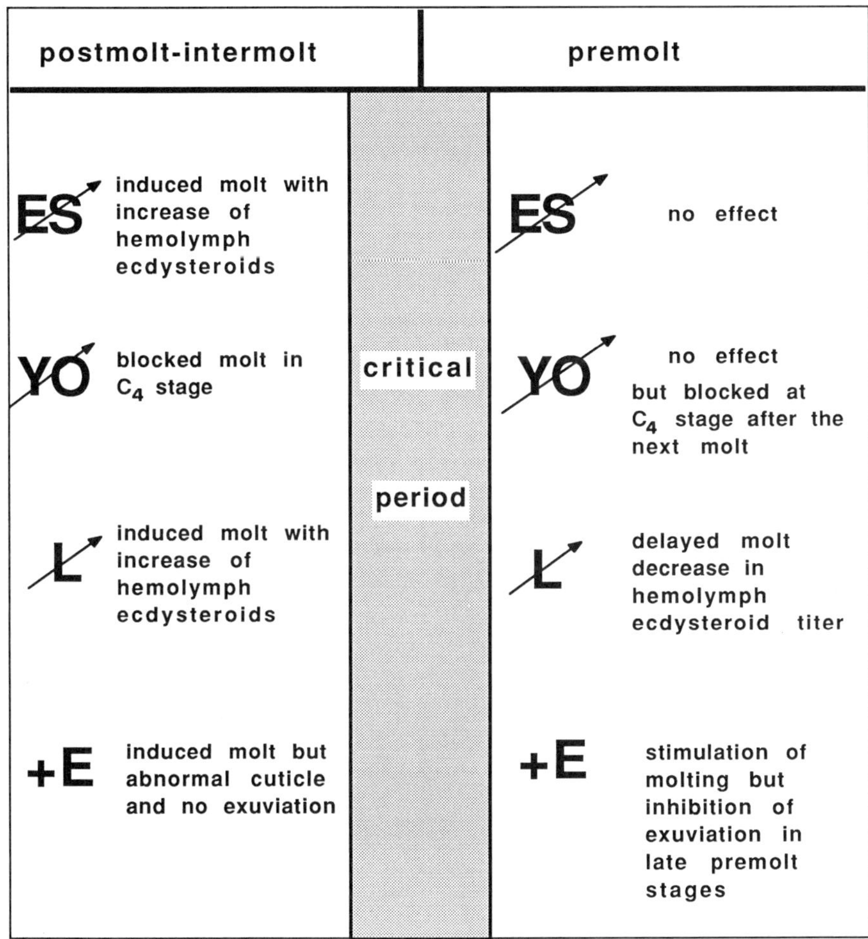

FIGURE 9.7 Effects of eyestalk removal (ES), Y-organs removal (YO), limb
autonomy (L), and exogenous ecdysteroid (+E) upon molt and hemolymph
ecdysteroid titers.

9.5.2.1. REMOVAL OF Y-ORGANS

Other than in *Pachygrapsus marmoratus* (Charmantier-Daures and DeReggi, 1980), removal of Y-organs (i.e., molting gland) or their degeneration (Charmantier et al., 1977) in crustaceans results in molt inhibition (Passano and Jyssum, 1963; Maissiat and Legrand, 1970; Blanchet, 1974; Burghause, 1975). When the organs are removed during intermolt (C_4), molting does not occur; when removal occurs during premolt (D_2), the animal is blocked in the C_4 stage after the next molt (Echalier, 1959). Considering that injection of 20E into such animals restores molting but not ecdysis (Blanchet, 1974; Charmantier and Trilles, 1973; Susuki, 1985), it is clear that this removal provokes a decrease in the titer of circulating ecdysteroids (Soumoff and Skinner, 1982).

9.5.2.2. REMOVAL OF EYESTALKS

Removal of eyestalks in the adult as well as in the larva mostly accelerates premolt initiation and provokes an increase in hemolymphatic ecdysteroids (Chang, 1985; Skinner, 1985). Implantation of X-organs (the neurosecretory system in the eyestalks of decapod crustaceans) into eyestalkless crabs restores molt inhibition (Passano, 1960), and injection of eyestalk extracts inhibits adult molt in *Carcinus* (Carlisle, 1957) and larval molts in lobster (Chang and Bruce, 1980; Snyder and Chang, 1986). Concomitant removal of eyestalks and Y-organs results in molt inhibition (Carlisle, 1957; Echalier, 1959; Démeusy, 1959; Passano and Jyssum, 1963). Hence, it can be assumed that the inhibitory action of eyestalks of X-organs proceeds via the Y-organs. In 1982, Soumoff and O'Connor, using an *in vitro* incubation technique, corroborated the statement that eyestalk extract inhibits secretion of ecdysteroids by the Y-organs.

In decapod crustaceans, the eyestalks accumulate a neurohormone or molt-inhibiting hormone (MIH) that acts directly on Y-organs by inhibiting ecdysteroid synthesis (Soumoff and O'Connor, 1982) from either cholesterol (Watson and Spaziani, 1985a,b) or ketodiol (Lachaise et al., 1988). This hormone appears to be a neuropeptide (Rao, 1965). A neuropeptide with an inhibitory effect on ecdysteroid synthesis by Y-organs has been isolated in *Carcinus maenas* by Webster and Keller (1986) and in the lobster by Chang et al. (1987). Soyez and Kleinholtz (1977) advanced the hypothesis that tryptophan or a tryptophan-yielding peptide might be responsible for MIH action. The putative neurohormone isolated from *Carcinus* has an inhibitory effect on ketodiol conversion by Y-organs (Lachaise et al., 1988) and by prothoracic glands of *Pieris brassicae* (P. Beydon, unpublished data). In competent Y-organs, 25-deoE formation is selectively inhibited by MIH or eyestalk extracts, suggesting that this neuropeptide controls both the nature and quantity of the hemolymph ecdysteroids.

However, while regulation of the hemolymph ecdysteroid titer is subject to the inhibitory action of MIH on ecdysteroid synthesis by Y-organs, MIH may act at other levels (Freeman and Bartell, 1976). This statement is supported by the finding that after injection of an eyestalk extract into the last abdominal segment of the shrimp *Palaemonetes pugio*, followed by a thoracoabdominal injection of 20E, a delay in setogenesis in the uropods, as compared with the accelerated setogenesis in the antennal scale, was observed. Freeman and Bartell conclude that there may be a possible competition between the molting hormone and MIH at the level of target tissues. However, the possibility also exists that MIH operates by accelerating 20E degradation or excretion. During anecdysis, crabs are blocked by MIH in the C_4 stage (Skinner, 1985). Considering that MIH action is reversible (Mattson and Spaziani, 1985a), it might be assumed that the Y-organs of such blocked crabs, when incubated *in vitro* (i.e., without MIH control), should convert ketodiol, but clearly they do not (Lachaise et al., 1986). These results, together with the occasional lack of molting initiation by eyestalk removal (Skinner, 1985), suggest the following possibilities: (1) a long-term delay of MIH action in the Y-organs of the donor crab; (2) a variable competency of Y-organs according to the molting stage of the animal and season; and (3) a regulation by factors other than MIH.

In chelicerates the regulation of the ecdysteroid titer remains poorly known. In myriapods, a brain gland was shown to contain a molt-inhibitory hormonal control (like the eyestalk in crustaceans) (Joly, 1966b). As in crustaceans, this factor acts during the intermolt period and in early premolt, but not in late premolt. Like crustaceans, myriapods possess molting glands (Seifert and Rosenberg, 1974), and it is likely that the brain glands act through them.

9.5.3. *Effects of Exogenous Ecdysteroid on Endogenous Titer*

The role of ecdysteroids in the arthropod molt has also been approached by using an exogenous hormonal supply (Fig. 9.7). The addition of exogenous ecdysteroids result in initiation of molt, although this is not an absolute rule. The animal frequently died at the time of exuviation. The effects of exogenous ecdysteroids depends on both the nature and the quantity of the hormone used and on the methods applied (injection, ingestion, or topical application); they are also influenced by external factors and the physiological state of the animal at the time of exogenous hormonal addition. The action of exogenous ecdysteroids shows itself in the initiation of molt and in ultrastructural changes at the level of target tissues.

9.5.3.1. METHOD OF APPLICATION AND NATURE OF ECDYSTEROIDS

The quantity of hormone used has a variable effect, depending on the animals, the treatment method, and the nature of the ecdysteroid tested. In certain chelicerates and crustaceans, a minimal dose (i.e., a dose below which the hormone is ineffective) and a maximal dose (i.e., a dose above which the hormone becomes toxic) have been established. Furthermore, dose-vs.-response rapidity correlations have been plotted (Krishnakumaran and Schneiderman, 1970; Herman, 1972; Hubschman and Armstrong, 1972; Jegla et al., 1972; Tighe-Ford and Vaile, 1972; Rao et al., 1973; Bonaric, 1976; Freeman and Bartell, 1976; Gilgan et al., 1977; Gilgan and Burns, 1979; Nair, 1980; Madhavan, 1981; Connat et al., 1983; Mellon and Greer, 1987). However, it is somewhat difficult to generalize owing to the heterogeneity of the concentrations and the nature of endogenous hormones that may sometimes be specific to the animal studied. Nevertheless, results from these studies show that the quantities utilized are in general unphysiologically high. In some cases, molt initiation is accomplished by very different doses depending on the type of hormone used or the method of introduction applied. In ticks, for which it is quite easy to supplement the blood food supply with ecdysteroids, a comparative analysis was carried out opposing the action of ecdysteroids by ingestion to topical application. In the tick *Ornithodoros moubata*, 10–20 ng per female of 22,25-dideoxyecdysone, when ingested, provokes a molt, whereas 5 ng per tick quantity applied topically has no effect (Connat et al., 1986a). In contrast, in the closely related species *Ornithodoros porcinus*, the reverse situation is observed: the quantity of ecdysone mixed with the blood food supply must be significantly higher than that applied topically to induce supermolting (Mango et al., 1976). In ticks, ingested ecdysteroids with an OH group at the C-22 position are inactivated by the digestive tract (see Section 9.4.2.). As a consequence, these animals are strongly resistant to ingestion of

E, 20E, PoA, and makisterone A (Connat et al., 1986a), but not to 22,25-dideoxyecdysone. Surprisingly, ketodiol, albeit devoid of an 22-OH group, is inactive when ingested by argasides at doses around 5 ng/ml (Diehl et al., 1986). Similarly, neither 22-deoxy-20-hydroxyecdysone nor poststerone had an effect when ingested at 5 and 10 ng/ml, respectively (Connat, 1987).

For a given ecdysteroid, there is little difference in activity in response to different methods of application: injection, topical application, or introduction into an aqueous phase (the most common method for aquatic crustaceans). E and 20E, which are the most widespread hormones in arthropods, have been the ones most tested for their action, but other ecdysteroids have been the subject of similar investigations. It can be assumed that according to their nature the action of ecdysteroids is variable and a circulating ecdysteroid precursor or metabolite does not necessarily have the potential for inducing a molt. In crustaceans, diverse sterols and ecdysteroids have been tested: (1) possible ecdysteroid precursors—cholesterol, triol [BSEA-1 (Krishnakumaran and Schneiderman, 1970)], and sitosterol; (2) circulating ecdysteroids—E, 20E, and PoA; (3) possible metabolites—inokosterone and 20E triacetate; and (4) a phytoecdysteroid—cyasterone.

In crayfish, neither cholesterol nor sitosterol is capable of inducing a molt in contrast to triol (Krishnakumaran and Schneiderman, 1970). In crustaceans, triol is not converted by tissues, and this could account for a specific action of this compound in crabs, as in ticks. In the crayfish *Procambarus*, the range of action of the ecdysteroid series is nearly 100% for 20E, 86% for inokosterone, 66% for ecdysone, 46% for PoA, and 30% for cyasterone (Krishnakumaran and Schneiderman, 1970).

Inokosterone, a possible metabolite of PoA in the crab *C. maenas* is also active in *Procambarus clarkii* (Fukutake et al., 1969) and *Homarus americanus* (Gilgan et al., 1977). In *Homarus*, inokosterone has an effect similar to 20E, although at much higher doses (Gilgan et al., 1977).

A pretreatment of spring animals by E prior to 20E injection, leads to a normal molt in the lobster, whereas 20E injection alone provokes death of the animal during the molt (Gilgan et al., 1977). 20E triacetate (a possible metabolite of 20E) has a positive effect on molt initiation, probably owing to the slow release of free 20E from acetates (Gilgan and Burns, 1976a,b). The results do not vary noticeably whether they are based on a single injection or several successive injections (or topical applications). Repeated injections generally seem to be better tolerated by the animals, since the dose given at each injection is smaller (Tighe-Ford and Vaile, 1972; Stevenson and Tschantz, 1973; Gilgan and Zinck, 1975; Madhavan, 1981). However, in the western rock lobster, *Palinurus longipes*, two successive injections were more effective in molt initiation than multiple ones (Dall and Barclay, 1977); this result suggests that successive injections can inhibit ecdysteroid synthesis, possibly owing to steroid feedback (Hopkins, 1986).

Various authors have compared the action of E and 20E. In the crayfish *Orconectes obscurus*, E and 20E have an equal effect (Warner and Stevenson, 1972). In the lobster, animals injected with 20E die during the initial stage of ecdysis, whereas those injected with E die in the later stages of ecdysis (Rao et al., 1973). Spring lobsters receiving both ecdysteroids passed through an accelerated premolt development, molted, and survived longer than those treated with a single ecdysteroid; E pretreatment seemed to inhibit the lethal effect of the preceding 20E and to permit some successful molts (Gilgan and Burns, 1977). In the amphipod *Orchestia gammarella*, the delaying response is shorter for 20E than for E; this might correspond to the time required for conversion of E into 20E by the tissues (Blanchet, 1972). In *Limulus* larvae, different ecdysteroids have been tested by injection (Herman, 1972; Jegla et al., 1972; Jegla and Costlow, 1979; Jegla, 1982). The doses used to obtain 50% molt at 25°C vary according to the ecdysteroids tested; the best results were achieved with cyasterone (a phytoecdysteroid) and PoA. The triol, active in ticks and crustaceans, is more active than E and 20E, which are the circulating hormones in *Limulus*.

9.5.3.2. BREEDING CONDITIONS

When introduction or hormone does not produce any effect, it may be that the animal is blocked due to unsuitable breeding conditions. Hence, restoration of suitable conditions should allow expression of the action. Other than insects, most arthropods studied are wild-caught animals that are maintained under laboratory conditions for studies. Molting makes animals very vulnerable, and the conditions in which they are kept might be an important factor in blocking molting. Temperature and photoperiod seem to have a variable effect on molt frequency depending on the animal considered. *Limulus*, whether alone or in a very crowded condition, and in the presence or absence of sand for burrowing, long or short photophase, and high or low temperature, reacts in a similar way to 20E injection; only starvation can modify the action of 20E (Herman, 1972).

The intermolt period of the first three instars of *Limulus* larvae was significantly longer at 10°C (Jegla and Costlow, 1982). Lobsters placed at temperatures ranging from 2° to 10°C, fail to respond to 20E injection, whereas control lobsters placed at 15°C enter premolt as soon as the hormone is injected (Gilgan and Burns, 1977). Gilgan and Burns conclude that there is an alteration of sensitivity to ecdysteroids by cold. Photoperiod has no effect on the molting frequency of *Balanus balanoides*, and temperature has little influence on the molting frequency when the animals are starved (Barnes and Stone, 1974). Handling stress depresses the ecdysteroid titer in the crab *Cancer antennarius*; this response is probably mediated by MIH release (Mattson and Spaziani, 1985b).

9.5.3.3. TEMPORAL ANECDYSIS

Juvenile chelicerates, like juvenile and adult myriapods or crustaceans, when caught in the wild and kept in suitable conditions, sometimes remain insensitive to ecdysteroid injection. It is likely that factors other than temperature and light intervene in blockage of the molt. In most of the animals studied, originating from regions with a seasonal cycle of temperature and photophase (spiders, crustaceans, or myriapods), there is an accelerated series of molts in spring and summer, and a slowing down, or even a complete stop, in autumn and winter. Molt interruption can also occur in females, which temporarily stop molting during the period of ovarian maturation and egg development. In temporarily blocked animals, initiation of molting by exogenous ecdysteroids occurs to a variable extent, depending on the species and the cause of blockage.

Termination of Diapause

Terrestrial animals, like chelicerates, whose development does not include any adult molt, have diapause as in insects. By injecting 20E into the spider, Bonaric (1977) successfully reinitiated molting even in winter, but the response to ecdysteroid injection is slower in animals caught in the wild from October to December than in spring and summer. In immature ixodids, applications of E, 20E, or 22,25-dideoxyecdysone terminated larval diapause (Wright, 1969; Sanassi and Subramoniam, 1972).

Reproduction Anecdysis Break

In adult female crustaceans, there is close interdependence between molting and the reproductive phenomena. The animal may either accomplish its complete reproductive cycle (ovarian development and egg laying) during the intermolt (e.g., in crabs) or lay its eggs immediately after molting (e.g., amphipod or prawn). In egg-carrying crustaceans, molting is blocked until hatching; it may also happen that ovarian maturation spreads over two intermolts (e.g., peracarids). In the case of the ovigerous and hence blocked females of the isopod *Armadillidium vulgare,* topical application of 20E induces molt and inhibits egg development, suggesting that anecdysis results from a lack of circulating ecdysteroids (Madhavan, 1981). In female *Carcinus* blocked in the reproductive period, the hemolymph yields little or no ecdysteroids (Lachaise et al., 1981), and Y-organs fail to convert ketodiol (Lachaise et al., 1988). Hence, there is strong evidence to suggest that termporarily blocked females can maintain a low level of circulating ecdysteroids by inhibiting their synthesis. Alternatively, it has been suggested that the action of the ecdysteroids is inhibited at the level of the target tissue or that all the hormones injected are immediately inactivated (Freeman and Bartell, 1976). In the ovigerous shrimp *Palaemonetes,* 20E does not induce molting, and Hubschman and Armstrong (1972) hypothesized that a molt-inhibiting factor would be sufficient to block the action of 20E.

Seasonal Anecdysis Break

In the crayfish *Orconectes obscurus* (Warner and Stevenson, 1972) and in the lobster (Gilgan et al., 1977), the effect of ecdysteroids differs, depending on whether the animal was injected during autumn or winter or during spring or summer. In the autumn–winter period, the injected lobsters die during early proecdysis or are refractory to ecdysteroid injection (Gilgan et al., 1977). In both *Carcinus* (Adelung, 1967) and *Procambarus simulans* (Lowe et al., 1968), no initiation of molting was observed after E or 20E injection. Male lobsters treated with 20E molted with the same frequency as did untreated controls (Gilgan, 1980). Crabs *(Carcinus maenas)* collected during the autumn–winter period are blocked at various intermolt stages (up to D_2), and eyestalk removal has no effect on molt initiation (Lachaise unpublished); in winter animals, Y-organs are incompetent to convert ketodiol and their hemolymph contains few ecdysteroids (Lachaise et al., 1988).

Moreover, in *Procambarus*, eyestalk removal allows molt acceleration by ecdysteroids (Lowe et al., 1968): MIH or another molt-inhibiting factor might act at the level of the target tissues, in addition to inhibiting ecdysteroid synthesis. These hypotheses seem to support those of Gilgan and Farquharson (1977), who showed that the lobster became more sensitive to exogenous 20E at the D stages (i.e., when MIH level is supposed to decrease).

9.5.3.4. TERMINAL ANECDYSIS

In animals definitively blocked in terminal anecdysis (Carlisle, 1957), i.e., chelicerates or more rarely crustacean adults, it nevertheless remains possible to restart the molt. This applies to the adult male of the crustacean *Sphaeroma*, where 20E injection triggers molting (Charmantier and Trilles, 1973, 1976). In argasid ticks, it is also possible to experimentally induce supermolting (Kitaoka, 1972; Mango et al., 1976; Ahmed and Bassal, 1982; Solomon et al., 1982; Connat et al., 1983; Campbell and Oliver, 1984; Pound et al., 1984; Diehl et al., 1986). In ixodid ticks, however, it is not possible to induce molt in the adult female by ecdysteroid injection (Connat, 1987).

9.5.3.5. SUCCESSIVELY MOLTING ANIMALS

The animals displaying successive molts in the spring–summer period are juvenile chelicerates, myriapods, and crustaceans. In these animals, the action of ecdysteroids is also variable, depending on the time when injection occurs with respect to the molt. The molting cycle is shortened when E or 20E are injected during the C period in crustaceans (Vernet, 1976; Spindler et al., 1980; Skinner, 1985). By contrast, at the D stage (i.e., late premolt), the ecdysteroids have an inhibitory effect in isopods (Graf, 1972; Maissiat and Graf, 1973) and amphipods (Blanchet, 1972), and no effect in shrimps (Hubschman and Armstrong, 1972; Touir and Charniaux-Cotton, 1974) and lobster (Rao et al., 1973). A comparative study of the action of ecdysteroids on diverse noninsect arthropods indicates that 20E injection triggers molting. Mostly, the animals tested die during exuviation (in crustaceans, Spindler et al., 1980; in ticks, Solomon et al., 1982; in myriapods, Scheffel et al., 1974; in spiders, Bonaric, 1986). In a few cases, ecdysteroid injections induce normal molts (Reidenbach, 1971; Warner and Stevenson, 1972; Herman, 1972; Gilgan et al., 1977). In decapods and

Limulus, molt is normal when the animals come from spring collections (Warner and Stevenson, 1972; Herman, 1972) or when the two hormones (E and 20E) are administered consecutively (Gilgan et al., 1977). Various authors have postulated the existence of an exuviation factor (Graf, 1972), probably synthesized in the Y-organs (Blanchet, 1974; Vernet, 1976; Skinner, 1985). The presence of PoA in the hemolymph of crustaceans could suggest a role of this hormone in the exuviation process.

9.6. Effects on Target Tissues

In contrast to the numerous studies devoted to the correlation between ecdysteroids (endogenous or exogenous) and molting, few investigations have focused on the mode of action of steroid hormones at the cellular or molecular level. The target tissues involved in ecdysteroid action are directly implicated in cuticle secretion, e.g., the epidermis. Ecdysteroids may have an indirect action on molting processes via the hepatopancreas, Y-organs, and neurosecretory cells, on gastrolith formation and calcification, and finally on tanning (Fingerman and Yamamoto, 1964).

9.6.1. *Morphological Modifications of the Epidermis*

In the tissue, cellular, and molecular molting processes, the epidermis secretes the new cuticle and intervenes in the resorption of the old one. This secretion implies morphological modifications, including retraction of the epidermis (apolysis) and growth of the epidermic cells, which increase in size and multiply during a limited period after apolysis and then regress during postmolt (Stevenson, 1985).

9.6.1.1. *IN VIVO* EXPERIMENTS

In two crustaceans *(Procambarus* and *Uca)* and three chelicerates *(Araneus cornutus, Dugesiella hentzi,* and *Limulus polyphemus),* the cuticle, secreted after 20E injection, appeared to be abnormal, particularly in its surface pattern and in the morphology of dactylus and claws for crustaceans and spines and bristles for spiders; moreover, animals that received higher doses of 20E secreted a thinner new cuticle than those that received lower doses (Krishnakumaran and Schneiderman, 1970).

9.6.1.2. *IN VITRO* EXPERIMENTS

Using *in vivo*–explanted fragments of larval antennae of the myriapod *Lithobius forficatus*, Scheffel (1983) showed the molt-stimulating effect of E and 20E. One-day exposure to 20E induced ecdysial responses in the isolated appendages and penis of the barnacle *Balanus eburneus*, whereas no change occurred in barnacles kept in seawater containing sitosterol or in controls (Cheung and Nigrelli, 1974). In crustaceans, MIH inhibits ecdysteroid synthesis, but also substantially inhibits 20E-stimulated apolysis in explants of barnacle mantle tissue over a culture period of 24 h, strongly suggesting that this hormone also acts on ecdysteroid target tissues (Freeman and Costlow, 1979).

9.6.2. *Biochemical Changes in the Epidermis*

The morphological changes in the epidermis that are associated with the molting process involve biochemical modifications, including synthesis of proteins and chitin.

9.6.2.1. PROTEINS AND NUCLEIC ACIDS

Initiation of apolysis of 20E was dependent on RNA and protein synthesis, as indicated by the inhibition of 20E-promoted apolysis by actinomycin D and cycloheximidine in mantle-tissue explants of the barnacle (Freeman and Costlow, 1979). Injected 20E caused a marked increase in the incorporation of radiolabeled leucine into epidermal proteins (McWhinnie and Mohrherr, 1970), in the synthesis of epidermal RNA (Stevenson, 1978; Dall and Barclay, 1979), and in the incorporation of amino acids into the hypodermis (McWhinnie et al., 1972). In isolated tissues of *Astacus leptodactylus*, 20E induced quantitative and qualitative changes in the hypodermis proteins in a dose- and time-dependent manner (Traub et al., 1987). These latter results, depending on the molting stage, suggested a reprogramming of protein biosynthesis from intermolt to the premolt stage (Traub et al., 1987). In the crayfish *Orconectes*, there is an increase in the incorporation of radiolabeled thymidine into epidermal cells in early premolt, suggesting increased DNA synthesis in the presence of ecdysteroid (Wittig and Stevenson, 1975; Stevenson, 1978). In contrast, Krishnakumaran and Schneiderman (1970) showed no incorporation of tritiated thymidine into epidermal-cell nuclei of the crayfish. In the western rock lobster, injected 20E caused marked increase in RNA synthesis but not in that of DNA, and Dall and Barclay (1979) concluded that the principal effect of the injected hormone is to make the epidermis secrete new cuticle by inducing RNA synthesis and not by stimulating cellular multiplication.

9.6.2.2.　CHITIN

By studying the incorporation of various chitin precursors, Willig and Keller (1973), Wittig and Stevenson (1974), and Armstrong and Stevenson (1978) observed, after 20E treatment, a significant incorporation of chitin precursor during the intermolt and early premolt. There are three enzymes in the pathway of chitin synthesis that seem the most likely to be affected by 20E: the enzyme that catalyses the synthesis of glucosamine 6-phosophate, uridine 5'-diphosphate-(UDP)–acetylglucosamine pyrophosphorylase, and chitin synthetase (Stevenson, 1978). 20E not only stimulates chitin synthesis in epidermal cells but also seems to affect resorption of the old cuticle: it stimulated the chitinolytic removal of n-acetylglucosamine from the cuticle of mantle epidermis explants of the barnacle *in vitro* (Freeman, 1979).

9.6.2.3.　PROTEIN KINASE

High kinase activity in the spider crab *Acanthonyx lunulatus* at ecdysis and a strong correlation between enhancement of the integumental kinase activity and the hemolymph ecdysteroid titer was reported by Chaix et al. (1981a). In the crayfish *Orconectes limosus*, injection of 20E into intermolt animals increased epidermal kinase activity, but *in vitro* incubation of an integumental extract with 20E showed no enhancement of cAMP levels in the tissue; this result suggests that the effect of 20E on epidermal kinase activity is not mediated by cAMP (Christ and Sedlmeier, 1987).

9.6.3. *Other Target Tissues and Calcification*

9.6.3.1. HEPATOPANCREAS

The hepatopancreas (midgut gland of crustaceans) seems to be a target tissue for molting hormone owing to the existence of specific receptor molecules (Gorell et al., 1972; Spindler et al., 1984). At the level of this tissue, incorporation of radiolabeled leucine into proteins increases in the presence of 20E, both *in vitro* in *Orconectes* (Gorell and Gilbert, 1969) and *in vivo* in *Palaemon* (Van Wormhoudt et al., 1978) and *Carcinus* (Adelung, 1971). However, in the crayfish, McWhinnie et al. (1972) showed that low doses of 20E increase the *in vivo* incorporation of aminoacid into the hypodermis but not into the hepatopancreas. In addition, it would then appear that the increase in lipid synthesis, characteristic of the early premolt hepatopancreas of the crab and crayfish, is not controlled by either 20E or the Y-organs (Bollenbacher et al., 1972a,b). In the midgut gland of the chelicerate *Limulus polyphemus*, 20E injections produced an early stimulation of protein synthesis, followed by a strong inhibition within 48 h (Winget and Herman, 1979). These seemingly controversial results can be explained by the presence in the eyestalks of factors stimulating or inhibiting the digestive enzyme synthesis. The disparate results obtained not only at different times throughout the molting cycle but also in the various seasons suggest that the stimulating factor is present in summer and the inhibiting factor in winter (McWhinnie and Mohrherr, 1970; Van Wormhoudt, 1973). The conflicting responses of spring–summer and autumn–winter animals has already been mentioned above in connection with the synthesis of ecdysteroids and responses to exogenous ecdysteroids. They suggest that in decapod crustaceans, at least, the regulation of these molt phenomena (among others) is achieved through neuropeptides, either stimulating or inhibiting, present in the eystalks.

9.6.3.2. NEUROSECRETORY CELLS AND Y-ORGANS

The neurosecretory cells of the myriapod *Lithobius forficatus* respond to the presence of 20E: addition of 20E to the hemolymph increased the frequency and amplitude of the electrical responses of the brain of *Lithobius* (Descamp and Lassalle, 1981), and 20E also increased the synthetic activity of the pars intercerebralis neurosecretory cells (Jamault-Navarro et al., 1983). In the crab *Cancer antennarius*, MIH released from the isolated ganglia of animals injected with 20E was significantly less than that released from ganglia of saline-injected controls (Mattson and Spaziani, 1986). In the Y-organs of *Portunus trituberculatus*, free ribosomes are more numerous in crabs after injection of 20E and the smooth ER and Golgi complexes are more developed (Taketomi and Hyodo, 1986).

9.6.3.3. OTHER TISSUES

Ecdysteroids also act on the ovaries (Gohar and Souty, 1984) and the androgenic glands (Hoffman and Foulks, 1973) of crustaceans by increasing protein and RNA synthesis respectively. Qualitative and quantitative changes in the hemolymph proteins occur during the molt cycle of *Acanthonyx lunulatus;* a new protein appears in the hemolymph when the ecdysone titer increases (Chaix et al., 1981b). During the period of proecdysis and ecdysis, a considerable amount of water is required for shedding the old cuticle, and water intake from the anus of crayfish is regulated by both eyestalk hormones and ecdysteroids (Muramoto, 1981). Depending on the concentration and length of exposure, 20E induced changes in the morphology, growth, and adhesiveness of tick cells isolated from the embryos of *Ripicephalus appendiculatus* and *Anocentor nitens.* Low concentrations stimulated, but higher concentrations inhibited, growth of the tick cell lines (Kurtti and Munderloh, 1983).

9.6.3.4. CALCIFICATION OF CRUSTACEAN CUTICLE

The calcification of preecdysial layers occurs after ecdysis, and the postecdysial layer calcifies as the old cuticle is shed. Calcification of the new carapace is achieved on the basis of exogenous calcium (yielded by water and food) and endogenous calcium sequestered before exuviation in special sites either common to all crustaceans (hepatopancreas or hemolymph) or particular to some of them (gastroliths, sternal plates, or posterior caeca). In aquatic crustaceans extracellular calcium is used, whereas in terrestrial animals endogenous calcium predominates. In the crayfish, calcium is converted in gastroliths after mobilization from the exoskeleton; 20E treatment results in gastrolith formation by animals in the intermolt stage, when the endogenous ecdysteroid titer is low (Krishnakumaran and Schneiderman, 1970; McWhinnie et al., 1972; Keller and Willig, 1976).

However, in eyestalkless animals or when higher doses of 20E are used, the rate of gastrolith formation is not increased (McWhinnie et al., 1972): ecdysteroids may control calcification, but their mechanism of action is probably indirect (Graf, 1972; Tighe-Ford and Vaile, 1972; Keller and Willig, 1976).

9.7. Effect on Regeneration

Like most invertebrates, chelicerates, crustaceans, and myriapods maintain morphogenetic potentialities—which are normally possessed only by the embryo or very young stages in vertebrates—throughout their life; e.g., they are capable of automizing or regenerating their appendages. First, after autotomy of the appendages, there is an increase in size and a migration of the epidermal cells. Next, mitoses appear, sometimes in a stepwise manner. Within several days, the epidermal cells have formed a layer and secreted a new cuticle. Several weeks following the loss of a limb, a blastema erupts through the scab. In addition to epidermal cells, which dedifferentiate and migrate to cover the autotomy plane, several types of cells participate in the formation of the blastema (Skinner, 1985).

9.7.1. *Regeneration and Molt*

In most cases, regeneration stimulates the molting processes in crustaceans (Spindler et al., 1980; Skinner, 1985). In crustaceans that are in anecdysis (but not in terminal anecdysis), the loss of numerous limbs (by induced autotomy in proecdysial animals), like eyestalk removal, triggers a precocious molt (Skinner and Graham, 1972; Fingerman and Fingerman, 1974; Rao, 1978). Moreover, provoked loss of limbs is used as a means of inducing a molt. This method is less stressful for the animal: following limb removal there is a normal molt (Skinner and Graham, 1970), whereas following eyestalk removal the animal dies just after or during molting (Skinner and Graham, 1972). The loss of many limbs exerts its most striking effects in animals with otherwise very long anecdysial stages; in animals undergoing rapid molt cycles, the molt-inducing effect is much less prominent (Chittleborough, 1975). However, autotomy of limbs from an adult anecdysial decapod crustacean results in precocious proecdysis and rapid regeneration; in some cases (crab larvae), the induced-accelerated ecdysis results in nonregeneration (McConaugha and Costlow, 1987). Precocious molts are initiated in the anecdysial or intermolt animal (when there is a low endogenous titer of ecdysteroids) by the loss of numerous limbs. In contrast, loss of one or more limbs from proecdysial animals (with a high endogenous ecdysteroid titer) delayed ecdysis. During early premolt, the initiation of secondary regenerates inhibits molting until these new secondary regenerates reach approximately the same size as the primary regenerates (Holland and Skinner, 1976). In late premolt, secondary regenerates of the animal failed to be initiated and ecdysis was not delayed (Holland and Skinner, 1976). All these results suggest the interdependence of circulatory endogenous ecdysteroids and regeneration.

9.7.2. *Regeneration and Ecdysteroids*

The interaction of regeneration and the molting cycle suggests a role of ecdysteroids in this process. When a molt is triggered by limb autotomy, there is an increase of endogenous ecdysteroids (Charmantier-Daures and DeReggi, 1980; Suzuki, 1985). In *Carcinus*, initiation of molting did not occur in Y-organectomized animals even in those that had autotomized pereopods (Bazin, 1973). Therefore, it can be concluded that the inactivation of molting by limb autotomy depends on ecdysteroid synthesis by the Y-organs, according to the classical scheme. In crustaceans in anecdysis, the limb autotomy factor that initiates molting is assumed to result in an increase of ecdysteroid titers via the Y-organs, whereas the effect of the limb autotomy factor that provokes inhibition of the ongoing molt is to reduce the ecdysteroid titer to the level found at the beginning of proecdysis (McCarthy and Skinner, 1977a,b).

For initiation and basal regeneration to occur a low titer of hemolymph ecdysteroids is required, whereas for proecdysial growth to happen a higher concentration of ecdysteroids is necessary. However, too high concentrations of ecdysteroids inhibit immediate regeneration and lead directly to a molt (Rao et al., 1972), regeneration being delayed to the following molt (autotomy without effect in late premolt). The molting stage and the critical ecdysteroid threshold, corresponding to reversal of the effect, vary according to the animal. In the crab *Gecarcinus lateralis*, this reversal of the effect can occur up to D_1: at that stage, regeneration still occurs after limb removal, although ecdysteroid concentrations are high (Tchernigovtzeff, 1965, 1974). The dramatic decrease of the hemolymph ecdysteroid titer, when autotomy occurs in premolt, is due to an acceleration of the processes of inactivation and excretion of ecdysteroids (McCarthy, 1980).

Exogenous ecdysteroids blocked the initiation of basal limb growth when administered to crabs exhibiting scar tissue without any visible limb; when injected into crabs with limb buds, exogenous ecdysteroids accelerated limb growth (Rao, 1978). Injection of ecdysteroids (mostly high titers) provokes inhibition of regeneration and stimulation of proecdysis in adult crustaceans (Krishnakumaran and Schneiderman, 1970; Rao et al., 1972; Noulin and Maissiat, 1974; Bazin, 1977a,b; Rao, 1978) and in larval crustaceans (McConaugha and Costlow, 1987). In contrast, 20E has no effect on the rate of limb regeneration in the prawn *Palaemon elegans* (Webster, 1983). In eyestalkless animals (with, in principle, a high titer of endogenous ecdysteroids), injection of 20E does not affect proecdysial limb growth (Hopkins et al., 1979) or delay limb regeneration (Flint, 1972).

9.8. Summary

Arthropods other than insects possess, like the latter, ecdysteroids which are transported by the hemolymph and metabolized by peripheral tissues. In crustaceans, these hormones are synthesized by the Y-organs. Classically, the ecdysteroid titer increases in the hemolymph of premolt animals, and there is a close correlation between this increase and the initiation of the molt. This results in the secretion of a new cuticle by the epidermis. Absence of cuticle secretion corresponds to a low level of hemolymph ecdysteroids. Although all the arthropods have a quite smaller hormonal control of molt by ecdysteroids, there are some noticeable between-group differences.

Most experiments dealing with arthropods other than insects were performed with wild-caught animals. The effects of exogenous ecdysteroids on them varied, depending on the season. This strongly suggests that environmental factors have a marked regulatory influence on both the synthesis and the mode of action of ecdysteroids.

Ecdysteroid precursors seem to be identical in crustaceans and in insects. In contrast, ticks fail to convert cholesterol into ecdysteroids. Hence, the question can be raised whether these animals, which are mainly ectoparasites and hematophagous, use a different pathway of ecdysteroid biosynthesis.

Adult molts occur in crustaceans, myriapods, and apterygote insects. Crustaceans and apterygote insects possess two major ecdysteroids (PoA and 20E), instead of one (20E) in chelicerates and pterygote insects. Regulation of the ecdysteroid titer in myriapods and crustaceans proceeds through a molt-inhibiting hormone, and not through a stimulating hormone as in pterygote insects. It is assumed that PoA and MIH might be involved in the continuation of molt in the adult stages.

Acknowledgments

I am indebted to P. Diehl for having supported this project. I wish to express deep gratitude to P. Harry, D. Lachaise, and G. Somme for help during the preparation of the manuscript. I also thank J. L. Connat, R. Lafont, and F. Xavier for helpful comments on the chapter. I gratefully acknowledge G. Carpentier for preparing the final version of the figures.

References

Adelung, D. 1967. Die Wirkung von Ecdyson bei *Carcinus maenas* L. und der Crustecdyson-titer während eines Häutungszyclus. Zool. Anz. Suppl. 30: 264–272.

Adelung, D. 1971. Untersuchungen zur Häutungsphysiologie der dekapoden Krebse am Beispiel der Strandkrabbe *Carcinus maenas*. Helgol. Wiss. Meeresunters. 22: 66–119.

Ahmed, S. H. and J. H. Bassal. 1982. Induction of adult tick molting by ecdysteroids. J. Egypt. Soc. Parasitol. 12: 383–387.

Armstrong, P. W. and J. R. Stevenson. 1979. The effect of ecdysterone on ^{14}C-N-acetyglucosamine incorporation into chitin in the crayfish *Orconectes obscurus* during the molt cycle. Comp. Biochem. Physiol. 63B: 63–65.

Barnes, H. and R. L. Stone. 1974. The effect of food, temperature and light-period (day-length) on moulting frequency in *Balanus balanoides* (L.). J. Exp. Mar. Biol. Ecol. 15: 275–284.

Bazin, F. 1973. Formation de blastèmes de régénération des péréiopodes autotomisés chez des crabes *Carcinus maenas* privés de leurs glandes de mue. C. R. Acad. Sci. Paris 276D: 2585–2588.

Bazin, F. 1977a. Effets d'injections d'ecdystérone sur la mue et la régénération chez le crabe *Carcinus maenas*. C. R. Acad. Sci. Paris 284D: 765–768.

Bazin, F. 1977b. Action inhibitrice de l'ecdystérone sur la régénération chez le crabe *Carcinus maenas*. C. R. Acad. Sci. Paris 284D: 1211–1214.

Behrens, W. and D. Bückmann. 1983. Ecdysteroids in the pycnogonid *Pycnogonum littorale* (Ström) (Arthropoda, Pantopoda). Gen. Comp. Endocrinol. 51: 8–14.

Bergamasco, R. and D. H. S. Horn. 1980. The biological activities of ecdysteroids and ecdysteroid analogues. Pp. 299–324 *in* J. A. Hoffmann (ed.), *Progress in Ecdysone Research*. Elsevier/North-Holland Publ., Amsterdam and New York.

Blanchet, M. F. 1972. Effets sur la mue et sur la vitellogénèse de la β-ecdysone introduite aux étapes A et D_2 du cycle d'intermue chez *Orchestia gammarella* Pallas (crustacé, amphipode): comparison avec les effets de la β- et de l'α-ecdysone aux autres étapes de l'intermue. C. R. Acad. Sci. Paris 274D: 3015–3018.

Blanchet, M. F. 1974. Etude du contrôle hormonal du cycle d'intermue et de l'exuviation chez *Orchestia gammarella* par microcautérisation des organes Y suivie d'injection d'ecdystérone. C. R. Acad. Sci. Paris 278D: 509–512.

Bollenbacher, W. E. 1974. *In vitro* secretions of arthropod ecdysial glands: correlation with the *in vivo* system. Paper, University of California, Los Angeles.

Bollenbacher, W. E., D. W. Borst, and J. D. O'Connor. 1972. Endocrine regulation of lipid synthesis in decapod crustaceans. Am. Zool. 12: 381–384.

Bollenbacher, W. E., S. M. Flechner, and J. D. O'Connor. 1972b. Regulation of lipid synthesis during early premolt in decapod crustaceans. Comp. Biochem. Physiol. 42B: 157–165.

Bonaric, J. C. 1976. Effects of ecdysterone on the molting mechanisms and duration of the intermolt period in *Pisaura mirabilis* Cl. Gen. Comp. Endocrinol. 30: 267–272.

Bonaric, J. C. 1977. Existence d'une diapause hivernale révelée par l'action de traitements hormonaux chez l'araignée *Pisaura mirabilis* Cl. (Pisauridae). C. R. Acad. Sci. Paris 284D: 1297–1300.

Bonaric, J. C. 1986. Moulting hormones. Pp. 111–118 *in* W. Nentwig (ed.), *Ecophysiology of Spiders*. Springer-Verlag, Berlin and New York.

Bonaric, J. C. and M. DeReggi. 1977. Changes in ecdysone levels in the spider *Pisaura mirabilis* nymphs (Araneae, Pisauridae). Experientia (Basel) 33: 1664–1665.

Bouvier, J., P. A. Diehl, and M. Morici. 1982. Ecdysone metabolism in the tick *Ornithodoros moubata* (Argasidae, Ixodoidea). Rev. Suisse Zool. 89: 967–976.

Bückmann, D., G. Starnecker, K. H. Tomaschko, E. Wilhelm, R. Lafont, and J. P. Girault. 1986. Isolation and identification of major ecdysteroids from the pycnogonid *Pycnogonum littorale* (Ström) (Arthropoda, Pantopoda). J. Comp. Physiol. 156B: 759–765.

Burghause, F. 1975. Das Y-organ von *Orconectes limosus* (Malacostraca, Astacura). Z. Morphol. Tiere 80: 41–57.

Butenandt, A. and P. Karlson. 1954. Über die Isolierung eines Metamorphosehormons der Insekten in kristallisierter Form. Z. Naturforsch. 9B: 389–391.

Campbell, J. D. and J. H. Oliver. 1984. Membrane feeding and developmental effects of ingested β-ecdysone on *Ornithodoros parkeri* (Acari: Argasidae).

Pp. 393–399 *in* D. A. Griffiths and C. E. Bowman (eds.), *Acarology VI*, Vol. 1. Ellis Horwood, Chichester, England.

Carlisle, D. B. 1957. On the hormonal inhibition of moulting in decapod Crustacea. II. The terminal anecdysis in crabs. J. Mar. Biol. Assoc. UK 36:291–307.

Carlisle, D. B. and R. O. Connick. 1973. Crustecdysone (20-hydroxyecdysone) site of storage in the crayfish *Orconectes propinquus*. Can. J. Zool. 51: 417–420.

Chaix, J. C., J. Marvaldi, and P. Mangeat. 1981a. Variation of epidermal, cyclic nucleotide–dependent, protein kinase activity during the molt cycle of the spider crab *Acanthonyx lunulatus*. Comp. Biochem. Physiol. 69B: 701–718.

Chaix, J. C., J. Marvaldi, and J. Seechi. 1981b. Variations of ecdysone titer and hemolymph major proteins during the molt cycle of the spider crab *Acanthonyx lunulatus*. Comp. Biochem. Physiol. 69B: 709–714.

Chang, E. S. 1985. Hormonal control of molting in decapod crustacea. Am. Zool. 25: 179–185.

Chang, E. S. and M. J. Bruce. 1980. Ecdysteroid titers of juvenile lobsters following molt induction. J. Exp. Zool. 214: 157–16.

Chang, E. S. and J. D. O'Connor. 1977. Secretion of α-ecdysone by crab Y-organs *in vitro*. Proc. Nat. Acad. Sci. USA. 74: 615–618.

Chang, E. S. and J. D. O'Connor. 1978. *In vitro* secretion and hydroxylation of α-ecdysone as a function of the crustacean molt cycle. Gen. Comp. Endocrinol. 36: 151–160.

Chang, E. S., M. J. Bruce, and R. W. Newcomb. 1987. Purification and amino acid composition of a peptide with molt-inhibiting activity from the lobster, *Homarus americanus*. Gen. Comp. Endocrinol. 65: 56–64.

Chang, E. S., B. A. Sage, and J. D. O'Connor. 1976. The quantitative and qualitative determinations of ecdysones in tissues of the crab, *Pachygrapsus crassipes*, following molt induction. Gen. Comp. Endocrinol. 30: 21–33.

Charmantier, G. and J. P. Trilles. 1973. Rétablissement de la mue par injection d'ecdystérone chez les mâles adultes, pubères de *Sphaeroma serratum* (crustacé, isopode). C. R. Acad. Sci. Paris 276D: 2561–2564.

Charmantier, G. and J. P. Trilles. 1976. Ecdystérone prémue et exuviation chez *Sphaeroma serratum* (Fabricius, 1787) (Crustacea, Isopoda, Flabellifera). Gen. Comp. Endocrinol. 28: 249–254.

Charmantier, G., M. Olle, and J. P. Trilles. 1977. Evolution du taux d'ecdystérone, dégénérescence des organes Y et sénescence chez les mâles pubères de *Sphaeroma serratum* (Fabricius, 1787) (Crustacea, Isopoda, Flabellifera). C. R. Acad. Sci. Paris 285D: 1487–1489.

Charmantier-Daures, M. and M. DeReggi. 1980. Aspects préliminaires des variations hémolymphatiques du taux d'ecdystéroïdes chez *Pachygrapsus marmoratus* (crustacé, décapode): influence de la régénération intensive et de l'ablation des organes Y. Bull. Soc. Zool. Fr. 105: 81–86.

Cheung, P. J. and R. F. Nigrelli. 1974. Ecdysterone-induced molting in isolated penis and appendages of barnacle *Balanus eburneus* Gould, kept in sterilized sea water. Am. Zool. 14(4): 1266 (abstract).

Chittleborough, R. G. 1975. Environmental factors affecting growth and survival of juvenile western rock lobsters *Palinurus longipes* (Milne-Edwards). Aust. J. Mar. Freshwater Res. 26: 117–196.

Christ, B. and D. Sedlmeier. 1987. Variations in epidermal cyclic nucleotide-dependent protein kinase activity during moult cycle of the crayfish *Orconectes limosus* and hormonal control of kinase activity by 20-hydroxyecdysone. Int. J. Biochem. 19(1); 79–84.

Connat, J. L. 1987. Aspects endocrinologiques de la physiologie du développement et de la reproduction chez les tiques. Doctoral thesis, University of Dijon, France.

Connat, J. L. and P. A. Diehl. 1986. Probable occurrence of ecdysteroid fatty acid esters in different classes of Arthropods. Insect Biochem. 16: 91–97.

Connat, J. L., P. A. Diehl, and M. Morici. 1984. Metabolism of ecdysteroids during the vittelogenesis of the tick *Ornithodoros moubata* (Ixodoidea, Argasidae): accumulation of apolar metabolites in the eggs. Gen. Comp. Endocrinol. 56: 100–110.

Connat, J. L., P. A. Diehl, and M. J. Thompson. 1986a. Possible inactivation of ingested ecdysteroids by conjugation with long-chain fatty acids in the female tick *Ornithodoros moubata* (Acarini: Argasidae). Arch. Insect Biochem. Physiol. 3: 235–252.

Connat, J. E., E. M. Dotson, and P. A. Diehl. 1987. Metabolism of ecdysteroids in the female tick *Amblyomma hebraeum* (Ixodiodea: Ixodidae): accumulation of free ecdysone and 20-hydroxyecdysone in the eggs. J. Comp. Physiol. 157B: 689–699.

Connat, J. L., R. Lafont, and P. A. Diehl. 1986b. Metabolism of [^{3}H]ecdysone by isolated tissues of the female ixodid tick *Amblyomma hebraeum* (Ixodoidea: Ixodidae). Mol. Cell. Endocrinol. 47: 257–267.

Connat, J. L., P. A. Diehl, H. Gfeller, and M. Morici. 1985. Ecdysteroids in females and eggs of the ixodid tick *Amblyomma hebraeum*. Int. J. Invertebr. Reprod. Dev. 8: 103–116.

Connat, J. L., P. A. Diehl, N. Dumont, S. Carminati, and M. J. Thompson. 1983. Effects of exogenous ecdysteroids on the female tick *Ornithodoros moubata:* induction of supermolting and influence on oogenesis. Z. Angew. Entomol. 96: 520–530.

Cox, B. L. 1960. Hormonal involvement in the molting process in the soft tick, *Ornithodoros turicata* Dugès. Thesis, University of Oklahoma, Norman.

Crosby, T. R. P. Evershed, D. Lewis, K. P. Wigglesworth, and H. Rees. 1986. Identification of ecdysone 22-long-chain fatty acyl esters in newly laid eggs of the cattle tick *Boophilus microplus*. Biochem. J. 240: 131–138.

Daig, K. and K. D. Spindler. 1980. In vitro uptake and retension of molting hormones in the integument of the crayfish, *Astacus leptodactylus*. Pp. 273–278 *in* E. Kurstak, K. Maramorosch, and A. Dübendorfer (eds.),

Invertebrates Systems In Vitro. Elsevier/North-Holland Publ., Amsterdam and New York.

Dall, W. and M. C. Barclay. 1977. Induction of viable ecdysis in the western rock lobster by 20-hydroxyecdysone. Gen. Comp. Endocrinol. 31: 323–334.

Dall, W., and M. C. Barclay. 1979. The effect of exogenous 20-hydroxyecdysone on levels of epidermal DNA and RNA in the western rock lobster. J. Exp. Mar. Biol. Ecol. 36: 103–110.

Delbecque, J. P., P. A. Diehl, and J. D. O'Connor. 1978. Presence of ecdysone and ecdysterone in the tick *Amblyomma hebraeum*. Koch. Experientia (Basel) 34: 1379–1380.

Démeusy, N. 1959. Pédoncules oculaires, glande de mue et appareil génital chez *Carcinus maenas* L. C. R. Acad. Sci. Paris 248D: 2652–2654.

Descamps, M. and B. Lassalle. 1981. Electrophysiological evidence for direct ecdysteroid action on the brain in *Lithobius forficatus* L. (Myriapoda: Chilopoda). Reprod. Nutr. Dév. 21: 681–687.

Diehl, P. A., J. L. Connat, and E. Dotson. 1986. Chemistry, function and metabolism of tick ecdysteroids. Pp. 165–193 *in* J. R. Sauer and J. A. Hair (eds.), *Morphology, Physiology, and Behavioural Biology of Ticks*. Wiley, New York.

Diehl, P. A., J. L. Connat, J. P. Girault, and R. Lafont. 1985. A new class of apolar ecdysteroid conjugates: esters of 20-hydroxecdysone with long-chain fatty acids in ticks. Int. J. Invertebr. Reprod. Dev. 8: 1–13.

Drach, P. 1939. Mue et cycle d'intermue chez les crustacés décapodes. Ann. Inst. Oceanogr. 19: 103–391.

Drach, P. and C. Tchernigovtzeff. 1967. Sur la méthode de détermination des stades d'intermue et son application générale aux crustacés. Vie Milieu Sér. A Biol. Mar. 18: 595–607.

Echalier, G. 1954. Recherches expérimentales sur le rôle de ''l'organe Y'' dans la mue de *Carcinus maenas* (L.) crustacé, décapode. C. R. Acad. Sci. Paris. 238D: 523–525.

Echalier, G. 1955. Rôle de l'organe Y dans le déterminisme de la mue de Carcinides *(Carcinus maenas)* (L.) (crustacés décapodes): expériences d'implantation. C. R. Acad. Sci. Paris 240D: 1581–1583.

Echalier, G. 1959. L'organe Y et le détermisme de la croissance et de la mue chez *Carcinus maenas* (L.) crustacé décapode. Ann. Sci. Nat. Zool. Biol. Anim. [Sér 12.] 1: 1–59.

El Bakary, Z., P. Porcheron, M. Morinière, and S. Fuzeau-Braesch. 1987. Mise en évidence et dosage d'ecdystéroïdes chez le scorpion *Leirus quinquestriatus*. C. R. Acad. Sci. Paris. 304: 453–456.

Faux, A., D. H. S. Horn, E. J. Middleton, H. M. Fales, and M. E. Lowe. 1969. Moulting hormones of a crab during ecdysis. Chem. Commun. 4: 175–176.

Fingerman, M. and S. W. Fingerman. 1974. The effects of limb removal on the rates of ecdysis of eyed and eyestalkless fiddler crabs, *Uca pugilator*. Zool. Jahrb. Physiol. 78: 301–309.

Fingerman, M. and Y. Yamamoto. 1964. Endocrine control of tanning in the crayfish exoskeleton. Science (Wash., DC) 144: 1462.

Flint, R. W. 1972. Effects of eyestalk removal and ecdysterone infusion on molting in *Homarus americanus*. J. Fish. Res. Bd. Canada 29: 1229–1233.

Freeman, J. A. 1979. Hormonal control of chitinolytic activity in the integument of *Balanus* amphitrite, *in vitro*. Comp. Biochem. Physiol. 65A: 13–17.

Freeman, J. A. and C. M. Bartell. 1976. Some effects of the molt-inhibiting hormone and 20-hydroxyecdysone upon molting in the grass shrimp, *Palaemonetes pugio*. Gen. Comp. Endocrinol. 28: 131–142.

Freeman, J. A. and J. D. Costlow. 1979. Hormonal control of apolysis in barnacle mantle tissue epidermis *in vitro*. J. Exp. Zool. 210: 333–346.

Fukutake, K., K. Morimoto, and K. Matsumoto. 1969. The effects of phytoecdysones upon the molt of the crayfish, *Procambarus*. Zool. Mag. 78: 482–483.

Gagosian, R. B., R. A. Bourbonniere, W. B. Smith, E. F. Couch, C. Blanton, and W. Novak. 1974. Lobster molting hormones: isolation and biosynthesis of ecdysterone. Experientia 30: 723–724.

Galbraith, M. N., D. H. S. Horn, E. J. Middleton, and R. J. Hackney. 1968. Structure of deoxycrustecdysone, a second crustacean moulting hormone. Chem. Commun. 3: 83–85.

Gersch, M. H., H. Eibisch, G. A. Böhm, and J. Koolman. 1979. Ecdysteroid production by the cephalic gland of the crayfish *Orconectes limosus*. Gen. Comp. Endocrinol. 39: 505–511.

Gilgan, M. W. 1980. The inhibition of normal molting in the adult male lobster *(Hormarus americanus)* by ecdysterone treatment. Comp. Biochem. Physiol. 65A: 207–209.

Gilgan, M. W. and B. G. Burns. 1976a. The successful induction of molting in the adult male lobster *(Homarus americans)* with a slow-release form of ecdysterone. Steroids 27: 571–581.

Gilgan, M. W. and B. G. Burns. 1976b. Molt induction in lobsters *(Homarus americanus)* by intramuscular injection of ecdysterone triacetate. Experientia (Basel) 33: 1114–1115.

Gilgan, M. W. and B. G. Burns. 1977. On the reduced sensitivity of the adult male lobster *(Homarus americanus)* to ecdysterone at reduced temperatures. Comp. Biochem. Physiol. 58A: 33–36.

Gilgan, M. W. and B. G. Burns. 1979. Ecdysterone triacetate-induced molting in the american lobster *(Homarus americanus)*. Comp. Biochem. Physiol. 64A: 125–131.

Gilgan, M. W. and T. E. Farquharson. 1977. A change in the sensitivity of adult male lobster *(Homarus americanus)* to ecdysterone on changing from intermolt to active premolt development. Comp. Biochem. Physiol. 58A: 29–32.

Gilgan, M. W. and M. E. Zinck. 1975. Response of the adult lobster *(Homarus americanus)* to graded and multiple doses of ecdysterone. Comp. Biochem. Physiol. 52A: 261–264.

Gilgan, M. W., T. E. Farquharson, and B. G. Burns. 1977. The effect of α-ecdysone, ecdysterone and inokosterone treatment, separately or in combinations, on premolt development and molting in adult male lobsters *(Homarus americanus)*. Comp. Biochem. Physiol. 56A: 43–49.

Gohar, M. and C. Souty. 1984. Action temporelle d'ecdystéroïdes sur la synthèse protéique ovarienne *in vitro* chez le crustacé isopode terrestre *Porcellio dilatatus* (Brandt). Reprod. Nutr. Dév. 24: 137–145.

Goodwin, T. W., D. H. S. Horn, P. Karlson, J. Koolman, K. Nakanishi, W. E. Robbins, J. B. Siddall, and T. Takemoto. 1978. Ecdysteroids: a new generic term. Nature 272: 122.

Gorell, T. A. and L. I. Gilbert. 1969. Stimulation of protein and RNA synthesis in the crayfish hepatopancreas by crustecdysone. Gen. Comp. Endocrinol. 13: 308–310.

Gorell, T. A., L. I. Gilbert, and J. D. Siddall. 1972. Binding proteins for an ecdysone metabolite in the Crustacean hepatopancreas. Proc. Nat. Acad. Sci. USA. 69: 812–815.

Graf, F. 1972. Action de l'ecdystérone sur la mue, la cuticule et le métabolisme du calcium chez *Orchestia cavimana* Heller (crustacé amphipode, talitridé). C. R. Acad. Sci. Paris 274D: 1731–1734.

Graf, F. and J. P. Delbecque. 1987. Ecdysteroid titers during the molt cycle of *Orchestia cavimana* (Crustacea, Amphipoda). Gen. Comp. Endocrinol. 65: 23–33.

Haag, T., C. Hetru, Y. Nakatani, B. Luu, L. Pichat, M. Audinot, and M. F. Meister. 1985. Synthesis of labelled ecdysone precursor. Part I. Tritium-labelled [$^{3}H_4$-22,23,24,25]-3β,14α-dihydroxy-5β-cholest-7-en-6-one. J. Labelled Comp. Radiopharmaceut. 22: 547–557.

Hampshire, F. and D. H. S. Horn. 1966. Structure of crustecdysone, a crustacean moulting hormone. J. Chem. Soc. Chem. Commun. pp. 37–38.

Herman, Wm. S. 1972. Molt initiation in response to phytoecdysones and low doses of animal ecdysone in the chelicerate arthropod, *Limulus polyphenus*. Gen. Comp. Endocrinol. 18: 301–305.

Hikino, H., Y. Ohizumi, and T. Takemoto. 1975. Detoxication mechanism of *Bombyx mori* against exogenous phytoecdysone ecdysterone. J. Insect Physiol. 21: 1953–1963.

Hoffman, D. L. and N. B. Foulks. 1973. Ecdysone-induced RNA synthesis in the androgenic glands of a shrimp. Am. Zool. 13(4): Abstract No. 110, pp. 1274–1275.

Hoffmann, J. A., M. Lagueux, C. Hetru, M. Charlet, and F. Goltzené. 1980. Ecdysone in reproductively competent female adults and in embryos of insects. Pp. 431–466 *in* J. A. Hoffmann (ed.), *Progress in Ecdysone Research*. Elsevier/North-Holland Publ., Amsterdam and New York.

Holland, C. A. and D. M. Skinner. 1976. Interactions between molting and regeneration in the land crab. Biol. Bull (Woods Hole) 150: 22–240.

Hopkins, P. M. 1986. Ecdysteroid titers and Y-organ activity during late anecdysis and proecdysis in the fiddler crab, *Uca pugilator*. Gen. Comp. Endocrinol. 63: 362–373.

Hopkins, P. M., D. E. Bliss, S. W. Sheehan, and J. R. Boyer. 1979. Limb growth-controlling factors in the crab *Gecarcinus lateralis*, with special reference to the limb growth-inhibiting factor. Gen. Comp. Endocrinol. 39: 192–207.

Hubschman, J. H. and P. W. Armstrong. 1972. Influence of ecdysterone on molting in *Palaemonetes*. Gen. Comp. Endocrinol. 18: 435–438.

Jamault-Navarro, C., R. Joly, and M. Descamps. 1983. Activation of neurosecretory protocerebral cells by 20-hydroxyecdysone in *Lithobius forficatus* L. (Myriapoda: Chilopoda). Gen. Comp. Endocrinol. 50: 36–42.

Jegla, T. C. 1982. A review of the molting physiology of the trilobite larva of *Limulus*. Pp. 83–101 *in* J. Bonaventure, C. Bonaventure, and S. Tesh (eds.), *Physiology and Biology of Horseshoe Crabs: Studies on Normal and Environmentally Stressed Animals*. Liss, New York.

Jegla, T. C. and J. D. Costlow. 1979. The *Limulus* bioassay for ecdysteroids. Biol. Bull (Woods Hole) 156: 103–114.

Jegla, T. C. and J. D. Costlow. 1982. Temperature and salinity effects on developmental and early posthatch stages of *Limulus*. Pp. 103–113 *in* J. Bonaventure, C. Bonaventure, and S. Tesh (eds.), *Physiology and Biology of Horseshoe Crabs: Studies on Normal and Environmentally Stressed Animals*. Liss, New York.

Jegla, T. C., J. D. Costlow, and J. Alspaugh. 1972. Effects of ecdysones and some synthetic analogs on horseshoe crab larvae. Gen. Comp. Endocrinol. 19: 159–166.

Jenkin, P. M. and H. E. Hinton. 1966. Apolysis in arthropod moulting cycles. Nature (Lond.) 211: 871.

Joly, R. 1966a. Contribution à l'étude du cycle de mue et de son déterminisme chez les myriapodes chilopodes. Bull. Biol. Fr. Belg. 3: 379–480.

Joly, R. 1966b. Etude expérimentale du cycle de mue et de sa régulation endocrine chez les myriapodes chilopodes. Gen. Comp. Endocrinol. 6: 519–533.

Joly, R., P. Porcheron, and F. Dray. 1979. Etude des variations du taux d'ecdystéroïdes au cours du cycle de mue dans l'hémolymphe de *Lithobius forficatus* L. (myriapode chilopode) par dosage radio-immunologique. C. R. Acad. Sci. Paris 288D: 243–246.

Juberthie-Jupeau, L., A. Strambi, M. De Reggi, and C. Juberthie. 1979. Presence of ecdysteroids and the variations of their level during the 1st adult stage of the myriapod *Hanseniella ivorensis* Juberthie-Jupeau and Kehe (Symphyla). Experientia (Basel) 35: 1406–1407.

Keller, R. and E. Schmidt. 1979. *In vitro* secretion of ecdysteroids by Y-organs and lack of secretion by mandibular organs of the crayfish following molt induction. J. Comp. Phsiol. 130: 347–353.

Keller, R. and A. Willig. 1976. Experimental evidence of the molt controlling function of the Y-organ of a macruran decapod, *Orconectes limosus*. J. Comp. Physiol. 108: 271–278.

King, D. S. and J. B. Siddall. 1969. Conversion of α-ecdysone to β-ecdysone by crustaceans and insects. Nature (Lond.) 221: 955–956.

Kitaoka, S. 1972. Effect of ecdysone on ticks, especially on *Ornithodoros moubata* (Acarina-Argasidae). Proc. 14th Int. Cong. Entomol., Australia, p. 272.

Koolman, J. 1982. Ecdysone metabolism. Insect Biochem. 12: 225–250.

Krishnakumaran, A. and H. A. Schneiderman. 1970. Control of molting in mandibulate and chelicerate arthropods by ecdysones. Biol. Bull. (Woods Hole) 139: 520–538.

Kuppert, P., M. Büchler, and K. D. Spindler. 1978. Distribution and transport of molting hormones in the crayfish, *Orconectes limosus*. Z. Naturforsch. 33C: 437–441.

Kurtti, T. J. and U. G. Munderloh. 1983. The effects of 20-hydroxyecdysone and juvenile hormone III on tick cells. J. Parasitol. 69: 1072–1078.

Lachaise, F. and R. Feyereisen. 1976. Ecdysone metabolism by different organs of *Carcinus maenas* L. incubated *in vitro*. C. R. Acad. Sci. Paris 283D: 1445–1448.

Lachaise, F. and J. A. Hoffmann. 1982. Ecdysteroids and embryonic development in the shore crab, *Carcinus maenas*. Hoppe-Seyler's Z. Physiol. Chem. 362: 521–529.

Lachaise, F. and R. Lafont. 1984. Ecdysteroid metabolism in a crab: *Carcinus maenas* L. Steroids 43: 243–260.

Lachaise, F., M. Hubert, S. G. Webster, and R. Lafont. 1988. Effect of molt-inhibiting hormone on ketodiol conversion by crab Y-organs. J. Insect Physiol. (in press).

Lachaise, F., M. F. Meister, C. Hétru, and R. Lafont. 1986. Studies on the biosynthesis of ecdysone by the Y-organs of *Carcinus maenas*. Mol. Cell. Endocrinol. 45: 253–261.

Lachaise, F., M. Goudeau, C. Hetru, C. Kappler, and J. A. Hoffmann. 1981. Ecdysteroids and ovarian development in the shore crab, *Carcinus maenas*. Hoppe-Seyler's Z. Physiol. Chem. 362: 521–529.

Lafont, R. and J. Koolman. 1984. Ecdysteroid metabolism. Pp. 196–226 *in* J. A. Hoffmann and M. Porchet (eds.), *Biosynthesis, Metabolism and Mode of Action of Invertebrate Hormones*. Springer-Verlag, Berlin and New York.

Leubert, F., H. Eibisch, and H. Scheffel. 1979. Isolation and characterization of the moulting hormone of *Lithobius forficatus* (L.) (Chilopoda). Zool. Jahrb. Abt. Allg. Zool. Physiol. Tiere 83: 334–339.

Lowe, M. E., D. H. S. Horn, and M. N. Galbraith. 1968. The role of crustecdysone in the moulting crayfish. Experientia (Basel) 24: 518–519.

Madhavan, K. 1981. Induction of precocious molting in the male and ovigerous female *Armadillidium vulgare* by topical application of 20-hydroxyecdysone. Gen. Comp. Endocrinol. 44: 28–36.

Maissiat, J. and F. Graf. 1973. Action de l'ecdystérone sur l'apolysis et l'ecdysis de divers crustacés isopodes. J. Insect Physiol. 19: 1265–1276.

Maissiat, J. and J. J. Legrand. 1970. Contribution à l'étude expérimentale du contrôle hormonal du cycle de mue chez les oniscoïdes *Porcellio dilatatus* et *Ligia oceanica*. C. R. Soc. Biol. Paris 164: 359–362.

Mango, C., T. R. Odhiambo, and R. Galun. 1976. Ecdysone and the supertick. Nature (Lond.) 260: 318–319.

Maroun, N. A. and K. A. Kamal. 1976. Biochemical and physiological studies of certain ticks (Ixodidae). Absence of sterol biosynthesis in *Dermacentor andersoni*. J. Med. Entomol. 13: 219–220.

Mattson, M. P. and E. Spaziani. 1985a. Characterization of molt-inhibiting action on crustacean Y-organ segments and dispersed cells in culture and a bioassay for MIH activity. J. Exp. Zool. 236: 93–101.

Mattson, M. P. and E. Spaziani. 1985b. Stress reduces hemolymph ecdysteroid levels in the crab: mediation by the eyestalks. J. Exp. Zool. 234: 319–323.

Mattson, M. P. and E. Spaziani. 1986. Evidence for ecdysteroid feedback on release of molt-inhibiting hormone from crab eyestalk ganglia. Biol. Bull. (Woods Hole) 171: 264–273.

McCarthy, J. F. 1979. Ponasterone A: a new ecdysteroid from the embryos and serum of brachyuran crustaceans. Steroids 34: 799–806.

McCarthy, J. F. 1980. Ecdysone metabolism and the interruption of proecdysis in the land crab, *Gecarinus lateralis*. Biol. Bull. 158: 91–102.

McCarthy, J. F. 1982. Ecdysone metabolism in premolt land crabs (*Gecarcinus lateralis*). Gen. Comp. Endocrinol. 47: 323–332.

McCarthy, J. F. and D. Skinner. 1977a. Proecdysial changes in serum ecdysone titers, gastrolith formation and limb regeneration following molt induction by limb autotomy and/or eyestalk removal in the land crab, *Gecarcinus lateralis*. Gen. Comp. Endocrinol. 33: 278–292.

McCarthy, J. F. and D. Skinner. 1977b. Interruption of proecdysis by autotomy of partially regenerated limbs in the land crab, *Gecarcinus lateralis*. Dev. Biol. 61: 299–310.

McCarthy, J. F. and D. Skinner. 1979a. Changes in ecdysteroids during embryogenesis of the blue crab, *Callinectes sapidus* Rathburn. Dev. Biol. 69: 627–633.

McCarthy, J. F. and D. Skinner. 1979b. Metabolism of α-ecdysone in intermolt land crabs (*Gecarcinus lateralis*). Gen. Comp. Endocrinol. 37: 250–263.

McConaugha, J. R. and J. D. Costlow. 1987. Role of ecdysone and eyestalk factors in regulating regeneration in larval crustaceans. Gen. Comp. Endocrinol. 66: 387–393.

McWhinnie, M. A. and C. J. Mohrherr. 1970. Influence of eyestalk factors, intermolt cycle and season upon ^{14}C-leucine incorporation into protein in the crayfish (*Orconectes virilis*). Comp. Biochem. Physiol. 34: 415–437.

McWhinnie, M. A., R. J. Kirchenberg, R. J. Urbansky, and J. E. Schwarz. 1972. Crustecdysone mediated changes in crayfish. Am. Zool. 12: 357–372.

Meister, M. F., J. L. Dimarcq, C. Kappler, C. Hetru, M. Lagueux, R. Lanot, B. Luu, and J. A. Hoffmann. 1985. Conversion of a radiolabelled ecdysone precursor, 2,22,25-trideoxyecdysone, by embryonic and larval tissues of *Locusta migratoria*. Mol. Cell. Endocrinol. 41: 27–44.

Mellon, DeF., Jr. and E. Greer. 1987. Induction of precocious molting and claw transformation in alpheid shrimps by exogenous 20-hydroxyecdysone. Biol. Bull. (Woods Hole) 172: 350–356.

Muramoto, A. 1981. Effects of eyestalk extracts and ecdysterone on water intake through the anus of the crayfish. Comp. Biochem. Physiol. 69A: 197–203.

Nair, V. S. K. 1980. Effects of ecdysterone on the molt initiation response in the Integument of *Jonespeltis splendidus* (Myriapoda: Diplopoda). Zool. Anz. 205: 81–89.

Nakaniski, K., M. Koreeda, M. L. Chang, and H. Y. Hsu. 1966. The structure of ponasterone A, an insect moulting hormone from the leaves of *Podocarpus narkaii* Hay. Chem. Commun. pp. 915–917.

Noulin, G. and J. Maissiat. 1974. Etude du rôle de l'organe Y et de l'effet de l'ecdystérone dans la régénération d'un appendice chez l'oniscoïde *Porcellio dilatatus*. J. Insect Physiol. 20: 1963–1974.

Passano, L. M. 1960. Molting and its control. Pp. 473–536 *in* T. H. Waterman (ed.), *The Physiology of Crustacea*. Academic Press, Orlando, Florida.

Passano, L. M. and S. Jyssum. 1963. The role of the Y-organ in crab proecdysis and limb regeneration. Comp. Biochem. Physiol. 9: 195–213.

Pound, J. M., J. H. Oliver, and R. H. Andrews. 1984. Induction of apolysis and cuticle formation in female *Ornithodoros parkeri* (Acari: Argasidae) by hemocoelic injections of β-ecdysone. J. Med. Entomol. 21: 612–614.

Rao, K. R. 1965. Isolation and partial characterization of the molt-inhibiting hormone of the crustacean eyestalk. Experientia (Basel) 21: 593–594.

Rao, K. R. 1978. Effects of ecdysterone, inokosterone and eyestalk ablation on limb regeneration in the fiddler crab, *Uca pugilator*. J. Exp. Zool. 203: 257–270.

Rao, K. R., S. W. Fingerman, and M. Fingerman. 1973. Effects of exogenous ecdysones on the molt cycles of fourth and fifth stage American lobsters, *Homarus americanus*. Comp. Biochem. Physiol. 44A: 1103–1120.

Rao, K. R., M. Fingerman, and C. Hays. 1972. Comparison of the abilities of α-ecdysone and 20-hydroxyecdysone to induce precocious proecdysis and ecdysis in the fiddler crab, *Uca pugilator*. Z. Vlg. Physiol. 76: 270–284.

Reidenbach, J. M. 1971. Action d'une ecdysone de synthèse sur la mue et le fonctionnement ovarien chez le crustacé isopode *Idotea Balthica* (Pallas). C. R. Acad. Sci. Paris 273D: 1614–1617.

Romer, F. and W. Gnatzy. 1981. Arachnid oenocytes: ecdysone synthesis in the legs of harvestmen (Opilionidae). Cell Tissue Res. 216: 449–453.

Sannasi, A. and T. Subramoniam. 1972. Hormonal rupture of larval diapause in the tick *Rhipicephalus sanguineus* (Lat.) Experientia (Basel) 28: 666–667.

Scheffel, H. 1969. Untersuchungen über die hormonal regulation von Häuntung und Anamorphose von *Lithobius forficatus* (L.) (Myriapoda, Chilopoda). Zool. Jahrb. Abt. Allg. Zool. Physiol. Tiere 74: 436–505.

Scheffel, H. 1983. *In vitro* studies on the moult stimulating effect of ecdysone and 20-hydroxyecdysone in centipedes. Zool. Jahrb. Abt. Allg. Zool.

Physiol. Tiere 87: 425–438.

Scheffel, H., C. Wilke, and W. Pollak. 1974. Die Wirkung von exogenem Ecdysteron auf Larven des Chilopoden *Lithobius forficatus*. Acta Entomol. Bohemoslov. 71: 233–238.

Seifer, G. and J. Rosenberg. 1974. Elektron mikroskopische Untersuchungen der Häutungsdrüsen lymphstränge von *Lithobius forficatus* L. (Chilopoda). Z. Morphol. Oekol. Tiere 78: 263–279.

Skinner, D. M. 1985. Molting and regeneration. Pp. 43–146 *in* D. E. Bliss and L. H. Mantel (eds.), *The Biology of Crustacea*. Academic Press, Orlando, Florida.

Skinner, D. M. and D. E. Graham. 1970. Molting in land crabs: stimulation by leg removal. Science (Wash., DC) 169: 383–385.

Skinner, D. M. and D. E. Graham. 1972. Loss of limbs as a stimulus to ecdysis in Brachyura (true crabs). Biol. Bull. (Woods Hole) 143: 222–233.

Snyder, M. J. and E. S. Chang. 1986. Effects of sinus gland extracts on larval molting and ecdysteroid titers of the American lobster, *Homarus americanus*. Biol. Bull. (Woods Hole) 170: 244–254.

Solomon, K. R., C. K. A. Mango, and F. D. Obenchain. 1982. Endocrine mechanisms in ticks: effects of insect hormones and their mimics on development and reproduction. Pp. 399–438 *in* F. D. Obenchain and R. Galun (eds.), *Physiology of Ticks*. Pergamon Press, Oxford and Elmsford, New York.

Soumoff, C. and J. D. O'Connor. 1982. Repression of Y-organ secretory activity by molt-inhibiting hormone in the crab *Pachygrapsus crassipes*. Gen. Comp. Endocrinol. 48: 432–439.

Soumoff, C. and D. M. Skinner. 1982. Effects of Y-organectomy and autotomy on circulating ecdysteroid levels and regeneration in the crab *Gecarcinus*. Am. Zool. 20: 1182.

Soyez, D. and L. H. Kleinholtz. 1977. Molt-inhibiting factor from the crustacean eyestalk. Gen. Comp. Endocrinol. 31: 233–242.

Spaziani, E. and S. B. Kater. 1973. Uptake and turnover of cholesterol-^{14}C in Y-organs of the crab *Hemigrapsus* as a function of the molt cycle. Gen. Comp. Endocrinol. 20: 534–549.

Spindler, K. D., L. Dinan, and M. Londerhausen. 1984. On the mode of action of ecdysteroids in crustaceans. Pp. 255–264 *in* J. A. Hoffmann and M. Porchet (eds.), *Biosynthesis, Metabolism and Mode of Action of Invertebrate Hormones*. Springer-Verlag, Berlin and New York.

Spindler, K. D., R. Keller, and J. D. O'Connor. 1980. The role of ecdysteroids in the crustacean molting cycle. Pp. 247–280 *in* J. A. Hoffmann (ed.), *Progress in Ecdysone Research*. Elsevier/North-Holland Publ., Amsterdam and New York.

Stevenson, J. R. 1978. The ecdysones and control of chitin sythesis in *Orconectes*. Freshwater Crayfish 4: 123–130.

Stevenson, J. R. 1985. Dynamics of the integument. Pp. 1–42 *in* D. E. Bliss and L. H. Mantel (eds.), *The Biology of Crustacea*. Academic Press, Orlando, Florida.

Stevenson, J. R. and J. A. Tschantz. 1973. Acceleration by ecdysterone of premoult substages in the crayfish. Nature (Lond.) 242: 133–134.

Suzuki, S. 1985. Effect of Y-organ removal on limb regeneration and molting in the terrestrial crab, *Sesarma haematocheir.* Gen. Comp. Endocrinol. 58: 202–210.

Taketomi, Y. and M. Hyodo. 1986. The Y-organ of the crab, *Portunus Trituberculatus:* effects of ecdysterone on the ultrastructure. Cell Biol. Int. Rep. 10: 367–374.

Tchernigovtzeff, C. 1965. Multiplication cellulaire et régénération au cours du cycle d'intermue des crustacés décapodes. Arch. Zool. Exp. Gén. 106: 377–497.

Tchernigovtzeff, C. 1974. Régénération et cycle d'intermue chez le crabe *Gecarcinus lateralis.* II. Situation du moment critique: incidence d'une régénération tardive sur le cours de la prémue. Arch. Zool. Exp. Gén. 115: 423–440.

Tighe-Ford, D. J. and D. C. Vaile. 1972. The action of crustecdysone on the cirripede *Balanus balanoides* (L.) J. Exp. Mar. Biol. Ecol. 9: 19–28.

Touir, A. and H. Charniaux-Cotton. 1974. Influence de l'introduction d'ecdystérone sur l'exuviation et la démarrage de la vitellogénèse chez la crevette *Lysmata seticaudata* Risso. C. R. Acad. Sci. Paris 278D: 119–122.

Traub, M., G. Gellissen, and K. D. Spindler. 1987. 20(OH)Ecdysone-induced transition from intermolt to premolt protein biosynthesis patterns in the hypodermis of the crayfish, *Astacus leptodactylus, in vitro.* Gen. Comp. Endocrinol. 65: 469–477.

Van Wormhoudt, A. 1973. Variations annuelles des activités enzymatiques digestives chez *Palaemon serratus.* C. R. Acad. Sci. Paris 277D: 369–373.

Van Wormhoudt, A., C. Bellon, and A. Le Roux. 1978. Influence des formations endocrines sur l'incorporation de leucine tritiée dans les protéines de l'hépatopancréas de la crevette *Palaemon serratus* Pennant. Gen. Comp. Endocrinol. 35: 263–273.

Vernet, G. 1976. Données actuelles sur le déterminisme de la mue chez les crustacés. Ann. Biol. 3–4: 155–188.

Warner, A. C. and J. R. Stevenson. 1972. The influence of ecdysones and eyestalk removal on the molt cycle of the crayfish *Orconectes obscurus.* Gen. Comp. Endocrinol. 18: 454–462.

Watson, D. R. and E. Spaziani. 1985a. Effect of eyestalk removal on cholesterol uptake and ecdysone secretion by crab *(Cancer antennarius)* Y-organs *in vitro.* Gen. Comp. Endocrinol. 57: 360–370.

Watson, D. R. and E. Spaziani. 1985b. Biosynthesis of ecdysteroids from cholesterol by crab Y-organs, and eyestalk suppression of cholesterol uptake and secretory activity, *in vitro.* Gen. Comp. Endocrinol. 59: 140–148.

Webster, S. G. 1983. Effects of exogenous ecdysterone upon moulting, proecdysial development, and limb regeneration in the prawn *Palaemon elegans.* Gen. Comp. Endocrinol. 49: 459–469.

Webster, S. G. and R. Keller. 1986. Purification, characterisation and amino acid composition of the putative moult-inhibiting hormone (MIH) of *Carcinus maenas* (Crustacea, Decapoda). J. Comp. Physiol. 156B: 617–624.

Willig, A. and R. Keller. 1973. Molting hormone content, cuticle growth and gastrolith growth in the molt cycle of the crayfish *Orconectes limosus*. J. Comp. Biochem. Physiol. 86: 377–388.

Willig, A. and R. Keller. 1976. Biosynthesis of α- and β-ecdysone by the crayfish *Orconectes limosus in vivo* and by its Y-organs *in vitro*. Experientia (Basel) 32: 936–937.

Winget, R. R. and W. S. Herman. 1976. Occurrence of ecdysone in the blood of the chelicerate arthropod, *Limulus polyphemus*. Experientia (Basel) 32: 1345–1346.

Winget, R. R. and W. S. Herman. 1979. Influence of molt cycle and β-ecdysone on protein synthesis in the chelicerate arthropod, *Limulus polyphemus*. Comp. Biochem. Physiol 62B: 119–122.

Wittig, K. P. and J. R. Stevenson. 1974. Modification of uridinediphosphoacetylglucosamine pyrophosphorylase by ecdysterone in the crayfish epidermis Am. Zool. 14: 1289.

Wittig, K. P. and J. R. Stevenson. 1975. DNA synthesis in the crayfish epidermis and its modification by ecdysterone. J. Comp. Physiol. 99: 279–286.

Wright, J. E. 1969. Hormonal termination of larval diapause in *Dermacentor albipictus*. Science (Wash., DC) 163: 390–391.

Zandee, D. I. 1964. Absence of sterol synthesis in some arthropods. Nature (Lond.) 202: 1335.

Zandee, D. I. 1966. Metabolism in the crayfish *Astacus astacus* L. III. Absence of cholesterol synthesis. Arch. Int. Physiol. Biochim. 74: 435–441.

Zandee, C. I. 1967. Absence of cholesterol synthesis as contrasted with the presence of fatty acid synthesis in some arthropods. Comp. Biochem. Physiol. 20: 811–822.

Metabolism of Insect Molting Hormones: Bioconversion and Titer Regulation

10

MALCOLM J. THOMPSON,
GUNTER F. WEIRICH, AND
JAMES A. SVOBODA

10.1.	Introduction	327
10.2.	Hydroxylation	331
10.3.	Epimerization	333
10.4.	Conjugation	335
	10.4.1. Phosphorylation/Acetylation	336
	10.4.2. Glucosylation	341
	10.4.3. Acylation	342
10.5.	Hydrolysis of Conjugates	343
10.6.	Formation of Ecdysteroid Acids	344
10.7.	Enzymes	347
	10.7.1. Monooxygenases (Hydroxylases)	347
	10.7.1.1. Ecdysone 20-Monooxygenase	347
	10.7.1.2. 2-Deoxyecdysone 2-Hydroxylase	350
	10.7.2. Ecdysone Oxidase, 3-Oxoecdysteroid 3α-Reductase, and 3-Oxoecdysteroid 3β-Reductase (Ecdysone 3-Epimerase System)	350
	10.7.3. ATP:Ecdysteroid Phosphotransferases	351
	10.7.4. Ecdysteroid–Phosphate Phosphohydrolase	352
10.8.	Summary	352
References		354

10.1. Introduction

The molting hormones (MHs), ecdysone and 20-hydroxyecdysone, were first isolated from pupae of the silkworm, *Bombyx mori* (Butenandt and Karlson, 1954; Karlson, 1956). Following the identification of ecdysone by chemical methods (Karlson et al., 1963) and X-ray diffraction spectroscopy (Huber and Hoppe, 1965), and of 20-hydroxyecdysone after its isolation from a seawater crayfish, *Jasus lalandei* (Hampshire and Horn, 1966), both ecdysteroids [a generic term for all compounds structurally related to ecdysone (Goodwin et al., 1978)] were isolated from pupae of the tobacco hornworm, *Manduca sexta* (Kaplanis et al., 1966a), 7 days after the larval–pupal ecdysis (peak titer). Later, a more polar ecdysteroid, present in small quantity, was identified as 20,26-dihydroxyecdysone (Thompson et al., 1967). This ecdysteroid turned out to be the major ecdysteroid in hornworms 12 days after the larval–pupal ecdysis, and smaller amounts of 3-epi-20-hydroxyecdysone, 20-hydroxyecdysone, 3-epi-20,26-dihydroxy-ecdysone, 3-epiecdysone, and ecdysone were also present at this stage (Kaplanis et al., 1979).

Thus, although the 3-epiecdysone and 3-epi-20-hydroxyecdysone were minor components at the time of peak titer, 5 days later, these two 3α-ecdysteroids quantitatively surpassed their 3β-isomers. 20,26-Dihydroxyecdysone, 3-epiecdysone, and 3-epi-20-hydroxyecdysone were found to be about one-tenth as active as ecdysone or 20-hydroxyecdysone and 3-epi-20,26-dihydroxyecdysone, and 1/300 as active as ecdysone (Thompson et al., 1967; Kaplanis et al., 1979) in the house fly assay (Kaplanis et al., 1966b). Thus, the biological activities strongly indicated that hydroxylation at C-26 and epimerization at C-3 are mechanisms for deactivating MHs.

The hydroxylation of ecdysone at C-20 to produce 20-hydroxyecdysone has been demonstrated in all species examined (Koolman and Karlson, 1985; Smith, 1985) and has been described as ''activation of the hormone.'' Although 20-hydroxylation of ecdysone is highly involved in titer regulation, in this chapter we will discuss mainly hydroxylations at positions other than at C-20.

To determine the mechanisms involved in MH titer regulation, the active hormones and their metabolites have to be isolated from insect sources (including excretory products) and identified. One of the major difficulties in such studies is that the active ecdysteroid(s) or hormone(s) may be present only in very minute amounts and only for a limited period of time, whereas their metabolites may be present in rather large quantities. Nevertheless, progress has been made and continues to be made, as a result of which we now have a better understanding of the numerous bioconversions of ecdysteroids occurring in various species, especially at those stages that constitute a closed system (eggs or pupae). The progress has been facilitated by using radiolabeled cholesterol as a precursor or examining the conversion of other labeled steroids to free and conjugated metabolites. The use of labeled precursors has made it possible to isolate previously undetected ecdysteroids and to determine their titer fluctuations during different periods of the life cycle. Only recently have techniques and methodologies been developed to isolate and analyze insect extracts directly for ecdysteroid acids and conjugates. With current instrumentation and techniques the structural elucidation of ecdysteroids is much less difficult than in the past. Publications during the last 5 years have uncovered the great variety of enzymatic bioconversions that regulate ecdysteroid titers in insects: hydroxylation, epimerization, conjugation, hydrolysis, and formation of ecdysteroid acids.

Several reviews concerned with regulation of ecdysteroid titers (Koolman and Karlson, 1985; Smith, 1985), with biosynthesis of ovarian ecdysteroids (Lagueux et al., 1984; Rees and Isaac, 1984), as well as with enzymes of ecdysone metabolism (Koolman, 1982; Lafont and Koolman, 1984) have been published in previous years. In the present chapter, we intend primarily to provide a comprehensive update and review of the more recent research. Figure 10.1 shows representative structures of neutral ecdysteroids and ecdysteroid acids, and Figure 10.2 those of ecdysteroid conjugates. The various bioconversions and enzymes involved in MH titer regulation will be discussed below in separate sections. Inevitably, this will make it somewhat difficult to visualize the interrelationships of the bioconversions in the control of ecdysteroid titers. However, several proposed schemes of metabolic pathways will be presented. Based on the observed accumulation of certain ecdysteroid metabolites in specific developmental stages and our current knowledge of biological activities, we can draw conclusions as to which biochemical reactions are most likely involved in controlling MH titers. However, we cannot rule out the possibility that compounds now assumed to be inactivation products may later turn out to have specific physiological functions.

Some of these biochemical transformations may be involved in more than one physiological function. For example, hydroxylations are essential for MH biosynthesis from cholesterol, and hydroxylation (at C-26) is also involved in MH inactivation. Formation of conjugates serves as an inactivation reaction, facilitates the storage of large quan-

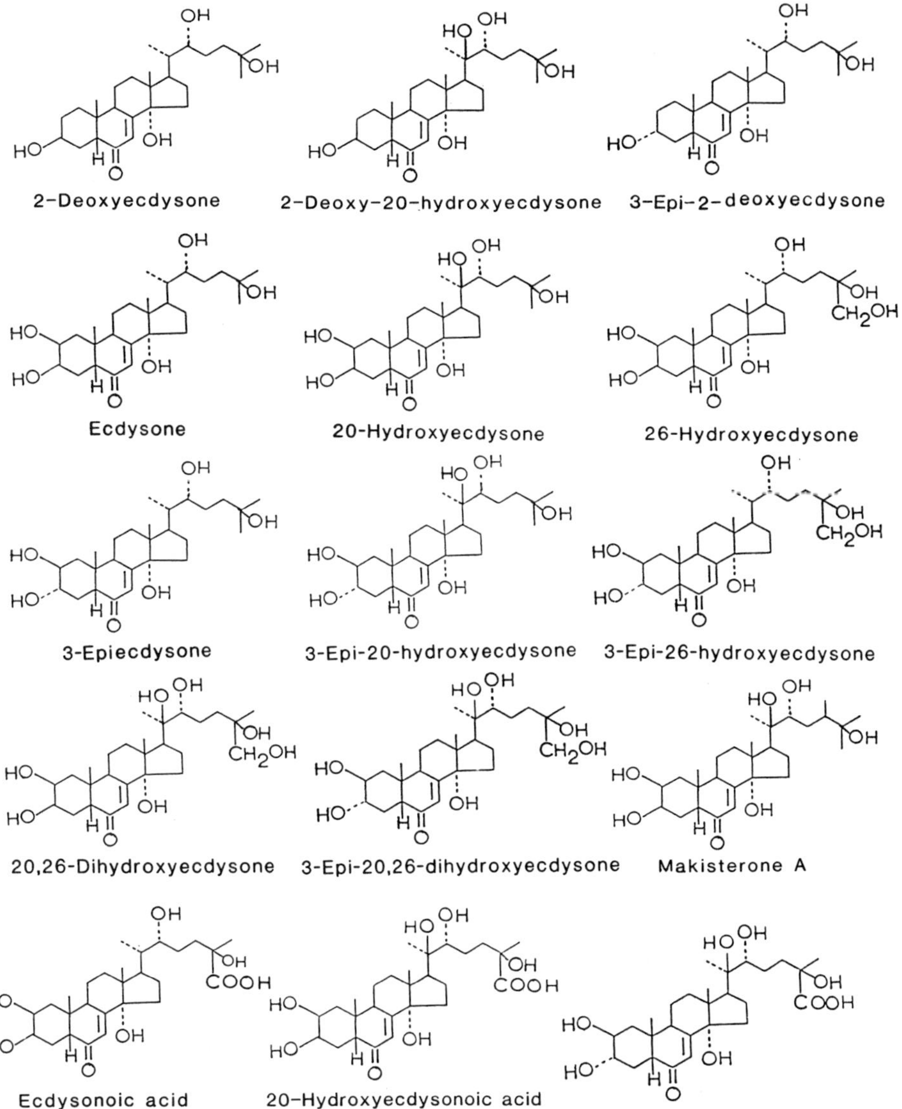

FIGURE 10.1. Representative structures of free ecdysteroids and ecdysteroid acids.

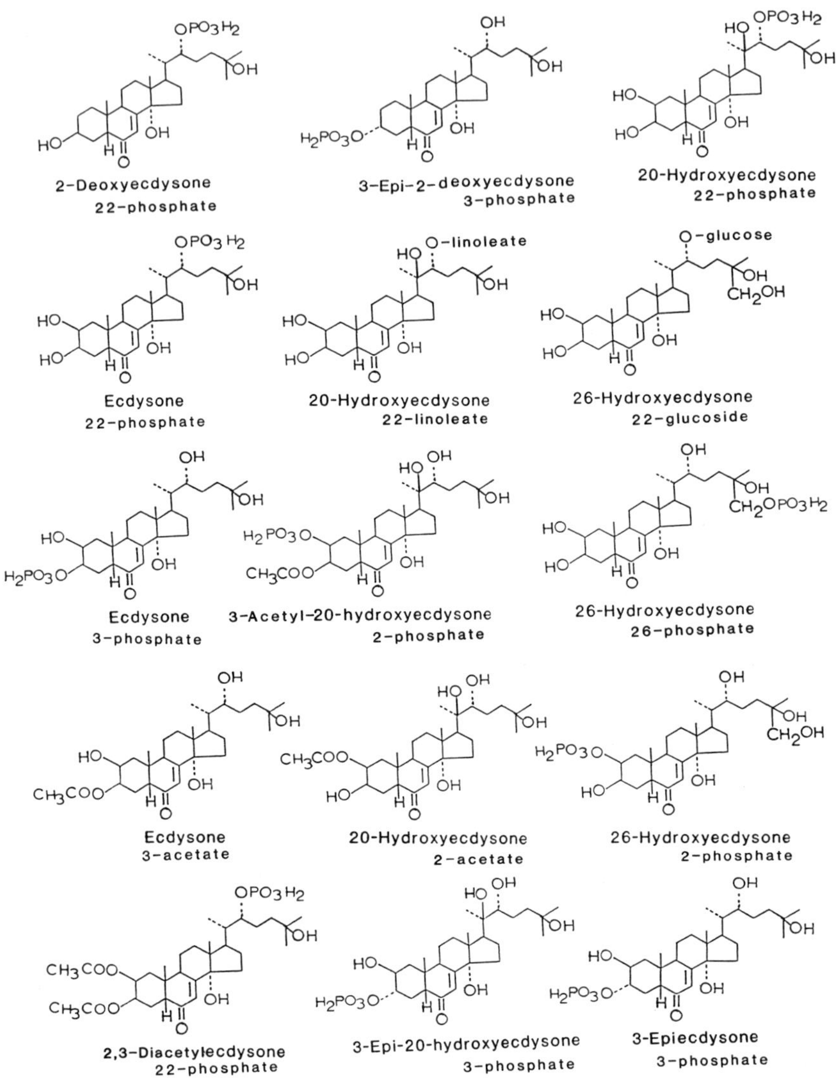

FIGURE 10.2. Representative structures of ecdysteroid conjugates.

tities of conjugated ecdysteroids in eggs of several species, and has also been proposed to enhance the overall synthesis of ecdysteroids by facilitating the removal of free steroids from the site of synthesis (Isaac et al., 1983a).

10.2. Hydroxylation

The discovery of 2-deoxyecdysone and 2-deoxy-20-hydroxyecdysone in *B. mori* ovaries (Ohnishi et al., 1981) and the presence of these ecdysteroids as conjugates in ovaries and newly laid eggs of the desert locust, *Schistocerca gregaria* (Isaac et al., 1981a), and the migratory locust, *Locusta migratoria* (Lagueux et al., 1984), suggest that in these insects the terminal step in the biosynthesis may be hydroxylation at C-2 for ecdysone and hydroxylation at either C-2 or C-20 for 20-hydroxyecdysone (Rees and Isaac, 1984). Either one of these reactions could be involved in titer regulation.

As mentioned above in Section 10.1, hydroxylation of ecdysone at C-20 to 20-hydroxyecdysone has been demonstrated in all species examined (Koolman and Karlson, 1985; Smith, 1985). *M. sexta* ovaries and eggs initially contain large amounts of 26-hydroxyecdysone 26-phosphate, part of which has been hydrolyzed in 1- to 18-h-old eggs to free 26-hydroxyecdysone (Kaplanis et al., 1980; Warren et al., 1986; Thompson et al., 1988). 26-Hydroxyecdysone is converted to 20,26-dihydroxyecdysone, the major free ecdysteroid in 48- to 64-h-old eggs (Thompson et al., 1988).

26-Hydroxylation is also a common reaction in insect ecdysteroid metabolism (Koolman, 1982), as indicated in Table 10.1. Although ecdysone appears to be a plausible precursor for 26-hydroxyecdysone, 26-hydroxylation of ecdysone has not been demonstrated directly (Thompson et al., 1986).

The conversion of ecdysone and 20-hydroxyecdysone to ecdysonoic and 20-hydroxyecdysonoic acid, respectively, probably proceeds via 26-hydroxyecdysone and 20,26-dihydroxyecdysone (see Section 10.6, below).

During pupal–adult development of *M. sexta* females, a clearance of ecdysone from the hemolymph and a concomitant rise in the titer of 20-hydroxyecdysone (peak titer days 9 and 10) has been shown. During the 4 days after the 20-hydroxyecdysone peak, 20-hydroxyecdysone disappears from the hemolymph, and as a result of 26-hydroxylation the concentration of 20,26-dihydroxyecdysone increases. 20,26-Dihydroxyecdysone is subsequently converted via oxi-

TABLE 10.1. Hydroxylation

Compound	Position	Product	Species	Source	Reference
2-Deoxyecdysone	C-2	Ecdysone	*Bombyx mori*	Ovaries	Ohnishi et al., 1981
			Schistocerca gregaria	Ovaries, eggs	Isaac et al., 1981a
			Locusta migratoria	Ovaries, eggs	Lagueux et al., 1984
2-Deoxy-20-hydroxy-ecdysone	C-2	20-Hydroxyecdysone	*Bombyx mori*	Ovaries	Ohnishi et al., 1981
			Schistocerca gregaria	Ovaries, eggs	Isaac et al., 1981a
			Locusta migratoria	Ovaries, eggs	Lagueux et al., 1984
26-Hydroxyecdysone	C-20	20,26-Dihydroxy-ecdysone	*Manduca sexta*	Eggs	Kaplanis et al., 1980
					Warren et al., 1986
					Thompson et al., 1988
Ecdysone	C-26	26-Hydroxyecdysone	*Manduca sexta*	Pupae	Warren and Gilbert, 1986
20-Hydroxyecdysone	C-26	20,26-Dihydroxy-ecdysone	*Manduca sexta*	Pupae	Thompson et al., 1967
					Kaplanis et al., 1979
				Hemolymph, gut	Warren and Gilbert, 1986
			Locusta migratoria	Eggs	Lagueux et al., 1984
				Adults, feces	Modde et al., 1984
			Calliphora vicina	Fat body	Lafont et al., 1983
			Schistocerca gregaria	Nymphs	Gibson et al., 1984
			Spodoptera littoralis	Pupae	Isaac et al., 1983b
			Pieris brassicae	Larvae	Beydon et al., 1987
3-Epi-20-hydroxy-ecdysone	C-26	3-Epi-20,26-dihydroxy-ecdysone	*Pieris brassicae*	Larvae	Beydon et al., 1987

dation to 20-hydroxyecdysonoic acid (Warren and Gilbert, 1986). Interestingly, the peak titers and the rates of titer fluctuations of ecdysone, 20-hydroxyecdysone, and 20,26-dihydroxyecdysone were similar for both hemolymph and gut, but higher levels of the ecdysteroid acids were localized in the gut.

All these results indicate that hydroxylation at C-26 is a means of controlling the titer of 20-hydroxyecdysone in many insect species (Table 10.1.). Analogous to the conversion of ecdysone to 20-hydroxyecdysone, the conversion of 26-hydroxyecdysone to 20,26-dihydroxyecdysone via C-20 hydroxylation, however, appears to be an activation process that occurs during embryonic development of *M. sexta* eggs (Warren et al., 1986; Thompson et al., 1988). This conclusion is based on the observation that 26-hydroxyecdysone is inactive in the housefly MH assay (Kaplanis et al., 1980) whereas 20,26-dihydroxyecdysone is about one-tenth as active as 20-hydroxyecdysone, and that in 48- to 64-h-old *M. sexta* eggs 20,26-dihydroxyecdysone is the major free ecdysteroid, which is subsequently converted to the less active 3-epi-20,26-dihydroxyecdysone and to the inactive 20-hydroxyecdysonoic acid (Thompson et al., 1988).

10.3. Epimerization

Thompson et al. (1974) first isolated 3-epi-20-hydroxyecdysone and 3-epiecdysone from meconium fluid of the tobacco hornworm. In the same year, 3-epiecdysone was also identified as the *in vitro* product of an enzyme system from *M. sexta* midgut (Nigg et al., 1974). Since then, other 3α-ecdysteroids (Table 10.2) have been isolated from *M. sexta* and a number of other species (Lafont and Koolman, 1984). In the conversion of 3β- to 3α-ecdysteroids, 3-dehydroecdysteroids are intermediates (see Section 10.7.2, below).

3-Epiecdysone, 3-epi-20-hydroxyecdysone, and 3-epi-20,26-dihydroxyecdysone have now been isolated and identified from various developmental stages of *M. sexta* (Svoboda and Thompson, 1985). These three ecdysteroids and the corresponding 3β-epimers have also been identified in extracts from *Pieris brassicae* pharate pupae that had been injected with [³H]ecdysone. The same extracts also contained ecdysonoic and 20-hydroxyecdysonoic acid and their 3α-epimers, 3-epiecdysonoic and 3-epi-20-hydroxyecdysonoic acid, as further metabolites (Lafont et al., 1980).

Epimerization is a very prominent reaction during the middle and final stages of embryogenesis of *M. sexta*. 3-Epi-20,26-dihydro-

TABLE 10.2. Epimerization

Compound	Position	Product	Species	Source	Reference
2-Deoxyecdysone	C-3	3-Epi-2-deoxyecdysone	*Schistocerca gregaria*	Eggs	Isaac et al., 1981b
					Rees and Issac, 1984
			Locusta migratoria	Eggs	Tsoupras et al., 1982
Ecdysone	C-3	3-Epiecdysone	*Manduca sexta*	Meconium	Thompson et al., 1974
				Midgut	Nigg et al., 1974
				Pupae	Kaplanis et al., 1979
			Pieris brassicae	pharate pupae	Lafont et al., 1980
20-Hydroxyecdysone	C-3	3-Epi-20-hydroxy-ecdysone	*Manduca sexta*	Meconium	Thompson et al., 1974
				Pupae	Kaplanis et al., 1979
			Pieris brassicae	Pharate pupae	Lafont et al., 1980
				Pupae,[a] larvae	Beydon and Lafont, 1987
20,26-Dihydroxy-ecdysone	C-3	3-Epi-20,26-dihydroxy-ecdysone	*Manduca sexta*	Pupae	Kaplanis et al., 1979
				Eggs	Kaplanis et al., 1980
					Thompson et al., 1988
				Larvae	Thompson et al., 1987a, 1988
			Pieris brassicae	Pharate pupae	Lafont et al., 1980
26-Hydroxyecdysone	C-3	3-Epi-26-hydroxy-ecdysone	*Manduca sexta*	Eggs	Kaplanis et al., 1979
					Thompson et al., 1988
				Larvae	Thompson et al., 1987a, 1988
Ecdysonoic acid	C-3	3-Epiecdysonoic acid	*Pieris brassicae*	Pharate pupae	Lafont et al., 1980
20-Hydroxyecdysonoic acid	C-3	3-Epi-20-hydroxy-ecdysonoic acid	*Pieris brassicae*	Pharate pupae	Lafont et al., 1980
				Larvae	Beydon and Lafont, 1987
			Manduca sexta	Pupae	Lozano et al., 1988
				Eggs, larvae	Thompson et al., 1988

[a] Newly ecdysed.

xyecdysone is first detected in 24- to 40-h-old eggs and reaches a maximum in 0- to 1-h-old larvae (2.9 μg/g of tissue). Around 96 h after oviposition, there appears to be a surge in the hydrolysis of the maternal ecdysteroid conjugate 26-hydroxyecdysone 26-phosphate to 26-hydroxyecdysone and subsequent epimerization to 3-epi-26-hydroxyecdysone. The latter ecdysteroid attains a level of 6.4 and 9.1 μg/g in 96-h-old eggs and 0- to 1-h-old larvae, respectively (Thompson et al., 1988). In fact, there is little or no free 3β-ecdysteroid in the young larvae (Thompson et al., 1987a). 3-Epi-20-hydroxyecdysonoic acid is first noticeable in 72- to 88-h-old eggs and accounts for 18.5% of the more polar metabolites in 0- to 1-h-old larvae. It could be derived via epimerization from 20-hydroxyecdysonoic acid since the level of 20-hydroxyecdysonoic acid decreases as the amount of 3-epi-20-hydroxyecdysonoic acid increases (Thompson et al., 1988).

3-Epi-2-deoxyecdysone 3-phospate is the major ecdysteroid in 8-day-old embryos of *L. migratoria* (Tsoupras et al., 1982) and a minor one in *S. gregaria* eggs (Isaac and Rees, 1984). Free 3-epi-2-deoxyecdysone, the most likely precursor of 3-epi-2-deoxyecdysone 3-phosphate, was first isolated from *S. gregaria* eggs (Isaac et al., 1981a).

Certainly, these results suggest that formation of the 3α-ecdysteroids from the more active 3β-ecdysteroids could be a means of regulating the MH titer.

10.4. Conjugation

Evidence for the conjugation of ecdysteroids was first reported in 1970 by Heinrich and Hoffmeister. Since that initial report, ecdysteroid conjugates have been detected as metabolites of labeled and unlabeled ecdysteroids in many species. Based only on hydrolysis with commercially available enzymes, the ecdysteroid conjugates have been designated as ecdysteroid glucosides, glucuronides, and sulfates (Koolman, 1982). However, the name under which an enzyme is marketed does not necessarily indicate the major activity contained therein (Weirich et al., 1986), and therefore some of the early reports require verification.

Formation of ecdysteroid conjugates apparently serves as a means of inactivation. Ecdysteroid conjugates are also utilized for storage of large quantities of ecdysteroids in insects eggs by mechanisms not yet understood. Ecdysteroids from these conjugates are released during embryogenesis, a developmental stage believed to be incapable of *de novo* ecdysteroid biosynthesis (Dinan and Rees, 1981; Lagueux et al.,

1981; Sall et al., 1983; and Isaac et al., 1983a). Conjugation could promote the overall synthesis of ecdysteroids by facilitating the removal of free steroids from the site of synthesis (Isaac et al., 1983b).

The first ecdysteroid conjugate isolated and identified was ecdysone 3-acetate from *S. gregaria* embryos (Isaac et al., 1981b). This was followed by the isolation and identification of the more polar 22-phosphates of ecdysone and 2-deoxyecdysone from the same source (Isaac et al., 1982). Today, the number of conjugates isolated and identified is nearly as large as the number of free ecdysteroids known to occur in insects (see Lafont and Koolman, 1984; Rees and Isaac, 1984). Although phosphoconjugation seems to be the predominant reaction in insect embryos and pupae, there are indications that glucosylation and acylation are also involved in regulating ecdysteroid titers.

10.4.1. *Phosphorylation/Acetylation*

Ecdysteroids can be phosphorylated at any one of the secondary alcohol positions C-2, C-3, or C-22, or at the C-26 primary alcohol position (Table 10.3). Thus far most of the ecdysteroid phosphates have been isolated from ovaries or newly laid eggs. The four major conjugates identified from eggs of *S. gregaria* are the 22-phosphates of ecdysone, 2-deoxyecdysone (Isaac et al., 1982; Isaac and Rees, 1985), 20-hydroxyecdysone, and 2-deoxy-20-hydroxyecdysone (Isaac et al., 1983a). During later stages of embryogenesis, five more ecdysteroid conjugates—3-acetyl-20-hydroxyecdysone 2-phosphate, 3-epi-2-deoxyecdysone 3-phosphate, 3-acetylecdysone 22-phosphate, 2-acetylecdysone 22-phosphate (Isaac and Rees, 1984), and 3-acetylecdysone 2-phosphate (Isaac et al., 1984)—are found. The last compound is rather unstable upon storage; it is converted to ecdysone 2-phosphate. Interestingly, both 3-acetylecdysone 2-phosphate and ecdysone 2-phosphate are less susceptible than ecdysone 22-phosphate to *in vitro* hydrolysis by an enzyme preparation from *S. gregaria* embryos (Isaac et al., 1984).

Analyses of ecdysteroids in *S. gregaria* eggs labeled maternally from [^{14}C]cholesterol showed that highly polar ecdysteroid derivatives predominated in all stages studied, but changes in their composition occurred between day 10 of development and hatching (day 17). During this period, the labeled polar conjugates of ecdysone 3-acetate (which represented 36% of the total conjugated steroids on day 17), 3-epi-2-deoxyecdysone, and ecdysteroid acids appeared.

TABLE 10.3. Phosphorylation/acetylation

Compound	Position	Product	Species	Source	Reference
2-Deoxyecdysone	C-22	2-Deoxyecdysone 22-phosphate	*Schistocerca gregaria*	Ovaries, eggs	Isaac et al., 1982, 1983a
			Locusta migratoria	Ovaries, eggs	Lagueux et al., 1984
Ecdysone	C-22	Ecdysone 22-phosphate	*Schistocerca gregaria*	Ovaries, eggs	Isaac et al., 1982, 1983a
			Locusta migratoria	Ovaries, eggs	Lagueux et al., 1984
	C-2,22	2-Acetylecdysone 22-phosphate	*Schistocerca gregaria*	Eggs	Isaac and Rees, 1984
	C-3,22	3-Acetylecdysone 22-phosphate			
	C-2,3	3-Acetylecdysone 2-phosphate			Isaac et al., 1984
	C-3	Ecdysone 3-acetate			Isaac et al., 1981b
					Isaac and Rees, 1985
				Nymphs	Gibson et al., 1984
	C-3	Ecdysone 3-phosphate	*Locusta migratoria*	Eggs	Lagueux et al., 1984
Ecdysone or ecdysone 2-phosphate	C-2,3,22	2,3-Diacetylecdysone 22-phosphate	*Locusta migratoria*	Eggs	Lagueux et al., 1984
3-Epi-2-deoxyecdysone	C-3	3-Epi-2-deoxyecdysone 3-phosphate	*Locusta migratoria*	Eggs	Tsoupras et al., 1982
			Schistocerca gregaria	Eggs	Isaac and Rees, 1984
3-Epiecdysone	C-3	3-Epiecdysone 3-phosphate	*Pieris brassicae*	Larvae[a]	Beydon and Lafont, 1987

TABLE 10.3. Phosphorylation/acetylation (continued)

Compound	Position	Product	Species	Source	Reference
2-Deoxy-20-hydroxy-ecdysone	C-22	2-Deoxy-20-hydroxyecdysone 22-phosphate	*Schistocerca gregaria*	Ovaries, eggs	Isaac et al., 1983b
20-Hydroxyecdysone	C-22	20-Hydroxyecdysone 22-phosphate	*Schistocerca gregaria*	Ovaries, eggs	Isaac et al., 1983b
	C-2	20-Hydroxyecdysone 2-acetate		Adults	Wilson and Lafont, 1986
	C-3	20-Hydroxyecdysone 3-acetate		Nymphs, adults	Modde et al., 1984
	C-2,3	3-Acetyl-20-hydroxyecdysone 2-phosphate	*Schistocerca gregaria*	Eggs	Isaac and Rees, 1984, 1985
			Locusta migratoria	Nymphs, adults	Modde et al., 1984
				Adults	Wilson and Lafont, 1986
3-Epi-20-hydroxyecdysone	C-3	3-Epi-20-hydroxyecdysone 3-phosphate	*Pieris brassicae*	Larvae[a]	Beydon and Lafont, 1987
26-Hydroxyecdysone	C-26	26-Hydroxyecdysone 26-phosphate	*Manduca sexta*	Ovaries, egs	Thompson et al., 1985
	C-2	26-Hydroxyecdysone 2-phosphate			Thompson et al., 1987b

[a] Red feces

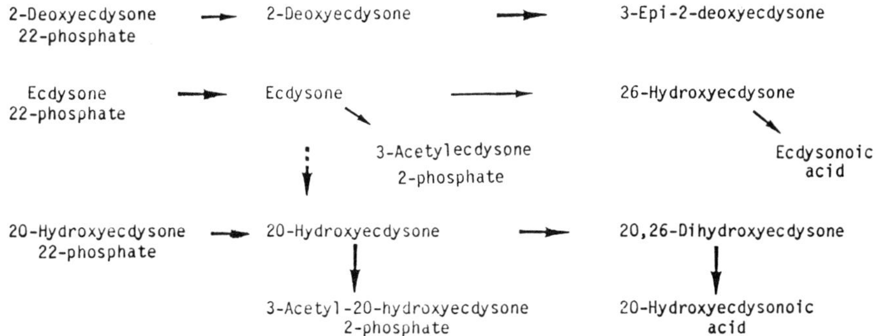

FIGURE 10.3. Proposed metabolic pathways of ecdysteroid 22-phosphates in developing eggs of *S. gregaria* (modified from Isaac and Rees, 1985).

[^{3}H]Ecdysone injected into eggs was metabolized to ecdysonoic acid, 3-acetylecdysone 2-phosphate, and—to a lesser extent—ecdysone 22-phosphate, 20-hydroxyecdysonoic acid, and possibly a phosphate ester of 20-hydroxyecdysone 3-acetate (Isaac and Rees, 1985). Based on these results, a metabolic scheme for the regulation of ecdysteroid levels in developing *S. gregaria* eggs was postulated (Fig. 10.3).

Ovarian ecdysteroids of *L. migratoria* are also conjugated at their C-22 position to phosphoric acid. For the *L. migratoria* conjugates, evidence was obtained that an adenosine moiety was covalently linked to the C-22 phosphates of 2-deoxyecdysone and ecdysone. However, the adenosine phosphates may only constitute a minor fraction of the 22-phosphates present in the eggs (Lagueux et al., 1984). During *L. migratoria* embryogenesis, the ecdysteroid 22-phosphates are hydrolyzed and 3-epi-2-deoxyecdysone is formed as the major metabolite of 2-deoxyecdysone. The concentrations of ecdysone conjugates also decrease during embryogenesis, and the main metabolites are ecdysone 3-phosphate and 2,3-diacetyl-22-phosphates of ecdysone and 20-hydroxyecdysone. It is not known whether the 2,3-diacetyl-22-phosphate of ecdysone is formed via acetylation of ecdysone 22-phosphate or acetylation and phosphorylation of free ecdysone. 20-Hydroxyecdysone is metabolized to 20,26-dihydroxyecdysone, 3-acetyl-20-hydroxyecdysone 22-phosphate, and other minor metabolites in *L. migratoria* at a later stage of embryonic development (Lagueux et al., 1984). Thus, after hydrolysis of maternal ecdysteroid conjugates, the titer of free ecdysteroids is first increased and then reduced again via reconjugation, predominantly acetylation and phosphorylation, according to a metabolic scheme (Fig. 10.4) proposed by Lagueux et al. (1984).

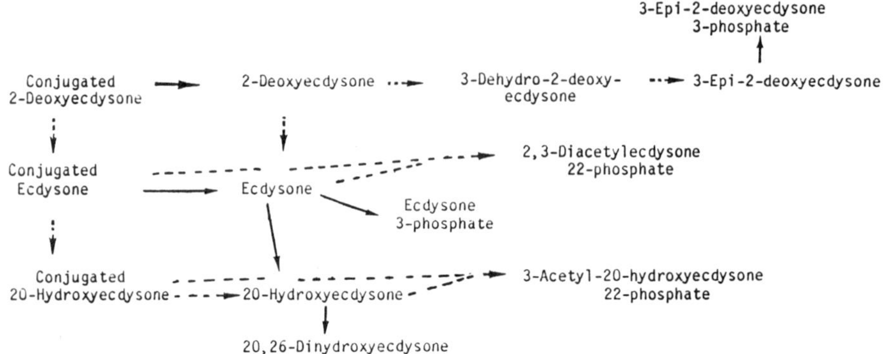

FIGURE 10.4. Proposed metabolic scheme of maternal ecdysteroids in eggs of *L. migratoria* during embryogenesis (modified from Lagueux et al., 1984).

M. sexta ovaries and newly laid eggs contain 26-hydroxyecdysone 26-phosphate as the major conjugate (Thompson et al., 1985) and 26-hydroxyecdysone 2-phosphate as a minor conjugate (Thompson et al., 1987b). It was concluded that 26-hydroxyecdysone 2-phosphate is an inactivation product. 26-Hydroxyecdysone 2-phosphate represents slightly less than 3% of total ecdysteroid conjugates in ovaries from 1-day-old adult females but more than 8% in ovaries from 4-day-old females. During oocyte maturation, 26-hydroxyecdysone 26-phosphate may possibly be converted in minute amounts to free 26-hydroxyecdysone, which in turn is phosphorylated to 26-hydroxyecdysone 2-phosphate. After initiation of embryogenesis the level of 26-hydroxyecdysone 2-phosphate, however, remains rather constant as the level of 26-hydroxyecdysone 26-phosphate decreases (Thompson et al., 1988). 26-Hydroxyecdysone 26-phosphate is metabolized to a number of metabolites (Fig. 10.5).

[^3H]Ecdysone injected into fifth-instar *S. gregaria* larvae was found to be metabolized and excreted as ecdysone 3-acetate, 20-hydroxyecdysone, and 20,26-dihydroxyecdysone, both free and as polar conjugates. Among the polar conjugates were 3-acetylecdysone 2-phosphate and 3-acetyl-20-hydroxyecdysone 2-phosphate (Gibson et al., 1984). The corresponding ecdysteroid acids were also detected. Early fifth-instar larvae (low endogenous ecdysteroid titer) excreted injected [^3H]ecdysone label mostly as free ecdysone and highly polar conjugates of ecdysone 3-acetate and 20-hydroxyecdysone. Late fifth-instar larvae (high endogenous ecdysteroid titer) excreted the radiolabel mostly as 20-hydroxyecdysone conjugate, and only a very small amount of ecdysone 3-acetate was detected (Gibson et al., 1984).

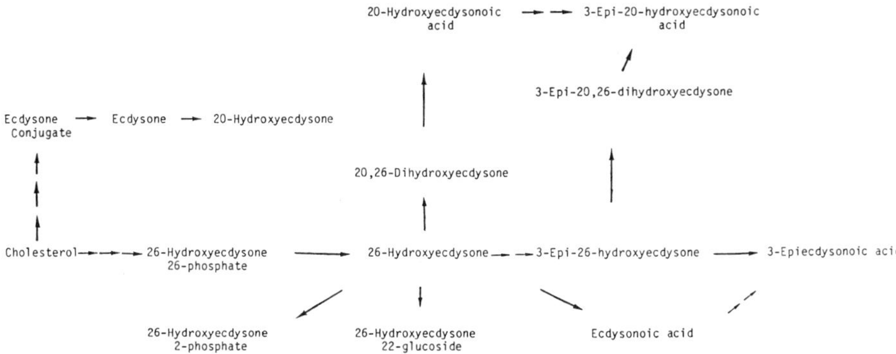

FIGURE 10.5. Metabolic scheme of ecdysteroids in *M. sexta* embryos and young larvae following incorporation of [^{14}C]cholesterol into ovarian ecdysteroids (Thompson et al., 1988).

The natural occurrence of ecdysteroid acetyl-phosphates during the later stages of embryogenesis in *L. migratoria* and *S. gregaria,* and the conversion of injected [^3H]ecdysone to acetylecdysteroid phosphates in fifth-instar *S. gregaria,* suggest that acetylation and phosphorylation are primary reactions for the inactivation of endogenous hormones in these two species.

10.4.2. *Glucosylation*

There is only limited evidence for glucosylation of ecdysteroids in insects. The first reported *in vitro* glucosylation of ecdysteroids by an insect tissue was the conversion of 20-hydroxyecdysone and ponasterone A into α-glucosides by transglucosylase in the fat body of *Calliphora vicina* (Heinrich and Hoffmeister, 1970). Similarly, when injected into *C. vicina* larvae, tritium-labeled ponasterone A was very quickly metabolized and appeared as the ecdysteroid "glucoside" in the hemolymph, fat body, and epidermis. In experiments on the metabolism of [^3H]22,25-dideoxyecdysone conducted with adult houseflies, *Musca domestica,* and pupae of *M. sexta* during termination of diapause, a mixture of polar ecdysteroids was found, which, on the basis of enzymatic hydrolysis, were assumed to be ecdysteroid sulfo- and glucoconjugates (Kaplanis et al., 1974).

The first unequivocal proof, however, of an ecdysteroid glucoside conjugate in an insect was obtained by the isolation and identification of 26-hydroxyecdysone 22-glucoside from young *M. sexta* larvae (Warren et al., 1986; Thompson et al., 1987a). The concentration of this conjugate increases from 2.7 μg/g in 24- to 40-h-old eggs to a maximum of 17 μg/g in 72- to 88-h-old eggs. The glucoside does not appear to be further metabolized and remains at this relatively high level in 0- to 1-h-old larvae (Thompson et al., 1988). It has been postulated that the glucosylating enzyme acts to remove free 26-hydroxyecdysone from the system (Warren et al., 1986).

Recently, ecdysone 25-β-D-glucopyranoside was identified as a major metabolite secreted into the culture medium following injection of [^3H]ecdysone into adults of the nematode *Parascaris equorum* (O'Hanlon et al., 1987). In future investigations of ecdysteroid conjugates of insects, more evidence may well be found for the formation of glucosides as a means of regulating ecdysteroid titers.

10.4.3. *Acylation*

Since most of the ecdysteroid acetates also contained a phosphate moiety, they were discussed in Section 10.4.1. In this section, we shall discuss those ecdysteroids acylated with long-chain fatty acids.

Acylation of ecdysteroids with long-chain fatty acids was first observed when radiolabeled ecdysone was injected into the hemocoel of nymphs of the tick *Ornithodoros moubata* (Bouvier et al., 1982). Apolar esterase-labile metabolites were also recovered from stage 5 nymphs after ingestion of high quantities of cold 20-hydroxyecdysone and identified as C-22 long-chain fatty acid esters of 20-hydroxyecdysone (Diehl et al., 1985). In studies of the metabolism of injected [^3H]ecdysone or [^3H]20-hydroxyecdysone in various species of ticks, spiders, scorpions, myriapods, crustaceans, and insects, most of these arthropods were found to convert the ecdysteroids to esterase-labile metabolites (Connat and Diehl, 1986). Based on reverse-phase high-performance liquid chromatography (HPLC), these metabolites were similar to the apolar conjugates of the tick (Connat et al., 1984). The apolar metabolites were most abundant in last-instar nymphs, young adult females, females with ootheca, and adult males of the German cockroach, *Blattella germanica*, in nymphs, male and female adults of the American cockroach, *Periplaneta americana*, and in pupae and adults of the yellow mealworm, *Tenebrio molitor* (Connat and Diehl, 1986). Incubation of various organs of adult female *Drosophila melanogaster* with [^3H]ecdysone also produced esterase-labile phila melanogaster with [³H]ecdysone also produced esterase-labile ecdysteroid metabolites (Dübendorfer and Maroy, 1986). In newly laid oothecae of *P. americana*, the bulk of the detectable ecdysteroids were also apolar conjugates (Slinger et al., 1986), which appeared to have properties similar to the ecdysteroid fatty acid esters identified in ticks (Connat et al., 1984; Wigglesworth et al., 1985).

To investigate the detoxification mechanism in larvae of the tobacco budworm, *Heliothis virescens*, a large amount of 20-hydroxyecdysone was orally administered. Four major metabolites, 22-linoleate, 22-palmitate, 22-oleate, and 22-stearate of 20-hydroxyecdysone, were isolated from frass (Kubo et al., 1987). Thus, it appears that esterification by long-chain fatty acids at the C-22 position of ecdysteroids may be a means of inactivation of ingested ecdysteroids in insects that feed on plants containing ecdysteroids and also a means of controlling the titer of endogenous hormones. The occurrence of ecdysteroid fatty acid esters in eggs of different classes of arthropods after injection of [³H]ecdysone could indicate a hormone-storage function (Connat and Diehl, 1986).

10.5. Hydrolysis of Conjugates

The ecdysteroid conjugates in ovaries and newly laid eggs of insects are apparently stored for controlled release during embryogenesis, a developmental stage most likely incapable of ecdysteroid biosynthesis. In most species studied, these conjugates are phosphates. Although, biosynthesis of ecdysteroids during later stages of embryogenesis is not precluded, the extremely large quantities of ecdysteroid conjugates in newly laid eggs strongly suggest that the free ecdysteroids observed during early embryonic development are released from these conjugates by the action of phosphatases.

Injection of [¹⁴C]cholesterol into *S. gregaria* females has been used to study the composition of ecdysteroids in eggs. Eggs at all stages examined contained primarily highly polar ecdysteroids, but their composition changed with the stages of development. For example, the concentration of a polar conjugate of ecdysone 3-acetate increased from 1.1% to 10.4% of the total ecdysteroids between day 10 and hatching (day 17) (Isaac and Rees, 1985). In separate studies, in which the [¹⁴C]ecdysteroid conjugates, consisting primarily of ecdysone 22-phosphate were extracted from newly laid eggs and injected into day-3 and day-13 eggs, the potential for *in vivo* hydrolysis of the ecdysteroid phosphates in *S. gregaria* embryos was demonstrated (Isaac and Rees, 1985).

The data obtained from *L. migratoria* eggs support the hypothesis that the four peaks of ecdysone and 20-hydroxyecdysone observed during embryogenesis result from the hydrolysis of the maternal ecdysteroid conjugates. Between days 3 and 5, the poorly immunoreactive maternal conjugates (22-phosphates of ecdysone and 2-deoxyecdysone) were converted to more immunoreactive, presumably free ecdysteroids (Lagueux et al., 1984).

During *M. sexta* embryogenesis, the maternal 26-hydroxyecdysone 26-phosphate is hydrolyzed to 26-hydroxyecdysone, which reaches its peak titer in 1- to 18-h-old eggs (Warren et al., 1986; Thompson et al., 1988). Throughout embryogenesis, there is a progressive appearance of other metabolites presumably derived from 26-hydroxyecdysone. Based on these studies, a metabolic scheme for the regulation of hormone titers in *M. sexta* embryos was proposed (Fig. 10.5) (Thompson et al., 1988). In contrast to these findings, the analysis of unfertilized *M. sexta* eggs yielded only the maternal ecdysteroids, the 2- and 26-phosphates of 26-hydroxyecdysone, and trace amounts of free 26-hydroxyecdysone (Feldlaufer et al., 1988).

26-Hydroxyecdysone 22-glucoside, the ecdysteroid accumulated by late *M. sexta* embryos, does not appear to be further metabolized either in embryos or in 0- to 1-h-old larvae. Whether it is hydrolyzed and used as a source of free hormone later in the young larvae or is a final inactivation product is not known at present (Thompson et al., 1988).

In conclusion, several lines of evidence suggest that ecdysteroid phosphates found in newly laid eggs are storage conjugates that can be hydrolyzed and used as a source of free hormone in embryonic development. Whether ecdysteroid phosphates found in insect larvae or pupae serve mainly as inactivation products or can also function as storage forms is not known. Further studies are needed to identify the ecdysteroid conjugates and the enzymes involved in their hydrolysis before their functions can be fully understood.

10.6.   Formation of Ecdysteroid Acids

The conversion of ecdysone and other ecdysteroids to their corresponding 26-acids (Table 10.4) is found in many species of insects and may be a pathway common to arthropods in general (Lafont and Koolman, 1984). The first step in this pathway is the hydroxylation at C-26.

TABLE 10.4. Occurrence of Ecdysteroid Acids in Insects: *In vivo* or *in vitro* Experiments

Ecdysteroid	Species	Source	Reference
Ecdysonoic	*Pieris brassicae*	Pupae	Beydon et al., 1981; Lafont et al., 1983
		Larvae[a]	Beydon and Lafont, 1987
	Schistocerca gregaria	Eggs	Isaac et al., 1983b
		Nymphs	Gibson et al., 1984
	Spodoptera littoralis	Pupae	Isaac et al., 1983b
	Manduca sexta	Eggs	Warren et al., 1986; Thompson et al., 1988
		Gut[b]	Warren and Gilbert, 1986
20-Hydroxyecdysonoic	*Pieris brassicae*	Pupae	Beydon et al., 1981; Lafont et al., 1983
		Larvae[a,c]	Beydon and Lafont, 1987
	Locusta migratoria	Nymphs	Lafont et al., 1983
	Calliphora vicina	Fat body	
	Schistocerca gregaria	Eggs	Isaac et al., 1983b
		Nymphs	Gibson et al., 1984
	Manduca sexta	Eggs	Thompson et al., 1988
		Gut[b]	Warren and Gilbert, 1986
		Pupae	Lozano et al., 1988
3-Epi-20-hydroxyecdysonoic	*Pieris brassicae*	Pharate pupae	Beydon and Lafont, 1987
		Larvae[a]	
	Manduca sexta	Eggs, larvae	Thompson et al., 1988
		Pupae	Lozano et al., 1988

[a] Green feces.

[b] Female during pupal–adult development.

[c] Red feces.

Pupae of *P. brassicae* are reportedly unable to form ecdysteroid conjugates (Beydon et al., 1981) but do form the acidic metabolites, ecdysonoic and 20-hydroxyecdysonoic acid (Lafont et al., 1983). The metabolism of ecdysteroids in the pharate pupae of *P. brassicae* leads to the accumulation of 3-epi-20-hydroxyecdysonoic acid as a major end product (Lafont et al., 1980). Following injection of [^3H]ecdysone into larvae of *P. brassicae*, labeled 20-hydroxyecdysone, 20,26-dihydroxyecdysone, ecdysonoic and 20-hydroxyecdysonoic acid as well as several other metabolites were isolated. Injection of [^3H]3-epi-20-hydroxyecdysone into *P. brassicae* larvae resulted in its conversion into 3-epi-20,26-dihydroxyecdysone and 3-epi-20-hydroxyecdysonoic acid (Beydon et al., 1987). In newly ecdysed pupae of *P. brassicae* that had been injected with [^3H]cholesterol 2 h after the last larval ecdysis, 3-epi-20-hydroxyecdysone and 3-epi-20-hydroxyecdysonoic acid were the major labeled ecdysteroids. Ecdysonoic, 20-hydroxyecdysonoic, and 3-epi-20-hydroxyecdysonoic acid were also detected in green feces collected during the larval feeding stage (Beydon and Lafont, 1987).

Ecdysonoic acid and 20-hydroxyecdysonoic acid were isolated and identified from late stages of embryogenesis of *S. gregaria* and from pupae of *Spodoptera littoralis* (Isaac et al., 1983b). Conversion to 20-hydroxyecdysonoic acid and, to a lesser extent, ecdysonoic acid are important metabolic routes for ecdysteroids in early fifth-instar *S. gregaria* (Gibson et al., 1984). Fifth-instar larvae of *L. migratoria* and fat body from late third-instar larvae of *C. vicina* were also shown to convert [^3H]ecdysone to [^3H]20-hydroxyecdysonoic acid (Lafont et al., 1983). Following injection of [^3H]ecdysone into male adult *L. migratoria*, 20-hydroxyecdysone, 20,26–dihydroxyecdysone, ecdysteroid acids, and other metabolites were isolated from feces (Modde et al., 1984).

The ecdysteroid acids—ecdysonoic, 20-hydroxyecdysonoic, 3-epi-20-hydroxyecdysonoic, and possibly 3-epiecdysonoic acid—also occur in the late stages of embryogenesis in *M. sexta* (Thompson et al., 1988). The 3-epiecdysteroid acids could be metabolites of 3-epi-26-hydroxyecdysone and 3-epi-20,26-dihydroxyecdysone, since these are the major free ecdysteroids in 96-h-old eggs (about 20 h before hatch) (Thompson et al., 1988) and in 0- to 1-h-old larvae of *M. sexta* (Thompson et al., 1987a). Alternatively, these acids could be derived via epimerization of ecdysonoic and 20-hydroxyecdysonoic acid. This view is supported by the observation that, as the level of 20-hydroxyecdysonoic acid decreases, there is a concurrent increase in the amount of 3-epi-20-hydroxyecdysonoic acid. In 8-day-old male pupae of *M. sexta*, injected as newly molted fifth-instar larvae with [^{14}C]cholesterol, 20-hydroxyecdysonoic and 3-epi-20-hydroxy-ecdysonoic acid (present in equal quantities) account for 48% of the total radioactivity of the polar ecdysteroid fraction (Loranzo et al., 1988).

Presently, in the absence of evidence for biological activity, these acids are considered inactivation products of the MHs (Beydon et al., 1981) and are involved in the regulation of 20-hydroxyecdysone and ecdysone titers in developing embryos (Isaac et al., 1983b). In like manner, ecdysteroid acid formation is involved in the titer regulation of 26-hydroxyecdysone and 20,26-dihydroxyecdysone in *M. sexta* embryos. Indeed, ecdysteroid acids may be key inactivation products common to insects at various developmental stages.

10.7. Enzymes

Much work has been done in recent years on ecdysone 20-monooxygenase. Ecdysone oxidase and 3-oxoecdysteroid 3-reductases, the enzymes involved in the 3-epimerization of ecdysteroids, have been studied in several laboratories, and as a result of these studies the 3-epimerization reaction is now better understood. 2-Deoxyecdysone 2-hydroxylase, ATP : ecdysteroid phosphotrans-ferase(s), and ecdysteroid-22-phosphate phosphohydrolase are new additions to the list of ecdysteroid-converting enzymes characterized *in vitro*.

10.7.1. *Monooxygenases (Hydroxylases)*

10.7.1.1. ECDYSONE 20-MONOOXYGENASE

Ecdysone 20-monooxygenase [ecdysone,hydrogen donor:oxygen oxi-doreductase (20-hydroxylating), EC 1.14.99.22], the enzyme system that converts ecdysone to 20-hydroxyecdysone, is the most extensively studied enzyme of ecdysone metabolism and has been the topic of two recent reviews (Weirich et al., 1984; Smith, 1985).

Ecdysone 20-monooxygenase is found in many insect tissues, but not in the prothoracic glands. The prothoracic glands carry the MH biosynthesis only up to ecdysone, and the 20-hydroxylation is carried out by other tissues. In adults of many insect species, the ovaries have been shown to synthesize 20-hydroxylated ecdysteroids (see Table 10.1), and in *Nauphoeta cinerea* follicle cells, ecdysone 20-hydroxylation has been demonstrated directly (Zhu et al., 1983). Thus, in contrast to the larval stages, in the adult insect the 20-hydroxylation is not always spatially separated from the earlier steps in the biosynthetic pathway for ecdysteroids.

Ecdysone 20-monooxygenase can be located in mitochondria and/or microsomes of insect tissues (Table 10.5; for further references, see Weirich et al., 1984; Smith, 1985). Wherever tested, ecdysone 20-monooxygenase was inhibited by inhibitors of cytochrome-P-450–dependent steroid oxidations (e.g., metyrapone, *p*-aminoglutethimide) and by carbon monoxide, and the carbon monoxide inhibition was (at least partially) reversed by 450-nm irradiation. Thus, ecdysone 20-monooxygenase is considered to be a cytochrome-P-450–containing enzyme.

In mitochondria and microsomes of mammalian tissues, the cytochromes P-450 are components of different multienzyme systems, consisting of NADPH-ferredoxin reductase, ferredoxin (an iron-sulfur protein), and cytochrome P-450 in mitochondria, and of NADPH–cytochrome P-450 reductase and cytochrome P-450 (in some instances supported by NADH–cytochrome b_5 reductase and cytochrome b_5) in microsomes (for references, see Weirich et al., 1984). Studies with *M. sexta* fat body mitochondria have provided evidence for the presence of a ferredoxin-type nonheme iron protein (Smith et al., 1980), and the functional involvement of an NADPH–cytochrome P-450 reductase in ecdysone 20-hydroxylation has been shown for *Diploptera punctata* midgut microsomes (Halliday et al., 1986). Thus, cytochrome P-450 enzyme systems of insects may be similar to mammalian systems in their composition.

In most insect species investigated, the activities of ecdysone 20-monooxygenase change during the course of development. These changes may reflect changing demands for the ecdysone 20-hydroxylation at different stages of development. In *L. migratoria*, injection of ecdysone has been shown to cause an increase of ecdysone 20-monooxygenase activity, and removal of the prothoracic glands led to a reduction of the enzyme activity (Feyereisen and Durst, 1980a,b). In kinetic experiments with midgut and fat body homogenates of *M. sexta* pupae, Smith et al. (1983) found that ecdysone 20-monooxygenase activity changes reflected changes in the number of available catalytic sites (measured as V_{max}) rather than in their affinity for the substrate ecdysone (measured as K_m). Hoggard and Rees (1988) have recently reported evidence, indicating that such

TABLE 10.5. Ecdysone 20-Monooxygenases

Species, stage	Tissue	Preparation	Apparant K_m (μM)	Reference
Diplopterea punctata last larval instar	Fat body	Whole tissue		Halliday et al., 1986
	Malpighian tubules	Whole tissue		
	Midgut	Microsomes	0.24–0.29	
Nauphoeta cinerea adult females	Ovaries	Follicle cells		Zhu et al., 1983
Schistocerca gregaria fifth instar	Malpighian tubules	Mitochondria	0.71	Greenwood and Rees, 1984
Manduca sexta last larval instar	Fat body	Homogenate	0.242	Smith et al., 1983
	Midgut	Homogenate	0.467	
	Midgut	Mitochondria, microsomes		Weirich et al., 1985
Spodoptera littoralis sixth larval instar	Fat body	Mitochondria, microsomes		Hoggard and Rees, 1988
	Malpighian tubules			
Pieris brassicae pupae	Wing disks	Microsomes	0.058	Blais and Lafont, 1986
Aedes aegypti adult females, blood-fed	Various abdominal tissues	Mitochondria, microsomes[a]	0.22–0.33[a]	Smith and Mitchell, 1986
Drosophila melanogaster wandering stage larvae	Whole body	Mitochondria, microsomes	0.064 0.099	Mitchell and Smith, 1986
Panulirus argus	Ovaries			
	Testes			
	Green glands	Mitochondria		James and Shiverick, 1984

[a] Preliminary evidence.

changes may be related to the reversible phosphorylation–dephosphorylation of some component(s) of the enzyme system in *S. littoralis.*

10.7.1.2. 2-DEOXYECDYSONE 2-HYDROXYLASE

The 2-deoxyecdysone 2-hydroxylase has so far been studied in only one species, *L. migratoria* (Kappler et al., 1986; Kabbouh et al., 1987). The enzyme was found in prothoracic glands, follicle cells, Malpighian tubules, fat body, and midgut. In Malpighian tubules and follicle cells, the hydroxylase was shown to be localized in mitochondria. Based on inhibition by metyrapone and piperonyl butoxide, and the oxygen requirement of the reaction, it has been suggested that 2-deoxyecdysone 2-hydroxylase is a cytochrome P-450 monooxygenase, although it is not inhibited by carbon monoxide.

10.7.2. *Ecdysone Oxidase, 3-Oxoecdysteroid 3α-Reductase, and 3-Oxoecdysteroid 3β-Reductase (Ecdysone 3-Epimerase System)*

Ecdysone oxidase catalyzes the oxygen-dependent conversion of ecdysone to 3-dehydroecdysone (3-oxoecdysone) (Koolman and Karlson, 1975, 1978). 3-Dehydroecdysone has been assumed to be the intermediate in the enzymatic conversion of ecdysone to 3-epiecdysone since this reaction was first reported to occur in an enzyme preparation from *M. sexta* midgut (Nigg et al., 1974) (Fig. 10.6). However, experimental proof for the two-step reaction has been obtained only in recent years.

Ecdysone oxidase [ecdysone : oxygen 3-oxidoreductase, EC 1.1.3.16] has first been found in high-speed supernatant of homogenates of *C. vicina* pupae (Koolman and Karlson, 1975) and more recently in enzyme extracts (postmicrosomal supernatants) of lepidopteran midgut (Blais and Lafont, 1984; Milner and Rees, 1985; Weirich et al., 1987). Lepidopteran midgut, however, contains not only ecdysone oxidase but also very active 3-oxoecdysteroid reductases and the cosubstrate(s) required for the enzymatic reduction, NADH and/or NADPH. Therefore, incubations of ecdysone or 20-hydroxyecdysone with such enzyme extracts yield only limited amounts of the 3-

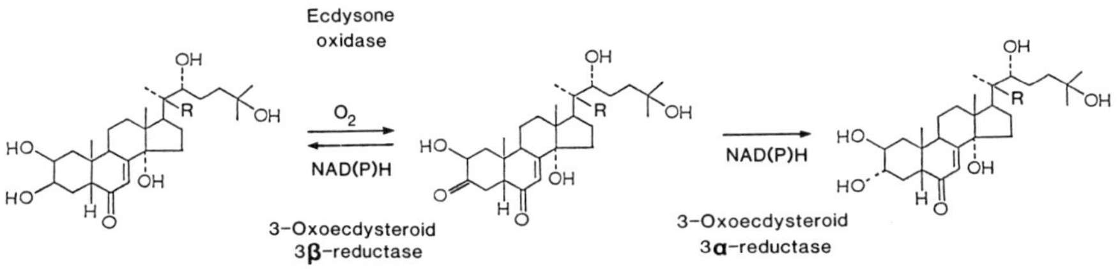

FIGURE 10.6. Enzymatic interconversions of 3β-ecdysteroids, 3-oxoecdysteroids, and 3α-ecdysteroids (R = H, ecdysone series; R = OH, 20-hydroxyecdysone series).

oxoecdysteroids or none at all (Milner and Rees, 1985; Weirich, Thompson, and Svoboda, unpublished data).

By incubating postmicrosomal supernatants of (mid)gut homogenates with 3-oxoecdysteroids, two types of 3-oxoecdysteroid 3-reductases have been detected in *P. brassicae* (Blais and Lafont, 1984), *S. littoralis* (Milner and Rees, 1985), and *M. sexta* (Weirich et al., 1987). The 3α-reductases convert the 3-oxoecdysteroids to the corresponding 3-epiecdysteroids, completing the irreversible epimerization sequence. The 3β-reductases convert the 3-oxoecdysteroids back to the original 3β-ecdysteroids (the substrates of the ecdysone oxidase reaction). Thus, 3β-ecdysteroids can be cycled through the ecdysone oxidase reaction and the 3β-reductase reaction. The reaction sequence for the 3-epimerization shown in Fig. 10.6 has been verified by isotope dilution experiments for *P. brassicae* (Blais and Lafont, 1984) and *S. littoralis* (Milner and Rees, 1985) and by kinetic analysis for *M. sexta* (Weirich et al., 1987).

10.7.3. *ATP:Ecdysteroid Phosphotransferases*

Although the number of isolated and identified ecdysteroid conjugates has been increasing in recent years (see Section 10.4), only one type of conjugate-forming enzyme in one species has been studied in some detail (Weirich et al., 1986).

Incubation of an 80,000g supernatant (cytosol) of midgut homogenate of *M. sexta* wandering larvae with ecdysone yielded highly polar metabolites (in addition to 3-epiecdysone, as outlined in the preceding section). By treatment with β-glucuronidase preparations from *Helix pomatia* or limpet (also containing sulfatase and acid phosphatase), the highly polar metabolites were converted to ecdysone and 3-epiecdysone. The hydrolysis by the limpet β-glucuronidase preparation was not inhibited by D-saccharic acid 1,4-lactone, a β-glucuronidase inhibitor, but was nearly 100% inhibited by PO_4^{3-} or NaF, inhibitors of the sulfatase and phosphatase activities present in the enzyme preparation. The highly polar metabolites were also hydrolyzed by human acid phosphatase preparations (free of sulfatase activity), and it was concluded that the metabolites were phosphoconjugates.

Four conjugates of ecdysone and two conjugates of 3-epiecdysone have been resolved by reversed-phase ion-pair HPLC. To account for this diversity, two hydroxyl groups of 3-epiecdysone have to be involved in conjugate formation, and at least three hydroxyl groups of ecdysone, most likely at C-2, C-3, and C-22.

The nature of the highly polar metabolites was further explored by determining the cosubstrate/cofactor requirements of the enzymes involved in their formation. After Sephadex G-25 filtration (removal of endogenous cosubstrates and cofactors), the cytosol had lost its potential for conjugate formation. The reaction was restored by addition of ATP and Mg^{2+}. These observations suggested that the enzymes involved in conjugate formation are ATP:ecdysteroid phosphotransferases.

10.7.4. *Ecdysteroid–Phosphate Phosphohydrolase*

The eggs of many insect species contain ecdysteroid conjugates of maternal origin (see Section 10.4). Ecdysone 22-phosphate is the major ecdysteroid conjugate in eggs of *S. gregaria* (Isaac et al., 1982). Isaac et al. (1983c) have obtained an enzyme preparation from day-15 *S. gregaria* embryos (total duration of embryogenesis 16–17 days) capable of hydrolyzing ecdysone 22-phosphate. This enzyme was most active at a pH below 5.0 and was found in the mitochondrial–lysosomal fraction. Hydrolysis of ecdysone 22-phosphate ([14C]-labeled) could also be shown *in situ* by injecting the conjugate into day-13 eggs (Isaac and Rees, 1985). These experiments have provided the first direct experimental evidence to show that embryonic phosphatase(s) can indeed hydrolyze ecdysteroid conjugates of maternal origin.

10.8. Summary

Based on the observed accumulation of certain ecdysteroids in specific developmental stages and our current knowledge of biological activities, it is apparent that hydroxylation, epimerization, conjugation, hydrolysis of conjugates, and formation of ecdysteroid acids are important biochemical reactions involved in the regulation of molting hormone (MH) titers in insects.

Hydroxylation of ecdysone at C-20 to give 20-hydroxyecdysone has been demonstrated in all species examined and has been described as "activation of the hormone." In some species, the terminal step in the biosynthesis may be hydroxylation at C-2 for ecdysone and hydroxylation at either C-2 or C-20 for 20-hydroxyecdsone. Either reaction can contribute to titer regulation. 26-Hydroxylation is a common reaction in ecdysteroid metabolism. Interestingly, 20,26-dihydroxyecdysone is about one-tenth as active as 20-hydroxyecdysone in the house fly MH assay; 26-hydroxyecdysone is inactive. The conversions of ecdysone and 20-hydroxyecdysone to ecdysonoic acid and 20-hydroxyecdysonoic acid, respectively, probably proceed via 26-hydroxyecdysone and 20,26-dihydroxyecdysone. Hydroxylation at C-26 is a means of controlling the titer of 20 hydroxyecdysone in many insect species.

Formation of 3α-ecdysteroids from the more active 3β-ecdysteroids is also a means of regulating MH titers. 3-Epiecdysone, 3-epi-20-hydroxyecdysone, and 3-epi-20,26-dihydroxyecdysone have been isolated from various developmental stages of *M. sexta* and a number of other species and are far less active than their corresponding 3β-epimers in the house fly assay. 3-Epi-26-hydroxyecdysone and 3-epi-20,26-dihydroxyecdysone are the major free ecdysteroids in 0- to 1-h-old *M. sexta* larvae. 3-Epi-2-deoxyecdysone 3-phosphate is the major ecdysteroid in 8-day-old embryos of *L. migratoria* and a minor one in *S. gregaria.*

Although phosphoconjugation seems to be the predominant reaction in insect embryos and pupae, there are indications that glucosylation and acylation are also involved in regulating ecdysteroid titers. In fact, analyses of ecdysteroids in *S. gregaria, L. migratoria,* and *M. sexta* indicate that the ecdysteroid conjugates in ovaries and newly laid eggs are phosphates. The analyses also suggest that the free ecdysteroids observed during early embryonic development are released from conjugates by the action of phosphatases. The free ecdysteroids are then reconjugated by acetylation, phosphorylation, and/or glucosylation.

Thus, hydroxylation, epimerization, conjugation, hydrolysis, and formation of ecdysteroid acids are means of controlling MH titers. The three proposed schemes of metabolic pathways of ecdysteroids presented here encompass all of these bioconversions except that formation of ecdysteroid acids was not observed for *L. migratoria* during embryogenesis. The evidence also suggests that ecdysteroid phosphates found in newly laid eggs are storage conjugates that can be hydrolyzed and used as a source of free hormone in embryonic development. Whether phophates found in insect larvae or pupae serve only as inactivation products or can also function as storage forms is not known. Ecdysteroid acids, however, appear to be key inactivation products common to insects at various developmental stages.

The most extensively studied enzyme of MH metabolism is ecdysone 20-monooxygenase, a mitochondrial or microsomal cytochrome P-450 enzyme. Another cytochrome P-450 monooxygenase, 2-deoxyecdysone 2-hydroxylase, has recently been demonstrated in *L. migratoria*. Ecdysone oxidase, previously known mainly from studies with *Calliphora vicina*, has now been found in several lepidopteran species. Ecdysone oxidase and 3-oxoecdysteroid 3α-reductase constitute the 3-epimerase system, and 3-oxoecdysteroid 3β-reductase catalyzes a reversal of the ecdysone oxidase reaction. ATP:ecdysteroid phosphotransferases have been extracted from *M. sexta* midgut; and ecdysone 22-phosphate–hydrolyzing enzyme(s), from *S. gregaria* embryos.

References

Beydon, P. and R. Lafont. 1987. Long-term cholesterol labeling as a convenient means for measuring ecdysteroid production and catabolism *in vivo:* application to the last larval instar of *Pieris brassicae*. Arch. Insect Biochem. Physiol. 5(2): 139–154.

Beydon, P., J.-P. Girault, and R. Lafont. 1987. Ecdysone metabolism in *Pieris brassicae* during the feeding last larval instar. Arch. Insect Biochem. Physiol. 4(2): 139–149.

Beydon, P., J. Claret, P. Porcheron, and R. Lafont. 1981. Biosynthesis and inactivation of ecdysone during the pupal–adult development of the cabbage butterfly, *Pieris brassicae* L. Steroids 38(6): 633–650.

Blais, C. and R. Lafont. 1984. Ecdysteroid metabolism by soluble enzymes from an insect: metabolic relationships between 3β-hydroxy-, 3α-hydroxy- and 3-oxoecdysteroids. Hoppe-Seyler's Z. Physiol. Chem. 365: 809–817.

Blais, C. and R. Lafont. 1986. Ecdysone 20-hydroxylation in imaginal wing discs of *Pieris brassicae* (Lepidoptera): correlations with ecdysone and 20-hydroxyecdysone titers in pupae. Arch. Insect Biochem. Physiol. 3(6): 501–512.

Bouvier, J., P. A. Diehl, and M. Morici. 1982. Ecdysone metabolism in the tick *Ornithodoros moubata* (Argasidae, Ixodoidae). Rev. Suisse Zool. 89: 967–976.

Butenandt, A. and P. Karlson. 1954. Über die Isolierung eines Metamorphose-Hormons der Insekten in kristallisierter Form. Z. Naturforsch. 9B: 389–391.

Connat, J.-L. and P. A. Diehl. 1986. Probable occurrence of ecdysteroid fatty acid esters in different classes of arthropods. Insect Biochem. 16(1): 91–97.

Connat, J. L., P. A. Diehl, and M. Morici. 1984. Metabolism of ecdysteroids during the vitellogenesis of the tick *Ornithodoros moubata* (Ixodoidea, Argasidae): accumulation of apolar metabolites in the eggs. Gen. Comp. Endocrinol. 56: 100–110.

Diehl, P. A., J. L. Connat, J. P. Girault, and R. Lafont. 1985. A new class of apolar ecdysteroid conjugates: esters of 20-hydroxy-ecdysone with long-chain fatty acids in ticks. Int. J. Invertebr. Reprod. Dev. 8: 1–13.

Dinan, L. N. and H. H. Rees. 1981. The identification and titres of conjugated and free ecdysteroids in developing ovaries and newly-laid eggs of *Schistocerca gregaria*. J. Insect Physiol. 27(1): 51–58.

Dübendorfer, A. and P. Maroy. 1986. Ecdysteroid conjugation by tissues of adult females of *Drosophila melanogaster*. Insect Biochem. 16(1): 109–113.

Feldlaufer, M. F., J. A. Svoboda, M. J. Thompson, and K. R. Wilzer. 1988. Fate of maternally-acquired ecdysteroids in unfertilized eggs of *Manduca sexta*. Insect Biochem. 18(2): 219–221.

Feyereisen, R. and F. Durst. 1980a. Development of microsomal cytochrome P-450 monooxygenases during the last larval instar of the locust, *Locusta migratoria*: correlation with the hemolymph 20-hydroxyecdysone titer. Mol. Cell. Endocrinol. 20: 157–169.

Feyereisen, R. and F. Durst. 1980b. Control of cytochrome P-450 monooxygenase in an insect by the steroid moulting hormone. Pp. 595–598 *in* M. J. Coon, A. H. Conney, R. W. Estabrook, H. V. Gelboin, J. R. Gillette, and J. P. O'Brien (eds.), *Microsomes, Drug Oxidations, and Chemical Carcinogenesis*. Vol. I. Academic Press, Orlando, Florida.

Gibson, J. M., R. E. Isaac, L. N. Dinan, and H. H. Rees. 1984. Metabolism of [^{3}H]ecdysone in *Schistocerca gregaria:* formation of ecdysteroid acids together with free and phosphorylated ecdysteroid acetates. Arch. Insect Biochem. Physiol. 1(4): 385–407.

Goodwin, T. W., D. H. S. Horn, P. Karlson, J. Koolman, K. Nakanishi, W. E. Robbins, J. B. Siddall, and T. Takemoto. 1978. Ecdysteroids: a new generic term. Nature (Lond.) 272: 122.

Greenwood, D. R. and H. H. Rees. 1984. Ecdysone 20-monooxygenase in the desert locust, *Schistocerca gregaria*. Biochem. J. 223: 837–847.

Halliday, W. R., D. E. Farnsworth, and R. Feyereisen. 1986. Hemolymph ecdysteroid titer and midgut ecdysone 20-monooxygenase activity during the last larval stage of *Diploptera punctata*. Insect Biochem. 16(4): 627–634.

Hampshire, F. and D. H. S. Horn. 1966. Structure of crustecdysone, a crustacean moulting hormone. Chem. Commun. pp. 37–38.

Heinrich, G. and H. Hoffmeister. 1970. Bildung von Hormonglykosiden als Inaktivierungsmechanismus bei *Calliphora erythrocephala*. Z. Naturforsch.

Hoggard, N. and H. H. Rees. 1988. Reversible activation–inactivation of mitochondrial ecdysone 20-monooxygenase: a possible role for phosphorylation–dephosphorylation. J. Insect Physiol. 34(7): 647–653.

Huber, R. and W. Hoppe. 1965. Zur Chemie des Ecdysons. VII. Die Kristall- und Molekülstrukturanalyse des Insektenverpuppungshormons Ecdyson mit der automatisierten Faltmolekülmethode. Chem. Ber. 98: 2403–2424.

Isaac, R. E. and H. H. Rees. 1984. Isolation and identification of ecdysteroid phosphates and acetylecdysteroid phosphates from developing eggs of the locust, *Schistocerca gregaria*. Biochem. J. 221: 459–464.

Isaac, R. E. and H. H. Rees. 1985. Metabolism of maternal ecdysteroid-22-phosphates in developing embryos of the desert locust, *Schistocerca gregaria*. Insect Biochem. 15(1): 65–72.

Isaac, R. E., N. P. Milner, and H. H. Rees. 1983b. Identification of ecdysonoic acid and 20-hydroxyecdysonoic acid isolated from developing eggs of *Schistocerca gregaria* and pupae of *Spodoptera littoralis*. Biochem. J. 213: 261–265.

Isaac, R. E., F. P. Sweeney, and H. H. Rees. 1983c. Enzymic hydrolysis of ecdysteroid phosphate during embryogenesis in the desert locust *(Schistocerca gregaria)*. Biochem. Soc. Trans. 11: 379–380.

Isaac, R. E., H. P. Desmond, and H. H. Rees. 1984. Isolation and identification of 3-acetylecdysone 2-phosphate, a metabolite of ecdysone, from developing eggs of *Schistocerca gregaria*. Biochem. J. 217: 239–243.

Isaac, R. E., H. H. Rees, and T. W. Goodwin. 1981a. Isolation of 2-deoxy-20-hydroxyecdysone and 3-epi-2-deoxyecdysone from eggs of the desert locust, *Schistocerca gregaria*, during embryogenesis. J. Chem. Soc. Chem. Commun. pp. 418–420.

Isaac, R. E., H. H. Rees, and T. W. Goodwin. 1981b. Isolation of ecdysone-3-acetate as a major ecdysteroid from the developing eggs of the desert locust, *Schistocerca gregaria*. J. Chem. Soc. Chem. Commun. pp. 594–595.

Isaac, R. E., M. E. Rose, H. H. Rees, and T. W. Goodwin. 1982. Identification of ecdysone-22-phosphate and 2-deoxyecdysone-22-phosphate in eggs of the desert locust *Schistocerca gregaria* by fast atom bombardment mass spectrometry and N.M.R. spectroscopy. J. Chem. Soc. Chem. Commun. pp. 249–251.

Isaac, R. E., M. E. Rose, H. H. Rees, and T. W. Goodwin. 1983a. Identification of the 22-phosphate esters of ecdysone, 2-deoxyecdysone, 20-hydroxyecdysone and 2-deoxy-20-hydroxyecdysone from newly laid eggs of the desert locust, *Schistocerca gregaria*. Biochem. J. 213: 533–541.

James, M. O. and K. T. Shiverick. 1984. Cytochrome-P-450-dependent oxidation of progesterone, testosterone, and ecdysone in the spiny lobster, *Panulirus argus*. Arch. Biochem. Biophys. 233: 1–9.

Kabbouh, M., C. Kappler, C. Hetru, and F. Durst. 1987. Further characterization of the 2-deoxyecdysone C-2 hydroxylase from *Locusta migratoria*. Insect Biochem. 17(8): 1155–1161.

Kaplanis, J. N., S. R. Dutky, W. E. Robbins, and M. J. Thompson. 1974. The metabolism of the synthetic ecdysone analog, 1α-^3H-22,25-bisdeoxyecdysone in relation to certain of its biological effects in insects. Pp. 161–175 *in* W. J. Burdette (ed.), *Invertebrate Endocrinology and Hormonal Heterophylly.* Springer-Verlag, Berlin and New York.

Kaplanis, J. N., M. J. Thompson, S. R. Dutky, and W. E. Robbins. 1979. The ecdysteroids from the tobacco hornworm during pupal–adult development five days after peak titer of molting hormone activity. Steroids 34(3): 333–345.

Kaplanis, J. N., M. J. Thompson, S. R. Dutky, and W. E. Robbins. 1980. The ecdysteroids from young embryonated eggs of the tobacco hornworm. Steroids 36(3): 321–336.

Kaplanis, J. N., M. J. Thompson, R. T. Yamamoto, W. E. Robbins, and S. J. Louloudes. 1966a. Ecdysones from the pupa of the tobacco hornworm, *Manduca sexta* (Johannson). Steroids 8(5): 605–623.

Kaplanis, J. N., L. A. Tabor, M. J. Thompson, W. E. Robbins, and T. J. Shortino. 1966b. Assay for ecdysone (molting hormone) activity using the house fly, *Musca domestica* L. Steroids 8(5): 625–631.

Kappler, C., M. Kabbouh, F. Durst, and J. A. Hoffmann. 1986. Studies on the C-2 hydroxylation of 2-deoxyecdysone in *Locusta migratoria.* Insect Biochem. 16(1): 25–32.

Karlson, P. 1956. Biochemical studies on insect hormones. *Vitam. Horm.* 14: 227–266.

Karlson, P., H. Hoffmeister, W. Hoppe, and R. Huber. 1963. Zur Chemie des Ecdysons. Liebigs Ann. Chem. 662: 1–20.

Koolman, J. 1982. Ecdysone metabolism. Insect Biochem. 12(3): 225–250.

Koolman, J. and P. Karlson. 1975. Ecdysone oxidase, an enzyme from the blowfly *Calliphora erythrocephala* (Meigen). Hoppe-Seyler's Z. Physiol. Chem. 356: 1131–1138.

Koolman, J. and P. Karlson. 1978. Ecdysone oxidase: reaction and specificity. Eur. J. Biochem. 89: 453–460.

Koolman, J. and P. Karlson. 1985. Regulation of ecdysteroid titer: degradation. Pp. 343–361 *in* G. A. Kerkut and L. I. Gilbert (eds.), *Comprehensive Insect Physiology, Biochemistry and Pharmacology,* Vol. 7. Pergamon Press, Oxford and Elmsford, New York.

Kubo, I., S. Komatsu, Y. Asaka, and G. DeBoer. 1987. Isolation and identification of apolar metabolites of ingested 20-hydroxyecdysone in frass of *Heliothis virescens* larvae. J. Chem. Ecol. 13(4): 785–794.

Lafont, R. and J. Koolman. 1984. Ecdysone metabolism. Pp. 196–226 *in* J. Hoffmann and M. Porchet (eds.), *Biosynthesis, Metabolism and Mode of Action of Invertebrate Hormones.* Springer-Verlag, Berlin and New York.

Lafont, R., P. Beydon, G. Somme-Martin, and C. Blais. 1980. High-performance liquid chromatography of ecdysone metabolites applied to the cabbage butterfly, *Pieris brassicae* L. Steroids 36(2): 185–207.

Lafont, R., C. Blais, P. Beydon, J.-F. Modde, U. Enderle, and J. Koolman.

1983. Conversion of ecdysone and 20-hydroxyecdysone into 26-oic derivatives is a major pathway in larvae and pupae of species from three insect orders. Arch. Insect Biochem. Physiol. 1(1): 41–58.

Lagueux, M., C. Sall, and J. A. Hoffmann. 1981. Ecdysteroids during embryogenesis in *Locusta migratoria*. Am. Zool. 21: 715–726.

Lagueux, M., J. A. Hoffmann, F. Goltzene, C. Kappler, G. Tsoupras, C. Hetru, and B. Luu. 1984. Ecdysteroids in ovaries and embryos of *Locusta migratoria*. Pp. 168–180 *in* J. Hoffmann and M. Porchet (eds.), *Biosynthesis, Metabolism and Mode of Action of Invertebrate Hormones*. Springer-Verlag, Berlin and New York.

Lozano, R., M. J. Thompson, J. A. Svoboda, and W. R. Lusby. 1988. Isolation of acidic and conjugated ecdysteroid fractions from *Manduca sexta* pupae. Insect Biochem. 18(2): 163–168.

Milner, N. P. and H. H. Rees. 1985. Involvement of 3-dehydroecdysone in the 3-epimerization of ecdysone. Biochem. J. 231: 369–374.

Mitchell, M. J. and S. L. Smith. 1986. Characterization of ecdysone 20-monooxygenase activity in wandering stage larvae of *Drosophila melanogaster*: evidence for mitochondrial and microsomal cytochrome-P-450-dependent systems. Insect Biochem. 16(3): 525–537.

Modde, J. F., R. Lafont, and J. A. Hoffmann. 1984. Ecdysone metabolism in *Locusta migratoria* larvae and adults. Int. J. Invertebr. Reprod. Dev. 7: 161–183.

Nigg, H. N., J. A. Svoboda, M. J. Thompson, J. N. Kaplanis, S. R. Dutky, and W. E. Robbins. 1974. Ecdysone metabolism: ecdysone dehydrogenase–isomerase. Lipids 9: 971–974.

O'Hanlon, G. M., O. W. Howarth, and H. H. Rees. 1987. Identification of ecdysone 25-*O*-β-D-glucopyranoside as a new metabolite of ecdysone in the nematode, *Parascaris equorum*. Biochem. J. 248(1): 305–307.

Ohnishi, E., T. Mizuno, N. Ikekawa, and T. Ikeda. 1981. Accumulation of 2-deoxyecdysteroids in ovaries of the silkworm, *Bombyx mori*. Insect Biochem. 11(2): 155–159.

Rees, H. H. and R. E. Isaac. 1984. Biosynthesis of ovarian ecdysteroid phosphates and their metabolic fate during embryogenesis in *Schistocerca gregaria*. Pp. 181–185 *in* J. Hoffmann and M. Porchet (eds.), *Biosynthesis, Metabolism and Mode of Action of Invertebrate Hormones*. Springer-Verlag, Berlin and New York.

Sall, C., G. Tsoupras, C. Kappler, M. Lagueux, D. Zachary, B. Luu, and J. A. Hoffmann. 1983. Fate of maternal conjugated ecdysteroids during embryonic development in *Locusta migratoria*. J. Insect Physiol. 29(6): 491–507.

Slinger, A. J., L. N. Dinan, and R. E. Isaac. 1986. Isolation of apolar ecdysteroid conjugates from newly-laid oothecae of *Periplaneta americana*. Insect Biochem. 16(1): 115–119.

Smith, S. L. 1985. Regulation of ecdysteroid titer: synthesis. Pp. 295–341 *in* G. A. Kerkut and L. I. Gilbert (eds.), *Comprehensive Insect Physiology,*

Biochemistry and Pharmacology, Vol. 7. Pergamon Press, Oxford and Elmsford, New York.

Smith, S. L. and M. J. Mitchell. 1986. Ecdysone 20-monooxygenase systems in a larval and an adult dipteran: an overview of their biochemistry, physiology and pharmacology. Insect Biochem. 16: 49–55.

Smith, S. L., W. E. Bollenbacher, and L. I. Gilbert. 1980. Studies on the biosynthesis of ecdysone and 20-hydroxyecdysone in the tobacco hornworm, *Manduca sexta*. Pp. 139–162 in J. A. Hoffmann (ed.), *Progress in Ecdysone Research*. Elsevier/North-Holland Publ., Amsterdam and New York.

Smith, S. L., W. E. Bollenbacher, and L. I. Gilbert. 1983. Ecdysone 20-monooxygenase activity during larval–pupal development of *Manduca sexta*. Mol. Cell. Endocrinol. 31: 227–251.

Svoboda, J. A. and M. J. Thompson. 1985. Steroids. Pp. 137–175 *in* G. A. Kerkut and L. I. Gilbert (eds.), *Comprehensive Insect Physiology, Biochemistry and Pharmacology*, Vol. 10. Pergamon Press, Oxford and Elmsford, New York.

Thompson, M. J., J. N. Kaplanis, W. E. Robbins, and R. T. Yamamoto. 1967. 20,26-Dihydroxyecdysone, a new steroid with moulting hormone activity from the tobacco hornworm, *Manduca sexta* (Johannson). Chem. Commun. pp. 650–653.

Thompson, M. J., J. A. Svoboda, M. F. Feldlaufer, and R. Lozano. 1986. The fate of radiolabeled steroids in ovaries and eggs of the tobacco hornworm, *Manduca sexta*. Lipids 21(1): 76–81.

Thompson, M. J., J. A. Svoboda, H. H. Rees, and K. R. Wilzer. 1987b. Isolation and identification of 26-hydroxyecdysone 2-phosphate: an ecdysteroid conjugate of eggs and ovaries of the tobacco hornworm, *Manduca sexta*. Arch. Insect Biochem. Physiol. 4(3): 183–190.

Thompson, M. J., J. A. Svoboda, R. Lozano, and K. R. Wilzer, Jr. 1988. Profile of labeled free and conjugated ecdysteroids and ecdysteroid acids during embryonic development of *Manduca sexta* (L.) following maternal incorporation of [^{14}C]cholesterol. Arch. Insect Biochem. Physiol. 7(3): 157–172.

Thompson, M. J., J. N. Kaplanis, W. E. Robbins, S. R. Dutky, and H. N. Nigg. 1974. 3-Epi-20-hydroxyecdysone from meconium of the tobacco hornworm. Steroids 24(3): 359–366.

Thompson, M. J., G. F. Weirich, H. H. Rees, J. A. Svoboda, M. F. Feldlaufer, and K. R. Wilzer. 1985. New ecdysteroid conjugate: isolation and identification of 26-hydroxyecdysone 26-phosphate from eggs of the tobacco hornworm, *Manduca sexta* (L.). Arch. Insect Biochem. Physiol. 2(3): 227–236.

Thompson, M. J., M. F. Feldlaufer, R. Lozano, H. H. Rees, W. R. Lusby, J. A. Svoboda, and K. R. Wilzer, Jr. 1987a. Metabolism of 26-[^{14}C]hydroxyecdysone 26-phosphate in the tobacco hornworm, *Manduca sexta* (L.), to a new ecdysteroid conjugate: 26-[^{14}C]hydroxyecdysone 22-

glucoside. Arch. Insect Biochem. Physiol. 4(1): 1–15.

Tsoupras, G., C. Hetru, B. Luu, M. Lagueux, E. Constantin, and J. A. Hoffmann. 1982. The major conjugates of ecdysteroids in young eggs and in embryos of *Locusta migratoria*. Tetrahedron Lett. 23: 2045–2048.

Warren, J. T. and L. I. Gilbert. 1986. Ecdysone metabolism and distribution during the pupal–adult development of *Manduca sexta*. Insect Biochem. 16(1): 65–82.

Warren, J. T., B. Steiner, A. Dorn, M. Pak, and L. I. Gilbert. 1986. Metabolism of ecdysteroids during the embryogenesis of *Manduca sexta*. J. Liq. Chromatogr. 9(8): 1759–1782.

Weirich, G. F., J. A. Svoboda, and M. J. Thompson. 1984. Ecdysone 20-monooxygenases. Pp. 227–223 *in* J. Hoffmann and M. Porche (eds.), *Biosynthesis, Metabolism and Mode of Action of Invertebrate Hormones*. Springer-Verlag, Berlin and New York.

Weirich, G. F., J. A. Svoboda, and M. J. Thompson. 1984. Ecdysone 20-monooxygenases. Pp. 227–233 *in* J. Hoffmann and M. Porche (eds.), *Biosynthesis, Metabolism and Mode of Action of Invertebrate Hormones*. Springer-Verlag, Berlin and New York.

Weirich, G. F., M. J. Thompson, and J. A. Svoboda. 1986. *In vitro* ecdysteroid conjugation by enzymes of *Manduca sexta* midgut cytosol. Arch. Insect Biochem. Physiol. 3(2): 109–126.

Weirich, G. F., M. J. Thompson, and J. A. Svoboda. 1987. Ecdysone 3-epimerization in *Manduca sexta*. Abstracts of 8th Ecdysone Workshop, Marburg, W. Germany, p. 123.

Wigglesworth, K. P., D. Lewis, and H. H. Rees. 1985. Ecdysteroid titre and metabolism to novel apolar derivatives in adult female *Boophilus microplus* (Ixodidae). Arch. Insect Biochem. Physiol. 2(1): 39–54.

Wilson, I. D. and R. Lafont. 1986. Thin-layer chromatography and high-performance thin-layer chromatography of [^3H]metabolites of 20-hydroxyecdysone. Insect Biochem. 16(1): 33–40.

Zhu, X. X., H. Gfeller, and B. Lanzrein. 1983. Ecdysteroids during oogenesis in the ovoviviparous cockroach *Nauphoeta cinerea*. J. Insect Physiol. 29: 225–235.

Note added in proof: In a study of ecdysone 20-monooxygenation in Musca domestica, Srivatsan et al. (1987) found the activity associated with mitochondria and microsomes, but only the activity in mitochondria was increased after treatment of larvae with ecdysone. From this they concluded that only the monooxygenase of mitochondria was involved in the physiological formation of 20-hydroxyecdysone. [Srivatsan, J., Kuwahara, T. and Agosin, M. 1987. The effect of alphaecdysone and phenobarbital on the alpha-ecdysone 20-monooxygenase of house fly larva. Biochem. Biophys. Res. Commun. 148:1075–1080.]

Mode of Action of Molting Hormones in Insects 11

G. KAEUSER, J. KOOLMAN, AND P. KARLSON

11.1.	Introduction	363
11.2.	Remarks on the Historical Background	363
11.3.	Original Model of the Mode of Action	364
11.4.	Isolation and Biochemical Characterization of Ecdysteroid Receptors	365
	11.4.1. Measurement of Steroid Receptors	372
11.5.	A Revised Model of the Mode of Action	374
	11.5.1. Methods for the Histochemical Detection of Ecdysteroid Receptors	375
	11.5.1.1. Autoradiography	375
	11.5.1.2. Immunohistology	376
	11.5.1.3. Photoaffinity Labeling	376
	11.5.2. A Caveat	377
11.6.	Ecdysteroid-Dependent Genes and Puffing	377
11.7.	Alternative Modes of Action of Ecdysteroids	378
11.8.	Summary	381
References		381

11.1. Introduction

Ecdysteroids function as hormones in arthropods and control molting and other processes related to molting. Control of these processes has been well studied both in insects and in crustaceans. More recently, other processes controlled by ecdysteroids have been studied: these hormones appear to be involved in vitellogenesis (for a review, see Hagedorn, 1983) and in embryogenesis (Lagueux et al., 1984).

The occurrence of ecdysteroids has been documented not only in arthropods but also in other taxa belonging to the Protostomia, such as platyhelminths, nemathelminths, annelids, and mollusks (Spindler, 1988). Whether ecdysteroids serve as hormones in all these proto-stomian organisms (Karlson, 1983) is as yet an open question.

Soon after the isolation of ecdysone (Butenandt and Karlson, 1954), Clever and Karlson (1960) discovered puff induction by this steroid hormone. The observation was used to formulate the currently accepted theory of the mode of action of steroid hormones (Karlson, 1961, 1963).

The mode of action of ecdysteroids has been reviewed in detail (Butterworth, 1980; Garen and Lepesant, 1980; O'Connor et al., 1980; Cherbas et al., 1980; Koolman and Spindler, 1983; Pongs, 1984; Spindler et al., 1984; Yund and Osterbur, 1985; Karlson, 1988). We will therefore focus in this chapter on areas where progress has been made in the last few years; special attention will be given to distribution of ecdysteroid receptor molecules within the cell, as well as to alternative modes of action of ecdysteroids.

11.2. Remarks on the Historical Background

Up to the 1940s, questions as to the mode of action of hormones were not thought to be an important problem in endocrinology. Endocrinologists in those days concentrated on other problems, such as the structural elucidation of hormone molecules and the identification of target tissues and biological functions. This changed in 1941, when Green proposed that hormones act as cofactors of enzymes. Although this hypothesis could not be confirmed, and later on turned out to be incorrect, it stimulated ideas and experimental study of the mode of action of hormones in general.

Another model formulated to explain the action of hormones was based upon their presumed control of cell permeability (cf. Karlson, 1965). In the case of ecdysteroids, the model is known as the Kroeger hypothesis: Kroeger (1963; see also Kroeger and Lezzi, 1966) explained the puffing phenomenon in terms of an interaction of ecdysteroids with ion pumps of cell membranes. This interaction would then change the K^+/Na^+ ratio and eventually result in puff formation and gene activation. The basic principle of the Kroeger hypothesis could not be substantiated by experimental evidence. However, recently a strong influence of K^+ ions on puffing phenomena has been found (M. Lezzi, personal communication, 1988). Whether this has physiological relevance for the control of transcription remains to be verified.

Using crystalline ecdysone, Clever and Karlson (1960) were able to induce puff formation in the giant chromosomes of *Chironomus tentans*. This experiment led to the formulation of a model for the mode of action fundamental to all steroid hormones: the hormone controls the process of transcription of specific genes in the cell nucleus (Karlson, 1963). Originally "repressor"-like factors were proposed (Karlson, 1965), as in the Jacob–Monod model of gene regulation in prokaryotes. A complete chain of events, starting with the interaction of the hormone with nuclear structures and ending with a definite biological effect, was established: the molting hormone interacted with chromatin and induced certain puffs. Synthesis of mRNA for DOPA (3,4-dihydroxyphenylalanine) decarboxylase was increased, which on translation yielded DOPA decarboxylase. This enzyme is involved in tanning of the cuticule and catalyzes an important step in molting (Sekeris, 1965).

In 1962, Jensen and Jacobsen found estradiol-binding proteins in target cells of this steroid hormone. These proteins were shown to be hormone receptors. Their presence allowed the target cells to respond to the hormone with high sensitivity and specificity. Both observations were combined, and a new model [explained in the next section (11.3)] was developed. Soon this model was regarded as the description of the general model of action of steroid hormones.

11.3. Original Model of the Mode of Action

Based on results of Clever and Karlson (1960) and Jensen and coworkers (1967), the steroid–receptor model can be described as follows:

The steroid hormone enters the cell and binds to a "cytoplasmic" receptor molecule. This leads to (or facilitates) a conformational change, a process termed "activation" of the receptor, by which the hormone receptor complex increases its affinity for specific nuclear structures. The cytoplasmic hormone–receptor complex then enters the nucleus of the target cell (it "translocates") and interacts with the chromatin; in giant chromosomes of insects, it induces puff formation as a visible sign of gene activity. As a result, RNA synthesis from a selected number of genes is increased. The newly synthesized RNA is processed by posttranscriptional modifications and, after its transfer into the cytoplasmic compartment, is translated into proteins. To test this model, receptor molecules had to be isolated and characterized. Also, their distribution within the cell had to be analyzed.

A receptor molecule in this respect is a protein with certain characteristics: it can bind to steroid as well as to DNA; the interaction with the steroid ligand is characterized by high affinity and steroid specificity; the interaction with DNA also appears to show high affinity and (nucleotide sequence) specificity. Moreover, the receptor is a rare molecule with only a few thousand molecules per cell.

11.4. Isolation and Biochemical Characterization of Ecdysteroid Receptors

Ecdysteroid receptors have been investigated in several tissues of the fruit fly *Drosophila melanogaster* (Table 11.1), in homogenates of the blow fly *Calliphora vicina*, and several tissues of three crab species. These analyses were possible, only when radiolabeled [^{3}H] ponasterone A (Fig. 11.1) became available. This hormone analogue has severalfold higher affinity for the receptor as compared with 20-hydroxyecdysone and ecdysone. Labeling of the receptor preparation with ponasterone A or a hormone analogue of similar affinity, such as muristerone A or kaladasterone (Fig. 11.1), should allow for purification and eventual isolation of the receptor molecule (Sage et al., 1986; Landon et al., 1988).

Generally, steroid hormone receptors have a high affinity for their specific steroids. It is two to three orders of magnitude lower in insects than in vertebrates. This difference is mainly due to a higher dissociation rate of the ecdysteroid–receptor complex. The half-life of the complex with ponasterone A at 4°C is between 27 and 29 h, and at around 20°C is less than 60 min (Sage et al., 1982; Lehmann and Koolman, 1988a). Interestingly, the crustacean ecdysteroid receptor has a somewhat higher affinity for ecdysteroids and an increased stability (Spindler et al., 1984). The kinetic instability of the ecdysteroid receptor complex from insects precludes an easy purification of the

TABLE 11.1. Published Biochemical Data on *Drosophila* Ecdysteroid Receptors[a]

	"Cytosolic" receptor				"Nuclear" receptor			
Tissue	K_D PoA	K_D 20	Binding sites per cell	MM (method)	$2K_D$ Pon A	$2K_D$ 20HOE	Binding sites per nucleus	Ref.
K_c cells	$2 * 10^{-9}$		$3.5 * 10^{-10}$/g protein		$3 * 10^{-9}$	$3 * 10^{-8}$		Maroy et al. (1978)
K_c cells	$3.4 * 10^{-9}$	$2.4 * 10^{-7}$	1200		$4.2 * 10^{-9}$	$2 * 10^{-7}$	800	Sage et al. (1982)
Imaginal disks	$3.3 * 10^{-9}$			250–480 kDa (GF)	$3.8 * 10^{-9}$		1000	Yund et al. (1978)
Imaginal disks					$3 * 10^{-9}$	$5 * 10^{-8}$	500–1000	Cherbas et al. (1984)
K_c cells	$3–4 * 10^{-9}$	$2 * 10^{-7}$	1000–2000					Cherbas et al. (1984)
Embryos	$6 * 10^{-9}$	$2 * 10^{-7}$						Cherbas et al. (1984)
Imaginal disks	$3.3 * 10^{-9}$		300–1000					Fristrom (1981)
K_c cells	$2.2 * 10^{-9}$		400		$1.6 * 10^{-8}$		700	Beckers et al. (1980)
Embryos					$6 * 10^{-9}$	$2 * 10^{-7}$		Osterbur and Yund (1982)
Salivary glands			1300	130kDa (SDS–PAGE)				Schaltmann and Pongs (1982)
K_c cells								Schaltmann and Pongs (1982)
K_c cells			100				200–300	Gronemeyer et al. (1983)
Fat body							9000–13000	Gronemeyer et al. (1983)
BII tumorous	10^{-10}		10 fmol/mg protein					Dinan (1985)
Blood cells	10^{-9}		30 fmol/mg protein					Dinan (1985)

[a] *Abbreviations:* K_D, dissociation constant (mol/litre); PoA, ponasterone A; 20E, 20-hydroxyecdysone; MM, molar mass; kDa, kilodaltion; GF, gelfiltration; SDS–PAGE, sodium dodecyl sulfate–polyacrylamide gelelectrophoresis.

receptor by following the radioactivity after initial labeling with tritiated hormone. At each purification step the receptor-containing fractions have to be identified by binding experiments. This inhibits progress in the purification and always lowers its yield. Another drawback is that no suitable affinity label has been found as yet, i.e., one which would allow a covalent labeling of the receptor and thereby facilitate its identification (such as steroid mesylates in vertebrate systems). Although photoaffinity labeling of ecdysteroid receptors appeared initially to be a very promising technique (Gronemeyer and Pongs, 1980), it later turned out to be less useful because of its very low yield and the problem of pseudoaffinity labeling (Gronemeyer et al., 1983; Gronemeyer, 1985; Sage et al., 1986).

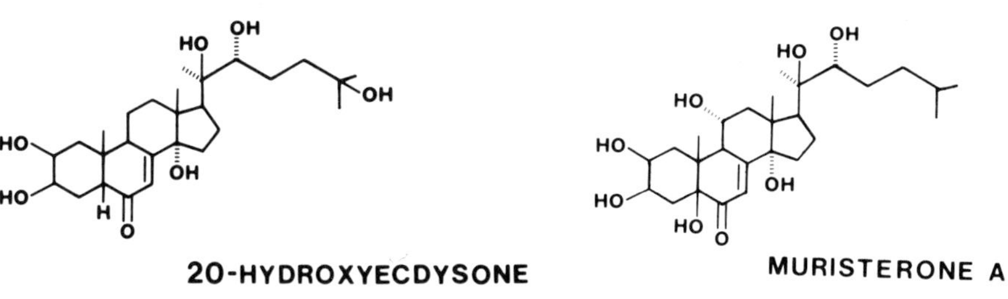

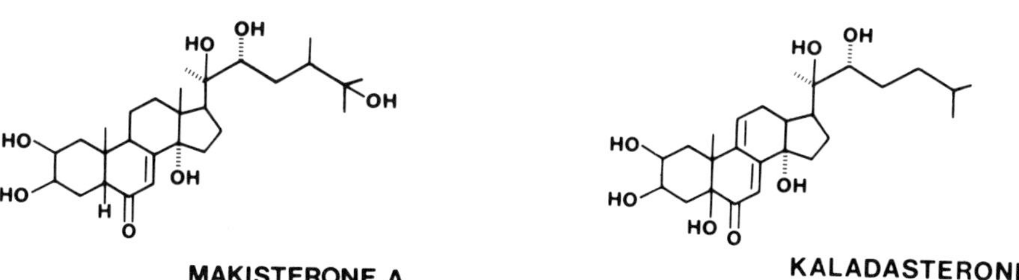

FIGURE 11.1. Chemical structure of selected ecdysteroids.

"Classical" biochemical techniques were employed to purify ecdysteroid receptors from insects: homogenates (from cell lines, from mass preparations of imaginal disks, or from whole organisms) were used as starting material to prepare a nuclear fraction. The nuclei were extracted with high-molarity salt buffers, and the extracts were desalted by dialysis. By the use of anion exchange chromatography and size exclusion chromatography, receptors could be purified (Lehmann and Koolman, 1988a). Other types of chromatography, such as affinity chromatography with immobilized ecdysteroid or DNA (Sage et a., 1986) may be used to further purify the hormone receptor. An approximately 10,000-fold purification appears to be necessary before the generation of monoclonal antibodies is feasible. This is still a future goal.

The results obtained in various laboratories working with insects are generally in accord (Table 11.1; cf. Lehmann and Koolman, 1988a): the ecdysteroid receptor is a protein with a molecular mass of approximately 100 kDa. Its ability to bind ecdysteroids has turned out to be extremly labile during preparative procedures. The K_D-values were deterimed to be approximately $10^{-9}M$ for ponasterone A, $10^{-7}M$ for 20-hydroxyecdysone, and $10^{-6}M$ for ecdysone (Fig. 11.2). As we have mentioned, these constants reflect a lower affinity between receptor and hormone than in vertebrates. The lower affinity, however, is compensated for by a higher concentration of steroid hormone in insect hemolymph as compared with vertebrate blood. In spite of the differences between the arthropods and vertebrates, this apparently leads to the same proportion of occupied receptors at peaks of hormone titer.

The experiments with ecdysteroid receptors do not provide any evidence for a stage- or tissue-dependent heterogeneity. A comparison of receptor affinities to ecdysteroids with circulating concentrations of ecdysteroids indicates that, in larval stages of insects, 20-hydroxyecdysone is the active form of the molting hormone (MH) and is bound by the receptor. Whether 26-hydroxyecdysone or other ecdysteroids found at high concentrations in insect embryos subserve the function of an MH for the embryonic stage remains to be determined (see Chapter 1 by Dorn in Part 2). In crustaceans ponasterone A has been detected (McCarthy, 1979). Since this ecdysteroid has a high affinity for crustacean ecdysteroid receptors (Londershausen and Spindler, 1981), it could well serve as an MH in Crustacea (see Chapter 9 in Part 2). In certain insect species mainly belonging to the orders Hemiptera and Hymenoptera, 20-hydroxyecdysone has been replaced by its C-24 methyl homologue, makisterone A, these organ-

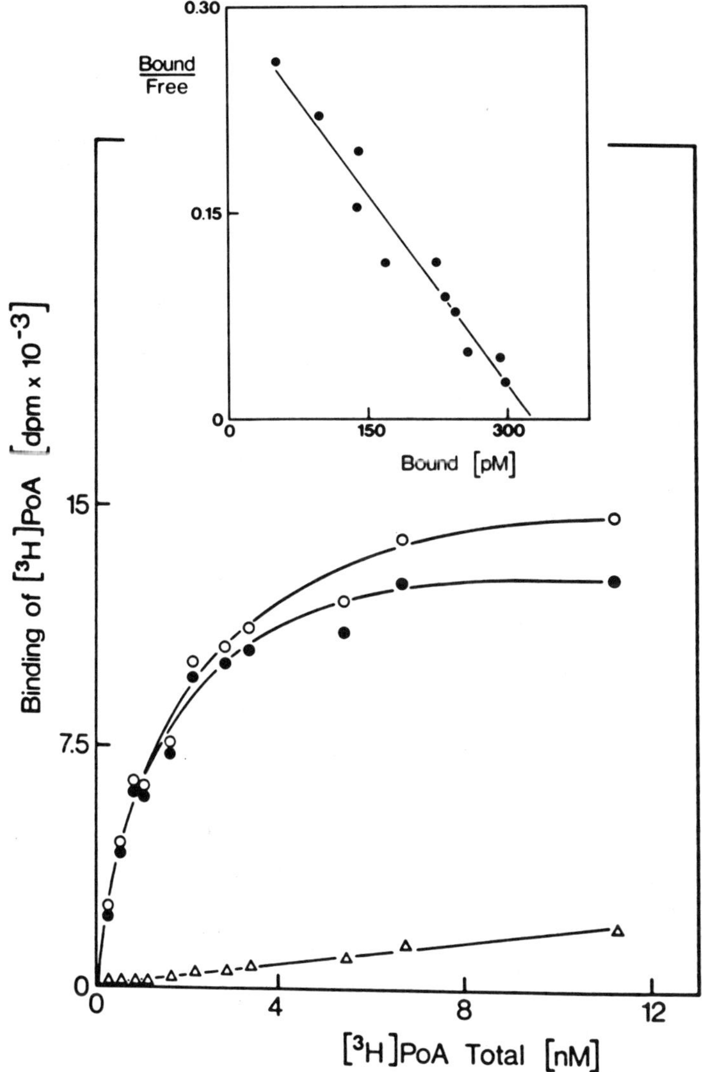

FIGURE 11.2. Saturation analysis of specific [^3H]ponasterone A binding. Purified nuclear extract from blowfly larvae was equilibrated at 20°C with increasing concentrations of tritiated ponasterone A. Specific binding (•) was determined as the difference between total binding (○) and nonspecific binding (Δ). The inset shows a Scatchard plot of the saturation binding data ($r = 0.974$). (From Lehmann and Koolman, 1988a.)

isms being unable to dealkylate the sterol precursor to cholesterol. In these cases, the ecdysteroid receptors may be expected to have changed their binding affinities appropriately (Feldlaufer and Svoboda, 1986).

Little is known about the nature of steroid binding and the conditions necessary for it. By use of mercurical reagents in receptor studies, Dinan and Spindler (1986) have found that the ecdysteroid receptor contains sulfhydryl (SH) groups that are involved in hormone binding.

Isolation and cloning of the steroid hormone receptor genes from vertebrates have indicated that steroid receptors in these animals (glucocorticoid, mineralocorticoid, and estrogen receptor) belong to one protein family (Loosfelt et al., 1986). The receptors, in principle, consist of three domains: the first is responsible for DNA binding; the second, for steroid binding; and the third, which was identified by monoclonal antibodies, probably has regulatory functions, but these are as yet ill defined (Gehring, 1987). The three domains show different amino acid sequence homologies between the various receptor types. The strongest homology is found between the DNA-binding domains, which might be interpreted as a hint for a general nucleic acid recognition mechanism of steroid hormone receptors. The observation of a DNA palindrome recognized by glucocorticoid and progesterone receptors appears to support this notion (Schütz, 1988).

So far, the three-dimensional structure of the steroid-binding domain has not been elucidated (by X-ray analysis) for any steroid receptor, since none of the receptors could be crystallized.

The analysis of ecdysteroid receptors still lags behind that of vertebrate steroid receptors. Thus, it is not known if the three-domain model of the vertebrate-type steroid receptor holds true for the ecdysteroid receptor as well, and whether the ecdysteroid receptor belongs to the protein family of vertebrate steroid receptors (a question that is of special relevance to the evolution of hormones).

The specificity of the ecdysteroid-binding site of ecdysteroid receptors has been investigated indirectly by a thorough comparison of biological activity with the chemical structure of a large number of ecdysteroids (Bergamasco and Horn, 1980). As a result of this study, three regions of the steroid molecule were defined that are relevant to its action: (i) the β("upper")-side of rings A and B is responsible for receptor binding; (ii) a region on the side chain with several hydroxyl groups at positions 20, 22, and 25 contributes to hydrophilic binding of the ecdysteroid to the receptor; and (iii) the hydroxy group at C-14 is relevant for biological activity.

The search for antiecdysteroids became successful only recently when the ability of certain brassinosteroids to compete with ecdysteroids for the binding site of ecdysteroid receptors was recognized (Lehmann and Koolman, 1988b). Brassinosteroids are known as plant hormones. When applied on intact insects, brassinosteroids inhibit molting and other ecdysteroid-dependent processes. These properties render brassinosteroids into interesting tools for the study of ecdysteroid effects and open the prospect for finding safer insecticides.

The application of modern methods of computer-aided drug design, which has already been used to investigate the vertebrate steroid receptor binding (Schmit and Rousseau, 1977; Duax et al., 1981), promises to yield even more information but has not yet been applied to ecdysteroid receptors.

For the transmission of the hormonal signal, the interaction of the steroid hormone receptor with chromatin is of equal importance as its interaction with the steroid. Apparently, the hormone receptors bind directly to sections of DNA near genes that are under the control of steroid hormones. Work on vertebrate receptors indicates that nucleotide sequences upstream of the gene contribute to its control (Payvar et al., 1981; Chandler et al., 1983; Scheidereit et al., 1983; Beato et al., 1986; Jantzen et al., 1987). For induction by glucocorticoids, a palindromic sequence of 15 base pairs is required. The progesterone receptor appears to interact with the same or a very related sequence (Strähle et al., 1987). Unpublished results indicate that, for ecdysteroids also, a 15-base-pair-long oligonucleotide confers inducibility to a nonregulated promoter (Schütz, 1988). This observation supports the hypothesis that the receptors of all steroid hormones share fundamental mechanisms and belong to a common protein family. Not yet clear is whether other nuclear components such as peptides, nonhistone chromosomal proteins, and small RNA molecules contribute to the specific binding of steroid hormone receptors to the DNA.

Ecdysteroid receptors were shown to bind to DNA *in vitro*, and after binding of the hormone the affinity to DNA increased (Lehmann, 1988). The ability of the receptor to bind to DNA is more labile than the ability to bind the hormone. We regard it as likely that the DNA binding of the ecdysteroid receptor is subject to an as yet unknown modification of physiological relevance.

11.4.1. *Measurement of Steroid Receptors*

Because of the lack of specific antibodies against ecdysteroid receptors, these molecules are usually identified and characterized by their ability to bind the labeled steroid ligand. Kinetic methods, such as saturation analysis or competition experiments, with radiolabeled ecdysteroids are employed. These methods follow complex protocols and may have certain drawbacks that can influence both the precision and relevance of the results obtained.

In the most common receptor assay, a preparation containing receptors (R) is incubated with a labeled ecdysteroid (predominately ^{3}H-side-chain-labeled ponasterone (A) (*E). This leads to the formation of an ecdysteroid–receptor complex (*ER):

$$*E + R \Longleftrightarrow *ER$$

Binding of a steroid to its receptor is, according to this model, a reversible event. Formally, it can be treated as a result of two processes: an ''association'' reaction, yielding the hormone–receptor complex,

$$*E + R \xrightarrow{K_a} *ER$$

and a ''dissociation'' reaction, resulting in free receptor and hormone molecules again,

$$*ER \xrightarrow{K_d} *E + R$$

here, K_A and K_D are the (kinetic) rate contants of association and dissociation.

When equilibrium between association and dissociation is reached, the overall reaction can be described by an ordinary law of mass action equation. Thus, for the association reaction; we have

$$K_A = \frac{[ER]}{[E] * [R]}$$

with [E], [R], and [ER] being the molar concentrations of the partners.

For the ''dissociation'' reaction the (thermodynamic) dissociation constants K_D can be calculated as

$$K_D = \frac{[E] * [R]}{[ER]} \tag{1}$$

which gives the following proportion between both thermodynamic constants (K_A being measured in litres/mole; K_D, in moles/litre):

$$K_A = \frac{1}{K_D}$$

Total concentration of receptors $[R_{tot}]$ is

$$[R_{tot}] = [R] + [ER] \tag{2}$$

Next, Eq. (1) and (2) may be combined, by writing Eq. (1) as

$$[ER] = \frac{1}{K_D} * [E] * [R] \tag{1'}$$

and Eq. (2) as

$$[R] = [R_{tot}] - [ER] \tag{2'}$$

Substituting Eq. (2') in (1'), we obtain

$$[ER] = \frac{1}{K_d} * [E] * ([R_{tot}] - [ER])$$

$$\frac{[ER]}{[E]} = \frac{1}{K_D} * ([R_{tot}] - [ER])$$

$$\frac{[ER]}{[E]} = -\frac{1}{K_D} * [ER] + \frac{1}{K_D} * [R_{tot}]$$

This equation describes a straight line with the slope giving the reciprocal of the dissociation constant K_D, while from the intercept of the X axis the total receptor concentration may be read (see Fig. 11.2).

This linearization is well known as the *Scatchard plot*. The concentration of the ecdysteroid–receptor complex [ER] is normally termed the ''bound'' fraction; the concentration of ecdysteroids [E], the ''free'' fraction. The Scatchard plot is obtained by relating the bound/free ratio with the concentration of bound material. The dissociation constant ($K_D = 1/K_A$) is usually termed the ''affinity'' of the receptor; the total receptor concentration (related to either the number of cells, or milligrams of protein or DNA, or millilitres of tissue extract) is known as the ''capacity'' of the receptor.

This simple method, however, will lead to severe misinterpretations of both affinity and capacity if certain prerequisites of the method are not strictly adhered to. The most important conditions are as follows:

(a) No processes must be occurring that change the concentrations of either hormone or receptor or both. Such changes could be due to the presence of endogeneous ecdysteroids or metabolism of ecdysteroids, as well as conversion of the receptor to a nonbinding state by denaturation, proteolytic degradation, or other enzymatic reactions.

(b) The interaction of receptor with ligand must be reversible, with association following second-order kinetics, dissociation showing first-order kinetics.

(c) The receptors present in the sample must behave homogeneously, i.e., they should have the same affinity toward the ligand used and should not influence each other; no cooperative effects should take place.

(d) Measurements of equilibrium constants must be made after equilibrium has been reached, which should be verified.

(e) Only a small proportion of the ligand must be bound.

There are a variety of other possible mistakes in this sort of assay, which are discussed in the literature (Rodbard, 1973; 1975; Chamness and McGuire, 1975; Braunsberg and Hammond, 1980; Klotz, 1982)

11.5. A Revised Model of the Mode of Action

A translocation of ecdysteroid receptors from cytosol to nuclei of target cells has not yet been observed. In this regard, ecdysteroid receptors appear to differ from vertebrate receptors (Bonner, 1982). However, in the last few years, the paradigm of a two-step process of steroid receptor action has also been intensively questioned in the vertebrate field (cf. Gorski, 1986). Different types of experiments suggested that the observation of "cytoplasmic" receptors was due to experimentally provoked artifacts. The discussion has not yet ended, but it appears that a sort of compromise is being worked out between the two opposing theories (cf. Clark, 1987).

By the use of indirect immunofluorescence, Pongs and his coworkers have demonstrated that ecdysteroids are present on the giant chromosomes of *Drosophila melanogaster* and *Chironomus tentans* (see Gronemeyer and Pongs, 1980; Gronemeyer et al., 1981; Dworniczak et al., 1983). Their experiments provided indirect evidence for the nuclear localization of ecdysteroid receptors. Direct evidence for the intracellular localization of ecdysteroid receptors was obtained only recently (H.J. Bidmon and J. Koolman, 1989). We have analyzed the intracellular localization of ecdysteroid receptors by two independent methods: autoradiography and immunohistochemistry. For autoradiography, larval tissues of the blow fly *Calliphora vicina* were incubated with [^3H] ponasterone A and subsequently processed, according to the thaw-mount technique developed by Stumpf and his group (Stumpf and Sar, 1975). For the immunohistological analysis, sections of blow fly tissues were incubated with an ecdysteroid-specific antiserum. The results from both approaches clearly demonstrated a nuclear localization of ecdysteroid receptors in a variety of target tissues. No indication of "cytosolic" receptors was found (Bidmon, 1988).

Ecdysteroid receptors from *Calliphora* also have been characterized biochemically (Lehmann and Koolman, 1988a). They were isolated from a nuclear extract. Again, no [^3H] ponasterone A–binding activity could be detected in the cytosolic supernatant. Thus, ecdysteroid receptors seem to have an exclusively nuclear localization. These experiments, of course, do not rule out the possibility that extra-nuclear "receptor" molecules exist which under the techniques employed might be unable to bind ecdysteroids.

11.5.1. *Methods for the Histochemical Detection of Ecdysteroid Receptors*

So far, antibodies against the ecdysteroid receptor are not available. Therefore, the distribution of ecdysteroid receptors within tissues must be analyzed by other methods. Three approaches have been used: autoradiography; immunohistology; and photoaffinity labeling.

11.5.1.1. AUTORADIOGRAPHY

The tissue is incubated with a limited concentration of radiolabeled ecdysteroid. To remove any unbound hormone, the tissue is washed extensively. The material is then frozen and sectioned. The distribution of radioactivity in the sections is determined by coating with a photographic emulsion. Exposure is continued in the frozen state and ranges between a few months and more than a year. The exposure is terminated by development of the photographic emulsion and staining of the tissue. The method, developed for vertebrate-type steroids by Stumpf and his group (see Stumpf and Sar, 1975), under certain conditions may give an exact image of the intracellular distribution and concentration of hormone receptors (Sheridan and Martin, 1987).

11.5.1.2. IMMUNOHISTOLOGY

With immunochemical experiments, results complementary to autoradiography can be obtained: the distribution of steroid hormone is visualized. Tissue is sectioned, treated with fixative, washed, and incubated with antiserum specific for ecdysteroids. By the chemical reaction of fixation, the endogenous ecdysteroids bound by the receptor become covalently linked to the receptor protein. This treatment, besides denaturing the proteins of the section, changes the three-dimensional structure of the receptor so that the covalently linked ecdysteroid is exposed and—at least to some extent—may be recognized by ecdysteroid-directed antibodies. The localization of these antibodies is detected by standard immunohistological methods, such as recognition with a second antibody against the steroid antibody and visualization by means of peroxidase/antiperoxidase staining. Also, fluorescent second antibodies can be used.

Histochemical methods employed for receptor localization of steroid hormones were recently discussed critically and the obtained results questioned (Underwood, 1987). We assume that the chemical structure of ecdysteroids (many functional groups!) tends to favor cross-linking to the receptor protein in comparison with vertebrate-type steroid hormones. Admittedly, a possible movement of ecdysteroids during fixation must be taken into consideration. This needs to be checked by appropriate control experiments (Bidmon, 1988). Moreover, in our experiments with the blow fly *C. vicina*, only one antiserum (DUL-3) out of eight proved to be suitable for the immunohistochemical detection of ecdysteroids.

11.5.1.3. PHOTOAFFINITY LABELING

This method was introduced into the field of ecdysteroids by Gronemeyer and Pongs (1980) and used to covalently link endogenous ecdysteroids to the receptor. By irradiation with ultraviolet (UV) light (315 nm), ecdysteroids are turned into a reactive state, leading to the formation of a covalent bond between ecdysteroid and neighboring molecules. Ideally, the molecule adjacent to the ecdysteroid is the hormone receptor. The ecdysteroid linked to the receptor is then detected by ecdysteroid-specific antibodies.

Although the method worked well with *Drosophila* and *Chironomus* giant chromosomes (Gronemeyer and Pongs, 1980; Gronemeyer et al., 1981; Dworniczak et al., 1983) and with ecdysteroid carrier protein

(Reum et al., 1982), its low efficiency (a small percentage of the hormone may be expected to be cross-linked to the protein) precluded its further use. Recent analyses with ecdysteroid receptors indicate that pseudoaffinity labeling (i.e., the labeling of any protein in solution with ecdysteroids) may prevail (Sage et al., 1986). This makes the method rather unattractive.

11.5.2. *A Caveat*

The main problem that arises in all the aforementioned assays is the diffusion of the steroid molecule during the analytic procedure. Therefore, any of these three methods for intracellular localization of ecdysteroids and ecdysteroid receptors must be accompanied by appropriate and extensive control experiments.

11.6. Ecdysteroid-Dependent Genes and Puffing

The mode of action of steroid hormones is a complex process that not only depends on the formation of the hormone–receptor complexes and their interaction with chromatin but also on interactions with other proteins. This is concluded from effects of ecdysteroids on the salivary glands of dipteran insects, especially *D. melanogaster*. The observations led to the so-called Ashburner model, which postulates a complex action of ecdysteroids on a set of tissue-specific genes (Ashburner et al., 1974). In combination with molecular biological data (Cherbas et al., 1986), the following picture emerges.

The action of ecdysteroids on salivary glands causes a sequential development of puffs, which can be divided into three major groups (Ashburner et al., 1974). The first comprises *intermold puffs*: a low ecdysteroid titer leads to an accumulation of their transcripts; by contrast, a high ecdysteroid titer results in regression of these puffs. The second group, called *early puffs*, is the result of an increased titer of ecdysteroids. Early puffs occur for a few hours only and are then replaced by the third group, *late puffs*. Using inhibitors of protein synthesis, Ashburner's group was able to show that the appearance of early and late puffs is regulated by the binding of the putative ecdysteroid–receptor complex *and* of protein(s) coded for by the early puffs. The initial binding of the hormone receptor complex activates the early puffs and leads to the synthesis of their coded protein(s).

These protein(s) inhibit the appearance of the early puffs, while promoting that of the late puffs. The late puffs, in contrast, are inhibited when the hormone receptor complex is attached to their chromatin; they are activated only after the increasing concentration of the early-puff product(s) has displaced the receptor.

A thorough analysis of some of the genes involved (Cherbas et al., 1986) led to the observation that one of the intermolt genes (*sgs-3* = salivary gland secretion protein No. 3) has a complex regulation pattern. It contains at least two control regions: one region (near the transcription site or internal to the gene) is responsible for the temporal and tissue specificity of the gene, whereas another region upstream of the gene is necessary for the normal level of expression.

The regulation of *sgs*-3 gene activity is not simply the result of a direct interaction of the gene with an ecdysteroid–receptor complex. Factors coded for by other genes are necessary for transcription of the *sgs* genes and for regression of the respective puff. In this case, it was even possible to separate the two phenomena of "puffing" and "transcription."

The analysis of the activation of early genes, indicated a similar complex pattern of control. The early puff 2B5, for example, codes for an important transcription-regulating factor, which is involved in (a) regression of intermolt puffs, (b) autoregulation of its own puff, and (c) activation of members of early and late gene sets. This is an indication of a special hierarchy of genes in the early gene set.

11.7. Alternative Modes of Action of Ecdysteroids

For nearly two decades, the elegant concept of receptor-mediated control of gene activity was believed to be *the* mode of action of ecdysteroids. In the last few years, however, alternative modes of steroid hormone action have been postulated for vertebrates as well as for insects. None of these models has yet been analyzed as deeply as has the chromatin–receptor model. From the field of vertebrates, two models should be mentioned: (1) In the oocytes of *Xenopus laevis*, a mechanism for the action of progesterone has been observed that depends on a membrane-bound steroid receptor. Binding of hormone regulates the adenylate cyclase system and eventually reinitiates meiosis (Sadler and Maller, 1985; Baulieu et al., 1985). (2) In rat brain, a direct interaction of steroid hormones with membrane-bound receptors of neurotransmitters and neuromodulators has been proved (Su et al., 1988). In the case of GABA (γ-aminobutyric acid) receptors, the steroids directly alter the response of the target cell to the transmitter action (Majewska et al., 1985).

The best-documented system, revealing a mode of action of ecdysteroids apparently independent of receptor interaction with chromatin, is based on experiments with fat body cells of the flesh fly *Sarcophaga peregrina* (Ueno et al., 1983; Ueno and Natori, 1984). Cryptic receptors for a hemolymph protein are controlled by 20-hydroxyecdysone.

Larval fat body of the flesh fly synthesizes a storage protein (generally called arylphorin) and releases it into the hemolymph, where it constitutes up to 75% of the hemolymph protein. In the pupal fat body, arylphorin is selectively taken up again and incorporated into protein granules. During later pupal development, this protein is used as a source of amino acids or of larger building blocks for the construction of adult structures during metamorphosis (Scheller, 1983). In the hemolymph of flies, arylphorin binds a large proportion of the circulating ecdysteroids and apparently serves as a transport protein for amphiphilic compounds (Enderle et al., 1983). Natori and coworkers (see Ueno et al., 1983; Ueno and Natori, 1984) have identified a receptor of this storage protein. The receptor is located on plasma membranes of *pupal* fat body cells. It has an M_r of 120,000. Plasma membranes of *larval* fat body cells already possess a cryptic receptor, which is inactive and slightly larger ($M_r = 125,000$) than the active pupal form. The binding of arylphorin to its receptor is Ca^{2+} dependent, shows a K_d of $4*10^{-9}M$, and is strictly pH dependent. Receptor molecules are repeatedly used for protein uptake. Natori's group demonstrated that the activation of cryptic larval receptor sites is induced by 20-hydroxyecdysone. By coincubation of larval fat body cells with 20-hydroxyecdysone and cycloheximide, they showed that the activation of the receptor is independent of protein synthesis. The biochemical basis for this process is not yet known.

In a second example, Natori's group was able to trace a similar phenomenon in the same tissue back to its biochemical origin. Incubation of fat body cells with radiolabeled phosphate and 20-hydroxyecdysone (with or without addition of cycloheximide) demonstrated that a protein of M_r 30,000, later identified as ribosomal protein S6, was selectively phosphorylated (Itoh et al., 1985). The physiological significance of this phosphorylation is not clear, but it was shown that it could be selectively inhibited by juvenile hormone. Juvenile hormone and 20-hydroxyecdysone apparently act antagonistically in this case (Itoh et al., 1987).

Over the years, more effects of ecdysteroids have been reported that cannot be explained by the classical model of ecdysteroid-induced alterations of transcription:

• Ito and Loewenstein (1965) found that ecdysone led within 1 h to a rise in mean nuclear membrane resistance of salivary gland cells of *Chironomus thummi.*

• Steel (1978) and Richter and Gersch (1981) observed that ecdysteroids changed the firing patterns of nerve cells. 20-Hydroxyecdysone, but not ecdysone, dampened the neuronal activity in NCC (nervus corporis cardiaci) II in *Rhodnius* (Steel, 1978), as well as in *Periplaneta* (Richter and Gersch, 1981), in a dose-dependent manner.

• Caveney and Blennerhassett (1980) detected changes in the intercellular resistance in the epidermis of *Tenebrio* upon exposure to 20-hydroxyecdysone. The membranes showed an increase in ionic conductance.

• Descamps and Lassalle (1981) presented electrophysiological evidence for a direct action of ecdysteroids on the brain of a myriapod.

• Spencer and Case (1984) found that chemoreceptors on the lateral antennulae of the crustacean *Panulirus interuptus* react to 20-hydroxyecdysone or ecdysone at concentrations of 10^{-12}–$10^{-13}M$. It appears as if in this case ecdysteroids act as phermones on the membrane receptors of sensory organs.

• Robert and colleagues (1986) measured fast muscular responses of the uterus of tsetse flies to treatment with ecdysteroids. Surprisingly, ecdysone and 20-hydroxyecdysone had opposite effects: ecdysone initiated phasic uterine contractions or enhanced the frequency of preexisting contractile activity, whereas 20-hydroxyecdysone decreased or abolished uterine contractions. Pharmacological investigations indicated that the ecdysteroids were acting on nervous connections between the uterus and ventral ganglionic chain.

• Lanot et al. (1987) observed that the meiosis of germinal vesicles of locusts, after its arrest in prophase, is continued after increase of 20-hydroxyecdysone—an effect that reminds one of the *Xenopus* model mentioned earlier.

Till now, these diverse reported observations have been viewed as isolated phenomena, and some of them may need confirmation by additional experiments. Nevertheless, they indicate that ecdysteroids have *extranuclear activities* that do not follow the classical scheme. We regard it as likely that membrane receptors for ecdysteroids exist, especially on nerve cells, which may control ion channels and/or intra-cellular enzymes such as protein kinases.

A field for new studies is opening up.

11.8. Summary

Ecdysteroids function as molting hormones (MHs) of arthropods and control, in addition, a variety of other developmental processes, such as vitellogenesis and embryogenesis. Like other steroid hormones, ecdysteroids act via receptors. Ecdysteroid receptors are intracellular proteins found in target cells of the hormone. They reveal all the characteristics of hormone receptors in terms of hormone specificity and affinity. The receptors have a nuclear localization and bind to DNA. The hormone apparently improves binding of the receptor to specific nucleotide sequences. As a result of a largely unknown process, specific genes are transcribed. Finally, mRNA is translated into specific proteins.

There is a slight heterogeneity of ecdysteroids functioning as molting hormones. While 20-hydroxyecdysone appears to be the major form of hormone in larval stages of most arthropod species, in some species makisterone A and ponasterone A have taken over this role. Other ecdysteroids perhaps have hormonal roles of their own. By use of brassinosteroids, which function as true antiecdysteroids by acting at the level of the hormone receptor, progress in the understanding of the classical mode of action of ecdysteroids may be expected.

Increasing evidence suggests that ecdysteroids also act in a nonclassical mode. The apparent target site is the plasma membrane, but details of this mechanism remain to be elucidated.

References

Ashburner, M., C. Chihara, P. Meltzer, and G. Richards. 1974. Temporal control of puffing activity in polytene chromosomes. Cold Spring Harb. Symp. Quant. Biol. 38: 655–662.

Baulieu, E. E., S. Schorderet-Slatkine, C. Le Goascogne, and J.-P. Blondeau. 1985. A membrane receptor mechanism for steroid hormones reinitiating meiosis in *Xenopus laevis* oocytes. Dev. Growth & Differ. 27: 223–231.

Beato, M., C. Scheidereit, P. Krauter, D. von der Ahe, S. Janich, A. C. B. Cato, G. Suske, and H. M. Westphal. 1986. Regulatory elements in steroid hormone inducible genes: structure and evolution of DNA sequences recognized by steroid hormone receptors. Pp. 121–141 *in* B. G. R. Reek, G. A. Goodwin, and P. Puigdomenech (eds.), *Chromosomal Proteins and Gene Expression*. Plenum Press. New York.

Beckers, Ch., P. Maroy, J. D. O'Connor, and H. Emmerich. 1980. Uptake and release of [^{3}H] ponasterone A by the K_c cell line of *Drosophila melanogaster*. Pp. 335–347 *in* J. A. Hoffmann (ed.), *Progress in Ecdysone Research*. Elsevier/North-Holland Publ., Amsterdam and New York.

Bergamasco, R. and D. H. S. Horn. 1980. The biological activities of ecdysteroids and ecdysteroid analogues. Pp. 299–324 *in* J. A. Hoffmann (ed.), *Progress in Ecdysone Research*. Elsevier/North-Holland Publ., Amsterdam and New York.

Bidmon, H.-J., 1988. Untersuchungen zur Ecdysonsynthese und Ecdysteroidrezeptorlokalisation in *Calliphora vicina*–Larven und einigen anderen Evertebraten. Dissertation, University of Giessen, W. Germany.

Bidmon, H.-J. and J. Koolman. 1989. Ecdysteroid-receptors located in the central nervous system of an insect. Experientia (Basel) 45:106–109.

Bonner, J. J. 1982. An assessment of the ecdysteroid receptor of *Drosophila*. Cell 30: 7–8.

Braunsberg, H. and K. D. Hammond. 1980. Practical and theoretical aspects in the analysis of steroid receptors. J. Steroid Biochem. 13: 1133–1145.

Butenandt, A. and P. Karlson. 1954. Über die Isolierung eines Metamorphose-Hormons der Insekten in kristallisierter Form. Z. Naturforsch. 9B: 389–391.

Butterworth, F. M. 1980. Ecdysone: introduction to the molecular mechanism of steroid action in insects. Pp. 248–254 *in* A. K. Roy and J. H. Clark (eds.), *Gene Regulation by Steroid Hormones*. Springer-Verlag, Berlin and New York.

Caveney, S. and M. G. Blennerhassett. 1980. Elevation of ionic conductance between insect epidermal cells by β-ecdysone *in vitro*. J. Insect Physiol. 26: 13–25.

Chamness, G. C. and W. L. McGuire. 1975. Scatchard plots: common errors in correction and interpretation. Steroids 26: 538–541.

Chandler, V. L., B. A. Maler, and K. R. Yamamoto. 1983. DNA sequences bound specifically by glucocorticoid receptor *in vitro* render a heterologous promoter hormone responsive *in vivo*. Cell 33: 489–499.

Cherbas, P., L. Cherbas, G. Demetri, M. Manteuffel-Cymborowska, C. Savakis, C. D. Yonger, and C.M. Williams. 1980. Ecdysteroid hormone effects on a *Drosophila* cell line. Pp. 278–308 *in* A. K. Roy and J. H. Clark (eds.), *Gene Regulation by Steroid Hormones*. Springer-Verlag, Berlin and New York.

Cherbas, L. J. W. Fristrom, and J. D. O'Connor. 1984. The action of ecdysone in imaginal discs and K_c cells of *Drosophila melanogaster*. Pp. 305–322 *in* J. A. Hoffmann, and M. Porchet (eds.), *Biosynthesis, Metabolism and Mode of Action of Invertebrate Hormones*. Springer-Verlag, Berlin and New York.

Cherbas, L., H. Benes, M. Bourouis, K. Burtis, A. Chao, P. Cherbas, M. Crosby, M. Garfinkel, G. Guild, D. Hogness, J. Jami, C. W. Jones, M. Koehler, J.-A. Lepesant, C. Martin, F. Maschat, P. Mathers, E. Meyerowitz, R. Moss, R. Pictet, J. Rebers, G. Richards, J. Roux, R. Schulz, W. Segraves, C. Thummel, and K. Vijayraghavan. 1986. Structural and functional analysis of some moulting hormone–responsive genes from *Drosophila*. Insect Biochem. 16: 241–248.

Clark, C. R. (ed.) 1987) *Steroid Hormone Receptors: Their Intracellular Localization*. Verlag Chemie, Weinheim, W. Germany.

Clever, U. and P. Karlson. 1960. Induktion von Puff-Veränderungen in den Speicheldrüsenchromosomen von *Chironomus tentans* durch Ecdyson. Exp. Cell Res. 20: 623–626.

Descamps, M. and B. Lassalle. 1981. Electrophysiological evidence for direct ecdysteroid action on the brain in *Lithobius forficatus*. Reprod. Nutr. Dev. 21: 681–687.

Dinan, L. 1985. Ecdysteroid receptors in a tumorous blood cell line of *Drosophila melanogaster*. Arch. Insect Biochem. Physiol. 2: 295–317.

Dinan, L. and K.-D. Spindler. 1986. The use of mercurial agents in the study of the mode of action of ecdysteroids. Insect Biochem. 16: 135–141.

Duax, W. L. J. F. Griffin, D. C. Rohrer, D. C. Swenson, and C. M. Weeks. 1981. Molecular details of receptor binding and hormonal action of steroids derived from x-ray crystallographic investigations. J. Steroid Biochem. 15: 41–47.

Dworniczak, B., R. Seidel, and O. Pongs. 1983. Puffing activities and binding of ecdysteroid to polytene chromosomes of *Drosophila melanogaster*. Europ. Mol. Biol. J. 2: 1323–1330.

Enderle, U., G. Käuser, L. Reum, K. Scheller, and J. Koolman. 1983. Ecdysteroids in the haemolymph of blowfly larvae are bound to calliphorin. Pp. 40–49 *in* K. Scheller (ed.), *The Larval Serum Proteins of Insects*. Thieme-Verlag, Stuttgart and New York.

Feldlaufer, M. F. and J. A. Svoboda. 1986. Makisterone A: a 28-carbon insect ecdysteroid. Insect Biochem. 16: 45–48.

Fristrom, J. W. 1981. *Drosophila* imaginal discs as a model system for the study of metamorphosis. Pp. 217–240 *in* L. I. Gilbert and E. Frieden (eds.), *Metamorphosis*. Plenum Press, New York.

Fristrom, J. W. and M. A. Yund. 1980. A comparative analysis of ecdysteroid action in larval and imaginal tissues of *Drosophila melanogaster*. Pp. 349–362 *in* J. A. Hoffmann (ed.), *Progress in Ecdysone Research*. Elsevier/North-Holland Publ., Amsterdam and New York.

Garen, A. and J. A. Lepesant. 1980. Hormonal control of gene expression and development by ecdysone in *Drosophila*. Pp. 255–262 *in* A. K. Roy and J. H. Clark (eds.), *Gene Regulation by Steroid Hormones*. Springer-Verlag, Berlin and New York.

Gehring, U. 1987. Steroid hormone receptors: biochemistry, genetics and molecular biology. Trends Biochem. Sci. 12: 399–402.

Gorski, J. 1986. The nature and development of steroid hormone receptors. Experientia 42: 744–750.

Green, D. E. 1941. Enzymes and trace substances. Adv. Enzymol. 1: 177–198.

Gronemeyer, H. 1985. Photoaffinity labelling of steroid hormone binding sites. Trends Biochem. Sci. 10: 264–267

Gronemeyer, H. and O. Pongs. 1980. Localization of ecdysterone on polytene chromosomes of *Drosophila melanogaster*. Proc. Nat. Acad. Sci. USA 77: 2108–2112.

Gronemeyer, H., H. Hameister, and O. Pongs. 1981. Photoinduced bonding of endogeneous ecdysterone to salivary gland chromosomes of *Chironomus tentans*. Chromosoma 82: 543–559.

Gronemeyer, H., P. Harry, and A. Alberga. 1983. A reappraisal of ecdysteroid binding in *Drosophila*. Mol. Cell. Endocrinol. 32: 171–178.

Hagedorn, H. H. 1983. The role of ecdysteroids in reproduction. Pp. 205–262 *in* G. A. Kerkut and L. I. Gilbert (eds.), *Comprehensive Insect Physiology, Biochemistry and Pharmacology*, Vol. 8. Pergamon Press, Oxford and Elmsford, New York.

Ito, S. and W. R. Loewenstein. 1965. Permeability of a nuclear membrane: changes during normal development and changes induced by growth hormone. Science (Wash., DC) 150: 909–910.

Itoh, K., K. Ueno, and S. Natori. 1985. Induction of selective phosphorylation of a fat body protein of *Sarcophaga peregrina* larvae by 20-hydroxyecdysone. Biochem. J. 227: 683–688.

Itoh, K., K. Ueno, and S. Natori. 1987. Counteraction by 20-hydroxyecdysone of the effect of juvenile hormone on phosphorylation of ribosomal protein S6. FEBS (Fed. Eur. Biochem. Soc.) Lett. 213: 85–88.

Jantzen, H.-M., U. Strähle, B. Gloss, F. Stewart, W. Schmid, M. Boshart, R. Miksicek, and G. Schütz. 1987. Cooperativity of glucocorticoid response elements located upstream of the tyrosine aminotransferase gene. Cell 49: 29–38.

Jensen, E. V. and H. I. Jacobson. 1962. Basic guides to the mechanism of estrogen action. Recent Prog. Horm. Res. 18: 387–408.

Jensen, E. V., T. Suzuki, T. Kawashima, W. E. Stumpf, P. W. Jungblut, and E. R. Desombre. 1967. A two-step mechanism for the interaction of estradiol with rat uterus. Proc. Nat. Acad. Sci. USA, 59: 632–638.

Karlson, P. 1961. Biochemische Wirkungsweise der Hormone. Dtsch. Med. Wochenschr. 86: 668–674.

Karlson, P. 1963. New concepts on the mode of action of hormones. Perspect. Biol. Med. 6: 203–214.

Karlson, P. (ed.) 1965. *Mechanisms of Hormone Action*. Thieme-Verlag, Stuttgart and New York.

Karlson, P. 1983. Why are so many hormones steroids? Hoppe-Seyler's Z. Physiol. Chem. 364: 1067–1087.

Karlson, P. 1988. The mode of action of ecdysone. Pp. 939–951 *in* A. Zabza, F. Sehnal, and D. L. Denlinger (eds.), *Endocrinological Frontiers in Physiological Insect Ecology*. Technical University of Wrocław Press, Wrocław, Poland (in press).

Klotz, I. M. 1982. Numbers of receptors sites from Scatchard graphs: facts and fantasies. Science (Wash., DC) 217: 1247–1249.

Koolman, J. and K.-D. Spindler. 1983. Mechanism of action of ecdysteroids. Pp. 179–201 *in* R. G. H. Downer and H. Laufer (eds.), *Endocrinology of Insects*. Liss, New York.

Kroeger, H. 1963. Chemical nature of the system controlling gene activity in insect cells. Nature (Lond.) 200: 1234–1235.

Kroeger, H. and M. Lezzi. 1966. Regulation of gene action in insect development. Annu. Rev. Entomol. 11: 1–22.

Lagueux, M., J. A. Hoffmann, F. Goltzene, C. Kappler, G. Tsoupras, C. Hetru, and B. Luu, B. 1984. Ecdysteroids in ovaries and embryos of *Locusta migratoria*. Pp. 168–180 *in* J. A. Hoffmann and M. Porchet (eds.), *Biosynthesis, Metabolism and Mode of Action of Invertebrate Hormones*, Springer-Verlag, Berlin and New York.

Landon, T. M., B. A. Sage, B. J. Seeler, and J. D. O'Connor. 1988. Characterization and partial purification of the *Drosophila* K_c cell ecdysteroid receptor. J. Biol. Chem. 263: 4693–4697.

Lanot, R., J. Thiebold, M. Lagueux, F. Goltzene, and J. A. Hoffmann. 1987. Involvement of ecdysone in the control of meiotic reinitiation of oocytes of *Locusta migratoria*. Dev. Biol. 121: 174–181.

Lehmann, M. 1988. Isolierung und Charakterisierung des Ecdysteroidrezeptors der Schmeissfliege, *Calliphora vicina*. Dissertation, University of Marburg, W. Germany.

Lehmann, M. and J. Koolman. 1988a. Ecdysteroid receptors of the blowfly, *Calliphora vicina*: partial purification and characterization of ecdysteroid binding. Mol. Cell. Endocrinol. 57: 239–249.

Lehmann, M. and J. Koolman, 1988b. Antiecdysteroid activity of brassinosteroids. Experientia (Basel) 44: 355–356.

Londershausen, M. and K.-D. Spindler. 1981. Characterization of cytoplasmic ecdysteroid receptors in the hypodermis of the crayfish, *Orconectes limosus*. Mol. Cell. Endocrinol. 24: 253–265.

Londershausen, M., P. Kuppert, and K.-D. Spindler. 1982. Ecdysteroid receptors: A comparison of cytoplasmic and nuclear receptors from crayfish hypodermis. Hoppe-Seyler's Z. Physiol. Chem. 363: 797–802.

Loosfelt, H., M. Atger, M. Misrahi, A. Guiochon-Mantel, C. Meriel, F. Logeat, R. Benarous, and E. Milgrom. 1986. Cloning and sequence analysis of rabbit progesterone–receptor complementary DNA. Proc. Nat. Acad. Sci. USA 83: 9045–9049.

Majewska, M. D., J.-C. Bisserbe, and R. L. Eskay. 1985. Glucocorticoids are modulators of GABA A receptors in brain. Brain Res. 339: 178–182.

Maroy, P., R. Dennis, C. Beckers, B. A. Sage, and J. D. O'Connor. 1978. Demonstration of an ecdysteroid receptor in a cultured cell line of *Drosophila melanogaster*. Proc. Nat. Acad. Sci. USA 75: 6035–6038.

McCarthy, J. P. 1979. Ponasterone A: a new ecdysteroid from the embryos and serum of brachyuran crustaceans. Steroids 34: 799–806.

O'Connor, J. D. 1983. Ecdysteroid and juvenile hormone receptors. Pp. 559–565 *in* G. H. Downer and H. Laufer (eds.), *Endocrinology of Insects*. Liss, New York.

O'Connor, J. D., P. Maroy, C. Beckers, R. Dennis, C. M. Alvarez, and B. A. Sage. 1980. Ecdysteroid receptors in cultured *Drosophila* cells. Pp. 263–277 *in* A. K. Roy and J. H. Clark (eds.), *Gene Regulation by Steroid Hormones*. Springer-Verlag, Berlin and New York.

Osterbur, D. L. and M. A. Yund. 1982. Ecdysteroid binding activity in embryos of *Drosophila melanogaster*. J. Cell. Biochem. 20: 277–282.

Payvar, F., O. Wrange, J. Carstedt-Duke, S. Okret, J. A. Gustavson, and K. Yamamoto. 1981. Purified glucocorticoid receptors bind selectively *in vitro* to cloned DNA fragment whose transcription is regulated by glucocorticoids *in vivo*. Proc. Nat. Acad. Sci. USA 78: 6628–6632.

Pongs, O. 1984. Ecdysteroid-regulated puffs and genes in *Drosophila*. Pp. 285–292 *in* J. A. Hoffmann and M. Porchet (eds.), *Biosynthesis, Metabolism and Mode of Action of Invertebrate Hormones*. Springer-Verlag, Berlin and New York.

Reum, L., G. Käuser, U. Enderle, and J. Koolman. 1982. A steroid-binding protein from insect haemolymph isolated by photoaffinity labelling and immunoadsorption. Z. Naturforsch. 37C: 967–976.

Richter, K. and M. Gersch. 1981. Electrophysiological activity of nervi corporis cardiaci I and II of *Periplaneta americana* in relation to the moulting cycle. Zool. Jahrb. Abt. Allg. Zool. Physiol. Tiere 85: 412–425.

Robert, A., A. Strambi, C. Strambi, and J. Gonella. 1986. Opposite effects of ecdysone and 20-hydroxyecdysone on *in vitro* uterus motility of a tsetse fly. Life Sci. 39: 2617–2622.

Rodbard, D. 1973. Mathematics of hormone–receptor interaction. I. Basic principles. Pp. 289–325 *in* B. W. O'Malley and A. R. Means (eds.), *Receptors for Reproductive Hormones*. Plenum Press, New York.

Rodbard, D. 1975. Theory of protein–ligand interaction. Methods Enzymol. 36: 3–15.

Sadler, S. E. and J. L. Maller. 1985. Inhibition of *Xenopus* oocyte adenylate cyclase by progesterone: a novel mechanism of action. Adv. in Cyclic Nucleotide Protein Phosphorylation Res. 19: 179–194.

Sage, B. A. and J. D. O'Connor. 1985. Measurement and characterization of ecdysteroid receptors. Methods Enzymol. 111: 458–468.

Sage, B. A., M. A. Tanis, and J. D. O'Connor. 1982. Characterization of ecdysteroid receptors in cytosol and naive nuclear preparations of *Drosophila* K_c cells. J. Biol. Chem. 257: 6373–6379.

Sage, B. A., D. H. S. Horn, T. M. Landon, and J. D. O'Connor. 1986. Alternative ligands for measurement and purification of ecdysteroid receptor in *Drosophila* K_c cells. Arch. Insect Biochem. Physiol., 3(Suppl. I): 25–33.

Schaltmann K. and O. Pongs. 1982. Identification and characterization of the ecdysterone receptor in *Drosophila melanogaster* by photoaffinity labeling. Proc. Nat. Acad. Sci. USA 79: 6–10.

Scheidereit, C., S. Geisse, H. M. Westphal, and M. Beato. 1983. The glucocorticoid receptor binds to defined nucleotide sequences near the promotor of mouse mammary tumor virus. Nature (Lond.) 304: 749–752.

Scheller, K. (ed.) 1983. *The Larval Serum Proteins of Insects*. Thieme-Verlag, Stuttgart and New York.

Schmit, J.-P. and G. G. Rousseau. 1977. Structure–activity relationships for glucocorticoids. II. Theoretical approach of molecular structures based on energy optimisation of a Westheimer model. J. Steroid Biochem. 8: 921–928.

Schütz, G. 1988. Control of gene expression by steroid hormones. Biol. Chem. Hoppe-Seyler 369: 77–86.

Sekeris, C. E. 1965. Action of ecdysone on RNA and protein metabolism in the blowfly, *Calliphora eythrocephala*. Pp. 149–168 *in* P. Karlson (ed.), *Mechanisms of Hormone Action*, Thieme-Verlag, Stuttgart and New York.

Sheridan, P. J. and P. M. Martin. 1987. Autoradiographic localization of steroid hormone receptors. Pp. 188–211 *in* C. R. Clark (ed.), *Steroid Hormone Receptors*. Verlag Chemie, Weinheim, W. Germany.

Spencer, M. and F. F. Case. 1984. Exogenous ecdysteroids elicit low-threshold sensory responses in spiny lobsters. J. Exp. Zool. 229: 163–166.

Spindler, K.-D. 1988. *Parasites and Hormones*. Springer-Verlag, Berlin and New York (in press).

Spindler, K.-D., L. Dinan, and M. Londershausen. 1984. On the mode of action of ecdysteroids in crustaceans. Pp. 255–264 *in* J. A. Hoffmann and M. Porchet, (eds.), *Biosynthesis, Metabolism and Mode of Action of Invertebrate Hormones*. Springer-Verlag, Berlin and New York.

Steel, C. G. H. 1978. Nervous and hormonal regulation of neurosecretory cells in the insect brain. Pp. 327–330 *in* P. J. Gaillard and H. H. Boer (eds.), *Comparative Endocrinology*, Elsevier/North Holland Publ., Amsterdam and New York.

Strähle, U., G. Klock, and G. Schütz. 1987. A DNA sequence of 15 base pairs is sufficient to mediate both glucocorticoid and progesterone induction of gene expression. Proc. Nat. Acad. Sci. USA 84: 7871–7875.

Stumpf, W. E. and M. Sar. 1975. Autoradiographic techniques for localizing steroid hormones. Methods Enzymol. 36: 135–136.

Su, T.-P., E. D. London, and J. H. Jaffe. 1988. Steroid binding at σ receptors suggests a link between endocrine, nervous, and immune systems. Science (Wash., DC) 240: 219–221.

Ueno, K. and S. Natori. 1984. Identification of storage protein receptor and its precursor in the fat body membrane of *Sarcophaga peregrina*. J. Biol. Chem. 259: 12107–12111.

Ueno, K., F. Ohsawa, and S. Natori. 1983. Identification and activation of storage protein receptor of *Sarcophaga peregrina* fat body by 20-hydroxyecdysone. J. Biol. Chem. 258: 12210–12214.

Underwood, J. C. E. 1987. Histochemical methods for steroid hormone receptor localization. Pp. 172–187 *in* C. R. Clark (ed.), *Steroid Hormone Receptors*. Verlag Chemie, Weinheim, W. Germany.

Yund, M. A. 1979. Specific binding of 20-hydroxyecdysone to nuclei of imaginal disks of *Drosophila melanogaster*. Mol. Cell. Endocrinol. 14: 19–35.

Yund, M. A. and D. L. Osterbur. 1985. Ecdysteroid receptors and binding proteins. Pp. 473–490 *in* G. A. Kerkut and L. I. Gilbert (eds.), *Comprehensive Insect Physiology, Biochemistry and Pharmacology*, Vol. 7. Pergamon Press, Oxford and Elmsford, New York.

Yund, M. A., D. S. King, and J. W. Fristrom. 1978. Ecdysteroid receptors in imaginal discs of *Drosophila melanogaster*. Proc. Nat. Acad. Sci. USA 75: 6039–6043.

Techniques for Identification and Quantification of Juvenile Hormones and Related Compounds in Arthropods

12

FRED. C. BAKER

12.1. Introduction   391

12.2. Structure and Stereochemistry of the JHs, JH Acids, and Methyl Farnesoate   395

12.3. Extraction and Purification of JHs and Related Compounds from Biological Samples   399

    12.3.1. A Few Words of Caution   399

    12.3.2. Extraction Methods and Solvents   400

    12.3.3. Thin-Layer Chromatography and Disposable Cartridges/Minicolumns   401

    12.3.4. High-Performance Liquid Chromatography   403

    12.3.5. Gas Chromatography   404

12.4. Detection, Identification, and Quantification of JHs and Related Compounds   406

    12.4.1. Bioassay   406

    12.4.2. Physicochemical Methods   409

        12.4.2.1. Analysis of Underivatized JHs   409

        12.4.2.2. Methods Employing Derivatization and GC with Electron Capture Detection   412

        12.4.2.3. Methods Employing Derivatization and GC/MS Detection   416

        12.4.2.4. Methyl Farnesoate and Other JH-Related Compounds   420

    12.4.3. *In Vitro* JH Biosynthesis   421

    12.4.4. Radioimmunoassay   424

        12.4.4.1. Preparation of Antibodies   425

12.4.4.2.   Radioligands   427
12.4.4.3.   Detection of JH *In Vivo* and *In Vitro*
    via RIA   427
12.4.5.   Other Methods   431
12.5.   Species in Which Qualitative and Quantitative JH
Measurements Have Been Made *In Vivo* Using
Physicochemical Methods   432
12.6.   Summary   433
Acknowledgments   438
References   439

12.1. Introduction

A breakthrough in the quest for isolating and identifying juvenile hormone (JH) came in 1956 when Williams discovered that abdomens of male cecropia moths (*Hyalophora cecropia*; Saturniidae) contain a factor that possesses very high JH biological activity. The active principle could be extracted into organic solvents, thus providing a clue to its chemical nature. Surprisingly, it was also claimed that the hormone was resistant to $0.2N$ acid, an observation, which is difficult to explain based on the (now) known structure of the JHs and their reactivity with dilute acids. It was a further 9 years until the isolation of pure JH from *H. cecropia* was reported (Röller and Bjerke, 1965). Two years later, Röller et al. (1967) published the structure of JH as the methyl ester of a dihomosesquiterpenoid epoxide (JH I, Fig. 12.1). The following year A. S. Meyer et al. (1968) isolated a second hormone (JH II, Fig. 12.1) from *H. cecropia*, occurring at one-fourth to one-seventh the level of JH I.

Classical transplantation experiments had indicated that the organs that secrete JH, the corpora allata (CA), from different species or

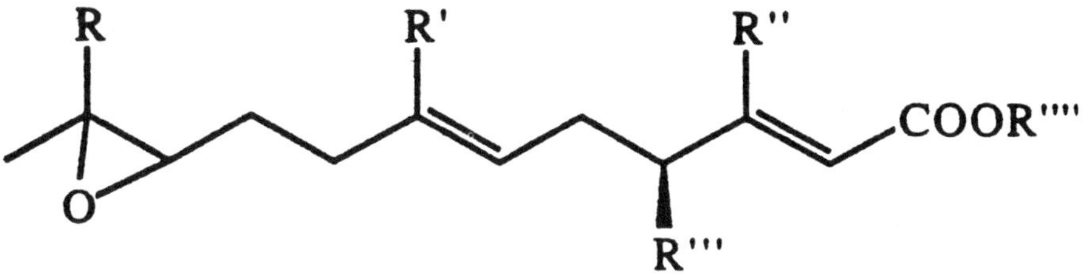

R=R'=R"=Me,   R'''=H,   (10*R*)-JH III
R=Et, R'=R"=Me,   R'''=H,   JH II
R=R'=Et, R"=Me,   R'''=H,   (10*R*,11*S*)-JH I
R=R'=R"=Et,   R'''=H,   JH O
R=R'=Et, R"=Me,   R'''=Me,   4-MeJH I

R""= Me for JH, R""= H for acids

FIGURE 12.1. Structures of the known JHs and corresponding JH acids. The trivial nomenclature for the more recently discovered JH O and 4Me–JH I (iso-JH 0) is based on the structures and nomenclature of the well-known JHs I, II, and III, which were named according to their chronological order of discovery.

different stages of the same species, elicited the same or similar response, suggesting the presence of a "universal" JH. Thus, the discovery of two JHs was surprising. This prompted the investigation of JHs in other insect species. At the same time, the unusual chemical structures of JH I and II piqued the interest of chemists, who developed methods to synthesize racemic and chiral JHs as well as many analogues. These small lipophilic molecules were able to penetrate the insect cuticle and elicit a JH effect resulting ultimately in death, or at least sterilization, of several insect species. It was thought there was great potential for these "juvenoids" as new generation pesticides (insect growth regulators, IGRs; see, e.g., Staal, 1975).

The observation that CA could be maintained *in vitro* in culture medium containing hemolymph (Röller and Dahm, 1970), allowed the investigation of JH biosynthesis in an environment conducive to simple isolation of pure JH. The discovery that the *S*-methyl group of methionine is the source of the methyl ester of JH I and II (Metzler et al., 1971) provided an excellent means of monitoring production of JH. Use of an *in vitro* technique very quickly led to discovery of a third JH (JH III; Fig. 12.1), produced along with JH II by CA from adult females of another lepidopteran, the tobacco hornworm moth, *Manduca sexta* (Judy et al., 1973b). Later, the *in vitro* technique was used to identify the JHs produced by miscellaneous insect species (see Section 12.4.3). While *in vitro* rates of JH biosynthesis by CA appeared to vary in a manner consistent with the developmental stage of donors, it was not clear that these data accurately reflected the *in vivo* JH fluctuations, or even if *in vitro* biosynthesis provided the correct qualitative JH profile. Furthermore, in those stages where CA were not accessible (e.g., most eggs), the only strategy was to isolate JHs from crude biological extracts. By the mid-1970s, several research groups were actively developing methods to detect and quantify the known JHs in biological samples. These included radioimmunoassay (RIA) and sophisticated physicochemical methods using various derivatization procedures to enhance JH detectability. Although many (or most) of these methods claim to have high sensitivity and selectivity, there has been considerable difficulty in obtaining consistent, reliable, and accurate identification and quantification of JHs *in vivo*. In fact, the literature is replete with misidentifications of JH. This is a reflection of the minute levels of JH present in most insect tissues and the difficulty of purifying these small amounts of the lipophilic JHs from a huge excess of lipid in the insect. As Schooley (1977) has astutely pointed out, "since physiological titers of these hormones are usually the order of a few nanograms per gram of tissue, separation of pure JH from

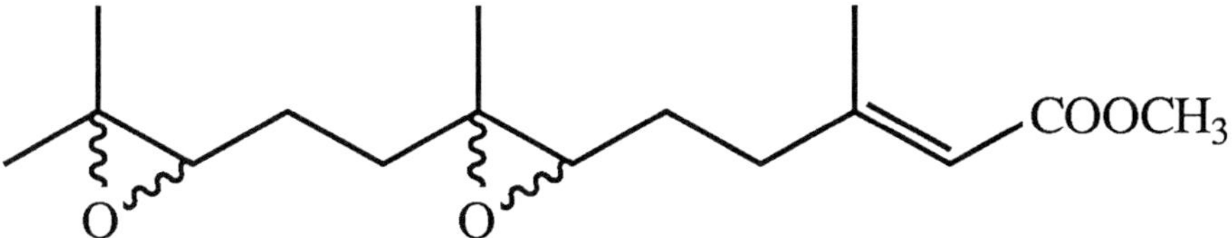

FIGURE 12.2. Methyl (2*E*,6*E*)-farnesoate.

chemically similar lipids can be a more challenging problem than structure determination.''

These problems aside, progress has been made using various techniques; the list of JHs has continued to expand. Using a rigorous purification scheme and a sensitive gas chromatographic/mass spectrometric (GC/MS) method developed to measure JH I, II, and III (Bergot et al., 1981a), Schooley's group at Zoecon isolated a new JH (JH 0; Fig. 12.1) from eggs of *M. sexta* (Bergot et al., 1980). Later, an isomer of JH 0 [4-methyl–JH I) Fig. 12.1] was also isolated from *M. sexta* eggs (Bergot et al., 1981d). To this group of five JHs may tentatively be added methyl farnesoate (MF); a simple acyclic sesquiterpenoid ester (Fig. 12.2) that is closely akin to JH III in structure and possesses JH activity (Wakabayashi et al., 1971). MF has been detected *in vivo* in certain crustaceans (Borst et al., 1987; Laufer et al., 1987) and is produced *in vitro* by the mandibular organ (see Chapter 2 by Borst and Laufer herein). Although MF appears to possess slight gonadotropic and morphogenetic activity in crustaceans (Borst et al., 1987; Laufer et al., 1987), it remains difficult to rule out the possibility that the compound is merely a prohormone that is epoxidized to JH III *in situ*. Very high levels of MF (and JH III) have been detected in embryos of *Nauphoeta cinerea* (Brüning et al., 1985) and are produced *in vitro* by CA dissected from embryos (Bürgin and Lanzrein, 1988). However, recent studies (Brüning and Lanzrein, 1987) suggest that MF is unable to substitute for the hormonal role of JH III in *N. cinerea*. Juvenile hormone acids (Fig. 12.1) are secreted by CA of certain male lepidopteran species and are later methylated in the male accessory glands (Weirich and Culver, 1979). JH acids are also secreted by CA of prepupal stages of *M. sexta* (Baker et al., 1987; Bhaskaran et al., 1986, 1987, 1988) and are methylated in the developing imaginal disks. Occurence of these ''prohormones'' in conjunction with the known JHs provides a further challenge to those developing methods for isolation, identification, and quantification of JHs from arthropod tissue.

Results gleaned during the period 1967–1975 from both *in vivo* and *in vitro* studies suggested a specific pattern regarding occurrence of JHs. It appeared (from the admittedly limited number of species investigated) that nonlepidopterans contained or biosynthesized only JH III whereas lepidopterans produced JH I, II, and III. However, in 1975 Lanzrein et al. reported the occurrence of JH I, II, and III in the cockroach *Nauphoeta cinerea*. They utilized a physicochemical method for identification of JHs, but quantification was via bioassay. This study was particularly interesting because the higher homologues (JH I, II) were predominant in nymphs whereas JH III was the major JH in adults, a result that led the authors to suggest differential roles for JH I and II vs. III as morphogenetic vs. gonadotropic hormones, respectively. Subsequently, there have been other reports of JH I and/or II in nonlepidopteran species. Several of these studies utilized physicochemical methods that lacked sufficient selectivity; others neither quantified the levels of JH nor gave unequivocal qualitative identification. Studies utilizing the RIA technique also began to emerge, and several of these reported the occurrence of JH I and/or II (immunoactive material) in nonlepidopteran species. Many of these "identifications" have been reexamined by other groups using very rigorous purification techniques and increasingly more specific methods of detection, particularly featuring gas chromatography/mass spectrometry (GC/MS). The latter results have indicated that only JH III is found in nonlepidopteran insects. Thus, we appear to be returning to the *status quo* of 1974/75. However, it must be kept in perspective that only a fraction of insect species have been examined for JH and many orders remain uninvestigated; therefore, the potential for discovery of ethyl-branched JHs or indeed different structural types in nonlepidopterans remains feasible. Ironically, each of the known JHs were first isolated and identified from stages other than the juvenile (nymphal or larval) morph. Perhaps this reflects the relatively low level (nanograms per gram or less) of JH(s) typically found in immature stages of most species investigated to date.

The subject of JH analysis was last reviewed in depth by Schooley (1977). During the ensuing years, interest in the physiological JH titers of miscellaneous insects and other arthropods has continued to expand. Many improvements have been made in the methods for isolation and detection of JHs. This is in part due to the increased availability of better and more sophisticated analytical instrumentation. However, introduction of simple apparatus, including disposable cartridges containing silica gel or reversed-phase packing materials, and the availability of very pure solvents have also contributed to the improvements. A commercial supply of unlabeled and radiolabeled JH I, II, and III have also facilitated methods development. Additionally, one cannot forget the wealth of knowledge gained by the many groups active in this field for >20 years.

While this chapter is devoted principally to description of techniques for identification and quantification of JHs in insects, techniques for detection of closely related compounds in insects and other arthropods will be discussed where appropriate. A comprehensive list of species and their corresponding *in vivo* JH(s) is provided in Section 12.5.

12.2. Structure and Stereochemistry of the JHs, JH Acids, and Methyl Farnesoate

It is useful to broach this subject, because it is intricately linked to the development of successful methods for extraction and quantification of JHs (and MF). For a more comprehensive treatment, the reader is referred to Chapter 2, herein) and earlier reviews (Schooley, 1977; Schooley and Baker, 1985).

The chemical structure of JH I (Fig. 12.1) was elucidated as methyl 10,11-epoxy-7-ethyl-3,11-dimethyl-2,6-tridecadienoate by means of MS analysis, nuclear magnetic resonance (NMR), and microderivatization techniques, using less than 300 μg of native material (Röller et al., 1967), a notable achievement at that time. Subsequently (Dahm et al., 1967), the structure was confirmed as the (2*E*,6*E*,10-*cis*)-isomer after synthesis of stereoisomers of known geometrical symmetry and comparisons by NMR and GC with natural product. The structure of JH I isolated from cecropia was also confirmed by A. S. Meyer et al. (1968). In addition, Meyer et al. reported the chemical structure of an additional bioactive compound, which they called JH II. The structure of JH II, methyl (2*E*,6*E*,10-*cis*)-10,11-epoxy-3,7,11-trimethyl-2,6-tridecadienoate, was determined using about 55 μg of natural product. Structure elucidation was aided by comparisons with methyl 10,11-epoxyfarnesoate, which had been synthesized earlier by van Tamelen et al. (1963) and later found to have potent JH biological activity by Bowers et al. (1965). Trost (1969) also mentioned that structure elucidation of JH I was aided by comparison to synthetic methyl 10,11-epoxyfarnesoate [methyl (2*E*,6*E*)-10,11-epoxy-3,7,11-trimethyl-2,6-dodecadienoate].

Meyer and Hanzmann (1970) subjected a mixture of cecropia JH I and II (9:1) to polarimetry measurements and obtained a positive rotation. The next year, Faulkner and Peterson (1971) and Loew and Johnson (1971) published syntheses of both (+)- and (−)-JH I. The (+)-JH I was nine times more bioactive on *Galleria mellonella* and about seven times more active on *Tenebrio molitor* than was (−)-JH I. Activity of the

(−)-enantiomer was largely due to contamination by (+)-isomer. Faulkner and Peterson (1971) synthesized optically active ketals of known absolute configuration as intermediates for synthesis of JH I. The (S)-ketal led to 11S-(+)-JH I, allowing assignment of its absolute configuration as 10R,11S (since knowing that the oxirane ring is cis allows assignment of the C-10 position as 10-R). The absolute configuration of JH I was later confirmed by Nakanishi et al. (1971) and by A. S. Meyer et al. (1971).

Corpora cardiaca–corpora allata (CC–CA) from adult female *M. sexta*, cultured in Grace's medium containing ^{14}C-labeled methionine as source of the JH methyl ester, produced two radiolabeled compounds with properties similar to known JHs (Judy et al., 1973b). The most abundant (8.7 μg from 738 CC–CA pairs) had a mass spectrum similar to JH II, whereas the less abundant product (5.3 μg) had a mass spectrum similar to that of methyl 10,11-epoxyfarnesoate. Thus, methyl 10,11-epoxyfarnesoate was discovered as a natural JH (JH III) some 8 years after synthetic material was found to possess JH activity (Bowers et al., 1965). [Subsequently, JH III was found to be the most ubiquitous of the JHs (see below).] Conversion of isolated JH II and III to corresponding 10,11-diols followed by periodate, then NaBH$_4$, treatment resulted in formation of the same primary alcohols, indicating a common skeleton from C$_1$–C$_{10}$ with the same 2*E*,6*E* geometry (Judy et al., 1973b).

The stereochemistry of biosynthetic JH II and III was investigated following conversion to their 10,11-diols (see Fig. 12.3 for JH II diol) and by subsequent derivatization with an optically active reagent. JH II diol (radiolabeled from methionine) eluted with the *threo*-glycol (from 10-*cis* epoxide), which was easily separable from the *erythro* isomer (from 10-*trans* epoxide). Reaction of 10,11-diols of both JH II and III with (R)-α-methoxy-α-trifluoromethylphenylacetyl chloride (MTPCl; Dale et al., 1969) produced the corresponding diastereomeric esters. The two optical antipodes of each racemic carrier were resolved by high-performance liquid chromatography (HPLC), but only the faster eluting peak contained radioactivity, implying pure optical isomers for both JH II and III secreted *in vitro*. The (R)-MTP ester of methyl 10,11-dihydroxyfarnesoate of known 10S configuration was analyzed and found to co-elute with the unnatural diastereomer; thus, natural JH III has the same (10R) configuration as JH I. The 10R,11S absolute configuration of JH II has never been rigorously established, but rather is based on relative HPLC elution orders of the JH II 10,11 diol derivative [10-(+)-MTP] compared to the similar derivative of JH III of

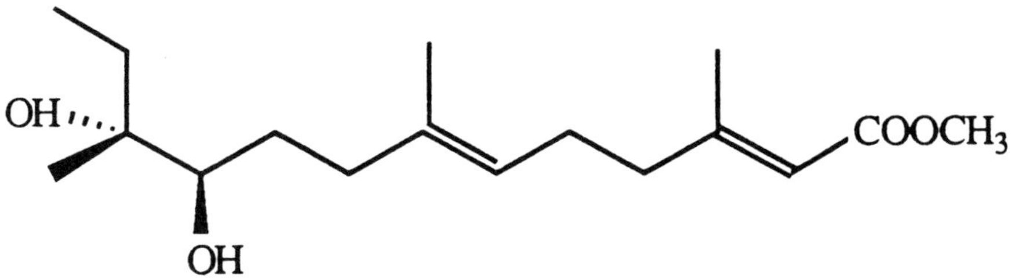

JH II threo diol

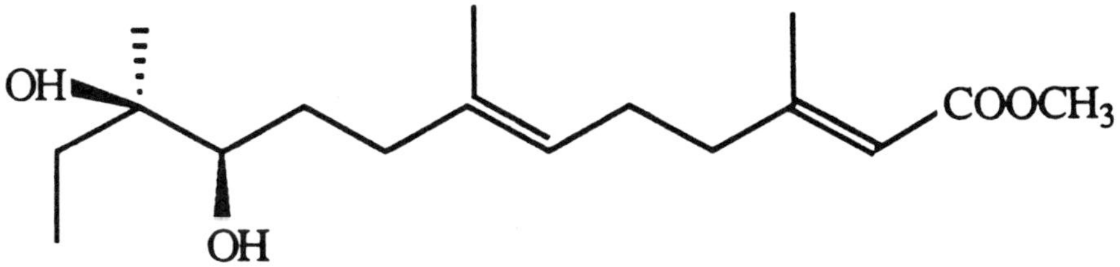

JH II erythro diol

FIGURE 12.3. Structures of *threo* and *erythro* JH II 10,11-diols.

known absolute configuration (Judy et al., 1973b; Jennings et al., 1975).

Structure elucidation of the trihomosesquiterpenoid JH 0 (Fig. 12.1; see Bergot et al., 1980) was completed with a mere 200 ng of material, isolated from *M. sexta* eggs, *cis* and *trans* isomers (Anderson et al., 1975). Catalytic hydrogenation (of 100 ng) afforded a tetrahydro derivative (*note:* no hydrogenolysis of oxirane structure; compare Röller et al., 1967) identical to the product formed from similar catalytic hydrogenation of synthetic 10-*cis*-JH 0. MS analysis of the tetrahydro derivative, in comparison to the tetrahydro derivative of JH III, provided strong evidence for the assignment of an extra methylene at the C-3 position. Capillary GC of the natural product showed retention time coincident with that of synthetic (2*E*,6*E*,10-*cis*)-JH 0, which in turn was resolved from the 2*E*,6*E*,10-*trans* isomer.

The discovery of JH 0 was closely followed by that of 4-methyl–JH I (4Me–JH I), also from *M. sexta* eggs (iso-JH 0; see Bergot et al., 1981d), which was detected as the d_3-methoxyhydrin derivative, in a fraction partially resolved by GC from the corresponding JH I derivative. About 50 ng of 4Me–JH I was obtained from 20 g of 48-h-old *M. sexta* eggs. The purified d_3-methoxyhydrin derivative afforded total electron impact (EI) and chemical ionization (CI) MS, which were almost identical to the MS of JH 0 except for slight differences in relative abundances of certain diagnostic ions. Further data were obtained from EI MS of underivatized 4Me–JH I (70–90 ng); the appearance of an ion at $m/z = 128$ identical to that seen for JH 0 verifies that the extra methylene is located at C-2, C-3, or C-4. Subsequent EI MS of the tetrahydroepoxymethyl ester confirmed the 4-methyl substitution: base peak at $m/z = 101$ and daughter fragment at $m/z = 69$ readily distinguished this derivative from the corresponding tetrahydro JH 0 (Bergot et al., 1980) and confirmed that the extra methyl group was absent from the C-2 or C-3 position and thus must occur at C-4. Comparison of natural product with synthetic ($2E,6E,10$-cis)-4Me–JH I (prepared as a mixture of the two pairs of enantiomers) showed idential EI and CI MS, as did corresponding tetrahydro derivatives. The ($4S$) absolute configuration of the natural product was deduced by using a combination of enzymatic synthesis and biotransformation to elaborate the 4Me–JH I structure *in vitro,* and taking advantage of the partial separation of diastereomers of chemically synthesized racemic 4Me–JH I on a 26-m Silar 10c glass capillary column (Koyama et al., 1987).

In those species of insects that secrete JH acid(s), structural confirmation of the free acid has been conducted simply by comparison of elution order on thin-layer chromatography (TLC) with samples of JH acid standards (e.g., see Bhaskaran et al., 1986) and subsequent conversion of the free acid to the methyl ester. It seems reasonable to assume that any JH acids secreted by insect CA have the same optical activity as the corresponding esters.

Naturally occurring MF in *Nauphoeta cinerea* embryos (Lanzrein et al., 1984) gave a mass spectrum identical to that of a pure sample of methyl ($2E,6E$)-farnesoate (Anderson et al., 1971). Methyl ($2E,6E$)-farnesoate was also identified (GC/MS identical to standard) in hemolymph of the crustaceans *Libinia emarginata* (Borst et al., 1987; Laufer et al., 1987) and *Homarus americanus* (Borst et al., 1987) and is produced *in vitro* by the mandibular organ (MO) of several crustacean species (Borst et al., 1987; Laufer et al., 1987; Tobe et al., 1989). The MO has been implicated in the regulation of crustacean reproduction (Le Roux, 1968) and is analogous to the CA of insects. The role of MF as a hormone or prohormone is perhaps more tenable in Crustacea than in Insecta, since JH III is only a minor constituent (and probably an artifact of sample processing) in the former class (Laufer et al., 1987).

12.3. Extraction and Purification of JHs and Related Compounds from Biological Samples

The techniques discussed can be applied to extraction and purification of lipophilic compounds structurally related to JHs, although some modifications may be necessary, depending on the polarity of the compound of interest. It should also be considered that the lipid content varies greatly according to species and developmental stage. Thus, it is difficult to develop a uniform protocol that can be applied equally to all arthropod species.

12.3.1. *A Few Words of Caution*

JHs often occur in biological samples at the parts-per-billion level or lower; thus, their isolation mandates an efficient and exhaustive purification procedure. Because the JHs are lipophilic compounds and most purification procedures are conducted with organic solvents, it is important to avoid introduction of impurities from contact with plastics, rubber tubing, serum stoppers, etc. If nitrogen gas is used to evaporate solvent from samples, the gas should be filtered and supplied via washed Teflon tubing rather than, e.g., Tygon or similar material containing a host of extractable contaminants. Any chemicals used in the purification procedure should be of high quality, and heat treated if possible. All-glass apparatus previously heat treated to 400°–500°C should be used to avoid contamination problems. The glassware should not be acid-washed as this could lead to decomposition of the JH 10,11-epoxide group. Neither should the glassware be silanized or siliconized prior to conducting extractions with organic solvents, since these treatments are unnecessary and lead to addition of contaminants. In laboratories that simultaneously utilize microgram or milligram levels of JHs, it is crucial to avoid cross-contamination of JH titer samples. By the same token, it is unwise to use microgram levels of JH "standards" to calibrate HPLC columns, etc., which would ultimately be used for purification of picogram or nanogram levels of JHs from biological samples. Non-JH markers or low mass of radiolabeled JH may be utilized to achieve the same goal. It is imperative to use high-purity solvents, particularly for later stages of purification, and to use dedicated equipment wherever possible. Finally, a blank sample should always be processed along with a batch of biological samples, as an indicator of potential false positive identifications.

12.3.2. *Extraction Methods and Solvents*

A variety of methods are available for the extraction of lipids from insects (Jackson and Armold, 1977). For isolation of JH from hemolymph, vigorous vortex mixing of the sample with solvent will suffice. Whole-body samples or eggs pose more difficulty, but these can usually be homogenized with solvent in an apparatus utilizing a mechanical shearing technique (e.g., Virtis homogenizer or Brinkmann Polytron). Alternatively, a glass homogenizer with glass or Teflon piston may be used (see, e.g., Hagenguth and Rembold, 1978). In those methods dedicated to quantitative determination of JH, it is important to accurately weigh the sample or measure its volume and to add a known amount of appropriate internal standard, preferably radio-labeled to allow estimation of recovery during the processing procedure. The ratio of mass of internal standard to mass of sample is a constant utilized in the ultimate analytical step to quantify each of the JHs. The accuracy of the end result depends upon several factors, including accurate weighing of sample and estimation of the amount of internal standard dispensed, complete homogenization of the sample, and stability of the internal standard and JHs (or related compound of interest) during storage. The addition of an inert material such as celite (heat treated) is advisable since it will aid subsequent filtration of the sample (Bergot et al., 1981a). Alternatively, samples may be centrifuged. Lyophilization of samples prior to solvent extraction has been employed by some groups (e.g., de Loof et al., 1976), although this is time consuming and may result in some loss of JHs due to volatility.

Numerous solvents and permutations of solvents have been utilized for the initial extraction and subsequent isolation of JHs from biological samples (see Table 12.1 and also references in Table 12.7, Section 12.5). Classical methods utilized diethyl ether to obtain JH active extracts of insects [e.g., the golden oil of *H. cecropia* (Williams, 1956)]. Gilbert and Schneiderman (1961) compared the ether-extractable mass of lipid from males and females of a large number of lepidopteran species. These studies showed a higher percentage of extractable oil in males vs. females of several species, and with levels as high as 37% of the fresh weight (e.g., male *Samia cynthia*). A useful purification can be achieved at this early step if the solvent is chosen carefully to effect complete extraction of the JHs without much of the extraneous lipid. Methanol extraction of whole-body or hemolymph samples, followed by partition between pentane and brine, resulted in a lipid extract that was usually only some 1–2% the mass of wet weight of sample (Schooley, 1977). Later (Bergot et al., 1981a), methanol was replaced

TABLE 12.1.   Solvents Utilized for Isolation of JHs from Insect Tissue

Solvent	Insect	Reference
Et_2O	*Hyalophora cecropia*	Williams, 1956
Et_2O:CH_3OH (1:1)	*Hyalophora cecropia*	Bieber et al., 1972
Et_2O:C_2H_5OH (6:1)	*Leptinotarsa decemlineata*	van Broekhoven et al., 1975
Et_2O:C_2H_5OH (5:1)	*Nauphoeta cinerea*	Lanzrein et al., 1975
EtOAc; followed by CH_3OH extraction of oil	*Manduca sexta*	Peter et al., 1976a
CH_3OH (CH_3CN); followed by pentane/ brine partitioning	Several spp.	Schooley et al., 1976; Bergot et al., 1981a
EtOAc:CH_3OH (1:1)	*Apis mellifera*	Hagenguth and Rembold, 1978
CH_3OH; followed by hexane partitioning	*Pieris brassicae*	Mauchamp et al., 1979a,b

by acetonitrile to effect a similar extraction of JHs. Acetonitrile extraction of the dried pentane phase offers a further useful purification step that efficiently removes JHs but does not extract poorly soluble glyceride esters of fatty acids, which comprise the major impurities (Bergot et al., 1981a). The extraction and partition methods used for JHs apply equally well to isolation of MF (see, e.g., Baker et al., 1988). Extraction of JH acids requires more polar extraction solvents (e.g., EtOAc). It is preferable to esterify JH acids to the corresponding JHs prior to additional purification and quantification (Baker et al., 1987, 1988). See Fig. 12.4 for the methodology used by Baker et al., (1988) to extract JH/JH acids from insect tissue, and the procedure utilized for subsequent purification and detection.

12.3.3.   *Thin-Layer Chromatography and Disposable Cartridges/Minicolumns*

Most reliable methods for JH isolation from whole-body tissue or hemolymph have employed one or more column chromatography or

Insect tissue (containing JHs and/or JH acids)
↓
a) Add CH_3CN, [^3H]internal standard (IS); homogenize, filter
↓
b) Partition filtrate (pentane/brine) → Extract brine layer (containing JH acids) with ethyl acetate
↓   ↓
c) Concentrate pentane layer   Add more [^3H]internal standard
↓   ↓
Crude lipid extract + JHs + IS   Concentrate ethyl acetate
↓   ↓
d) Extract oil (5X with CH_3CN)   Treat with CH_2N_2/ether
↓   ↓
e) Apply to C_{18} Sep-Pak; elute with CH_3CN **JHs + other esters**
↓   ↓
f) Alumina filtration; elute with ether   Evaporate, redissolve in CH_3CN, go to step b
↓
g) d_4-methanol/H^+
↓

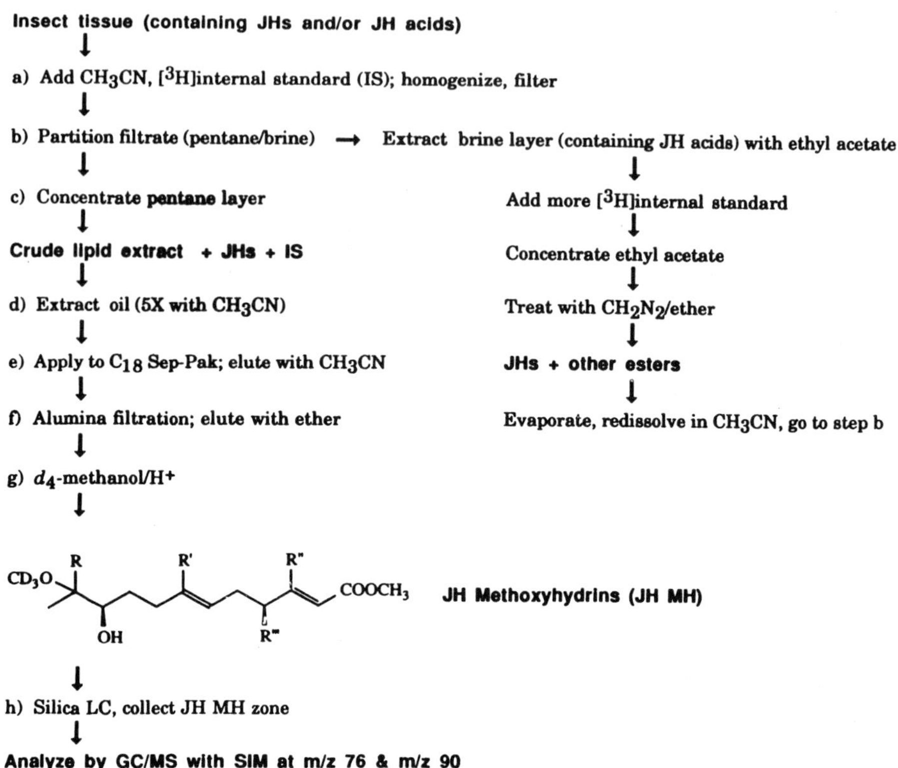

↓
h) Silica LC, collect JH MH zone
↓
Analyze by GC/MS with SIM at m/z 76 & m/z 90

FIGURE 12.4. Method for extraction, purification, and quantification of JHs and JH acids from arthropod tissue (from Baker et al., 1988).

thin-layer chromatography (TLC) steps. Schooley (1977) tabulated data from several studies that compared the effectiveness of TLC and column chromatography using various absorbents, including silica and alumina. These methods continued in popularity (see, e.g., Hagenguth and Rembold, 1979; Huibregste-Minderhoud et al., 1980a; Mauchamp et al., 1979a; Strambi et al., 1981—for TLC of JH 10,11-diol derivative). TLC is time consuming, however, and may lead to poor recovery, and inadvertent addition of contaminants (from binder) to the sample. Furthermore, it is often difficult to delineate the JH zone following TLC in the presence of a large mass of contaminating lipids. Epoxidation of MF may occur during TLC, leading to artifactual production of JH III (G. E. Pratt et al., unpublished results). Column chromatography is a preferable and more rapid method, particularly if disposable minicolumns are utilized. The availability of commercially prepared, standardized cartridges containing either normal-phase or reversed-phase adsorbents has greatly facilitated workup of samples. While cartridges are typically fabricated of plastic, this does not constitute a contamination problem provided they are prewashed and used sufficiently early in a purification scheme. Bergot et al. (1981a) found that a C_{18} Sep-Pak cartridge (Waters Associates) effectively removed residual sterols and di- and triglycerides from crude acetonitrile extracts of insect tissue. This step was followed by alumina filtration in order to remove free fatty acids from JH. This latter step used heat-treated Pasteur pipettes as a convenient minicolumn. Together, the two minichromatography steps achieved a 10^3–10^4 purification factor. The same method has been applied to purification of MF from insect or crustacean tissues (Borst et al., 1987; Baker et al., 1988). Mauchamp et al. (1981a) and Temin et al. (1986) also used C_{18} Sep-Pak cartridges for purification of JH from *Pieris brassicae* and *Locusta migratoria*, respectively, although these authors preferred to use methanol in place of acetonitrile. Rembold and Lackner (1985) used an alternative commercial product, Bond Elut LiChroprep RP-8 (Analytichem International) to effect a similar purification. They also employed a subsequent alumina filtration step in their procedure.

12.3.4. *High-Performance Liquid Chromatography*

Although time consuming, the HPLC technique offers high resolution and great flexibility, and consequently has been utilized as the penultimate step in most methods developed for analysis of JH in biological samples. One notable exception to this is the recently developed method of Rembold and Lackner (1985), which relies heavily on use of disposable cartridges instead. The inherent advantages of the HPLC technique can be further enhanced if JH(s) is converted prior to HPLC to a derivative with different polarity (see, e.g., Bergot et al., 1981a; Strambi et al., 1981, and Sections 12.4.2.2 and 12.4.2.3, below, for rationale concerning derivatization). Normal-phase HPLC is most desirable if this step precedes analysis by a physicochemical method [e.g., GC/MS or GC/ECD (electron-capture detection)] since the collected JH fractions require no additional extraction steps. Reversed-phase HPLC may be more suitable if this is to be followed by RIA detection, although the level of organic solvent usually prohibits direct use for RIA without further dilution. The use of highest purity solvents and uncontaminated glassware etc. is of the utmost importance at the HPLC step, and for dissolution of samples prior to the final analysis.

While collection of JHs (or derivatives) as a group leads to accumulation of large volumes of eluent, containing many prospective contaminants, this method is preferable to collection of individual fractions corresponding to each JH (or derivative). The latter approach can lead to loss of the desired fraction(s) if slight variations in R_T occur during analysis. Furthermore, accuracy of JH quantification is seriously jeopardized once internal standard and component JHs are separated from one another. In a group separation, the JH-containing zone can be delineated by use of suitable chromophoric marker compounds [levels of JH present in biological samples are usually too low to allow direct detection, e.g., by ultraviolet (UV) absorbance]. It is important to determine that these markers do not interfere with subsequent analysis of the JHs. At the beginning of each series of HPLC analyses, it is prudent to ensure that the system is performing optimally and especially that the desired JH fraction does elute within the boundaries of the marker compounds, since small variations in solvent composition may drastically affect the separation (unpublished work in our laboratory, utilizing a silica Spheri-5 column and Et_2O/CH_2Cl_2, 1/2 water saturated). However, at this crucial step it is important not to introduce contaminants (e.g., by injecting a large mass of unlabeled JHs, or derivatives, as markers). The use of high-specific-activity radiolabeled JHs (or derivatives) is the most feasible method. It is also a good practice to analyze samples in order of anticipated increasing JH content, if known, beginning with the procedure blank. After collection of desired JH fractions (several millilitres), it is necessary to concentrate the samples and transfer them to small vials for subsequent analysis (typically GC in most physicochemical procedures). It is important to ensure that caps with septa coated with Teflon or similar inert material are used for sealing samples of storage prior to analysis. RIA procedures require JHs to be in aqueous solution containing a substantial level of protein to allow subsequent precipitation of antibody–ligand complexes. Addition of protein [e.g., bovine serum albumen (BSA)] no doubt improves the solubility of JHs in aqueous solution. However, the inherently poor solubility of JHs or derivatives in aqueous solution is one of the drawbacks to using the RIA technique.

12.3.5. *Gas Chromatography*

In most physicochemical methods for JH identification and quantification, GC constitutes the ultimate separation step and probably provides better resolution of JH (or derivatives) from impurities than any prior purification steps. It is advisable to use a dedicated syringe and analytical column, and to inject standards, followed by a solvent blank, prior to analysis of samples, to ensure that no inadvertent contamination occurs. JHs are not very stable under most GC conditions and may decompose to produce several contaminants, which could hamper identification and quantification (see Section 12.3.2). Thus, it is preferable that JHs be derivatized at some prior stage in the purification scheme. Derivatization of JH plays an important part in enhancement of detectability and selectivity; however, depending upon the degree of purification before and after the derivatization step, many contaminants may still be present. This depends, to some extent, on the lipid content of the biological samples under scrutiny: samples of *Oncopeltus fasciatus* eggs and *Anastrepha suspensa* larvae contain prodigious levels of lipid, which complicate purification of the JH fraction (unpublished work in our laboratory). The use of derivatizing reagents that are nonspecific (e.g., those which react with —OH groups) can lead to other (derivatized) impurities, many of which are not removed by further purification steps. Thus, in the final GC analysis many peaks will be observed, some of which may co-elute with one or other of the JH derivatives. However, the very strength of the GC method lies in the wide range of different phases available. It is important to question all positive identifications and reanalyze samples on a column containing a phase of different selectivity (Huibregste-Minderhoud et al., 1979). The use of temperature programming also offers another dimension for improving resolution. Biological methods generally lack this depth of exhaustive analysis.

Capillary columns offer improved resolution compared with typical 2-m packed glass columns. However, little, if any, improvement in the limit of detection of JHs was observed with a capillary column vs. 2-m packed column (F. C. Baker and G. C. Jamieson, unpublished observations). Furthermore, the packed columns may be interchanged more easily than glass capillary columns and in addition are much less expensive. A wide range of stationary phases have been used for GC analysis of JHs and various derivatives; the most popular include SE-30, OV-17, and OV-1. Dunham et al. (1975) have pointed out that the use of a slightly basic stationary phase [e.g., phenyldiethanolamine succinate) (PDEAS)] is more desirable for analysis of underivatized JHs.

12.4. Detection, Identification, and Quantification of JHs and Related Compounds

12.4.1. *Bioassay*

Bioassay of biological extracts for JH activity played a crucial role in the many studies that paved the way for subsequent purification and identification of the natural JHs. Several key features of JH action relevant to bioassay are discussed by Schneiderman (1972). Thus, JHs (and many analogues and other JH active substances) readily penetrate the cuticle of insects, and can be bioassayed via topical treatment. Since JHs are lipophilic, extracts are often concentrated and then diluted in oil to minimize metabolism following application or injection (see Bjerke and Röller, 1974; Wigglesworth, 1969, 1985). The many available methods for bioassay have been reviewed in detail by Staal (1972) and Bjerke and Röller (1974). There is almost no limit to the variety of insect species that can be utilized. However, the most sensitive and consequently the most utilized of JH bioassays are the wax tests (see de Wilde et al., 1968; Schneiderman, 1972; Staal, 1972; Bjerke and Röller, 1974—and references therein) in which extracts are concentrated, dissolved in paraffin oil, and then mixed with a known mass of melted paraffin wax on a microscope slide. After cooling, the solid is cut into equal sections and melted on the abraded or punctured mesonotum region of pupae from *Antheraea polyphemus* (Schneiderman and Gilbert, 1958), *Hyalophora cecropia* (Fisher and Sanborn, 1964), or, more commonly, *Galleria mellonella* (de Wilde et al., 1968). The appearance of patches of pupal cuticle over the wounded area, following the adult molt, is scored as a positive JH response. As little as 5 pg of JH I or II can be detected using this technique (see Table 12.2), but JH III is much less active. Topical application of juvenoids in acetone to *Galleria* pupae has also been used for bioassay (Henrich et al., 1973); although this is a simpler procedure, sensitivity is sacrificed. *Tenebrio molitor* has also been utilized extensively for JH bioassay, including topical treatment (Bowers and Thompson, 1963; Henrick et al., 1973), injection (Yamamoto and Jacobson, 1962), or cuticle tests (Karlson and Nachtigall, 1961; Bjerke and Röller, 1974). These *T. molitor* bioassays are much less sensitive than the wax tests for the known JHs (see Table 12.2; also Dahm and Röller, 1970; Dahm et al., 1976). However, Röller and Bjerke (1965) successfully utilized a *Tenebrio* pupal injection bioassay to monitor purification of JH from

TABLE 12.2.   Relative Activities of JH I, II, and III Expressed as *Galleria* or *Tenebrio* Units (GU and TU, respectively)

JH	$GU/\mu g$	$TU/\mu g$	GU/TU
I	2×10^{5}	8×10^{2}	2.5×10^{2}
II	2×10^{5}	30	6.7×10^{3}
III	2×10^{3}	0.5	$4 \ \times 10^{4}$

SOURCE: Modified from Dahm et al. (1976).

NOTE: JH biological activity of JH O and 4Me–JH I has not been tested using the *Galleria* wax test or by injection into *Tenebrio* pupae; however, these compounds are known to possess substantial JH activity when testing using other bioassays (see Bergot et al., 1980, 1981d).

cecropia abdomens, as did Röller et al., (1967) in the study leading to the first isolation and identification of JH (JH I). Historically, *Rhodnius prolixus* has figured prominently in JH research, as exemplified by the many early studies by Wigglesworth (e.g., see his 1969 paper for recent use of the *Rhodnius* bioassay).

The latter bioassays are all aimed at measuring the morphogenetic effect of a JH or extract. However, it should be stressed that many other bioassays are designed to monitor JH gonadotropic hormone activity. Typically, these measure the effects of excess JH on egg development (oocyte length or egg number; see, e.g., Baker et al., 1983) and/or viability of eggs (see, e.g., Masner et al., 1968). Alternatively, rescue by JH of egg development in animals that have been allatectomized offers another bioassay procedure applicable to a wide range of species.

In their isolation of JHs from *H. cecropia*, A. S. Meyer et al. (1965) used a battery of injection and topical bioassay procedures including the wax test(s) and *Tenebrio* topical and injection tests. Later A. S. Meyer et al. (1968) identified JH II (in addition to JH I) in purified extracts from *H. cecropia* bioassayed with the *Galleria* wax test, a result in part attributable to the superior sensitivity of this bioassay procedure compared to the *Tenebrio* assay used by Röller et al. (1967). This, in fact, brings out some important points: a useful bioassay should be simple, sensitive, rapid, specific, and quantitative. Unfortunately, as far as JH bioassays are concerned, available methods fall short in several of these respects, but particularly regarding specificity

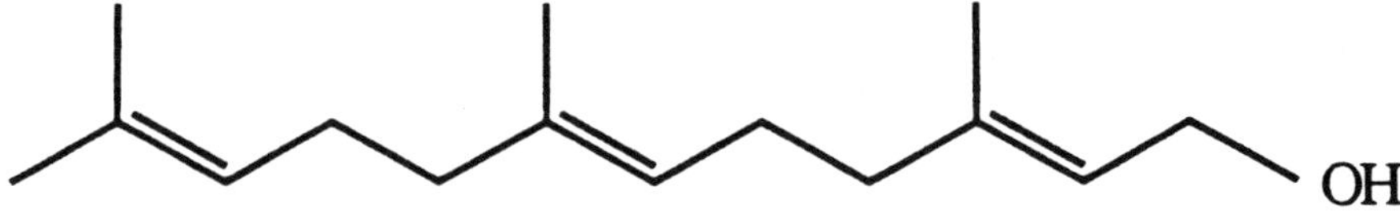

FIGURE 12.5. Farnesol.

and quantification. The known JHs are not equipotent in a given bioassay. Furthermore, the relative potency of the JHs varies according to the insect species tested. This is exemplified in the case of *Galleria* wax test vs. *Tenebrio* injection bioassay (Table 12.2) where JH III has very low activity in the *Tenebrio* test. On the contrary, farnesol (Fig. 12.5), a natural product recognized as possessing JH activity (Wigglesworth, 1961; Schmialek, 1963; Bowers and Thompson, 1963; Schneiderman and Gilbert, 1963; Bowers and Thompson, 1963; Schneiderman and Gilbert, 1964), is as potent as cecropia oil when both are assayed via the *Tenebrio* topical test but is considerably less potent than cecropia oil in the *Galleria* wax test (A. S. Meyer et al., 1965). Compare also the relative biological activity of extracts from several species measured in terms of *Galleria* units vs. Cecropia units (Dahm et al., 1981).

In many JH bioassays (measuring either morphogenetic or gonadotropic effects), JH I is much more active than JH III. In some instances, this has led to the assumption that JH I is therefore the natural JH of the species under test. There is no basis for this reasoning. In this context, it should be pointed out that while JH I is a much more active morphogenetic hormone than JH III when applied to *Tenebrio molitor* (see Table 12.2), JH III is the natural JH of *T. molitor* (Trautmann et al., 1974a).

Several hemipteran species have been utilized for testing of miscellaneous biological extracts and JH analogues. The famous "paper factor" (Slama and Williams, 1965), later identified and named juvabione (Bowers et al., 1966), was active on the bug *Pyrrhocoris apterus* but was inactive on other bugs, including *Oncopeltus fasciatus* and *Rhodnius prolixus*, and the lepidopteran *H. cecropia*. Furthermore, cecropia oil (JH I/II) was inactive when tested on *P. apterus* (Slama and Williams, 1965), whereas synthetic (RS)-JH III was active on all hemipterans (Bowers et al., 1966). These apparent inconsistencies serve to emphasize that hemipterans appear heterogeneous with regard to response to JH active extracts and further underline the lack of specificity of the various bioassays. Dorn (1975) and Rankin and Riddiford (1978) reported the occurrence of very high levels of JH biological activity in extracts from *O. fasciatus*. However, little, if any, of the known JHs, JH acids, or related compounds could be detected (by a physicochemical method) in three strains of *O. fasciatus* (Baker et al., 1988). Observations that certain components of crude biological extracts can lead to false positive JH effects (e.g., fatty acids; see Bowers and Thompson, 1963), while positive JH effects may be masked by toxic impurities in the extract, further cloud the reliability of bioassays. Carlisle and Ellis (1968) have pointed out that olive oil (a common vehicle for injection of biological extracts) is not inert in JH bioassays.

Manduca sexta has been featured prominently in lepidopteran endocrine research, and it was appropriate that a JH bioassay be developed for this species. The neck ligation bioassay (Truman et al., 1973)–or its more convenient equivalent utilizing the naturally JH-deficient black mutant (Fain and Riddiford, 1975)—has found acceptance in some laboratories. However, the sensitivity of this bioassay is not as good as the wax tests. Staal (1977) has also developed several feeding and topical JH bioassays utilizing *M. sexta.*

Owing to the occurence of more than one JH in lepidopteran species and because of the pitfalls and semiquantitative nature of bioassay methods, it has become more desirable to develop new methods to identify and quantify the various JHs. However, there are several instances in which JH bioassays continue to play significant roles:

(a) Testing of JH analogues for potential commercial develoment (see Staal, 1972, 1975; Henrick, 1982; Henrick and Staal, 1988)

(b) Testing biological extracts containig JH active ingredients thought to be of novel structure (i.e., not containing the known JHs), and monitoring of activity during purification of active principle

(c) Testing of biological extracts for JH content when access to more sophisticated (expensive!) analytical equipment is limited or not feasible

Other alternatives to expensive physicochemical methods are discussed in Sections 12.4.3 and 12.4.4.

12.4.2. *Physicochemical Methods*

12.4.2.1. ANALYSIS OF UNDERIVATIZED JHs

The existence of more than one JH mandated the development of methods that would allow both qualitative and quantitative analysis of a mixture of JHs in biological samples. Mass spectrometric methods,

TABLE 12.3.   Gas Chromatographic/Mass Spectrometric Methods for JH Analysis

Mode	Ions monitored	Internal standard	Reference
EI	m/z 114 vs. m/z 117 (IS)	[^{3}H]JH I radioisotope dilution; only JH I monitored	Bieber et al., 1972
EI	Multiple (m/z 114, particularly characteristic)	[^{3}H]JH I	Trautmann et al., 1974a
CI	Multiple	None	Lanzrein et al., 1975
CI	$m/z = 263, 280, 295$	[^{3}H]JH I	Mauchamp et al., 1979b
CI	$m/z = 235, 249, 263, 277$	[^{3}H]Iso-JH II ethyl ester	Mauchamp et al., 1981a
CI	Several	JH I ethyl ester; JH III	Mauchamp et al., 1981b, 1984
CI	$m/z = 235, 249, 263$	[$5 - {}^2$H$_2$]JH III	Camps et al., 1987

usually in conjunction with GC, had already figured prominently in the first identification and structure confirmation of JH I and II (see Section 12.2). Thus, it was not surprising that more general methods were developed based on these techniques, several of which are shown in Table 12.3. However, it is important to point out that the JHs per se are not ideal candidates for gas chromatography because of their reactive oxirane ring. Gas chromatography of JHs results in poor recovery, tailing peaks, and generation of breakdown products unless special precautions are taken (see Dunham et al., 1975). A. S. Meyer et al. (1968) used a specially designed metal-free injector to circumvent thermal metal catalysis. Furthermore, the electron impact mass spectra of underivatized JHs are complex and thus not conducive to high sensitivity or selectivity. Several groups have utilized GC/MS in the chemical ionization mode (see Table 12.3) to improve selectivity.

Bieber et al. (1972) developed a GC/MS radioisotope dilution method (utilizing [^{3}H]JH I as tracer) to quantify JH I in *H. cecropia.* This technique was not utilized for JH analysis in any other species; furthermore, the method had a very poor limit of detection

(LOD ≅ 200 ng) and did not detect JH II in *H. cecropia*. Trautmann et al. (1974a) used a capillary GC method to quantify JH I, II, and III in several insects. They found only JH III in *Melolontha melolontha* and no JH in *Tenebrio molitor* (larvae or young adults), *Vanessa io*, or *Musca domestica*. Subsequently, Trautmann et al. (1974b) detected JH III in mature adult (8- to 16-day-old) *T. molitor* and several species from three other orders (see Table 12.7, in Section 12.5, below). With the exception of *Schistocerca nitens*, the occurrence of JH III was confirmed by GC/MS analysis, by monitoring several ions in the upper mass range. This radioisotope dilution method typically required kilogram levels of insects to achieve a LOD of ~0.1 ng/g. Nevertheless, the results achieved using this method have stood the test of time and have, in several instances, been confirmed by recent physicochemical methods (see Section 12.5).

One result of particular significance was the detection of only JH III in adult *Nauphoeta cinerea* and two other cockroaches (Trautmann et al., 1974b). Additional work using bioassay methods confirmed that JH I or II were not present in these species. However, Lanzrein et al. (1975) reported that hemolymph of *N. cinerea* larvae contained predominantly JH I and II, whereas adults contained mainly JH III. They used CI mass spectrometry (which typically provides fewer fragment ions than EI–MS) for the qualitative identification, but quantification was by *Galleria* wax test. Unfortunately, ~200-ng levels of JH I, II, and III standards were used to "calibrate" the HPLC column prior to purification of biological extracts. Consequently, it is highly possible that this resulted in contamination of the extracts, particularly since a septum-type injector notorious for "memory effects" was used in this study. The occurrence of *only* JH III in all stages of *N. cinerea* was confirmed in a further study (Baker et al., 1984) that utilized the derivatization and GC/MS method of Bergot et al. (1981a) for detection (see Section 12.5).

Madhavan et al. (1981) used a method similar to that by Lanzrein et al. (1975) to measure JH in the primitive apterygote *Thermobia domestica* (adults). After processing ~1.25 kg of insects, Madhavan et al. claimed the occurrence of JH I but no quantification was provided. Later, Baker et al. (1984), using ~5g samples, showed that *T. domestica* nymphs or adults contained ~1ng JH III/g but none of the other known JHs. Insufficient details were provided by Madhavan et al., but it is possible that contamination of sample with JH I standard was responsible for detection of this compound by GC/MS. In a perplexing study, Girard et al. (1976) extracted 60 kg of *Musca domestica* but were unable to detect any of the known JHs by GC/MS. However, they "estimated" that JH I was present at 0.01 ng/g of animals. Mauchamp et al. (1979b) were unable to obtain reliable results using EI selective ion monitoring (SIM) at $m/z = 114$ (used earlier by Bieber et al., 1972) to identify JHs in *Pieris brassicae* hemolymph. Instead, Mauchamp et al. (1979b) used capillary GC, and CI mass spectrometry with methane, isobutane, or ammonia. The latter method gave the best results—mainly high mass fragment ions, including $[MH]^+$ and $[MH-15]^+$. They utilized $[^3H]JH$ I as internal standard and reported the presence of JH I in *P. brassicae* hemolymph. The minimum detectable quantity (MDQ) was ~ 10 and ~ 100 pg for JH standards and biological samples, respectively. Again, this method suffered from the drawback of using relatively high levels of JH ($\cong 100$ ng) to calibrate the HPLC column used to purify biological samples. Later, Mauchamp et al. (1981b, 1984) used JH I ethyl ester or JH III as the internal standard in an otherwise similar method, again aimed only at JH titer determination in *P. brassicae*. A novel alternative method employing HPLC/MS was described by Mauchamp et al. (1981a). They used a micro HPLC column (flow rate of 50 $\mu l/min$) and a splitter device. The method detected 10-pg JH standards, and JH I was again reported in *P. brassicae*. However, this method was not used subsequently.

12.4.2.2. METHODS EMPLOYING DERIVATIZATION AND GC WITH ELECTRON CAPTURE DETECTION

In parallel with early methods employing GC/MS for detection of underivatized JHs, attention turned to the conversion of JHs to derivatives that were stable to GC analysis and more amenable to detection (summarized in Table 12.4). The conversion of JHs to electrophoretic organohalide derivatives provided a means to greatly enhance their sensitivity. In combination with GC and electron capture detection (ECD; see Schooley, 1977, and Zlatkis and Poole, 1981, for principles of the ECD technique), this appeared to be an ideal, relatively inexpensive approach to detection of JHs at picogram levels. This technique was used by Judy et al. (1973b) in an attempt to demonstrate the presence of the newly discovered JH III in larval *M. sexta* hemolymph. Extracts from the latter were hydrolyzed ($HClO_4$) in order to convert JHs to their corresponding 10,11-diols (Fig. 12.6); these in turn were reacted with trifluoroacetic anhydride to form the 10,11-bistrifluoroacetates, which were purified and analyzed by GC/ECD. Results confirmed the presence of JH III but did not give conclusive results for JH I and II [*M. sexta* larvae were later found to contain

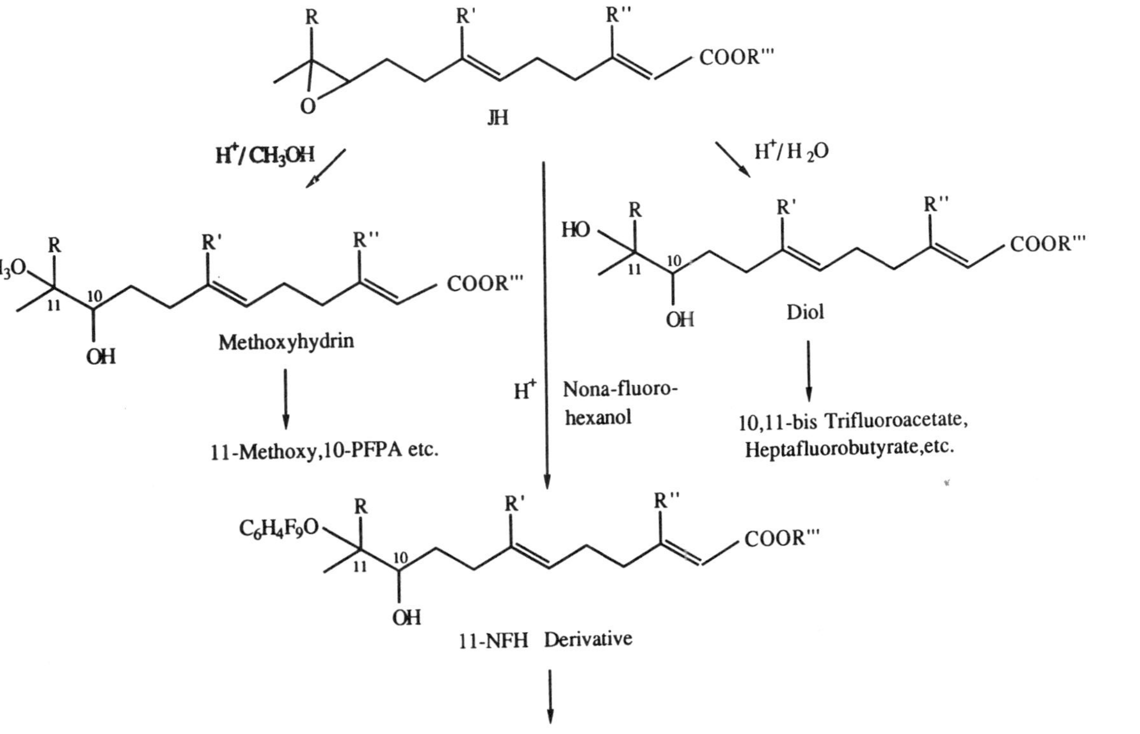

FIGURE 12.6. Scheme showing conversion of JHs to alkoxyhydrins or diols, and subsequent derivatives detected by GC/ECD or GC/MS with SIM.

TABLE 12.4.   Gas Chromatography with Electron Capture Detection for JH Analysis

JH derivative	Internal standard	References
10,11-Bistrifluoroacetate	None	Judy et al., 1973b
10,11-Bisheptaflurobutyrate	None	Ajami, 1974; van Broekhoven et al., 1975
11-Methoxy-10-pentafluoro-phenoyxyacetate	n-Propyl ester analogue of [^3H]JH I	Schooley et al., 1976; Bergot et al., 1976
11-Methoxy-10-(2,4-dichloro-benzoate)	[^3H]2E,6E,10-*trans* JH 0	Peter et al., 1976a
11-Methoxy-10-heptafluoro-butyrate	[^3H]*JH* I (isotope dilution)	Hagenguth and Rembold, 1978, 1979
11-Methyoxy-10-heptafluoro-butyrate capillary GC	10-*trans*-JH II and the ethyl ester of the 2Z isomer of JH III	Huibregtse-Minderhoud et al., 1979, 1980a,b

mainly JH I and II and little JH III (D. A. Schooley et al., unpublished results; Baker et al., 1987), underlying the lack of specificity of this GC/ECD method]. The aforementioned authors pointed out that opening of the JH oxirane ring to form the 10,11-diol is an erratic reaction at microgram or submicrogram levels, perhaps due to intramolecular cyclization of JH; furthermore, the bistrifluoroacetate derivatives were hydrolytically unstable. Using an improved procedure, the same group (Bergot et al., 1976; Schooley et al., 1976) first converted JHs to their 11-methoxy-10-hydroxy (methoxyhydrin) derivatives, which were further reacted to give the electrophoretic 10-pentafluorophenoxyacetates (Fig. 12.6). This method also incorporated a radiolabeled internal standard (n-propyl ester of [^{3}H]JH I). The sensitivity of the GC/ECD technique is very high (~ 1 pg with a signal-to-noise ratio of about 7:1), but specificity is poor and a large number of interfering products (especially esters) are a severe drawback to the technique. Nevertheless, using 5- to 20-g whole-body or 1- to 10-ml hemolymph samples, Schooley et al. (1976) confirmed the presence of JH II and III in adult female *M. sexta* (see also Judy et al., 1973b; Peter et al., 1976a). Low titers of JH I, II, and III were reported for *Musca domestica* and *Sarcophaga bullata,* but these could not be confirmed later using a GC/MS method (F. C. Baker et al., unpublished results).

Meanwhile, Ajami (1974) utilized GC/ECD of the bistrifluoroacetate derivative of JH I diol in an attempt to monitor specific activity of radiolabeled JH I formed *in vivo* from radiolabeled precursors. In 1975, van Broekhoven et al. utilized the bisheptafluorobutyrate derivative of JH III diol (Fig. 12.6) for GC/ECD quantification of JH III in hemolymph from *Leptinotarsa decemlineata.* These authors noted that the detection response was not linear over a wide range of concentrations but was acceptable between 0 and 100 pg. However, in practice the limit of detection with biological samples was only ~ 1 ng/g (1 ng/ml) owing to the high degree of impurity of samples.

Peter et al. (1976a,b) detected JHs (as their 11-methoxy-10-[2,4-dichlorobenzoates] via GC/ECD) in *M. sexta* hemolymph. They claimed an impressive limit of detection of $\leqslant 0.003$ ng/ml and quantified JH I, II, and III in fourth- and fifth-stadium larvae. They also demonstrated the absence of all JHs in wandering fifth-stadium larvae, and detected predominantly JH II and III in adult females but predominantly I and II in adult males. However, the purification scheme prior to GC/ECD required four TLC steps and one HPLC step, and thus was too unwieldy for routine use. The method was notable in that a novel radiolabeled JH analogue was used, [^{3}H-*methyl*](2*E*,6*E*,10-*trans*)-JH 0 (4 years prior to the actual discovery of (2*E*,6*E*,10-*cis*) JH 0 in *M. sexta* eggs!)

Hagenguth and Rembold (1978) converted JH methoxyhydrin derivatives to the 10-heptafluorobutyrate esters for GC/ECD detection and quantification of JH III in all developmental stages of the honey bee. [10-^{3}H]JH I was used as internal standard; JH III was the major, if not only, JH detected, and the levels of JH III detected were consonant with estimates of JH via bioassay. No details were given as regards precision and accuracy of the method. Essentially, the same method was described in a later publication by the same authors (Hagenguth and Rembold, 1979); however, the degree of contamination from miscellaneous sources, including solvents, greatly compromises the technique.

Huibregste-Minderhoud et al. (1979, 1980a,b) also detected JHs as their 11-methoxy-10-heptafluorobutyrate esters but used alternative internal standards (see Table 12.4). Furthermore, they substituted a capillary column for a packed GC column. Using this technique, they detected only JH III in *Locusta migratoria* (see Section 12.4.4) and essentially only JH III in *Leptinotarsa decemlineata*. While 2.5 pg of the JH methoxyhydrin heptafluorobutyrate derivatives could be detected with a signal-to-noise ratio of 10:1, the limit of detection of JH I, II, and III in biological samples was only 0.5, 0.3, and 0.2 ng, respectively.

12.4.2.3. METHODS EMPLOYING DERIVATIZATION
AND GC/MS DETECTION

As an alternative to GC/ECD methods, which may suffer from insufficient selectivity, Rembold et al. (1980) and Bergot et al. (1980) published new GC/MS methods for quantification of JHs in biological samples (see Table 12.5). These rely on the use of the mass spectrometer in the selected ion monitoring mode (SIM) as a sensitive, selective, yet versatile detector. Rembold et al. (1980) reacted JHs with nonafluorohexanol/0.03% HClO$_4$ to form the 11-nonafluorohexoxy-10-hydroxy (NFH) derivatives (Fig. 12.6), which produce a few major fragments of high mass (particularly $m/z = 305$ and 319 for JH III and I, respectively; see Fig. 12.7). They used the ethyl ester analogue of JH III as the internal standard. The method could detect only 2 pmol of JH III–NFH derivative with SIM at $m/z = 305$; glass capillary GC improved specificity. They reported the occurrence of only JH III in worker and queen honeybees when this procedure was used. Curiously, this included the detection of −0.5-ng JH III/g in worker pupae

TABLE 12.5. GC/MS Methods for Analysis of JH Derivatives

JH Derivative	Mode	Ions monitored	Internal standard	Reference
11-Nonafluorohexoxy-10-OH	EI	$m/z = 305, 319$	Ethyl ester analogue of JH III	Rembold et al., 1980
11-Deuteromethoxy-10-OH	EI	$m/z = 76, 90$	[^{3}H]Iso-JH II ethyl ester	Bergot et al., 1980, 1981a,c
Several, including 11-nonafluoroctoxy-10-heptafluorobutyrate	EI	$m/z = 333, 347$	Ethyl ester analogues of JH I and III	Rembold, 1981
11-Methoxy-10-nonafluorohexyl-dimethylsilyl	EI	$m/z = 376, 390$	Ethyl ester analogues of JH I and III	Bührlen et al., 1984; Rembold and Lackner, 1985

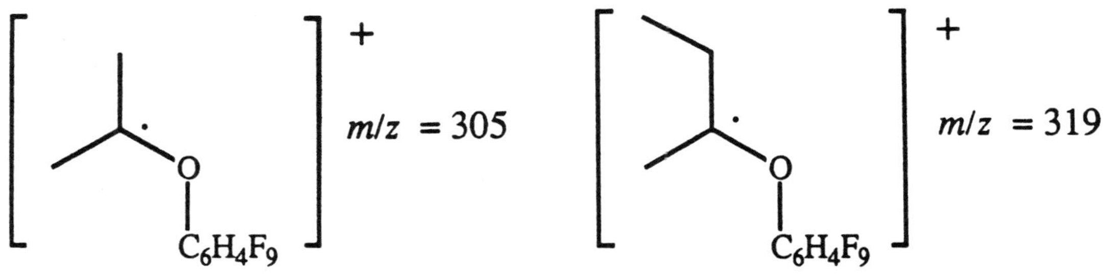

FIGURE 12.7. Fragments with $m/z = 305$ and 319 from JH III and JH I 11- nonafluorohexoxy-10-hydroxy (NFH) derivatives, respectively.

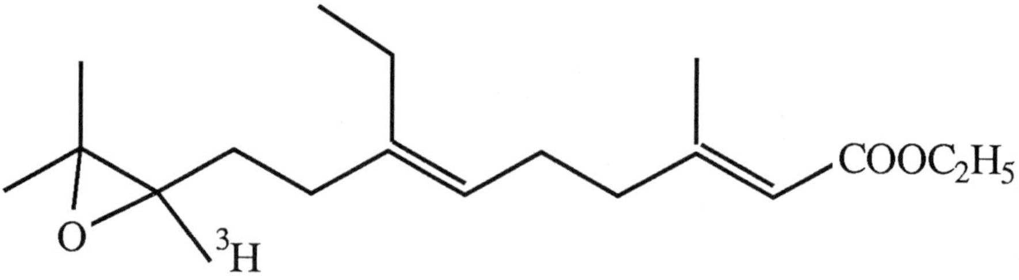

FIGURE 12.8. [10-³H]Iso JH II ethyl ester used as internal standard by Bergot et al. (1981a).

(day 1). One minor disadvantage of the technique is that an approximately 20% attack of the NFH can occur at the C-10 rather than C-11 position of JH III.

Bergot et al. (1980) discovered a new JH [JH 0; methyl (2*E*,6*E*,10-*cis*)-10,11-epoxy-3,7-diethyl-11-methyl-2,6-tridecadienoate] in *M. sexta* eggs. This discovery was highly dependent upon their recently developed GC/MS with SIM method, whose details were published the following year (Bergot et al., 1981a). Important features of this method include the use of [10-³H]iso-JH II ethyl ester (Fig. 12.8) of known specific activity as internal standard (IS), and conversion of the JHs and IS to the corresponding 11-d_3-methoxy-10-hydroxy derivatives. The latter derivatives of all of the known JHs give an $m/z=76$ (JH III, IS) or $m/z=90$ (JH 0, I, II, and the as yet undiscovered 4Me–JH I; see Bergot et al., 1981b) as the major fragment ion in the electron impact (EI) mode. These ions carry between 32% and 47% of the total ion current, and this affords a sensitive method of detection. Methanol is less desirable than deuteromethanol for the derivatization step because the derivatives give rise to fragments with $m/z=73$ or 87 instead of $m/z=76$ or 90. The $m/z=73$ ion (trimethylsilyl ion) commonly occurs as a constituent of "column bleed" from GC stationary phase containing silicone. Bergot et al. (1981a) went to great pains to monitor the accuracy and precision of the method, as well as JH recovery. Limits of detection were typically ~0.01 ng/g using 2-m packed columns and 5–10 g of biological tissue.

Rembold (1981) utilized alternative derivatives for detection of JHs. These included the 11-nonafluoroctoxy-10-OH or 11-trideca-fluorononoxy-10-OH derivatives, which gave fragments with higher mass than those of their earlier method (Rembold et al., 1980). For better resolution of derivatives by capillary GC/MS, Rembold (1981) suggested esterification of the 10-OH function by reacting with hepta-fluorobutyryl iodide. Samples of *Locusta migratoria* were processed for detection of JHs as their 11-nonafluoroctoxy-10-heptafluorobutyrate derivatives and found to contain only JH III (see Section 12.4.4). In 1984, Bührlen et al. introduced yet another variation: they detected JHs as their 11-methoxy-10-nonafluorohexyldimethylsilyl derivatives. Further details of this procedure were provided in a later publication (Rembold and Lackner, 1985), along with an extensive purification scheme for JHs and their derivatives. Their identification of JHs in miscellaneous insect species confirmed the presence of only JH III in nonlepidopteran species (see Section 12.5, below).

A prime advantage of the GC/MS with SIM technique is the additional confirmatory evidence for a particular JH structure that can be accumulated during analysis. For example, other important ions may be monitored and the ratios of the ions compared to those from analysis of the JH standard under similar conditions. Thus, Baker et al. (1984) monitored the relative ratios of ions with $m/z=76$, 193, and 225 to confirm that the JH (detected as deuteromethoxyhydrin) of *Nauphoeta cinerea* was JH III (Fig. 12.9). A similar method was used to

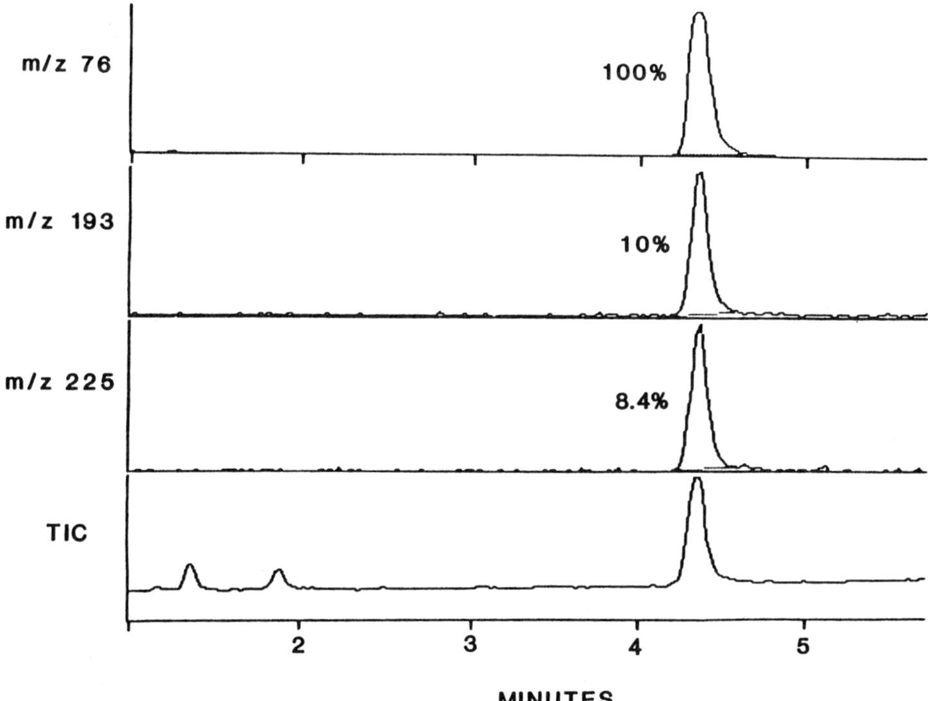

FIGURE 12.9. Selected ion chromatogram of the 11-methoxy-d_3-10-hydroxy derivative formed from JH III in day 34 *Nauphoeta cinerea* embryos (from Baker et al., 1984). TIC − total ion chromatogram.

confirm JH III in the aphid *Megoura viciae* (Hardie et al., 1985). If the JH derivative is present at a sufficiently high level, its full mass spectrum can be observed. As pointed out earlier, analysis of samples on more than one column (with different properties) is also very important in order to rule out false positive identifications. This technique proved useful during analysis of *Drosophila melanogaster* samples, which contained an impurity that co-eluted with JH I (Sliter et al., 1987).

The use of the propyl ester analogue of JH I as IS, in conjunction with methanol extraction and storage in methanol, can result in trans-esterification to the methyl ester and thus apparent contamination of the samples with JH I. This reaction has been observed even in samples stored in the freezer and led Bergot et al. to modify their earlier (1976) extraction procedure (and to substitute CH_3CN for CH_3OH). In addition, Bergot et al. (1981a) synthesized an alternative IS, which possessed a JH carbon skeleton different to any of the known JHs (i.e., iso-JH II ethyl ester) and which could not undergo trans-esterification to a known JH. Mauchamp et al. (1984) reviewed the status of GC/MS methods for detection of JHs and included some interesting comments regarding use of an IS and sensitivity of detection. Most recently Camps et al. (1987) reported the synthesis of racemic $5[-^2H_2]$JH III and its use as an internal standard during quantification of JH III in *Blattella germanica*.

12.4.2.4. METHYL FARNESOATE AND OTHER JH-RELATED COMPOUNDS

Methyl farnesoate (MF), the precursor of JH III in nonlepidopterans (Pratt and Tobe, 1974), is present in substantial quantities in embryos of *Nauphoeta cinerea* (Lanzrein et al., 1984). Recently, it was detected (by a physicochemical method employing GC/MS) in the hemolymph of the crustacean *Libinia emarginata* (Laufer et al., 1987). Later (Borst et al., 1987), MF was found in several other crustaceans; an IS (ethyl $[10-^3H]$farnesoate) was used to improve the precision of the quantification. A high level of methyl $(2E,6E)$-farnesoate has been reported from the plant *Cyperus iria* and several other *Cyperus* species (see Toong et al., 1988, and references therein). Further details of the method for quantifying methyl farnesoate and several closely related compounds (JH acids, farnesol, JH III 10,11-diol) are described by Baker et al. (1988). Farnesol is a fairly common constituent of biological samples, while JH III 10,11-diol has been detected as a secretion product by *Locusta migratoria* CA (Gadot et al., 1987).

12.4.3.   In Vitro *JH Biosynthesis*

In 1970, Röller and Dahm demonstrated JH I biosynthesis by brain–CC–CA complexes from *H. cecropia* male pupae incubated for a long term in Grace's medium containing sterile heat-treated hemolymph from pharate adult *H. cecropia.* In retrospect, this was a somewhat peculiar result since later Peter et al. (1981) showed that CA from adult male *H. cecropia* secrete JH acids that are finally methylated in the accessory glands where they are stored. Subsequently, it was discovered that the *S*-methyl moiety of methionine is the *in vivo* source of the methyl ester group of the JHs (Metzler et al., 1971). This led to general use of *S*-methyl radiolabeled methionine to monitor JH biosynthesis *in vitro.* Judy et al. (1973b) maintained CC–CA from adult female *M. sexta* in culture medium containing [*methyl*-^{14}C] methionine for many days, and isolated JH II and III from extracts. Similar studies with CA from the grasshopper *Schistocerca vaga* (Judy et al., 1973a) and *Tenebrio molitor* (Judy et al., 1975) resulted in isolation of only JH III, in keeping with those JHs observed *in vivo* (Trautmann et al., 1974b). Jennings et al. (1975) detected both JH I and II produced *in vitro* by CC–CA from the lepidopteran *Heliothis virescens.*

Pratt and Tobe (1974) and Tobe and Pratt (1974) streamlined the *in vitro* method substantially. Utilizing (2*E*,6*E*)-[^{3}H]farnesoic acid or (2*E*,6*E*,10-*cis*)-dihomofarnesoic acid, and [*methyl*-^{14}C]methionine, they were able to show efficient formation of dual-labeled JH III or I by CA of *Schistocerca gregaria* during 3-h incubations in a specially prepared culture medium. Afterward (Tobe and Pratt, 1975; Pratt et al., 1975), [*methyl*-^{14}C]methionine was utilized as a marker to monitor spontaneous rates of JH biosynthesis in CA from *Schistocerca gregaria* and *Periplaneta americana,* respectively, a method which has continued in popularity and is commonly referred to as the radiochemical assay (RCA) for *in vitro* JH biosynthesis.

In these and many related studies, the products were isolated by solvent extraction of the incubation medium, subjected to TLC, and quantified by liquid scintillation counting. Later, the method for quantification was also streamlined by simply extracting incubation medium with isooctane and subjecting the latter to liquid scintillation counting (Feyereisen, 1985a,b). This method has the drawback that minuscule levels of aqueous phase (containing relatively high levels of radiolabeled methionine) may contaminate the solvent extract and might lead to erroneously high values. It is perhaps advisable to filter isooctane extracts through mini silica gel columns and elute them with ether/pentane to avoid this problem of methionine carryover (F. C. Baker, unpublished results). For those species that secrete several JHs (e.g., *M. sexta* larvae or adults), it is desirable to separate the individual products via HPLC. Hamnett and Pratt (1978) used a more sophisticated method (automated capillary column radio gas chromatography) to delineate *in vitro* products from *Periplaneta americana* CA. Other variations employed for analyzing products from *in vitro* JH biosynthesis include GC/MS (Mauchamp et al., 1985; Trabalon et al., 1987) and RIA procedures (e.g., Granger et al., 1979; Nariki et al., 1986). However, RIA methods utilizing "specific antibodies" to quantify a mixture of JHs (unresolved by HPLC) should be treated with caution.

In general, it can be said that the identity of the JHs produced *in vitro* mirror those found *in vivo* [cf. Table 12.6 (here) and Table 12.7 (in Section 12.5)]. CC–CA from nonlepidopteran species produce only JH III *in vitro* unless certain homosesquiterpenoid precursors are added to the medium (see Pratt and Tobe, 1974) or under extremely forcing conditions (Feyereisen and Farnsworth, 1988). Lepidopteran CC–CA, which normally secrete more than one JH, do one strictly produce these JHs in the same proportions as they are found *in vivo* (cf. Baker et al., 1987, and Granger et al., 1982a, for larvae; Schooley et al., 1984, and Kramer and Law, 1980 for adults).

Observed differences in JH biosynthetic rates by CA isolated from animals of known age have been compared with *in vivo* JH titers from animals of corresponding age. It appears that at least in some instances there is a good correlation between relative *in vitro* biosynthetic rates and corresponding *in vivo* titers (e.g., Tobe et al., 1985; Couillaud et al., 1985) and thus activity of CA *in vitro* probably reflects activity *in vivo*. However, many artifacts can affect the accuracy of *in vitro* results, e.g., asymmetry in rates of hormone biosynthesis between right and left glands (Tobe, 1977); biosynthesis of MF could be quantified incorrectly as JH. Recently, Gadot et al. (1987) described the biosynthesis of JH III 10,11-diol by CA from *Locusta migratoria* and another group (Richard et al., 1989) have demonstrated the biosynthesis of a new JH (JH B_3:methyl 6,7;10,11-bisepoxy-3,7,11-trimethyl-(2E)-dodecenoate) by ring glands from *Drosophila melanogaster*. Although there is no substitute for *in vivo* JH titer determinations, the *in vitro* JH biosynthesis technique can give useful qualitative and quantitative information and can be used to study complicated hormonal interactions (Granger et al., 1987). The *in vitro* technique has been indispensable for studying the mode of JH biosynthesis (see Schooley and Baker, 1985, for a review; also Brindle et al., 1987, 1988) and regulation of JH biosynthesis (Feyereisen, 1985).

The CA from certain adult male moths (e.g., *H. cecropia* or *M. sexta*) and prepupal stages of *M. sexta*, and perhaps other lepidopterans (see Bhaskaran, 1988), secrete JH acids. Thus, the *in vitro RCA* technique

TABLE 12.6. Identification of JHs (and MP) Produced by CA *In Vitro*

Order	Insect	JH	Reference
Lepidoptera	*Hyalophora cecropia*	I	*Röller* and Dahm, 1970[a]
		I,II	Dahm et al., 1976
	Manduca sexta	II,III	Judy et al., 1973b; Dahm et al., 1976
		I,II,III	Kramer and Law, 1980
	Heliothis virescens	I,II	Jennings et al., 1975
	Galleria mellonella	II	Dahm et al., 1976
	Bombyx mori	I,II	Kmura and Sakurai, 1984[b]
	Samia cynthia	I,II,III	Brindle et al., 1988
Orthoptera	*Schistoceca vaga*	III	Judy et al., 1973a
	Schistocerca gregaria	III	Pratt et al., 1975
	Gastrimargus africanus	III	Judy et al., 1976
	Locusta migratoria	III	Mauchamp et al., 1985[c]
	Melanoplus sanguinipes	III	McCaffery and McCaffery, 1983
	Acheta domesticus	III	Strambi et al., 1984b[b]
	Gryllus bimaculatus	III	Koch and Hoffmann, 1985
	Teteogryllus commodus	III	Ruegg et al., 1986
Dictyoptera	*Periplaneta americana*	III	Müller et al., 1975; Pratt et al., 1975
	Periplaneta fuliginosa	III	Peter et al., 1976a
	Diploptera punctata	III	Tobe and Stay, 1977
	Leucophaea madera	III	Koeppe et al., 1980
	Nauphoeta cinerea	III,MF	Lanzrein et al., 1980
	Blaberus discoidalis	III	Dahm et al., 1976
Coleoptra	*Tenebrio molitor*	III	Judy et al., 1975
	Leptinotarsa decemlineata	III	Schooneveld et al., 1979
Hemiptera	*Dysdercus fasciatus*	MF	Feldlaufer et al., 1982
		III	Bowers et al., 1983
	Nezara viridula	III	Bowers et al., 1983
	Oncopeltus fasciatus	III	Bowers et al., 1983
Diptera	*Calliphora vomitoria*	III	Trabalon et al., 1987[c]
	Culex pipiens	III	Readio et al., 1988
Isoptera	*Macrotermes subhyalinus*	III	Lanzrein et al., 1978
	Zootermopsis angusticollis	III	Greenberg and Tobe, 1974

[a] Did not use radiolabeled methionine.

[b] RIA.

[c] GC/MS.

NOTE: All other studies used radiolabeled methionine (RCA).

from these animals. Nevertheless, alternative methods may be applied [including RIA detection of JH acids, GC/MS analysis of methylated products, or use of radiolabeled propionate as a precursor of JH I/II acids (Bhaskaran et al., 1988)].

In crustaceans, the mandibular organ (MO) is thought to perform an equivalent function to the CA of insects. The *in vitro* radiochemical assay has been applied to study of JH biosynthesis by MOs from several crustaceans including *Libinia emarginata* (Laufer et al., 1987), *Homarus americanus* (Borst et al., 1987), and *Scylla serrata* (Tobe et al., 1989). From these studies, it is clear that MF or related compounds (including farnesoic acid) are produced by the MO *in vitro*, but insect JHs are minor products and perhaps artifacts of the isolation procedure (see Laufer et al., 1987). CA from embryos of the cockroach *Nauphoeta cinerea* produce very high levels of MF (Lanzrein et al., 1984), and CA from other insect species also produce detectable levels of MF as well as JH III (Pratt et al., 1984; Couillaud et al., 1988).

12.4.4. *Radioimmunoassay*

The RIA technique appeared to be a rapid, cheap, and very sensitive alternative to physicochemical methods for quantification and identification of JHs both *in vivo* and *in vitro*. In practice, however, the technique has been less than satisfactory, and in many instances the results obtained by this technique could not be confirmed by more rigorous physicochemical methods. Unlike ecdysteroids, which have been quantified routinely from insect tissue via RIA, the JHs typically occur at much lower levels. Because of their lipophilic nature, they are much more difficult than ecdysteroids to purify from other lipoids and only poorly soluble in aqueous solution, factors which affect the performance of the RIA. Cross-reactivity between the various JHs of nominally JH-specific antibody can lead to erroneous results. Furthermore, the cross-reactivity can vary depending on the radiolabeled ligand used in the binding studies. This cross-reactivity for various JHs may be unimportant in those (the majority) of species that contain only JH III. Often JH antibodies will cross-react with JH acid or related metabolites/precursors of JH. This can be disadvantageous for identification purposes but may provide useful new information that might be missed using alternative (e.g., physicochemical) methods. Provided that the various drawbacks of the RIA technique are considered carefully, the method can be utilized to advantage for some studies, particularly unpurified *in vitro* products containing only JH III, or even detection of carefully purified extracts of JH III from whole animals or hemolymph. Baehr et al. (1987) recently developed an enzyme-linked immunoassay as an alternative to RIA for JH and JH diol. The sensitivity of the method is only 0.2 ng JH diol/ml. So far, the method has not been applied to analysis of biological samples.

12.4.4.1. PREPARATION OF ANTIBODIES

Since JH itself is not immunogenic, methods for producing antisera have relied on derivatization of JH and subsequent conjugation to protein (especially human serum albumin, HSA; see Fig. 12.10 and the review of Granger and Goodman, 1983). For a general treatise on RIA methodology, the reader is referred to Chard (1987). Lauer et al. (1974) were the first to describe a method for producing antibodies to insect JH III (Fig. 12.10). However, the antibodies produced exhibited low binding affinities and were not suitable for quantification of JH in biological samples. Furthermore, the antibodies produced were not specific for JH III but cross-reacted with JH II and I as well as various derivatives formed from modification of the C-1 terminus of JH III. The lack of specificity for the C-11 substituent is probably related to instability of the immunogen C-10,11 epoxy group and its modification to 10,11-diol after injection into rabbits. Consequently, a mixed poulation of antibodies may have been produced to both JH and JH diol. Baehr et al. (1976) produced antibodies to JH I by a procedure similar to that of Lauer et al. (1974), but they reduced the level of antigen injected and the frequency of booster injections, and consequently obtained antibodies with considerably higher binding affinity. Later, antibodies were produced to JH II and III (Baehr et al., 1979).

Strambi et al. (1981) developed an alternative method to produce JH antibodies. They opted for conversion of JH I to the corresponding 10,11-diol and its subsequent conjugation to HSA via the C-10 hydroxyl moiety (Fig. 12.10). The resulting antibodies are specific for the JH ester group but cannot differentiate between JH I, II, and III; therefore, an initial HPLC purification of extract is necessary to separate JHs if more than one JH is present. Strambi et al. used smaller amounts of antigen than did Baehr et al. (1976) and more frequent booster injections, and obtained antiserum of high binding affinity. This technique measures JHs in biological samples after first converting them to the corresponding diols. JH diols are more water soluble than JH and can be more readily separated from interfering lipids, which may cross-react with antibodies. In addition, the JH diols appear to be more immunoreactive than JHs. However, JH diols are natural metabolites of JHs, and therefore JH quantification by this method could give overestimates of JH content. A major drawback to this and other RIA procedures is the lack of a suitable internal standard enabling one to estimate recovery of JHs following the several purification steps. Strambi et al. (1981) claimed 40% recovery of

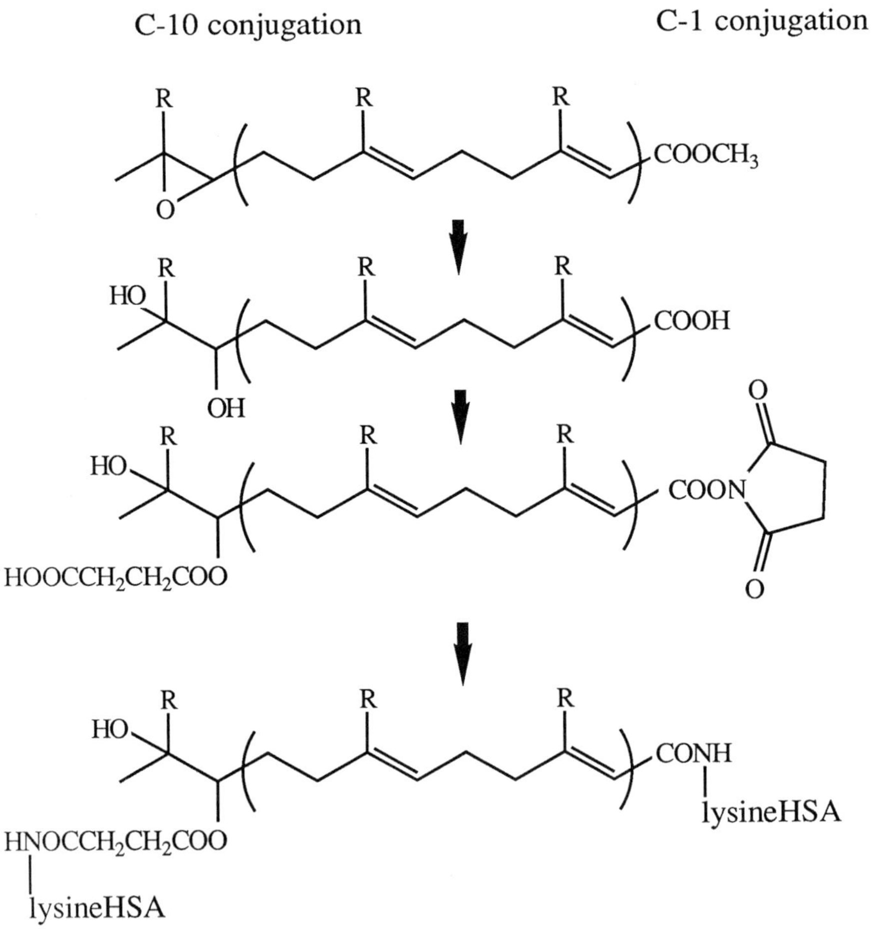

FIGURE 12.10. Procedure for conjugation of JH via C-1 or C-10 for antibody preparation (Strambi et al., 1981; Lauer et al., 1974; Baehr et al., 1976; Granger et al., 1982a,b).

radiolabeled JH I processed through their methodology, and later this value was revised downward. This value was factored into their quantification of JHs. However, it is unlikely that recovery is consistent for all samples, particularly if spillage occurs.

12.4.4.2. RADIOLIGANDS

Commercially available [^{3}H]JH of high specific activity has been utilized extensively in the JH RIA procedure. However, to further enhance the sensitivity of the technique, various radiolabeled JH analogues of even higher specific activity have been prepared (Fig. 12.11). Methodology similar to that used in derivatization and conjugation of JHs for corresponding antigen preparation was used to prepare radiolabeled ligands. As discussed by Granger and Goodman (1983), the use of a particular radioligand will affect the cross-reactivity observed with a given antiserum by selecting for a specific antibody subpopulation. This must be considered when one is conducting cross-reactivity studies.

12.4.4.3. DETECTION OF JH *IN VIVO* AND *IN VITRO* VIA RIA

Baehr et al. (1979, and references therein) reported the occurrence of JH III only in the dermapteran *Labidura riparia*, and considerable levels of JH II and/or JH I plus III in hemolymph from nymphs and/or larvae of several insect species, including *Locusta migratoria, Schistocerca gregaria, Rhodnius prolixus,* and *Apis mellifera*. The same group reported the occurrence of JH I, II, and III in larvae or pupae of *Leptinotarsa decemlineata* (Deleurance et al., 1979, 1980) and JH I and III in *Thermobia domestica* (Bitsch et al., 1985). However, the occurrence of JH I and/or II in each of these species has been refuted by other groups using physicochemical detection methods (see Bergot et al., 1981c; de Kort et al., 1982; also Sections 12.4.3, and 12.5). At a roundtable discussion during the Second International CNRS Symposium, Strasbourg, 1983, a group of scientists active in the area of JH quantification via RIA or physicochemical methods resolved to compare the reliability of their various methods. A "neutral" party dispensed samples containing a mixture of JH standards, hemolymph extracts, and products from *in vitro* JH biosynthesis, and these were distributed to seven groups (three utilizing RIA techniques and four using physicochemical methods). Unfortunately, owing to many

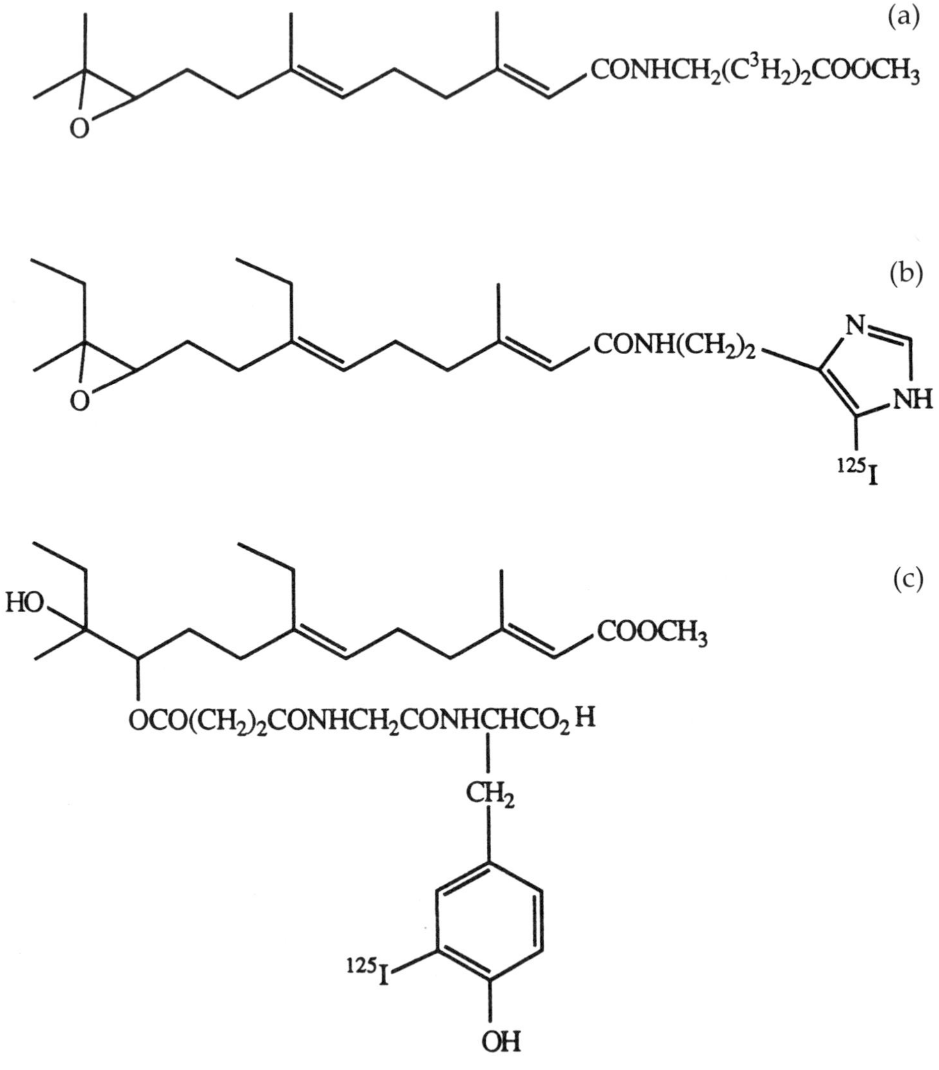

FIGURE 12.11. High-specific-activity radiolabeled ligands used in JH RIA: (a) Lauer et al. (1974); (b) Baehr et al. (1976); (c) Strambi et al. (1981).

difficulties (including postal strikes and disagreement concerning the appropriate levels of JH to be dispensed for analysis!), no concrete conclusions were drawn from this study and two groups failed to supply results. However, the qualitative and quantitative results obtained for mixtures of JH standards were reasonably close to theoretical values for most of the groups, and all groups reported similar levels of JH III in the *in vitro* biosynthetic sample. Certainly, none of the values differed as widely as might have been expected based on the comparison of *in vivo* JH titers measured by RIA vs. physicochemical methods, as described on page 427.

Major pitfalls in the RIA technique include lack of specificity of the JH antibody and cross-reactivity of JH antibody with other lipid contaminants even after partial purification of JHs. Unfortunately, RIA methods are less flexible than physicochemical methods as regards discerning contaminants. Baehr et al. (1981) suggest that the discrepancies in results obtained by their RIA procedure vs. those from other groups using physicochemical methods arc related to pulse-like secretion of JHs and use of relatively small hemolymph samples (e.g., from individual animals) and different strains of animals. This does little to explain the vastly different qualitative JH profiles obtained by the two methods. In defending their RIA procedure, Baehr et al. (1981) suggest that the validity of the RIA is proven by correlation of results with physiological events. It is certainly true that lack of specificity of the RIA procedure can lead to interesting discoveries (e.g., JH acids) that do correlate to physiological events and would be overlooked by specific physicochemical or biochemical methods. However, it is important that RIA procedures be validated by physicochemical procedures, particularly as regards qualitative JH determination and also to avoid misidentification of JH metabolites or precursors as JH. To date, RIA techniques appear unsuited for qualitative JH determination unless initial HPLC purification of the individual JHs is performed. However, as for physicochemical and bioassay methods, it is important that the JHs used to calibrate the HPLC column do not subsequently contaminate the biological samples under study, leading to false positive identifications.

The RIA procedure of Strambi et al. (1981) has also been utilized to measure JH titers in several species. Thus, Gharib et al. (1983) and Gharib and de Reggi (1983) detected JH I, II, and III, and measured their modulation during egg development in *Bombyx mori*. The levels of all three hormones were surprisingly high (>1 μg/g, on a molar basis similar to corresponding ecdysteroid titers). Plantevin et al. (1984) measured JH titers in several stages of *Galleria mellonella*. They detected JH I, II, and III not only in larvae but surprisingly also in pupae. No JH was detected in pupae of another lepidopteran, *M. sexta*, when a physicochemical method of analysis was used (D. A. Schooley et al., unpublished results). The procedure of Strambi et al. (1981) was also used by Guilvard et al. (1984) to measure JH fluctuation in the mosquitoes *Aedes detritus* and *A. caspius.* JH III was found to be the predominant JH in agreement with studies on other *Aedes* species by groups using physicochemical methods (see Section 12.5). However, the levels of JH III determined by RIA were very much higher, and Guilvard et al. (1984) mentioned that JH I and II accounted for less than 5% of total JH, suggesting that they were present. Physicochemical methods did not detect *any* JH I or II in *Aedes aegypti* (Baker et al., 1983; Shapiro et al., 1986). A thorough study by de Kort et al. (1985) showed that the RIA procedure of Strambi et al. (1981) gave comparable results to the physicochemical method of Bergot et al. (1981a) when applied to analysis of *Leptinotarsa decemlineata* hemolymph. However, the limit of detection of the RIA method was disappointingly low (1.9 ± 1.0 ng/ml) and, considering the simpler end-point analysis, relatively few (36) samples per person per week could be processed by this method. Strambi et al. (1984a) also determined JH and ecdysteroid titers simultaneously in hemolymph of individual animals of *Bombus hypnorum* and *B. terrestris.* They found that JH III was the only detectable JH in males and females of both species, with levels as high as 80 ng/ml.

Several groups have utilized RIA procedures in preference to the simpler radiochemical assay (see Section 12.4.3, above) to quantify JHs produced by corpora allata *in vitro.* The relative purity of *in vitro* products lends itself to RIA detection, although moderate levels of protein are necessary to effect precipitation of antibody—radioligand complexes. Paradoxically, the "advantage" of the RIA technique in this instance is its lack of specificity, since, in addition to JHs, it is able to quantify JH acids (and related precursors/metabolites depending upon antibody) unlike the radiochemical assay. Granger's group has utilized this technique to study JH biosynthesis by *M. sexta* CC–CA (see Granger et al., 1979, 1982a,b) and interendocrine control by 20-hydroxyecdysone (Watson et al., 1986). Their results indicated that not only were rates of JH biosynthesis higher in CC–CA from wandering and prepupal *M. sexta* than day-0 fifth-stadium animals but considerable levels of JH III (or JH III acid) were produced in fifth-stadium animals. Physicochemical methods (D. A. Schooley et al., unpublished results; Peter et al., 1976a,b; Baker et al., 1987) have shown that JH I and II are the predominant JHs in larval *M. sexta* and that JH III is minimal. Similarly, *in vitro* JH biosynthesis using the RCA detected predominantly JH I and II in CC–CA from early fifth-stadium larvae (Kramer and Kalish, 1984) but little or no JHs produced by CC–CA from prepupal *M. sexta* (S. J. Kramer and F. C. Baker, unpublished results). Taking into consideration the JH activity of JH I acid described by Bhaskaran et al. (1980), subsequent experiments (Sparagana et al., 1984, 1985; Bhaskaran et al., 1987) showing biosynthesis of the JH carbon skeleton but absence of methyl transferase at the prepupal stages, and finally the detection of JH acid prior to JH in prepupal *M. sexta in vivo* (Baker et al., 1987), it is probable that JH acid is the predominant species detected by Granger et al. (1982a,b). However, it is difficult to rationalize their detection of high levels of JH acid RIA activity in fractions isolated by HPLC as JH (esters). It is also still difficult to explain the relatively high levels of JH III/JH III acid detected by the RIA technique of Granger et al. (1982a,b). Unfortunately, their method does not detect JH II, the predominant JH homologue detected in *M. sexta* by other techniques. Using a similar JH antibody, Caruelle et al. (1979) claimed that CA from fifth-stadium nymphs of *Locusta migratoria migratorioides* synthesize JH I, II, and III *in vitro. In vitro* studies, utilizing the RCA or GC/MS, with CA from *Locusta migratoria* and other orthopterans detect *only* JH III unless immediate homologous JH precursors are provided in the incubation medium (see Section 12.4.3). Recent physicochemical methods have detected only JH III in all insect orders studied except Lepidoptera (see Schooley et al., 1984). Nariki et al., (1986) reported *in vitro* JH I biosynthesis by CA from *Bombyx mori*, detected by an RIA technique.

12.4.5. *Other Methods*

Because of the generally low level at which JHs occur in insects and the difficulty of isolating pure material, classical chemical techniques such as nuclear magnetic resonance (NMR) spectrometry, infrared-(IR), and ultraviolet (UV) spectroscopy are not used routinely for identification of JHs. Rather, these techniques have been utilized in studies describing the initial discovery or isolation of a new JH and the associated proof of structure. However, at certain developmental stages in some species (e.g., *Diploptera punctata, Hyalophora cecropia,* or *Nauphoeta cinerea*) levels of JH detected are so high that the UV absorbance of the α, β-unsaturated ester moiety can be utilized to give an estimate of level of JH present. In practice, the UV absorbance is monitored during an HPLC purification step, rather than in a spectrophotometer, and is only feasible if samples have been processed through a rigorous purification scheme. Thus, microgram amounts of JH III (as the d_3-methoxyhydrin derivative) isolated from d5 adult female *D. punctata* hemolymph could be observed in the HPLC purification step prior to GC/MS analysis (F. C. Baker, unpublished results; Tobe et al., 1985). Recently, the very high levels (>100 μg/g) of JH III (as d_3-MH) isolated from the plant *Cyperus iria* were quantified easily by UV absorbance during HPLC of purified extracts (Toong et al., 1988). Detection at $\cong 220$ nm allows the quantification of <100 ng of JH, but at this level high purity of the sample is crucial. Simple HPLC analysis of crude insect extracts (e.g., Wright and Thomas, 1983; Sasagawa and Kuwahara, 1988) cannot be regarded as a reliable method for either identification or quantiication of JHs. In many studies, HPLC analysis of biological extracts has been preceded by analysis of JH standards in order to calibrate the column prior to collection of specific zones for further analysis. However, on most occasions, the level of JH standards utilized to calibrate the HPLC column far outweight the levels of JH(s) present in the biological samples. Thus, there is great potential for cross-contamination of biological samples with JH. In order to circumvent this problem, the use of unrelated marker compounds or radiolabeled JHs of high specific activity an be recommended for conducting such column calibration procedures. Alternatively, the use of nanogram levels of JH with detection at $\cong 220$ nm should result in minimal cross-contamination problems (see, e.g., Baehr et al., 1981).

12.5. Species in Which Qualitative and Quantitative JH Measurements Have Been Made *In Vivo* Using Physicochemical Methods

The list of species in which JH identification and quantification has been reported continues to grow. However, the literature is replete with many equivocal determinations. Therefore, it is deemed important to attempt to report the results that appear reliable. While it could be argued that those results represent the present author's personal bias, it is hoped that the omission of incorrect or inconclusive results will eliminate further propagation of these data. Because the number of studies are numerous, the salient features are more easily presented in tabular form (Table 12.7). Many of the references have already been discussed in Section 12.4.2. Some of the results derive from studies that are still unpublished. A high proportion of the JH identifications stem from studies utilizing the method of Bergot et al. (1981a)—no doubt because this method has been used for analysis of more than 2300 samples encompassing >30 species.

12.6. Summary

The identify of the JHs of some 100 insect species has been elucidated by a combination of methods discussed here. While this number still reflects only a minuscule fraction of the known insects, and indeed only a few insect orders are represented, certain conclusions can be drawn from the results so far:

(1) JH III is the predominant insect JH (particularly insect gonadotropic hormone).

(2) Higher homologs of JH are produced only by lepidopterans.

(3) The identity of the JH in many flies and bugs is still unresolved.

(4) MF or related compounds may be important equivalents of JH in crustaceans.

In vitro and RIA techniques will probably be used increasingly to identify JHs and study the phsiology of JH secretion in miscellaneous species. Although it is unlikely that use of physicochemical methods for JH quantification will increase markedly because of the high cost of such techniques, it is to be hoped that one or more of these methods will be employed to conduct important structure confirmation studies where appropriate. This is particularly important since it is apparent that the secretion of JH in some species is much more complicated than originally envisaged. JH acids are secreted by certain lepidopterans at the prepupal stage of development, and methylation occurs in th imaginal disks of these animals. Many adult male moths secrete JH acids, which are methylated and stored in the accessory sex organs. While it is known that this JH may be transferred to females during mating, its function in the males (and even in the females) remains a mystery.

Just as the use of the *in vitro* technique led to the discovery of JH III (and JH II) in *M. sexta*, it is clear that CA or equivalent organs (ring glands from dipterans; mandibular organs from crustaceans) secrete materials (other than, or in addition to, JHs) that may possess some JH activity. Thus, *Locusta migratoria* CA appears to secrete JH 10,11-diol (Gadot et al., 1987), *Drosophila melanogaster* ring glands secrete a putative new JH (JHB$_3$; Richard et al., 1989), and the crab *Scylla serrata* secretes farnesoic acid and a lesser amount of MF (Tobe et al., 1989). These variations further complicate the task of detecting and identifying JH or JH-like products and mandate the use of several different methods.

TABLE 12.7. Species in Which Qualitative and Quantitative JH Determinations Have Been Made *In Vivo* Using Physicochemical Methods

Species	Stages examined	JH(s)	Method	Reference
Lepidoptera				
Manduca sexta	larvae, adults	I,II,III	GC/ECD	Peter et al., 1976a; Schooley et al., 1976
	eggs	0,I	GC/MS	Bergot et al., 1980
	eggs	II,I,0 4-MeJH I	GC/MS	Bergot et al., 1981d
	larvae	0,I,II,III	GC/MS	Baker et al., 1987
Attacus atlas	adults	II,I	GC/MS	Baker et al., 1985
Trichoplusia ni	larvae	III,II,I	GC/MS	Schooley et al., 1984; Jones et al., in preparation
Danaus plexippus	adults	III,II,I	GC/MS	Lessman et al., in press
Heliothis virescens	adults, eggs	II,I,0	GC/MS	Bergot et al., 1981a,d
Hyalophora gloveri	adults	I,II	GC/MS	Dahm and Röller, 1970
Hyalophora cecropia	adults, eggs	II,I,0	GC/MS	Bergot et al., 1981c,d
Galleria mellonella	larvae	II	GC/MS	Rembold and Lackner, 1985;
			GC/MS	Sehnal and Rembold, 1985
Pieris brassicae	larvae	I	CI GC/MS	Mauchamp et al., 1979a,b
	larvae, pupae	I	CI GC/MS	Mauchamp et al., 1981a
Samia cynthia	larvae	I,II,III	GC/ECD	Schooley et al., 1976
Diatraea grandiosella	larvae, pupae	I,II,III	GC/ECD	Schooley et al., 1976; Bergot et al., 1976

Coleoptera

Species	Stage		Method	Reference
Leptinotarsa decemlineata	adults	III	GC/MS	Trautmann et al., 1974b; Peter et al., 1976b
	adults, eggs	III	GC/MS	de Kort et al., 1982
Diabrotica undecempunctata	adults	III	GC/MS	Schooley et al., 1984
Tenebrio molitor	adults	III	GC/MS	Trautmann et al., 1974b; F. C. Baker, unpublished results
Melolontha melolontha	eggs	III	GC/MS	Trautmann et al., 1974a
	adults	III	GC/MS	Trautmann et al., 1976
Epilachna varivestis	adults	III	GC/MS	Rembold and Lackner, 1985

Orthopera

Species	Stage		Method	Reference
Locusta migratoria	nymphs, adults	III	GC/MS	Bergot et al., 1981b
	adults	III	GC/MS	Rembold, 1981; Rembold and Lackner, 1985
	eggs	III	GC/MS	Temin et al., 1986; Pener et al., 1986
Teleogryllus commodus	all	III	GC/MS	Loher et al., 1983
Taeniopoda eques	adults	III	GC/MS	Loher et al., 1983
Schistocerca gregaria	adults	III	GC/MS	Trautmann et al., 1984b
Schistocerca vaga	nymphs	III	GC/MS	Schooley et al., unpublished results
Gastrimargus africanus	nymphs	III	GC/MS	Schooley et al., unpublished results
Melanoplus sanguinipes	nymphs	III	GC/MS	Schooley et al., unpublished results

Dictyoptera

Species	Stage		Method	Reference
Blatella germanica	adults	III	GC/MS	Camps et al., 1987
Diploptera punctata	nymphs, adults	III	GC/MS	Tobe et al., 1985
Nauphoeta cinerea	adults	III	GC/MS	Trautmann et al., 1974b
	all	III	GC/MS	Baker et al., 1984
Blatta orientalis	adults	III	GC/MS	Trautmann et al., 1974b
Leucophaea maderae	adults	III	GC/MS	Trautmann et al., 1974b
Periplaneta americana	adults	III	GC/MS	Edwards et al., 1987

TABLE 12.7.   (continued)

Species	Stages examined	JH(s)	Method	Reference
Diptera				
Aedes aegypti	larvae, adults	III	GC/MS	Baker et al., 1983
Drosophila melanogaster	pupae	III	GC/MS	Rembold and Lackner, 1985
	all	III	GC/MS	Rembold, 1985; Bownes and Rembold, 1987; Sliter et al., 1987
Drosophila hydei	all	III	GC/MS	Klages et al., 1981; Bührlen et al., 1984
	larvae	III	GC/MS	Rembold and Lackner, 1985
Musca domestica	adults	none	GC/MS	F. C. Baker et al., unpublished results
Sarcophaga bullata	adults	none	GC/MS	F. C. Baker et al., unpublished results
Calliphora vicina	larvae., adults	none	GC/MS	F. C. Baker et al., unpublished results
Anastrepha suspensa	larvae, adults	III	GC/MS	Lawrence et al., in press
Hemiptera				
Oncopeltus fasciatus	eggs, adults	none	GC/MS	Baker et al., 1988

Homoptera				
Megoura viciae	adults	III	GC/MS	Hardie et al., 1985
Aphis fabae	adults	III	GC/MS	Hardie et al., 1985
Thysanura				
Thermobia domestica	eggs, adults	III	GC/MS	Baker et al., 1984
Hymenoptera				
Chelonus sp.	larvae	III	GC/MS	GC/MS. Jones et al., in preparation Bühler et al., 1985
Apis mellifera	adults	III	GC/MS	Trautmann et al., 1974b
	workers	III	GC/MS	Trautmann et al., 1976
	larvae, pupae, adults	III	GC/ECD	Hagenguth and Rembold, 1978
	larvae, pupae	III	GC/MS	Rembold et al., 1980; Rembold and Hagenguth, 1981; Rembold, 1986
	adults	III	GC/MS	Rembold and Lackner, 1985
Scaptotrigona postica	larvae	III	GC/MS	Rembold and Lackner, 1985
Melipona quadrifasciata	adults	III	GC/MS	Rembold and Lackner, 1985
Biosteres longicaudatus	larvae	III	GC/MS	Lawrence et al., in press
Isoptera				
Macrotermes subhyalinus	queens	III	GC/MS	D. R. Meyer et al., 1976

Greatest advances in method development remain to be made in the areas of RIA and in purification of JHs prior to analysis. Several groups are active in the development of better, more sensitive methods for RIA detection of JHs, including development of monoclonal antibodies and use of enantiomerically pure JH for antibody production. A major feat would be the development of a sensitive RIA method that also incorporates an appropriate internal standard (IS) to monitor recovery of JH during the procedure. Presently, this creates a problem because of the cross-reactivity of any labeled JH IS with JH antibody.

It is expected that additional studies using existing or improved methods will expand the list of species whose JH identity has been determined. However, the trend is toward more sophisticated investigations with an emphasis on physiological or developmental biology. In several species, the temporal fluctuation (quantitative and/or qualitative) of JHs during the entire insect life cycle is known (e.g., *Manduca sexta*, *Nauphoeta cinerea*, or *Galleria mellonella*). More detailed scrutiny of JH fluctuation in conjunction with ecdysteroid levels, egg development, etc. have been performed in a few species (e.g., fifth-stadium *M. sexta*, embryonic *Nauphoeta cinerea*, or adult female *Diploptera punctata*; see Baker et al., 1987; Lanzrein et al., 1984; Tobe et al., 1985). More sophisticated studies have measured JH levels in insects after surgical procedures were performed in an effort to find the method (neural, humoral, etc.) by which JH levels are regulated, e.g., during egg laying (Loher et al., 1983, 1987). Finally, several studies have measured the effect of juvenoids or anti-JH agents on the suppression of *in vivo* JH titer in a few insect species (Edwards et al., 1983, 1987; Quistad et al., 1985; Baker et al., 1987; Pener et al., 1986); the use of physicochemical methods is particularly crucial for these types of studies.

Acknowledgments

I thank Dr. Gary B. Quistad for helpful comments, Dr. David A. Schooley for help with the computer-drawn figures, and Ms. Noreen Collins for her skillful typing of the manuscript and assistance in compiling the references and indexes. I am grateful to the National Science Foundation for partial financial support (DMB-8518299).

References

Ajami, A. M. 1974. Approaches to the biosynthesis of the cecropia C_{18}-juvenile hormone. J. Insect Physiol. 20: 2497–2511.

Anderson, R. J., C. A. Henrick, J. B. Siddall, and R. Zurflüh. 1971. Stereoselective synthesis of the racemic C-17 juvenile hormone of cecropia. J. Am. Chem. Soc. 94: 5379–5386.

Anderson, R. J., V. L. Corbin, G. Cotterrell, G. R. Cox, C. A. Henrick, F. Schaub, and J. B. Siddall. 1975. Stereoselective addition of organocopper reagents to acetylenic esters and amides: synthesis of juvenile hormone analogs. J. Am. Chem. Soc. 97: 1197–1204.

Baehr, J. C., P. Pradelles, and F. Dray. 1979. A radioimmunological assay for naturally occurring insect juvenile hormones using iodinated tracers: its use in the analysis of biological samples. Ann. Biol. Anim. Biochim. Biophys. 19: 1827–1836.

Baehr, J. C., J. P. Caruelle, P. Porcheron, and P. Cassier. 1981. Quantification of juvenile hormones using a radioimmunological assay in the analysis of biological samples. Pp. 47–57 *in* G. E. Pratt and G. T. Brooks (eds.), *Juvenile Hormone Biochemistry*. Elsevier, Amsterdam and New York.

Baehr, J. C., J. Casas, A. Messeguer, P. Pradelles, and J. Grassi. 1987. Enzyme immunoassays of JH III and JH III diol using acetylcholinesterase as tracer: an alternative to radioimmunoassay. Insect Biochem. 17: 929–937.

Baehr, J. C., P. Pradelles, C. Lebreux, P. Cassier, and F. Dray. 1976. A simple and sensitive radioimmunoassay of insect juvenile hormone using an iodinated tracer. FEBS (Fed. Evr. Biochem. Soc.) Lett. 69: 123–128.

Baker, F. C., G. C. Jamieson, B. Morallo-Rejesus, and D. A. Schooley, 1985. Identification of the juvenile hormones from adult *Attacus atlas*. Insect Biochem. 15: 321–324.

Baker, F. C., L. W. Tsai, C. C. Reuter, and D. A. Schooley. 1987. *In vivo* fluctuation of JH, JH acid, and ecdysteroid titer, and JH esterase activity, during development of fifth-stadium *Manduca sexta*. Insect Biochem. 17: 989–996.

Baker, F. C., L. W. Tsai, C. C. Reuter, and D. A. Schooley. 1988. The absence of significant levels of the known juvenile hormones and related compounds in the milkweed bug, *Oncopeltus fasciatus*. Insect Biochem. (in press).

Baker, F. C., H. H. Hagedorn, D. A. Schooley, and G. Wheelock. 1983. Mosquito juvenile hormone: identification and bioassay activity. J. Insect Physiol. 29: 465–470.

Baker, F. C., B. Lanzrein, C. A. Miller, L. W. Tsai, G. C. Jamieson, and D. A. Schooley. 1984. Detection of *only* JH III in several life-stages of *Nauphoeta cinerea* and *Thermobia domestica*. Life Sci. 35: 1553–1560.

Bergot, B. J., M. Ratcliff, and D. A. Schooley. 1981a. Method for quantitative determination of the four known juvenile hormones in insect tissues using gas chromatography-mass spectroscopy. J. Chromatogr. 204: 231–244.

Bergot, B. J., D. A. Schooley, and C. A. D. de Kort. 1981b. Identification of JH III as the principal juvenile hormone in *Locusta migratoria*. Experientia (Basel) 37: 909–910.

Bergot, B. J., G. C. Jamieson, M. A. Ratcliff, and D. A. Schooley. 1980. JH zero: new naturally occurring insect juvenile hormone from developing embryos of the tobacco hornworm. Science (Wash., DC) 210: 336–338.

Bergot, B. J., D. A. Schooley, G. M. Chippendale, and C.-M. Yin. 1976. Juvenile hormone titer determinations in the southwestern corn borer, *Diatraea grandiosella*, by electron capture–gas chromatography. Life Sci. 18: 811–820.

Bergot, B. J., G. C. Jamieson, M. A. Ratcliff, and D. A. Schooley. 1981c. Identification of a new insect juvenile hormone, and a method for JH titer determination using gas chromatography–mass spectroscopy. Pp. 173–182 *in* M. Kloza, F. Sehnal, A. Zabza, J. J. Menn, and B. Cymborowsky (eds.), *Regulation of Insect Development and Behaviour: International Conference* Wrocław Technical University Press, Wrocław, Poland.

Bergot, B. J., F. C. Baker, D. C. Cerf, G. Jamieson, and D. A. Schooley, 1981d. Qualitative and quantitative aspects of juvenile hormone titers in developing embryos of several insect species: discovery of a new JH-like substance extracted from eggs of *Manduca sexta*. Pp. 33–45 *in* G. E. Pratt and G. T. Brooks (eds.), *Juvenile Hormone Biochemistry*. Elsevier, Amsterdam and New York.

Bhaskaran, G., S. P. Sparagana, P. Barrera, and K. H. Dahm. 1986. Change in corpus allatum function during metamorphosis of the tobacco hornworm *Manduca sexta:* regulation at the terminal step in juvenile hormone biosynthesis. Arch. Insect Biochem. Physiol. 3: 321–338.

Bhaskaran, G., G. de Leon, B. Looman, P. D. Shirk, and H. Röller. 1980. Activity of juvenile hormone acid in brainless, allatectomized diapausing cecropia pupae. Gen. Comp. Endocrinol. 42: 129–133.

Bhaskaran, G., K. H. Dahm, G. L. Jones, K. Peck, and S. Faught. 1987. Juvenile hormone acid synthesis and HMG–CoA reductase activity in corpora allata of *Manduca sexta* prepupae. Insect Biochem. 17: 933–937.

Bhaskaran, G., S. P. Sparagana, K. H. Dahm, P. Barrera, and K. Peck. 1988. Sexual dimorphism in juvenile hormone synthesis by corpora allata and in juvenile hormone acid methyltransferase activity in corpora allata and accessory sex glands of some Lepidoptera. Int. J. Invertebr. Reprod. Dev. 13: 87–100.

Bieber, M. A., C. C. Sweeley, D. J. Faulkner, and M. R. Petersen. 1972. Purification and quantitative determination of cecropia juvenile hormone. Anal. Biochem. 47: 261–272.

Bitsch, C., J. C. Baehr, and J. Bitsch. 1985. Juvenile hormones in *Thermobia domestica* females: identification and quantification during biological cycles and after precocene application. Experientia (Basel) 41: 409–410.

Bjerke, J. S. and H. Röller. 1974. Assays for juvenile hormone. Pp. 130–139 *in* W. J. Burdette (ed.), *Invertebrate Endocrinology and Hormonal Heterophylly*. Springer-Verlag, Berlin and New York.

Borst, D. W., H. Laufer, M. Landau, E. S. Chang, W. A. Hertz, F. C. Baker, and D. A. Schooley. 1987. Methyl farnesoate and its role in crustacean reproduction and development. Insect Biochem. 17: 1123–1127.

Bowers, W. S. and M. J. Thompson. 1963. Juvenile hormone activity: effects of isoprenoid and straight-chain alcohols on insects. Science (Wash., DC) 142: 1469–1470.

Bowers, W. S., M. J. Thompson, and E. C. Uebel. 1965. Juvenile and gonadotropic hormone activity of 10,11-epoxyfarnesenic acid methyl ester. Life Sci. 4: 2323–2331.

Bowers, W. S., H. M. Fales, M. J. Thompson, and E. C. Uebel. 1966. Juvenile hormone: identification of an active compound from balsam fir. Science (Wash., DC) 154: 1020–1021.

Bowers, W. S., P. A. Marsella, and P. H. Evans. 1983. Identification of an hemipteran juvenile hormone: *in vitro* biosynthesis of JH III by *Dysdercus fasciatus*. J. Exp. Zool. 228: 555–559.

Bownes, M. and H. Rembold. 1987. The titre of juvenile hormone during the pupal and adult stages of the life cycle of *Drosophila melanogaster*. Eur. J. Biochem. 164: 709–712.

Brindle, P. A., D. A. Schooley, L. W. Tsai, and F. C. Baker. 1988. Comparative metabolism of branched-chain amino acids to precursors of juvenile hormone biogenesis in corpora allata of lepidopterous vs. nonlepidopterous insects. J. Biol. Chem. 263: 10653–10657.

Brindle, P. A., F. C. Baker, L. W. Tsai, C. C. Reuter, and D. A. Schooley. 1987. Sources of propionate for the biogenesis of ethyl-branched insect juvenile hormones: role of isoleucine and valine. Proc. Nat. Acad. Sci. USA. 84: 7906–7910.

Brüning, E. and B. Lanzrein. 1987. Function of juvenile hormone III in embryonic development of the cockroach, *Nauphoeta cinerea*. Int. J. Invertebr. Reprod. Dev. 12: 29–44.

Brüning, E., A. Saxer, and B. Lanzrein. 1985. Methyl farnesoate and juvenile hormone III in normal and precocene-treated embryos of the ovoviviparous cockroach *Nauphoeta cinerea*. Int. J. Invertebr. Reprod. Dev. 8: 269–278.

Bühler, A., T. N. Hanzlik, and B. D. Hammock. 1985. Effects of parasitization of *Trichoplusia ni* by Chelonus sp. Physiol. Entomol. 10: 383–394.

Bührlen, U., H. Emmerich and H. Rembold. 1984. Titer of juvenile hormone III in *Drosophila hydei* during metamorphosis determined by GC–MS–MIS. Z. Naturforsch. 39C: 1150–1154.

Bürgin, C. and B. Lanzrein. 1988. Stage-dependent biosynthesis of methyl farnesoate and juvenile hormone III and metabolism of juvenile hormone III in embryos of the cockroach, *Nauphoeta cinerea*. Insect Biochem. 18: 3–9.

Camps, F., J. Casas, F.-J. Sanchez, and A. Messeuguer. 1987. Identification of juvenile hormone III in the hemolymph of *Blattella germanica* adult females by gas chromatography–mass spectrometry. Arch. Insect Biochem. Physiol. 6: 181–189.

Carlisle, D. B., and P. E. Ellis. 1968. Insect hormones: olive oil is not an inert vehicle for hormone injection into locusts. Science (Wash., DC) 162: 1393–1394.

Caruelle, J. P., J.-C. Baehr, and P. Cassier. 1979. Synthèse d'hormones juvéniles *in vitro* pour les *Corpora allata* de larves du 5e *stade de Locusta migratoria migratorioides* (R. et F.) (Insecte, Orthopteroide). C. R. Acad. Sci. Paris 288D: 1107–1110.

Chard, T. 1987. *An Introduction to Radioimmunoassay and Related Techniques*, 3rd ed. Elsevier, Amsterdam and New York.

Couillaud, F., B. Mauchamp, and A. Girardie. 1985. Regulation of juvenile hormone titer in African locust. Experientia (Basel) 41: 1165–1167.

Couillaud, F., B. Mauchamp, A. Girardie, and S. de Kort. 1988. Enhancement by farnesol and farnesoic acid of juvenile-hormone biosynthesis in induced low-activity locust corpora allata. Arch. Insect Biochem. Physiol. 7: 133–143.

Dahm, K. H. and H. Röller. 1970. The juvenile hormone of the giant silk moth *Hyalophora gloveri* (Strecker). Life Sci. 9: 1397–1400.

Dahm, K. H., B. M. Trost, and H. Röller. 1967. The juvenile hormone. V. Synthesis of the racemic juvenile hormone. J. Am. Chem. Soc. 89: 5292–5294.

Dahm, K. H., G. Bhaskaran, M. G. Peter, P. D. Shirk, K. R. Seshan, and H. Röller. 1976. On the identity of the juvenile hormone in insects. Pp. 19–47 *in* L. I. Gilbert (ed.), *The Juvenile Hormones*. Plenum Press, New York.

Dahm, K. H., G. Bhaskaran, M. G. Peter, P. D. Shirk, K. R. Seshan, and H. Röller. 1981. The juvenile hormones of cecropia. Pp. 183–198 *in* M. Kloza, F. Sehnal, A. Zabza, J. J. Menn, and B. Cymborowski(eds.), *Regulation of Insect Development and Behaviour: International Conference.* Wrocław Technical University Press, Wrocław, Poland.

Dale, J. A., D. L. Dull, and H. S. Mosher. 1969. α-Methoxy-α-trifluoromethylphenylacetic acid, a versatile reagent for the determination of enantiomeric composition of alcohols and amines. J. Org. Chem. 34: 2543–2549.

de Kort, C. A. D., B. J. Bergot, and D. A. Schooley. 1982. The nature and titer of juvenile hormone in the Colorado potato beetle, *Leptinotarsa decemlineata*. J. Insect Physiol. 28: 471–474.

de Kort, C. A. D., A. B. Koopmanschap, C. Strambi, and A. Strambi. 1985. The application and evaluation of a radioimmunoassay for measuring juvenile hormone titres in Colorado beetle haemolymph. Insect Biochem. 15: 771–775.

Deleurance, S., J.-C. Baehr, P. Porcheron, and P. Cassier. 1979. Dosages radioimmunologiques des ecdystéroides et des hormones juvéniles au cours des stades larvaires III et IV, nymphal et imaginal de *Leptinotarsa decemlineata* Say (Coléoptère Chrysomélides). C. R. Acad. Sci. Paris 289D: 1089–1092.

Deleurance, S., J.-C. Baehr, P. Porcheron, and P. Cassier. 1980. Diapause et évolution des taux des ecdystéroides et des hormones juvéviles chez les nymphes et les imagos de *Leptinotarsa decemlineata* Say. C. R. Acad. Sci. Paris 290D: 367–370.

deLoof, A., K. Madhavan, J. E. Girard, H. A. Schneiderman, and J. Meinwald. 1976. A novel method for the quantitive extraction of juvenile hormones from insects including the silverfish, *Thermobia domestica*. J. Insect Physiol. 22: 829–832.

de Wilde, J., G. B. Staal, C. A. D. de Kort, A. de Loof, and G. Baard. 1968. Juvenile hormone titer in the haemolymph as a function of photoperiodic treatment in the adult Colorado beetle (*Leptinotarsa decemlineata* Say). Koninkl. Nederl. Akademie Van Wetenschappen–Amsterdam. Reprinted from Proc. series 71C: 321–326.

Dorn, A. 1975. Structure and function of the embryonic corpus allatum of *Oncopeltus fasciatus* Dallas (Insecta, Heteroptera). Verh. Dtsch. Zool. Ges. 67: 85–89.

Dunham, L. L., D. A. Schooley, and J. B. Siddall. 1975. A survey of the chromatographic analysis of natural insect juvenile hormones and the insect growth regulator, Altosid® . J. Chromatogr. Sci. 13: 334–336.

Edwards, J. P., B. J. Bergot, and G. B. Staal. 1983. Effects of three compounds with anti–juvenile hormone activity and a juvenile hormone analogue on endogenous juvenile hormone levels in the tobacco hornworm, *Manduca sexta*. J. Insect Physiol. 29: 83–89.

Edwards, J. P., J. Chambers, N. R. Price, and J. P. G. Wilkins. 1987. Action of a juvenile hormone analogue on the activity of *Periplaneta americana* corpora allata *in vitro* and on juvenile hormone III levels *in vivo*. Insect Biochem. 17: 1115–1118.

Fain, M. J. and L. M. Riddiford. 1975. Juvenile hormone titers in the hemolymph during late larval development of the tobacco hornworm, *Manduca sexta* (L). Biol. Bull. (Woods Hole) 149: 504–521.

Faulkner, D. J. and M. R. Peterson. 1971. Synthesis of C-18 *Cecropia* juvenile hormone to obtain optically active forms of known absolute configuration. J. Am. Chem. Soc. 93: 3766–3767.

Feldlaufer, M. F., W. S. Bowers, D. M. Soderlund, and P. H. Evans. 1982. Biosynthesis of the sesquiterpenoid skeleton of juvenile hormone 3 by *Dysdercus fasciatus* corpora allata *in vitro*. J. Exp. Zool. 223: 295–298.

Feyereisen, R. 1985a. Regulation of juvenile hormone titer: synthesis. Pp. 391–429 *in* G. A. Kerkut and L. I. Gilbert (eds.), *Comprehensive Insect Physiology, Biochemistry and Pharmacology*, Vol. 7. Pergamon Press, Oxford and Elmsford, New York.

Feyereisen, R. 1985b. Radiochemical assay for juvenile hormone III biosynthesis *in vitro*. Methods Enzymol. 111B: 530–539.

Feyereisen, R. and D. E. Farnsworth. 1988. Forced synthesis of trace amounts of juvenile hormone II from propionate by corpora allata of a juvenile hormone III—producing insect. Experientia (Basel) 44: 47–49.

Feyereisen, R. and S. S. Tobe. 1981. A rapid partition assay for routine analysis of juvenile hormone release by insect corpora allata. Anal. Biochem. 111: 372–375.

Fisher, F. M. and R. C. Sanborn. 1964. *Nosema* as a source of juvenile hormone in parasitized insects. Biol. Bull. (Woods Hole) 126: 235–252.

Fuzeau-Braesch, S., G. Nicolas, J.-C. Baehr, and P. Porcheron. 1982. A study of hormonal levels of the locust *Locusta migratoria cinerescens* artifically changed to the solitary state by a chronic CO_2 treatment of one minute per day. Comp. Biochem. Physiol. 71A: 53–58.

Gadot, M., A. Goldman, M. Cojocaru, and S. W. Applebaum. 1987. The intrinsic synthesis of juvenile hormone-III diol by locust corpora allata *in vitro*. Mol. Cell. Endocrinol. 49: 99–107.

Gharib, B. and M. de Reggi. 1983. Changes in ecdysteroid and juvenile hormone levels in developing eggs of *Bombyx mori*. J. Insect Physiol. 29: 871–876.

Gharib, B., M. de Reggi, J.-L. Connat, and J.-C. Chaix. 1983. Ecdysteroid and juvenile hormone changes in *Bombyx mori* eggs, related to the initiation of diapause. FEBS (Fed. Eur. Biochem. Soc.) Lett. 160: 119–123.

Gilbert, L. E. and H. A. Schneiderman. 1961. The content of juvenile hormone and lipid in Lepidoptera: sexual differences and developmental changes. Gen. Comp. Endocrinol. 1: 453–472.

Girard, J. E., K. Madhavan, T. C. McMorris, A. de Loof, J. Chong, V. Arunachalam, H. A. Schneiderman, and J. Meinwald. 1976. Identification of a juvenile hormone from *Musca domestica*. Insect Biochem. 6: 347–350.

Granger, N. A. and W. G. Goodman. 1983. Juvenile hormone radioimmunoassays theory and practice. Insect Biochem. 13: 333–340.

Granger, N. A., S. M. Neimiec, L. I. Gilbert, and W. E. Bollenbacher. 1982a. Juvenile hormone synthesis *in vitro* by larval and pupal corpora allata of *Manduca sexta*. Mol. Cell Endocrinol. 28: 587–604.

Granger, N. A., S. M. Niemiec, L. I. Gilbert, and W. E. Bollenbacher. 1982b. Juvenile hormone III biosynthesis by the larval corpora allata of *Manduca sexta*. J. Insect Physiol. 28: 385–391.

Granger, N. A., L. R. Whisenton, W. P. Janzen, and W. E. Bollenbacher. 1987. Interendocrine control by 20-hydroxyecdysone of the corpora allata of *Manduca sexta*. Insect Biochem. 17: 949–953.

Granger, N. A., W. F. Bollenbacher, R. Vince, L. T. Gilbert, J. C. Baehr, and F. Dray. 1979. *In vitro* biosynthesis of juvenile hormone by the larval corpora allata of *Manduca sexta*: quantification by radioimmunoassay. Mol. Cell. Endocrinol. 16: 1–17.

Greenberg, S. L. and S. S. Tobe. 1984. The *in vitro* activity of corpora allata from neotenic queens of the primitive termite *Zootermopsis angusticollis*. Gen. Comp. Endocrinol. 53: 489.

Guilvard, E., M. de Reggi, and J.-A. Rioux. 1984. Changes in ecdysteroid and juvenile hormone titers correlated to the initiation of vitellogenesis in two *Aedes* species (Diptera, Culicidae). Gen. Comp. Endocrinol. 53: 218–223.

Hagenguth, H. and H. Rembold. 1978. Identification of juvenile hormone 3 as the only JH homolog in all developmental stages of the honey bee. Z. Naturforsch. 33C: 847–850.

Hagenguth, H. and H. Rembold. 1979. Identification and quantification of insect juvenile hormones as their 10-heptafluorobutyryloxy-11-methoxy derivatives by combination of high-performance liquid chromatography and electron capture gas–liquid chromatography. J. Chromatogr. 170: 175–184.

Hamnett, A. F. and G. E. Pratt. 1978. Use of automated capillary column radio gas chromatography in the identification of insect juvenile hormones. J. Chromatog. 158: 387–399.

Hardie, J., F. C. Baker, G. C. Jamieson, A. D. Lees, and D. A. Schooley. 1985. The identification of an aphid juvenile hormone, and its titre in relation to photoperiod. Physiol. Entomol. 10: 297–302.

Henrick, C. A. 1982. Juvenile hormone analogs: structure–activity relationships. Pp. 315–402 *in* J. R. Coats (ed.), *Insecticide Mode of Action.* Academic Press, Orlando; Florida.

Henrick, C. A. and G. B. Staal. 1988. Stereoisomerism in juvenile hormone structure–activity relationships. Pp. 303–326 *in* E. J. Ariëns, J. J. S. van Rensen, and W. Welling (eds.), *Stereoselectivity of Pesticides.* Elsevier, Amsterdam and New York.

Henrick, C. A., G. B. Staal, and J. B. Siddall. 1973. Alkyl 3,7,11-trimethyl-2,4-dodecadienoates, a new class of potent insect growth regulators with juvenile hormone activity. J. Agric. Food Chem. 21: 354–359.

Huibregtse-Minderhoud, L., A. C. van der Kerk-van Hoof, P. Wijkens, H. W. A. Biessels, and C. A. Salemink. 1979. A method for the quantitative determination of juvenile hormones in insects. Recl. Trav. Chim. Pays-Bas Belg. 98: 371–424.

Huibregtse-Minderhoud, L., A. C. van der Kerk-van Hoof, P. Wijkens, H. W. A. Biessels, and C. A. Salemink. 1980a. Glass capillary column gas chromatographic method for simultaneous quantitative determination of insect juvenile hormones at the picogram level: a comparitive study of various halogenated derivatives. J. Chromatogr. 196: 425–434.

Huibregtse-Minderhoud, L., M. A. M. van den Hondel-Franken, A. C. van der Kerk–van Hoof, H. W. A. Biessels, C. A. Salemink, D. J. van der Horst, and A. M. Th. Beenakkers. 1980b. Quantitative determination of the juvenile hormones in the haemolymph of *Locusta migratoria* during normal development and after implantation of corpora allata. J. Insect Physiol. 26: 627–631.

Jackson, L. L. and M. T. Armold. 1977. Insect lipid analysis. Pp. 171–206 *in* R. B. Turner (ed.), *Analytical Biochemistry of Insects.* Elsevier, Amsterdam and New York.

Jennings, R. C., K. J. Judy, D. A. Schooley, M. S. Hall, and J. B. Siddall.

1975. The identification and biosynthesis of two juvenile hormones from the tobacco budworm moth (*Heliothis virescens*). Life Sci. 16: 1033–1040.

Judy, K. J., D. A. Schooley, M. S. Hall. B. J. Bergot, and J. B. Siddall. 1973a. Chemical structure and absolute configuration of a juvenile hormone from grasshopper corpora allata *in vitro*. Life Sci. 13: 1511–1516.

Judy, K. J., D. A. Schooley, L. L. Dunham, M. S. Hall, B. J. Bergot, and J. B. Siddall. 1973b. Isolation, structure, and absolute configuration of a new natural insect juvenile hormone from *Manduca sexta*. Proc. Nat. Acad. Sci. USA 70: 1509–1513.

Judy, K. J., D. A. Schooley, R. G. Troetschler, R. C. Jennings, B. J. Bergot, and M. S. Hall. 1975. Juvenile hormone production by corpora allata of *Tenebrio molitor in vitro*. Life Sci. 16: 1059–1066.

Judy, K. J., D. A. Schooley, B. J. Bergot, M. S. Hall, R. C. Jennings, and R. G. Troetschler. 1976. *In vitro* studies on insect juvenile hormone structure and biosynthesis. Colloq. Int. Cent. Nat. Rech. Sci. No. 251 (*Actualités sur les hormones d'invertebrés, Villeneuve d'Ascq, France, 1975*): 451–457.

Karlson, P. and M. Nachigall. 1961. Ein biologischer Test zur quantitativen Bestimmung der Juvenilhormon-aktivität von Insektenextrakten. J. Insect Physiol. 7: 210–215.

Kimura, M. and S. Sakurai. 1984. Identification of the juvenile hormone from the silkworm, *Bombyx mori*. Zool. Sci. Proc. 55th Annu. Meet. Zool. Soc. Jpn.

Klages, G., H. Emmerich, and H. Rembold. 1981. Identification and titer of juvenile hormones in the fruitfly. P. 207 *in* M. Kloza, F. Sehnal, A Zabza, J. J. Menn, and B. Cymborowski (eds.), *Regulation of Insect Development and Behaviour: International Conference*. Wrocław Technical University Press, Wrocław, Poland.

Koch, P. B. and K. H. Hoffmann. 1985. Juvenile hormone and reproduction in crickets, *Gryllus bimaculatus* De Geer: corpus allatum activity (*in vitro*) in females during adult life cycle. Physiol. Entomol. 10: 173–182.

Koeppe, J. K., S. G. Brantley, and M. M. Nijhout. 1980. Structure, haemolymph titre, and regulatory effects of juvenile hormone during ovarian maturation in *Leucophaea maderae*. J. Insect Physiol. 26: 749–753.

Koyama, T., K. Ogura, F. C. Baker, G. C. Jamieson, and D. A. Schooley. 1987. Synthesis and absolute configuration of 4-methyl juvenile hormone I (4-Me JH I) by a biogenetic approach: a combination of enzymatic synthesis and biotransformation. J. Am. Chem. Soc. 109: 2853–2854.

Kramer, S. J. and F. Kalish. 1984. Regulation of the corpora allata in the black mutant of *Manduca sexta*. J. Insect Physiol. 30: 311–316.

Kramer, S. J. and J. H. Law. 1980. Control of juvenile hormone production: the relationship between precursor supply and hormone synthesis in the tobacco hornworm, *Manduca sexta*. Insect Biochem. 10: 569–575.

Kramer, S. J. and G. B. Staal. 1981. *In vitro* studies on the mechanism of action of anti–juvenile hormone agents in larvae of *Manduca sexta*. Pp.

425–437 *in* G. E. Pratt and G. T. Brooks (eds.), *Juvenile Hormone Biochemistry*. Elsevier, Amsterdam and New York.

Lanzrein, B., V. Gentinetta, and M. Lüscher. 1978. *In vitro* juvenile hormonsynthese durch corpora allata der termite *Macrotermes subhyalinus*. Rev. Suisse Zool. 85: 10.

Lanzrein, B., M. Hashimoto, V. Parmakovich, and K. Nakanishi. 1975. Identification and quantification of juvenile hormones from different developmental stages of the cockroach *Nauphoeta cinerea*. Life Sci. 16: 1271–1284.

Lanzrein, B., H. Imboden, C. Bürgin, E. Brüning, and H. Gfeller. 1984. On titers, origin, and functions of juvenile hormone III, methyl farnesoate, and ecdysteroids in embryonic development of the ovoviviparous cockroach *Nauphoeta cinerea*. Pp. 454–465 *in* J. Hoffmann and M. Porchet (eds.), *Biosynthesis, Metabolism and Mode of Action of Invertebrate Hormones*. Springer-Verlag, Berlin and New York.

Lauer, R. C., P. H. Solomon, K. Nakanishi, and B. F. Erlanger. 1974. Antibodies to insect C_{16}-juvenile hormone. Experientia (Basel) 30: 558–560.

Laufer, H., D. Borst, F. C. Baker, C. Carrasco, M. Sinkus, C. C. Reuter, L. W. Tsai, and D. A. Schooley. 1987. Identification of a juvenile hormone–like compound in a crustacean. Science (Wash., DC) 235: 202–205.

Lessman, C. A., W. S. Herman, D. A. Schooley, L. W. Tsai, B. J. Bergot, and F. C. Baker. 1989. Detection of juvenile hormones I, II, and III in adult monarch butterflies (*Danaus plexippus*). Insect Biochem. (in press).

Le Roux, A. 1968. Description d'organes mandibulaires nouveaux chez les crustaces decapodes. C. R. Acad. Sci. (Paris) 266D: 1414–1417.

Lawrence, P. O., F. C. Baker, L. W. Tsai, C. A. Miller, D. A. Schooley, and L. Geddes. 1989. JH III levels in superparasitized pharate pupae of *Anastrepha suspensa* (Diptera: Tephritidae) and in larvae of the parasitic wasp *Biosteres* longicaudatus (Hymenoptera: Braconidae). Arch. Insect Biochem. Physiol. (in press).

Loew, P. and W. S. Johnson. 1971. The synthesis of the optically active form of the C-18 *cecropia* juvenile hormone. J. Am. Chem. Soc. 93: 3765–3766.

Loher, W., D. A. Schooley, and F. C. Baker. 1987. Influence of the ovaries on JH titer in *Teleogryllus commodus*. Insect Biochem. 17: 1099–1102.

Loher, W., L. Ruzo, F. C. Baker, C. A. Miller, and D. A. Schooley. 1983. Identification of the juvenile hormone from the cricket, *Teleogryllus commodus*, and juvenile hormone titre changes. J. Insect Physiol. 29: 585–589.

Madhavan, K., T. Ohta, W. S. Bowers, and H. A. Schneiderman. 1981. Identification of a juvenile hormone of a primitive insect, the firebrat, *Thermobia domestica* (Thysanura). Am. Zool. 21: 733–735.

Masner, P., K. Slama, and V. Landa. 1968. Natural and synthetic materials with insect hormone activity. IV. Specific female sterility effects produced by a juvenile hormone analogue. J. Embryol. Exp. Morphol. 20: 25–31.

Mauchamp, B., R. Lafont, and D. Jourdain. 1979a. Mass fragmentographic analysis of juvenile hormone 1 levels during the last larval instar of *Pieris brassicae*. J. Insect Physiol. 25: 545–550.

Mauchamp, B., R. Lafont, and P. Krein. 1981a. Analysis of juvenile hormones by high-performance liquid chromatography coupled with mass spectrometry. Pp. 21–31 *in* G. E. Pratt and G. T. Brooks (eds.), *Juvenile Hormone Biochemistry*. Elsevier, Amsterdam and New York.

Mauchamp, B., M. Zander, and R. Wolff. 1984. The qualitative and quantitative determination of juvenile hormones by mass spectrometry. Pp. 363–372 *in* J. Hoffmann and M. Porchet (eds.), *Biosynthesis, Metabolism and Mode of Action of Invertebrate Hormones*. Springer-Verlag, Berlin and New York.

Mauchamp, B., F. Couillaud, and C. Malosse. 1985. Gas chromatography–mass spectroscopy analysis of juvenile hormone released by corpora allata. Anal. Biochem. 145: 251–256.

Mauchamp, B., R. Lafont, M. Hardy, and D. Jourdain. 1979b. Analysis of insect juvenile hormones by gas chromatography–mass spectrometry: problems of sample preparation and choice of detection procedure. Biomed. Mass Spect. 6: 276–281.

Mauchamp, B., R. Lafont, J.-L. Pennetier, and J. Doumas. 1981b. Detection and quantification of the juvenile hormone 1 during the post-embryonic development of *Pieris brassicae* L. Pp. 199–206 in M. Kloza, F. Sehnal, A. Zabza, J. J. Menn, and B. Cymborowsky (eds.), *Regulation of Insect Development and Behaviour: International Conference*. Wrocław Technical University Press, Wrocław, Poland.

McCaffery, A. R. and V. A. McCaffery. 1983. Corpus allatum activity during overlapping cycles of oocyte growth in adult female *Melanoplus sanguinipes*. J. Insect Physiol. 29: 259–266.

Metzler, M., K. H. Dahm, D. Meyer, and H. Röller. 1971. On the biosynthesis of juvenile hormone in the adult cecropia moth. Z. Naturforsch. 26B: 1270–1276.

Metzler, M., D. Meyer, K. H. Dahm, H. Röller, and J. B. Siddall. 1972. Biosynthesis of juvenile hormone from 10-epoxy-7-ethyl-3,11-dimethyl-2,6-tridecadienoic acid in the adult cecropia moth. Z. Naturfosch. 27B: 321–322.

Meyer, A. S. and E. Hanzmann. 1970. The optical activity of the cecropia juvenile hormones. Biochem. Biophys. Res. Commun. 41: 891–893.

Meyer, A. S., H. A. Schneiderman, and L. I. Gilbert. 1965. A highly purified preparation of juvenile hormone from the silk moth *Hyalophora cecropia* L. Nature (Lond.) pp. 272–275.

Meyer, A. S., E. Hanzmann, and R. C. Murphy. 1971. Absolute configuration of *cecropia* juvenile hormone. Proc. Nat. Acad. Sci. USA. 68: 2312–2315.

Meyer, A. S., H. A. Schneiderman, E. Hanzmann, and J. H. Ko. 1968. The two juvenile hormones from the cecropia silk moth. Proc. Nat. Acad. Sci. USA 60: 853–860.

Meyer, D. R., B. Lanzrein, M. Lüscher, and K. Nakanishi. 1976. Isolation and identification of a juvenile hormone (JH) in termites. Experientia (Basel) 32: 773.

Müller, P. J., P. Masner, K. H. Trautmann, M. Suchy, and H.-K. Wipf. 1975. The isolation and identification of juvenile hormone from cockroach corpora allata *in vitro*. Life Sci. 15: 915–921.

Nakanishi, K., D. A. Schooley, M. Koreeda, and J. Dillon. 1971. Absolute configuration of the C_{18} juvenile hormone: application of a new circular dichroism method using tris(dipivaloylmethanato)praseodymium. Chem. Comm. pp. 1235–1236.

Nariki, M., T. Nakamachi, and T. Ohatki. 1986. JH-secreting activity of corpora allata from 4th and 5th instar larvae of *Bombyx mori*. Proc. 57th Annu. Meet. Zool. Soc. Jpn. 13: 1092.

Pener, M. P., D. Dessberg, P. Lazarovici, C. C. Reuter, L. W. Tsai, and F. C. Baker. 1986. The effect of a synthetic prococene on juvenile hormone III titre in late *Locusta* eggs. J. Insect Physiol. 32: 853–857.

Peter, M. G., K. H. Dahm, and H. Röller. 1976a. The juvenile hormones in blood of larvae and adults of *Manduca sexta* (Joh.). Z. Naturforsch. 31C: 129–131.

Peter, M. G., P. D. Shirk. K. H. Dahm, and H. Röller. 1981. On the specificity of juvenile hormone biosynthesis in the male *cecropia* moth. Z. Naturforsch. 36C: 579–585.

Peter, M. G., H. W. A. Biessels, K. R. Seshan, H. Röller, G. Bhaskaran, and K. H. Dahm. 1976b. Identification of juvenile hormones in insects. Abstr. Pap. 172nd Am. Chem. Soc. Nat. Meet., San Francisco, 1976, Abstr. No. 40 (pesticide chemistry).

Plantevin, G., M. de Reggi, and C. Nardon. 1984. Changes in ecdysteroid and juvenile hormone titers in the hemolymph of *Galleria mellonella* larvae and pupae. Gen. Comp. Endocrinol. 56: 218–230.

Pratt, G. E. and S. S. Tobe. 1974. Juvenile hormones radiobiosynthesised by corpora allata of adult female locusts *in vitro*. Life Sci. 14: 575–586.

Pratt, G. E. and R. J. Weaver. 1975. Juvenile hormone biosynthesis by cultured cockroach corpora allata. J. Endocrinol. 64: 67P.

Pratt, G. E., R. C. Jennings, and R. J. Weaver. 1984. The influence of a P_{450} inhibitor on methyl farnesoate levels in cockroach corpora allata, *in vitro*. Insect Biochem. 14: 609–614.

Pratt, G. E., S. S. Tobe, R. J. Weaver, and J. R. Finney. 1975. Spontaneous synthesis and release of C_{16} juvenile hormone by isolated corpora allata of female locust *Schistocerca gregaria* and female cockroach *Periplaneta americana*. Gen. Comp. Endocrinol. 26: 478–484.

Quistad, G. B., D. C. Cerf, S. J. Kramer, B. J. Bergot, and D. A. Schooley. 1985. Design of novel insect anti–juvenile hormones: allylic alcohol derivatives. J. Agric. Food Chem. 33: 47–50.

Rankin, M. A. and L. M. Riddiford, 1978. Significance of haemolymph juvenile hormone titer changes in timing of migration and reproduction in adult *Oncopeltus fasciatus*. J. Insect Physiol. 24: 31–38.

Readio, J., K. Peck, R. Meola, and K. H. Dahm. 1988. Corpus allatum activity (*in vitro*) in female *Culex pipiens* during adult life cycle. J. Insect Physiol. 34: 131–135.

Reimbold, H. 1981. Modulation of JH III–titer during the gonotrophic cycle of

Locusta migratoria, measured by gas chromatography–selected ion monitoring mass spectrometry. Pp. 11–20 *in* G. E. Pratt and G. T. Brooks (eds.), *Juvenile Hormone Biochemistry*. Elsevier, Amsterdam and New York.

Rembold, H. 1985. Mikroanalytische Bestimmung von Juvenilhormonen. Separatdruck Chimia 39: 348–354.

Rembold, H. 1986. Control of imaginal development by juvenile hormone. Pp. 59–68 *in* M. Porchet, J.-C. Andries, and A. Dhainaut (eds.), *Advances in Invertebrate Reproduction*, Vol. 4. Elsevier, Amsterdam and New York.

Rembold, H. and H. Hagenguth. 1981. Modulation of hormone pools during postembryonic development of the female honey bee castes. Pp. 427–440 *in* M. Kloza, F. Sehnal, A. Zabza, J. J. Menn, and B. Cymborowski (eds.), *Regulation of Insect Development and Behaviour: International Conference*. Wrocław Technical University Press, Wrocław, Poland.

Rembold, H. and B. Lackner. 1985. Convenient method for the determination of picomole amounts of juvenile hormone. J. Chromatogr. 323: 355–361.

Rembold, H., H. Hagenguth, and J. Rascher. 1980. A sensitive method for detection and estimation of juvenile hormones from biological samples by glass capillary combined gas chromatography–selected ion monitoring mass spectrometry. Anal. Biochem. 101: 356–363.

Röseler, P.-F. and I. Röseler. 1978. Studies on the regulation of the juvenile hormone titre in bumblebee workers, *Bombus terrestris*. J. Insect Physiol. 24:707–713.

Röller, H. and J. S. Bjerke, 1965. Purification and isolation of juvenile hormone and its action in lepidopteran larvae. Life Sci. 4: 1617–1624.

Röller, H. and K. H. Dahm. 1970. The identity of juvenile hormone produced by corpora allata *in vitro*. Naturwissenschaften 57: 454–455.

Röller, H., K. H. Dahm, C. C. Sweeley, and B. M. Trost. 1967. The structure of the juvenile hormone. Angew. Chem. Int. Ed. Engl. 6: 179–180.

Richard, D. S., S. W. Applebaum, T. J. Sliter, F. C. Baker, D. A. Schooley, C. C. Reuter, V. C. Henrich, and L. I. Gilbert. 1989. Juvenile hormone bisepoxide biosynthesis *in vitro* by the ring gland of *Drosophila melanogaster*: a putative juvenile hormone in the higher Diptera. Proc. Nat. Acad. Sci. USA 86:1421-1425.

Ruegg, R. P., S. S. Tobe, and W. Loher. 1986. Juvenile hormone biosynthesis during egg development in the cricket, *Teleogryllus commodus*. J. Insect Physiol. 32: 517–521.

Sasagawa, H. and Y. Kuwahara. 1988. Quantitative determination method for juvenile hormone III titer in honeybee haemolymph by high-performance liquid chromatography. Agric. Biol. Chem. 52: 1295–1297.

Schmialek, P. 1963. Über Verbindungen mit Juvenilhormonwirkung. Z. Naturforsch. 18B: 516–519.

Schneiderman, H. A. 1972. Insect hormones and insect control. Pp. 3–27 *in* J. J. Menn and M. Beroza (eds.), *Insect Juvenile Hormones*. Academic Press, Orlando, Florida.

Schneiderman, H. A. and L. I. Gilbert. 1958. Substances with juvenile hormone activity in crustacea and other invertebrates. Biol. Bull. (Woods Hole) 115: 530–535.

Schneiderman, H. A. and L. I. Gilbert. 1964. Control of growth and development in insects. Science 143: 325–333.

Schooley, D. A. 1977. Analysis of the naturally occurring juvenile hormones: their isolation, identification, and titer determination at physiological levels. Pp. 241–287 *in* R. B. Turner (ed.), *Analytical Biochemistry of Insects.* Elsevier, Amsterdam and New York.

Schooley, D. A. and F. C. Baker. 1985. Juvenile hormone biosynthesis. Pp. 363–389 *in* G. A. Kerkut and L. I. Gilbert (eds.), *Comprehensive Insect Physiology, Biochemistry and Pharmacology.* Pergamon Press, Oxford and Elmsford, New York.

Schooley, D. A., K. J. Judy, B. J. Bergot, M. S. Hall, and J. B. Siddall. 1973. Biosynthesis of the juvenile hormones of *Manduca sexta:* labeling pattern from mevalonate, propionate, and acetate. Proc. Nat. Acad. Sci. USA 70: 2921–2925.

Schooley, D. A., K. J. Judy, B. J. Bergot, M. S. Hall, and R. C. Jennings. 1976. Determination of the physiological levels of juvenile hormones in several insects and biosynthesis of the carbon skeletons of the juvenile hormones. Pp. 101–117 *in* L. I. Gilbert (ed.), *The Juvenile Hormones.* Plenum Press, New York.

Schooley, D. A., F. C. Baker, L. W. Tsai, C. A. Miller, and G. C. Jamieson. 1984. Juvenile hormones 0, I, and II exist only in Lepidoptera. Pp. 373–383 *in* J. Hoffmann and M. Porchet (eds.), *Biosynthesis, Metabolism and Mode of Action of Invertebrate Hormones.* Springer-Verlag, Berlin and New York.

Schooneveld, H., S. J. Kramer, H. Privee, and A. van Huis. 1979. Evidence of controlled corpus allatum activity in the adult Colorado potato beetle. J. Insect Physiol. 25: 449–453.

Sehnal, F. and H. Rembold. 1985. Brain stimulation of juvenile hormone production in insect larvae. Experientia (Basel) 42: 684–685.

Shapiro, A. B., G. D. Wheelock, H. H. Hagedorn, F. C. Baker, L. W. Tsai, and D. A. Schooley. 1986. Juvenile Hormone and juvenile hormone esterase in adult females of the mosquito *Aedes aegypti.* J. Insect Physiol. 32: 867–877.

Slama, K. and C. M. Williams. 1965. Juvenile hormone activity for the bug *Pyrrhocoris apterus.* Proc. Nat. Acad. Sci. USA 54: 411–414.

Sliter, T. J., B. J. Sedlak, F. C. Baker, and D. A. Schooley. 1987. Juvenile hormone in *Drosophila melanogaster:* Identification and titer determination during development. Insect Biochem. 17: 161–165.

Sparagana, S. P., G. Bhaskaran, and P. Barrera. 1985. Juvenile hormone acid methyl transferase activity in imaginal discs of *Manduca sexta* prepupae. Arch. Insect Biochem. Physiol. 2: 191–202.

Sparagana, S. P., G. Bhaskaran, K. H. Dahm, and V. Riddle, 1984. Juvenile hormone production, juvenile hormone esterase, and juvenile hormone acid methyltransferase in corpora allata of *Manduca sexta.* J. Exp. Zool. 230: 309–313.

Staal, G. B. 1972. Biological activity and bioassay of juvenile hormone analogs. Pp. 69–94 *in* J. J. Menn and M. Beroza (eds.), *Insect Juvenile Hormones.* Academic Press, Orlando, Florida.

Staal, G. B. 1975. Insect growth regulators with juvenile hormone activity. Annu. Rev. Entomol. 20: 417–460.

Staal, G. B. 1977. Essential differences between natural juvenile hormones and juvenile hormone analogs elucidated by use of a substitution assay. Scr. Varia Pont. Acad. Sci. 41: 333–348.

Strambi, A., C. Strambi, P.-F. Röseler, and I. Röseler. 1984a. Simultaneous determination of juvenile hormone and ecdysteroid titers in the hemolymph of bumblebee prepupae (*Bombus hypnorum* and *B. terrestris*). Gen. Comp. Endocrinol. 55: 83–88.

Strambi, C., J. P. Delbecque, and J. L. Connat. 1984b. Identification by high-pressure liquid chromatography and radioimmunoassay of JH-III in *Acheta domesticus*. Insect Biochem. 14: 719–723.

Strambi, C., A. Strambi, M. L. de Reggi, M. H. Hirn, and A. Delaage. 1981. Radioimmunoassay of insect juvenile hormones and of their diol derivatives. Eur. J. Biochem. 118: 401–406.

Temin, G., M. Zander, and J.-P. Roussel. 1986. Physico-chemical (GC–MS) measurements of juvenile hormone III titres during embryogenesis of *Locusta migratoria*. Int. J. Invertebr. Reprod. Dev. 9: 105–112.

Tobe, S. S. 1977. Asymmetry in hormone biosynthesis by insect endocrine glands. Can. J. Zool. 55: 1509–1514.

Tobe, S. S. and G. E. Pratt. 1974. The influence of substrate concentrations on the rate of insect juvenile hormone biosynthesis by corpora allata of the desert locust *in vitro*. Biochem. J. 144: 107–113.

Tobe, S. S. and G. E. Pratt. 1975. Corpus allatum activity *in vitro* during ovarian maturation in the desert locust, *Schistocerca gregaria*. J. Exp. Biol. 62: 611–627.

Tobe, S. S. and B. Stay. 1977. Corpus allatum activity *in vitro* during the reproductive cycle of the viviparous cockroach, *Diploptera punctata* (Eschscholtz). Gen. Comp. Endocrinol. 31: 138–147.

Tobe, S. S., R. P. Ruegg, B. A. Stay, F. C. Baker, C. A. Miller, and D. A. Schooley. 1985. Juvenile hormone titre and regulation in the cockroach *Diploptera punctata*. Experientia (Basel) 41: 1028–1034.

Toong, Y. C., D. A. Schooley, and F. C. Baker. 1988. Isolation of insect juvenile hormone III from a plant. Nature (Lond.). 333: 170–171.

Trabalon, M., M. Campan, J. C. Baehr, and B. Mauchamp. 1987. *In vitro* biosynthesis of juvenile hormone III by the corpora allata of *Calliphora vomitoria* and its role in ovarian maturation and sexual receptivity. Experientia (Basel) 43: 1113–1115.

Trautmann, K. H., A. Schuler, M. Suchy, and H.-K. Wipf. 1974a. A. method for the qualitative and quantitative determination of three natural insect juvenile hormones. Z. Naturforsch. 29C: 161–168.

Trautmann, K. H., P. Masner, A. Schuler, M. Suchy, and H.-K. Wipf. 1974b. Evidence of the juvenile hormone methyl (2*E*,6*E*)-10,11-epoxy-3,7,11-trimethyl-2,6-dodecadienoate (JH-3) in insects of four orders. Z. Naturforsch. 29C: 757–759.

Trautmann, K. H., M. Suchy, P. Masner, H.-K. Wipf, and A. Schuler. 1976. Isolation and identification of juvenile hormones by means of a radioactive isotope dilution method: evidence for JH III in eight species from four orders. Pp. 118–130 *in* L. I. Gilbert (ed.), *The Juvenile Hormones.* Plenum Press, New York.

Trost, B. M. 1969. The juvenile hormone of *Hyalophora cecropia.* Life Sci. 3: 120–130.

Truman, J. W., L. M. Riddiford, and L. Safranek. 1973. Hormonal control of cuticle coloration in the tobacco hornworm, *Manduca sexta:* basis of an ultrasensitive bioassay for juvenile hormone. J. Insect Physiol. 19: 195–203.

van Broekhoven, L. W., A. C. van der Kerk–van Hoof, and C. A. Salemink. 1975. Gas chromatographic determination of subnanogram amounts of the juvenile hormone methyl (2*E*,6*E*)-10,11-epoxy-3,7,11-trimethyl-2,6-dodecadienoate (JH III) in insect material by electron capture detection. Z. Naturforsch. 30C: 726–729.

van Tamelen, E. E., A. Storni, E. J. Hessler, and M. Schwartz. 1963. The biogenetically patterned *in vitro* cyclization of farnesyl acetate. J. Am. Chem. Soc. 85: 3295–3296.

Wakabayashi, N., M. Schwarz, P. E. Sonnet, R. M. Waters, R. E. Redfern, and M. Jacobson. 1971. Compounds related to insect juvenile hormones. Bull. Soc. Entomol. Suisse 44: 131–139.

Watson, R. D., L. R. Whisenton, W. E. Bollenbacher, and N. A. Granger. 1986. Interendocrine regulation of the corpora allata and prothoracic glands of *Manduca sexta.* Insect Biochem. 16: 149–155.

Weirich, G. F. and M. G. Culver. 1979. *S*-Adenosylmethionine: juvenile hormone acid methyltransferase in male accessory reproductive glands of *Hyalophora cecropia* (L). Arch. Biochem. Biophys. 198: 175–181.

Wigglesworth, V. B. 1961. Some observations on the juvenile hormone effect of farnesol in *Rhodnius prolixus* Stål (Hemiptera). J. Insect Physiol. 7: 73–78.

Wigglesworth, V. B. 1969. Chemical structure and juvenile hormone activity: comparative tests on *Rhodnius prolixus.* J. Insect Physiol. 15: 73–94.

Wigglesworth, V. 1985. Historical perspectives. Pp. 1–24 *in* G. A. Kerkut and L. I. Gilbert (eds.), *Comprehensive Insect Physiology, Biochemistry and Pharacology,* Vol. 7. Pergamon Press, Oxford and Elmsford, New York.

Williams, C. M. 1956. The juvenile hormone of insects. Nature (Lond.) 178: 212–213.

Wright, J. E. and B. R. Thomas. 1983. Boll weevil: determination of ecdysteroids and juvenile hormones with high-pressure liquid chromatography. J. Liquid Chromatogr. 6: 2055–2066.

Yamamoto, R. T. and M. Jacobson. 1962. Juvenile hormone activity of isomers of farnesol. Nature (Lond.) 196: 908–909.

Zlatkis, A. and C. F. Poole (eds.). 1981. Electron capture, theory, and practice in chromatography. J. Chromatogr. Libr. No. 20.

Methods for Ecdysteroid Analysis

13

RENÉ LAFONT AND

PHILIPPE BEYDON

13.1.   Introduction   457
13.2.   Structures of Arthropod Ecdysteroids   457
13.3.   Physical Properties of Ecdysteroids   464
        13.3.1.   Ultraviolet and Infrared Spectrometry   464
        13.3.2.   Mass Spectrometry   465
        13.3.3.   Nuclear Magnetic Resonance
                  Spectrometry   465
        13.3.4.   Optical Rotation Properties   467
13.4.   Basic Reactions   468
        13.4.1.   Derivatization Reactions   468
            13.4.1.1.   Formation of Acetates   468
            13.4.1.2.   Formation of Acetonides   470
            13.4.1.3.   Formation of Methyl Esters   470
            13.4.1.4.   Reduction (for 3 = Oxo Compounds)   470
            13.4.1.5.   Formation of Fluorescent
                        Derivatives   471
        13.4.2.   Hydrolysis of Conjugates   472
            13.4.2.1.   Enzymatic Hydrolysis   472
            13.4.2.2.   Chemical Hydrolysis   473
13.5.   Assays   473
        13.5.1.   Bioassays   473
        13.5.2.   Immunoassays   474
13.6.   Methods for Extraction and Separation   479
        13.6.1.   Extraction Procedures   479
        13.6.2.   Separation Procedures   479
            13.6.2.1.   Thin-Layer Chromatography   479
            13.6.2.2.   Droplet Countercurrent
                        Chromatography   483
            13.6.2.3.   Gas–Liquid Chromatography   483
            13.6.2.4.   High-Performance Liquid
                        Chromatography   485

	13.6.2.5. Supercritical Fluid Chromatography	488
	13.6.2.6. High-Voltage Paper Electrophoresis	489
13.7.	Practice of Ecdysteroid Analysis	490
	13.7.1. Identification of Ecdysteroids Whose Reference is Available	490
	13.7.2. Identification of New Ecdysteroids from Rich Materials	491
	13.7.3. Identification of Ecdysone Metabolites Produced *In Vivo*	493
	13.7.4. Identification of Ecdysone Metabolites Produced *In Vitro*	494
	13.7.5. Utilization of Cholesterol Labeling for the Isolation of Ecdysteroids	495
	13.7.6. Quantification of Ecdysteroids by Reversed-Phase HPLC/RIA	497
	13.7.7. Quantification of Ecdysteroids After Long-Term Cholesterol Labeling	499
13.8.	Summary	501
	Acknowledgments	502
	References	502

13.1. Introduction

Ecdysteroids have been detected in all groups of arthropods, where they control both development and reproduction. They represent a large family, comprising about 50 compounds from invertebrates (zooecdysteroids) and more than 100 isolated from plants (phytoecdysteroids, some of which are identical with zooecdysteroids). Since the initial isolation of ecdysone by Butenandt and Karlson (1954) and the elucidation of its structure about 10 years later (Huber and Hoppe, 1965), many methods for ecdysteroid analysis have been developed. This methodological progress has led to the design of very efficient routine procedures [e.g., high-performance liquid chromatography/radioimmunoassay (HPLC/RIA)] and of more sophisticated techniques with the general improvement of analytical methods. The topic of this chapter has been covered by several previous review articles to which the reader is referred for more specific information (Morgan and Poole, 1976; Koreeda and Teicher, 1977; Morgan and Wilson, 1980, 1988; Lafont et al., 1980b; Hoffmann and Hétru, 1983; Horn and Bergamasco, 1985; Rees and Isaac, 1985; Lafont 1987).

13.2. Structures of Arthropod Ecdysteroids

Arthropod ecdysteroids constitute a large group of compounds, many of which are listed in Table 13.1. The formulas of major representatives are given in Fig. 13.1. Some of these compounds are found in a great many species: for instance, 20-hydroxyecdysone is present in almost all species tested, except those containing exclusively C_{28} ecdysteroids, which, in turn, contain its 24-methyl homologue, makisterone A. By contrast, other compounds are found in only a few species, even occasionally a single species (at least at the present time).

Ecdysteroids may be classified according to their polarity as members of apolar, medium polar, or very polar compounds. *Polar compounds* comprise nonionic ecdysteroids = glucosides; ionic ecdysteroids = ecdysonoic acids and phosphate esters. *Medium-polarity compounds* comprise ecdysone and 20-hydroxyecdysone (or C_{28} homologues); immediate precursors (one −OH less than ecdysone); metabolites bearing one additional −OH group (26-OH); 3-dehydroecdysteroids, 3-epimers, and conjugates like acetates or glycolates. *Apolar compounds* comprise various precursors (e.g., ketodiol) and conjugates (C-22 fatty acylester). Of course, all these compounds have rather different properties (chemical and biological) and require specific methods for their analysis.

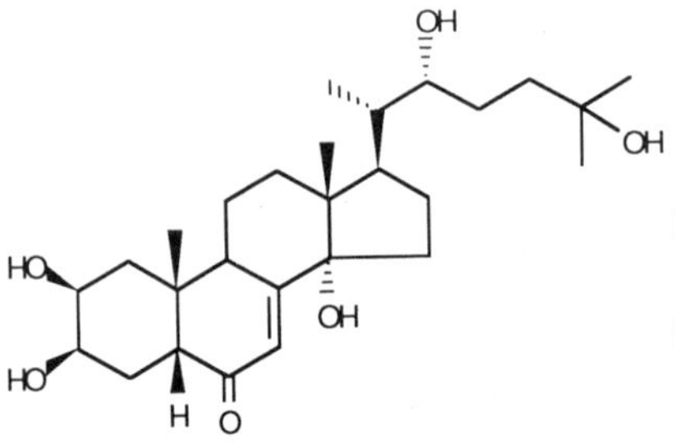

ECDYSONE

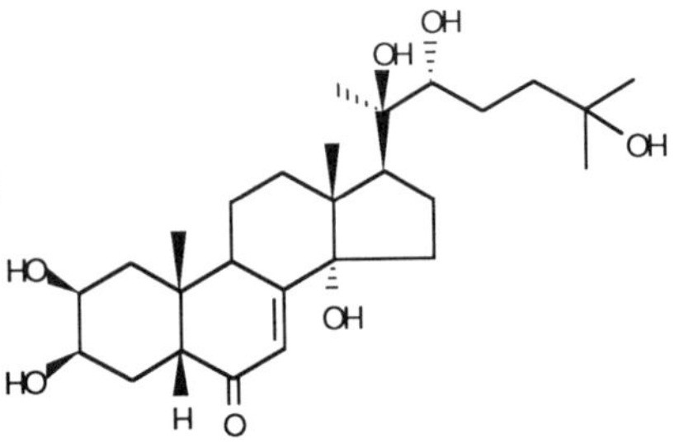

20-HYDROXYECDYSONE

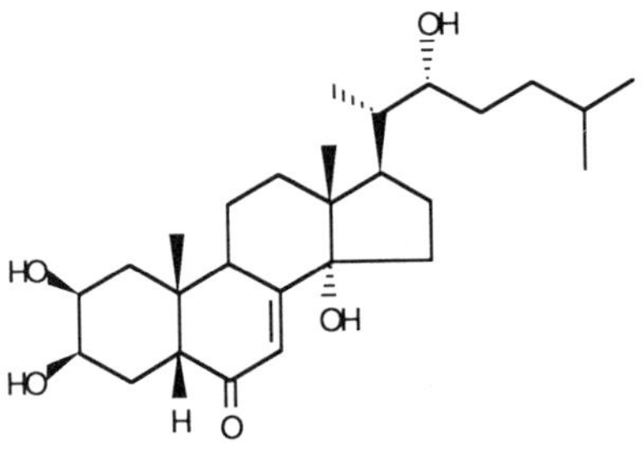

25-DEOXYECDYSONE

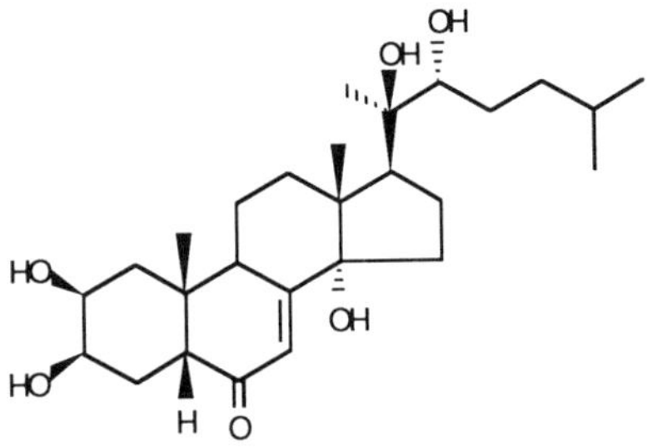

PONASTERONE A

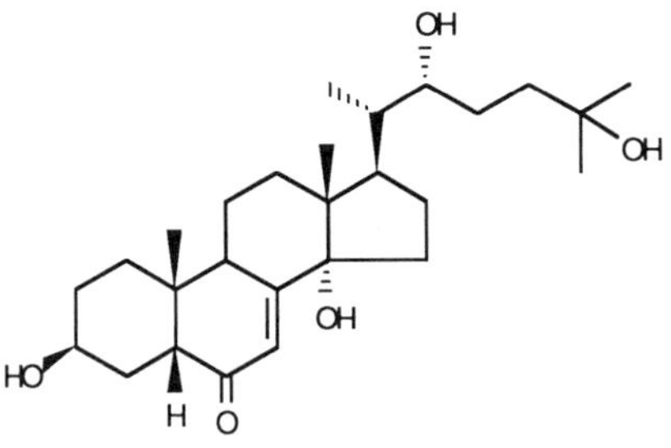

2-DEOXY-ECDYSONE

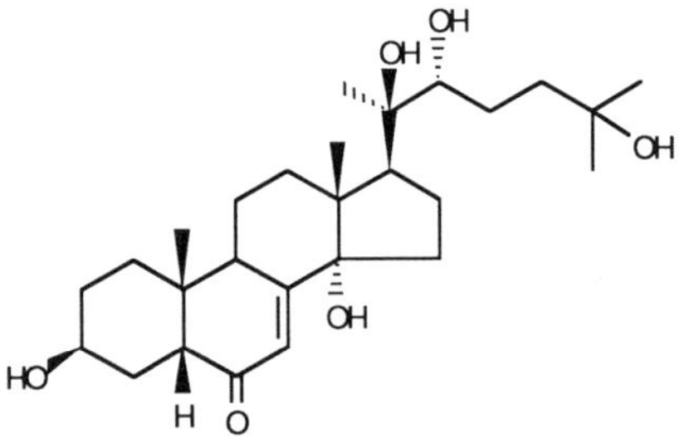

2-DEOXY-20-HYDROXYECDYSONE

FIGURE 13.1. Structures of the major ecdysteroids isolated from arthropods.

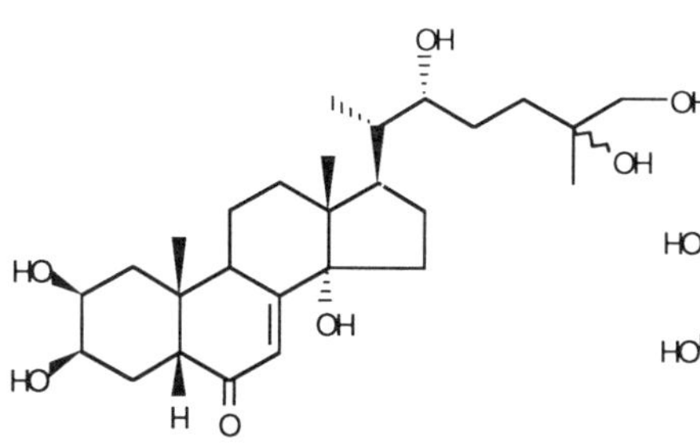

OH
OH
26-HYDROXY-ECDYSONE 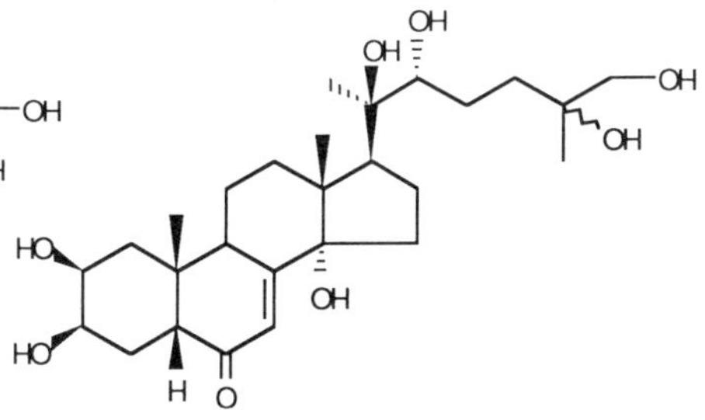
OH
OH
OH
20,26-DIHYDROXYECDYSONE 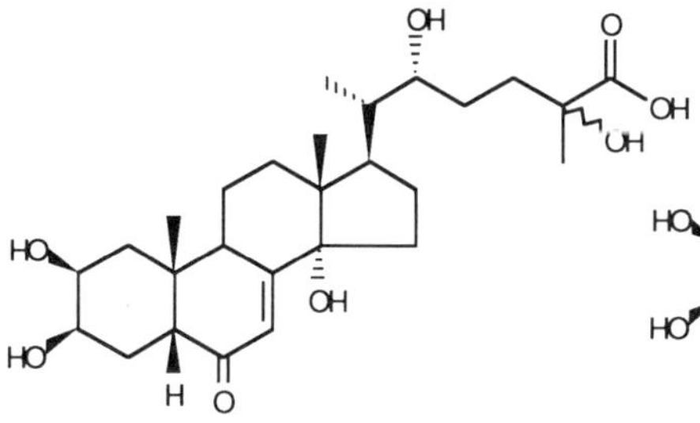
OH
O
OH
OH
ECDYSONOIC ACID 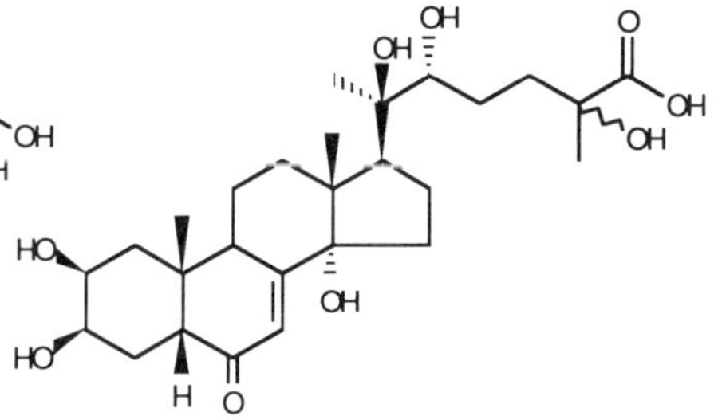
OH
O
OH
OH
20-HYDROXYECDYSONOIC ACID 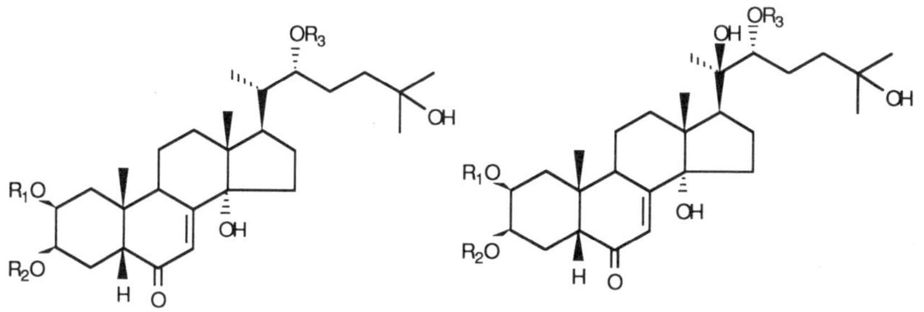
OR3
OH
OR3
OH
ECDYSONE and 20-HYDROXYECDYSONE CONJUGATES
(ACETATES, ACYL ESTERS, PHOSPHATES, ...)

TABLE 13.1. Ecdysteroids isolated from Arthropods

Compounds	Species	Reference
20-Hydroxyecdysone and its precurors		
20-Hydroxyecdysone (= crustecdysone, ecdysterone, β-ecdysone)	*Jasus lalandei*	Hampshire and Horn, 1966
	Bombyx mori	Hocks and Wiechert, 1966
		Hoffmeister and Grützmacher, 1966
Ecdysone (= α-ecdysone)	*Bombyx mori*	Butenandt and Karlson, 1954
2-Deoxyecdysone	*Bombyx mori*	Ohnishi et al., 1977
2-Deoxy-20-hydroxyecdysone	*Jasus lalandei*	Galbraith et al., 1968
	Schistocerca gregaria	Isaac et al., 1981a
2,22-Dideoxy-20-hydroxyecdysone	*Bombyx mori*	Ikekawa et al., 1980
2,14,22,25-Tetradeoxyecydsone (5β-ketol)	*Locusta migratoria*	Hétru et al., 1978
2,22,25-Trideoxyecdysone (5β-ketodiol)	*Locusta migratoria*	Hétru et al., 1978
22-Deoxy-20-hydroxyecdysone	*Pycnogonum litorale*	Bückman et al., 1986
25-Deoxyecdysone	*Carcinus maenas*	Lachaise et al., 1986
25-Deoxy-20-hydroxyecdysone (= ponasterone A)	*Gecarcinus lateralis*	McCarthy, 1979
26-Hydroxyecdysteroids (and ecdysonoic acids)		
26-Hydroxyecdysone	*Manduca sexta*	Kaplanis et al., 1973
20,26-Dihydroxyecdysone	*Manduca sexta*	Thompson et al., 1967
22-Deoxy-20,26-dihydroxyecdysone	*Pyconogonum litorale*	Bückman et al., 1986

Inokosterone	*Callinectes sapidus*	Faux et al., 1969
Ecdysonoic acid and	*Pieris brassicae*	Lafont et al., 1983
20-Hydroxyecdysonoic acid	*Locusta migratoria*	Modde et al., 1984
	Schistocerca gregaria	Isaac et al., 1983
	Spodoptera littoralis	Isaac et al., 1983

3-Dehydro- and 3-epiecdysteroids

3-Dehydroecdysone	*Calliphora vicina*	Karlson et al., 1972
3-Dehydro-20-hydroxyecdysone	*Calliphora vicina*	Koolman and Splindler, 1977
3-Epi-2-deoxyecdysone	*Schistocerca gregaria*	Isaac et al., 1981a
3-Epiecdysone	*Manduca sexta*	Kaplanis et al., 1979
3-Epi-20-Hydroxyecdysone	*Manduca sexta*	Thompson et al., 1974
3-Epi-26-hydroxyecdysone	*Manduca sexta*	Kaplanis et al., 1980
3-Epi-20,26-dihydroxyecdysone	*Manduca sexta*	Kaplanis et al., 1979

Acyl esters

Ecdysone 3-acetate	*Schistocerca gregaria*	Isaac et al., 1981b
20-Hydroxyecdysone 3-acetate	*Locusta migratoria*	Modde et al., 1984
20-Hydroxyecdysone 22-acetate	*Pycnogonum litorale*	Bückman et al., 1986
	Drosophila melanogaster	Maroy et al., 1988
20,26-Dihydroxyecdysone 22-acetate	*Pycnogonum litorale*	Bückman et al., 1986
20-Hydroxyecdysone 22-glycolate	*Pycnogonum litorale*	Bückman et al., 1986
Ecdysone 22-palmitate	*Heliothis armigera*	Robinson et al., 1987
	Boophilus microplus	Crosby et al., 1986
Ecdysone 22-linoleate, 22-oleate,	*Boophilus microplus*	Crosby et al., 1986
22-palmitoleate, 22-stearate		

TABLE 13.1. Ecdysteroids isolated from Arthropods (continued)

Compounds	Species	Reference
20-Hydroxyecdysone 22-linoleate,	*Ornithodoros moubata*	Diehl et al., 1985
22-oleate, 22-palmitate, 22-stearate	*Heliothis virescens*	Kubo et al., 1987
Glucosides		
26-Hydroxyecdysone 22-glucoside	*Manduca sexta*	Thompson et al., 1987a
Phosphate esters and nucleotides		
Ecdysone 22-phosphate and	*Schistocerca gregaria*	Isaac et al., 1982a
20-Hydroxyecdysone 22-phosphate	*Locusta migratoria*	Tsoupras et al., 1982b
2-Deoxyecdysone 22-phosphate	*Schistocerca gregaria*	Isaac et al., 1982a
2-Deoxy-20-hydroxyecdysone 22-phosphate	*Locusta migratoria*	Tsoupras et al., 1982b
26-Hydroxyecdysone 2-phosphate	*Manduca sexta*	Thompson et al., 1987b
26-Hydroxyecdysone 26-phosphate	*Manduca sexta*	Thompson et al., 1985b

Compound	Species	Reference
Ecdysone 22-adenosine monophosphate	*Locusta migratoria*	Tsoupras, 1982a; Hétru et al., 194
Ecdysone 22-(N^6-isopentenyl)-adenosine monophosphate	*Locusta migratoria*	Tsoupras et al., 1983
3-Epiecdysone 3-phosphate	*Locusta migratoria*	Tsoupras, 1982
3-Epi-20-hydroxyecdysone 3-phosphate	*Pieris brassicae*	Beydon et al., 1987b
Double conjugates		
Ecdysone 3-acetate 2-phosphate	*Schistocerca gregaria*	Isaac et al., 1984
20-Hydroxyecdysone 3-acetate 2-phosphate	*Locusta migratoria*	Modde et al., 1984
20-Hydroxyecdysone 3-acetate 22-phosphate	*Locusta migratoria*	Tsoupras et al., 1982b
Side-chain cleavage product		
Poststerone	*Calliphora stygia*	Galbraith et al., 1968
C_{28} compounds		
Makisterone A	*Callinectes sapidus*	Faux et al., 1969
	Oncopeltus fasciatus	Kaplanis et al., 1975
20-Deoxymakisterone A	*Drosophila melanogaster*	Redfern, 1984

13.3. Physical Properties of Ecdysteroids

13.3.1. *Ultraviolet and Infrared Spectrometry*

Ecdysteroids are characterized by the presence of a 7-ene-6-one chromophore that renders them easily detectable by their ultraviolet (UV) absorbance (γ_{max}=242–243 nm in ethanol; $\epsilon \approx 12,000$). These values are only slightly changed by the various substitutions on the steroid nucleus or on the side chain. Upon treatment with HCl/ MeOH, the 14-OH is readily removed, which results in the appearance of two dehydration products corresponding respectively to a 7,14-dien-6-one (γ_{max}=295 nm) and a 8,14-dien-6-one (γ_{max}=248 nm) (Fig. 13.2). This property has often been used to assess the presence of the 6-one/7-ene/14-OH system (Nakanishi, 1971; Horn, 1971).

Ecdysteroids show usual infrared (IR) absorption bands for C=C (1612 cm^{-1}), C=O (1640–1667 cm^{-1}), and −OH (3330–3400 cm^{-1}) corresponding to common characteristics. Only some compounds share additional bands corresponding to particular groups; for instance, this is the case with 3-dehydroecdysteroids, which show an additional band at 1715 cm^{-1} (Koolman and Spindler, 1977). This technique is in fact of limited use for the characterization of ecdysteroids.

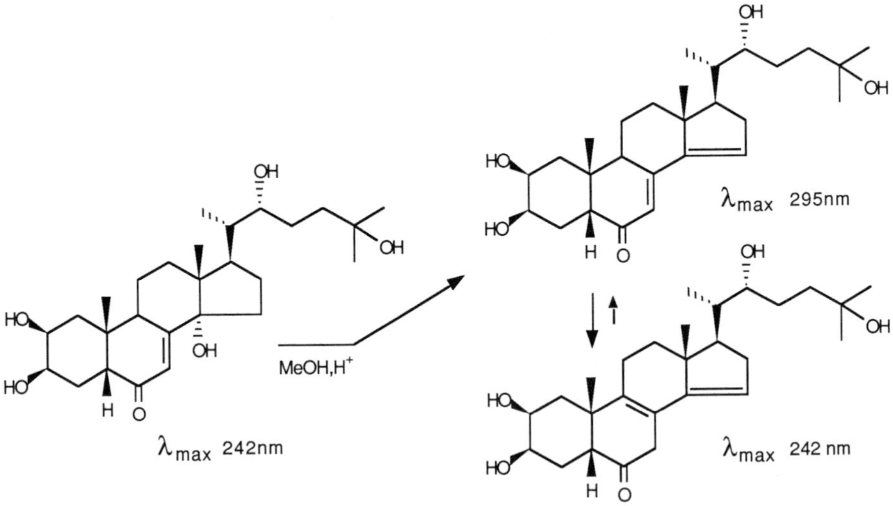

FIGURE 13.2. Dehydration of the ecdysone molecule under acidic treatment. (After Nakanishi, 1971.)

13.3.2. *Mass Spectrometry*

Several techniques are available, which include, for instance, electron impact (EI), chemical ionization (CI), and fast-atom bombardment (FAB) (and also several more "exotic" procedures). EI gives at best very weak M^+ peaks, but a large range of fragments, which are useful for structural elucidation, whereas CI (with ammonia, methane or isobutane) gives larger molecular (M^+) or pseudomolecular (MH^+) ions and thus allows the easy determination of the molecular weight of ecdysteroids. FAB is particularly well suited for poorly volatile or fragile compounds, e.g., ionic conjugates (like phosphate esters) and ecdysonoic acids with negative ion detection (see Rees and Isaac, 1985). Of course, FAB works equally well with all ecdysteroids; however, it requires higher amounts of ecdysteroids than do EI or CI.

Classically, ecdysteroids undergo several losses of water molecules and side-chain cleavage between C-20 and C-22 (especially if they possess a 20,22-diol) and between C-17 and C-20 (Fig. 13.3). Fragmentation patterns provide a direct information about the position (nucleus vs. side chain) of extra hydroxyl groups and side-chain branching at C-24 (Nakanishi, 1971).

13.3.3. *Nuclear Magnetic Resonance Spectrometry*

NMR spectroscopy is the method of choice for assessing definitely the structure of ecdysteroids. Among the several NMR procedures available, two are of general use: proton (^{1}H-NMR) and carbon (^{13}C-NMR). ^{31}P-NMR use is limited to phosphate conjugates (Hétru et al., 1985; Rees and Isaac, 1985). To obtain a high-quality ^{1}H-NMR spectrum with a modern apparatus requires about 50 μg of a pure compound (Fig. 13.4), and this is generally sufficient to determine classical changes like epimerization, appearance of an extra $-OH$ group, or the position involved in a conjugation process (Table 13.2). More important changes would need ^{13}C-NMR studies, which require milligram amounts. Complete NMR spectra of ecdysone and 20-hydroxyecdysone have been published recently (Kubo et al., 1985b; Girault and Lafont, 1987), and general tables are given in several recent reviews (Hoffmann and Hétru, 1983; Horn and Bergamasco, 1985; Rees and Isaac, 1985). In many cases, a minimal amount of information regarding methyl signals and a few characteristic proton signals can be sufficient, provided that it is used in conjunction with other independent data (e.g., derivatization), to assess the identity of

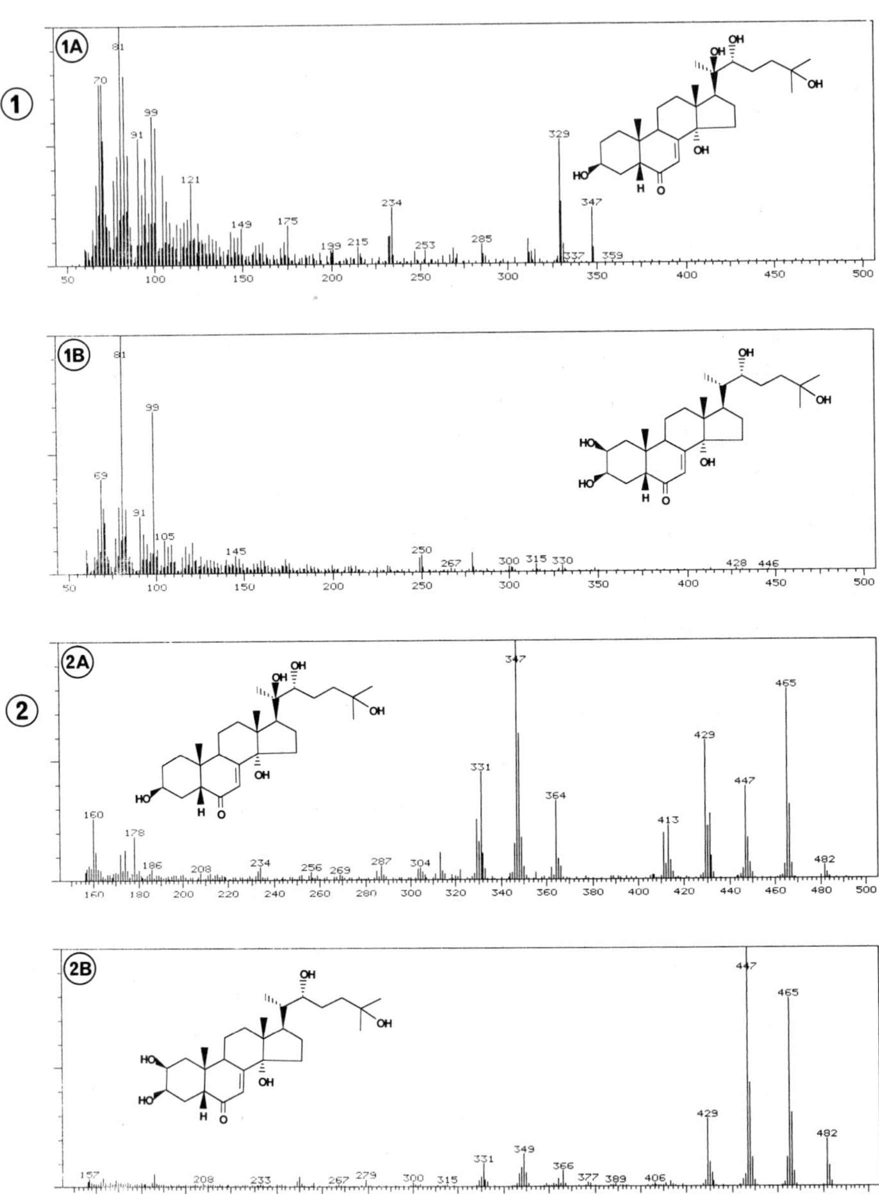

FIGURE 13.3. Comparison of the mass spectra (EI) of two ecdysteroids having the same molecular mass: (1B,2B) ecdysone; (1A, 2A) 2-deoxy-20-hydroxy-ecdysone.

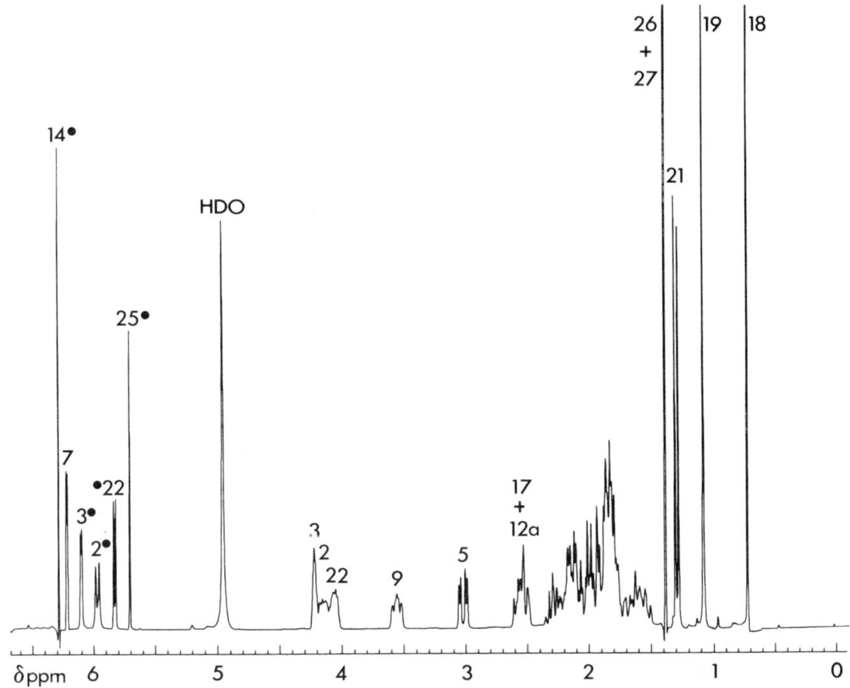

FIGURE 13.4. ^{1}H-NMR spectrum of ecdysone in C_5D_5N (Brucker 250 MHz): • = OH signals: (Courtesy of Prof. J.-P. Girault.)

a new ecdysteroid. Most NMR data have been obtained by using pyridine as solvent, except very apolar compounds (apolar conjugates or acetate derivatives, which are more soluble in chloroform) and polar conjugates (analyzed in water). Of course, many other solvents can be used, e.g., methanol, dimethyl sulfoxide (DMSO), and methylene chloride. The use of such solvents may be of interest for specific compounds in connection with either solubility problems or the masking of signals by those of classical solvents. For ^{1}H-NMR studies with modern machines, the water solubility of most common ecdysteroids is high enough to obtain excellent spectra, and this solvent presents several interesting features (Girault and Lafont, 1988).

13.3.4. Optical Rotation Properties

Optical rotation properties have been used to assess stereochemistry at C-5, C-20, and C-22 in early studies and are still of interest for the

TABLE 13.2. Proton NMR Data for Selected Ecdysteroids (δ in ppm)

	Methyl signals in pyridine-d$_5$[a]			
Compound	*18-Me*	*19-Me*	*21-Me*	*26,27-Me*
Ecdysone	0.70	1.05	1.25(d)	1.38, 1.38
20-Hydroxyecdysone	1.20	1.07	1.56(s)	1.38, 1.38
2-Deoxyecdysone	0.74	1.05	1.28(d)	1.38, 1.38
25-Deoxyecdysone	0.72	1.06	1.25(d)	0.83, 0.83
26-Hydroxyecdysone	0.74	1.08	1.23(d)	−, 1.47
3-Dehydroecdysone	0.74	1.06	1.32(d)	1.40, 1.40
3-Epiecdysone	0.75	1.07	1.23(d)	1.40, 1.40

	Proton signals in CD$_3$OD[b]			
Compound[c]	*H-2*	*H-3*	*H-7*	*H-22*
Ecydsone	3.85(m)	3.96(m)	5.81(d)	3.61(m)
20-Hydroxyecdysone	3.84(m)	3.95(m)	5.80(d)	3.33(m)
2-Deoxyecdysone	−	−	5.80(d)	−
Ecdysone 2-PH	4.37(m)	4.18(m)	5.81(d)	−
Ecdysone 22-PH	3.86(m)	3.97(m)	5.81(d)	4.20(m)
Ecdysone 3-AC-2-PH	−	−	5.82(d)	3.58(m)

[a] From Horn and Bergamasco (1985).
[b] After Rees and Isaac (1985).
[c] AC = acetate; PH = phosphate.

characterization of ecdysteroids isolated from new biological materials.
Details about these methods can be found in a review by Nakanishi
(1971).

13.4. Basic Reactions

13.4.1. *Derivatization Reactions*

13.4.1.1. FORMATION OF ACETATES

Acetylation is a common and easy reaction of ecdysteroids that is generally used for the identification of labeled metabolites. The ecdysone
molecule bears several hydroxyl groups, which can react with acetic

anhydride in pyridine to give a complex mixture of derivatives. The respective reactiveness of the $-OH$ groups is $2 > 22 > 3 \gg 25$. Thus, under short-term conditions mainly 2-OAc and 2,22-diOAc are formed, whereas upon longer treatment 2,3,22-triOAc accumulates (Galbraith and Horn, 1969). Acetylation of the tertiary 25-OH needs strictly anhydrous reagents and a long reaction time at room temperature (24–48 h). Upon warmer conditions (e.g., 90°C), the molecule may undergo dehydration of tertiary alcohol groups [thus $\Delta 24(25)$ and/or $\Delta 17(20)$ for 20-hydroxyecdysteroids] (Horn and Bergamasco, 1985).

Any change in the ecdysone molecule may result, in drastic modifications of the acetate pattern; thus, the addition of a primary alcohol function (26-OH) or epimerization at C-3 will introduce very reactive positions. As a consequence, acetylation is often performed on mixtures of a labeled compound to identify and an unlabeled reference. It is thus possible to verify by thin-layer chromatography (TLC) that all acetates comigrate (−"fingerprint"). Even better, if HPLC is used with monitoring of both ultraviolet (UV) absorbance *and* radioactivity, it can easily be verified that the relative proportion of individual acetates also is identical for both labeled and unlabeled compounds (Fig. 13.5).

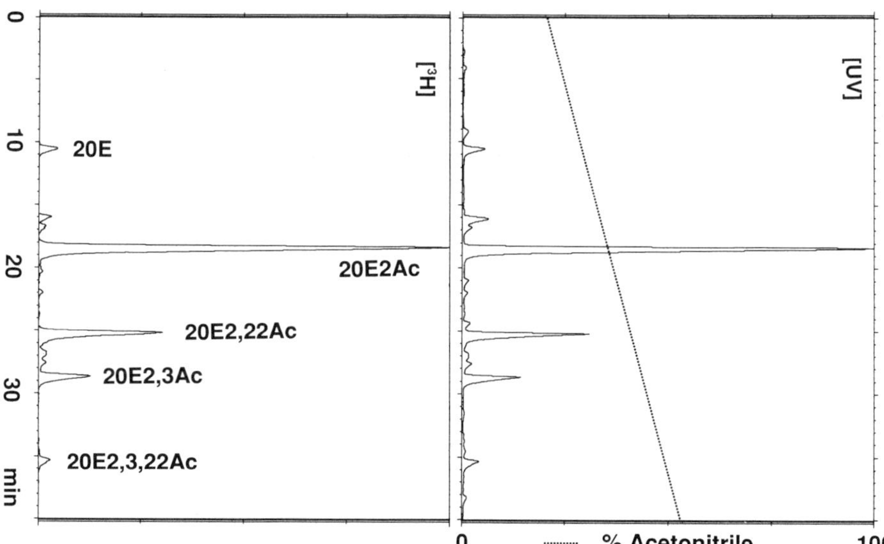

FIGURE 13.5. HPLC analysis of 20-hydroxyecdysone acetates. Operating conditions: column Spherisorb ODS-2 (250 mm long, 4.6 mm i.d.) eluted with a linear gradient 20–70% acetonitrile in 0.1% TFA. Flow rate: 1 ml/min. On-line radioactivity monitoring used a Flo-One model IC (Radiomatic Instruments).

13.4.1.2. FORMATION OF ACETONIDES

Reaction of ecdysteroids with anhydrous acetone in the presence of a catalyst (*p*-toluenesulfonic acid or anhydrous copper sulfate) leads to the formation of characteristic derivatives, acetonides, which involve vicinal *cis*-diols. Thus, 2,3 and 20,22 (if a 20-OH is present) acetonides may be formed, and (as the reaction is not complete), for instance, 20-hydroxyecdysone, will give a mixture of three derivatives: 2,3-monoacetonide, 20,22-monoacetonide, and the diacetonide. The 20,22 position is much more reactive than the 2,3 one. The reaction is used to characterize labeled metabolites in the following ways: 3-epimers and 3-oxo compounds do not react at C-2,3; 26-OH-ecdysteroids have an additional 25,26 diol and thus can give diacetonides (in the 20-deoxy series) or triacetonides (in the 20-OH series); when one (or, of course, both) of the −OH groups of a *cis*-diol is involved in a conjugation reaction, no further acetonide can be formed (e.g., 22-esters of 20-hydroxyecdysone; see Crosby et al., 1986).

Acetonides are unstable under mild acidic conditions and easily again give their parent compounds upon acidic hydrolysis.

13.4.1.3. FORMATION OF METHYL ESTERS

This reaction concerns only ecdysonoic acids. Their reaction with diazomethane in methanol will form nonionic methyl esters, which can be run on normal-phase HPLC. This reaction is used for the characterization of ecdysonoic acids among polar metabolites; only carboxylic acids react, whereas conjugates remain unchanged (Lafont et al., 1983; Rees and Isaac, 1985).

13.4.1.4. REDUCTION (FOR 3-OXO COMPOUNDS)

3-Dehydroecdysteroids can be characterized by their reduction into 3α- and/or 3β-OH compounds. This reduction can be performed either by enzymatic means [reductases present in the cytosol of many tissues (Blais and Lafont, 1984; Milner and Rees, 1985)] or by chemical means with sodium borohydride; the latter method was used in the preparation of 3-epimers of ecdysone and 20-hydroxyecdysone (Dinan and Rees, 1978).

13.4.1.5. FORMATION OF FLUORESCENT DERIVATIVES

This topic has been studied from several aspects. Ecdysteroids placed in a 50% H_2SO_4 solution form (unidentified) fluorescent derivatives (Takemoto et al., 1967) that can be used for a quantitative fluorometric assay (Gilgan and Zinck, 1972; Mayer and Svoboda, 1978; Koolman, 1980). The interest of researchers lies in the high sensitivity of fluroescence detection, which allows measurements of nanogram amounts—provided that the compounds are pure enough (e.g., as HPLC fractions; see Koolman, 1980).

A second type of reaction has been designed to prepare (specific!) fluorescent derivatives used in chromatographic detection. Two different derivatization reactions have been described so far. The first used phenanthreneboronic acid and HPTLC/HPLC separation (Poole et al., 1978), but it has received no further application. The second was more recently designed and uses 1-anthroyl nitrile as reagent (Kubo and Komatsu, 1986). It forms derivatives at C-2 (Fig. 13.6) and

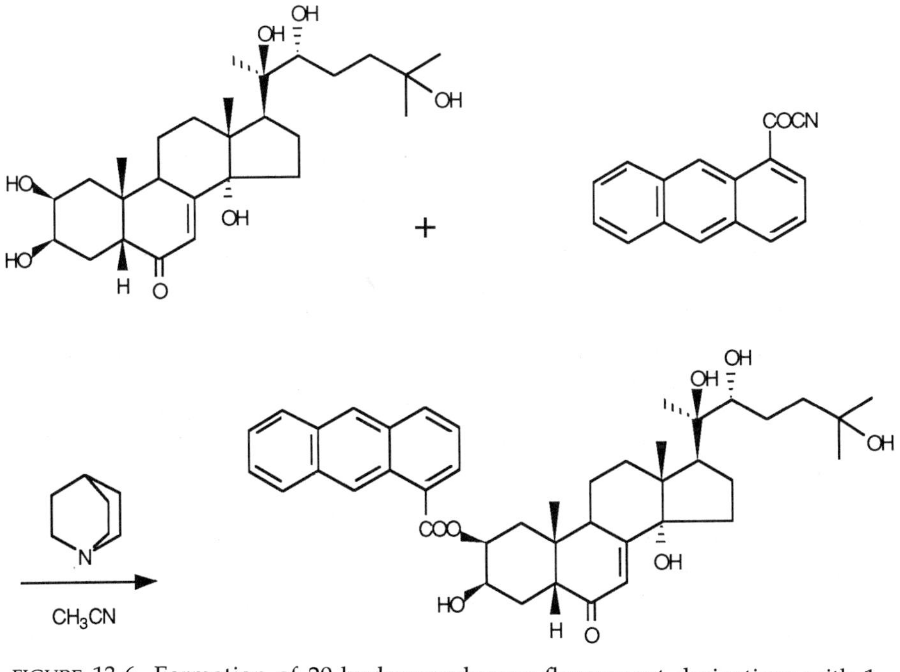

FIGURE 13.6. Formation of 20-hydroxyecdysone fluorescent derivatives with 1-anthroyl nitrile used as reagent. (After Kubo and Komatsu, 1986.)

was applied to extracts of termite queens. Kubo and Komatsu report a detection limit of about 10 pg, but in fact the method was used with a large sample (one termite queen!) containing rather high ecdysone concentrations, of which a small aliquot was thereafter analyzed by HPLC. If the method would work equally well with small samples containing low hormone concentrations, it could really compete with HPLC–RIA. Note, however, that the aforementioned reactions are not really specific for ecdysteroids and that many compounds might therefore interfere. Nevertheless, it appears that they improve sensitivity (if not specificity) when compared with UV absorbance.

13.4.2. *Hydrolysis of Conjugates*

Metabolism of ecdysteroids, as regards either excretion or storage mechanisms, proceeds mainly through conjugation reactions (see Table 13.1 for the nature of conjugates). It is often of interest to identify the genin (aglycone) in such conjugates, which can be achieved after cleavage of the ester link.

13.4.2.1. ENZYMATIC HYDROLYSIS

Enzymatic hydrolysis can be performed using a complex mixture of hydrolases contained in snail digestive juice (Sannasi and Karlson, 1974) or purified enzymes (α/β-glucosidases, phosphatases, sulfatases, β-glucuronidase, or esterases). Snail juice is widely used and is especially active toward ecdysteroid phosphates, but it also cleaves the other esters upon long-term incubation (usually overnight, sometimes 48 h or even more). Nevertheless, its esterase activity is low; therefore, for the hydrolysis of apolar conjugates, esterase (e.g., from pig liver) is preferred (Diehl et al., 1985). Some crude preparations of snail juice may contain noticeable amounts of ecdysteroids (Romer, 1979), and this must be checked for in experiments using immunoassays after hydrolysis. Purified enzyme preparations do not contain ecdysteroids. It must be noted that such preparations are not actually at all pure and, as they are used in large quantities, no firm conclusion can be drawn from their effectiveness regarding the nature of conjugates. More precise conclusions can be drawn after simultaneous use of inhibitors of some enzyme activities (Weirich et al., 1986).

13.4.2.2. CHEMICAL HYDROLYSIS

Chemical methods can be used for the cleavage of conjugates. However, they are of limited use (except for chemical syntheses). A classical method is the treatment with potassium carbonate in methanol/water for the cleavage of carboxylic esters (acetates or fatty acyl esters). Note that this method does not work with 22-esters in the 20-deoxy series (e.g., esters of ecdysone), and that upon prolonged conditions some equilibration between 5-β/5-α forms may take place (Horn, 1971).

A second method has been described, i.e., solvolysis under acidic conditions (refluxing in anhydrous dioxane containing 1% acetic acid; see King, 1972). This method was used under the assumption that the conjugates were sulfate esters. Perhaps owing to the easy degradation of ecdysteroids under acidic conditions (see Section 13.2.1), this method has had only very limited application.

13.5. Assays

Measurement of ecdysteroids can be performed either on crude samples or after a chromatographic separation. Of course, chromatographic methods use their own detectors and in several cases [e.g., GLC/ECD (gas–liquid chromatography/electron-capture detection), GLC/MS, HPLC/UV; see section 13.6], allow a direct qualitative and quantitative determination of ecdysteroid content. Nevertheless, off-line methods are of wide use when the amounts to be measured are low or simply when the required sophisticated equipment is not available. In these cases, bioassays and better immunoassays represent invaluable tools (which of course can also be used with crude samples as a rapid screening method).

13.5.1. *Bioassays*

Bioassays were used to a large extent by insect physiologists for quantifications and isolations of ecdysteroids. At the present time, more practical assays are used for the same purposes. However, bioassays are still useful for determining the biological activity of new ecdysteroids and for structure–activity studies (Bergamasco and Horn, 1980). Ecdysteroids can be assayed for their biological activity using *in vivo* or *in vitro* procedures (Table 13.3). Note that the relative activity of the different ecdysteroids may strongly vary according to the test animal and also the type of assay for a given species.

The historical work of Fraenkel (1935) was the basis for the most commonly used *in vivo* bioassay: the dipteran assay (the *Calliphora* or *Musca* test). It measures the ability to induce pupation of a larval abdomem that has been isolated by a ligation. The *Chilo* dipping test (Sato et al., 1968) was used to a large extent as a means of screening for the presence of ecdysteroids in various plants. More recently, a feeding bioassay for lepidopterous larvae (Kubo et al., 1981), an *Oncopeltus* assay (Kelly et al., 1981), and a *Limulus* assay (Jegla and Costlow, 1979) were developed.

In vitro assays include puffing assays on salivary gland polytene chromosomes, induction of morphological changes in *Drosophila* cell lines, the ability to induce evagination or cuticle formation with imaginal disks, ability to induce mRNA of protein P1 with fat body from thermosensitive *Drosophila* mutant *ecd*[1] (*ecd-ts*).

In vivo bioassays require a large supply of synchronous test animals, which must be injected and scored for biological effect. The sensitivity of the assay is on the order of 5- to 50-ng 20-hydroxyecdysone per abdomen in the *Calliphora* test (Thomson et al., 1970) and 5- to 6-ng 20-hydroxyecdysone per abdomen in the *Musca* test (Kaplanis et al., 1966). In *in vivo* assays, the observed hormonal activity is the net result of many variable contributing factors, which include distribution rates, transport, metabolism, inactivation rates, and intrinsic molting hormone activity (MH) of active compound present during the bioassay. For detecting intrinsic MH activity, short-term *in vitro* tests are preferable, because the metabolism of the tested ecdysteroids is generally very limited or even cannot be detected.

13.5.2. *Immunoassays*

The specific binding proteins used in immunoassay (IA) are antibodies obtained by immunization with ecdysteroids (Hirn and Delaage, 1980; Delaage et al., 1982). Ecdysone is not directly immunogenic. It must therefore be coupled to a suitable protein before immunization in order to be rendered immunogenic. Generally, ecdysone is derivatized to create a carboxylic function that can react with the amino groups of protein carriers [bovine serum albumin (BSA), human serum albumin (HSA), or thyroglobulin].

TABLE 13.3. Bioassays for Ecdysteroids

Test	Biological response	Limit	Duration	Reference
In vivo tests				
Calliphora test	Cuticular darkening of isolated abdomens	10^{-8} g/insect	2–3 days	Karlson, 1956 Fraenkel and Zdarek, 1970
Musca test	Cuticular darkening of isolated abdomens	5×10^{-9} g/insect	2–3 days	Kaplanis et al., 1966
Chilo dipping test	Apolysis and sclerotization of isolated abdomens	5×10^{-4} M in methanol topic application	2–3 days	Sato et al., 1968
Limulus test	Molting	2×10^{-11} g/crab	5–10 days	Jegla and Costlow, 1979
Oncopeltus test	Molting	1×10^{-6} g/insect	5 days	Kelly et al., 1981
Lepidopterous Larvae test	Molting cycle failure	10 ppm in artificial diet	A few days	Kubo et al., 1981
In vitro tests				
Drosophila salivary glands	Induction of early puffs of giant chromosomes	10^{-8} M	5–10 min	Richards, 1978
Drosophila imaginal disks	Evagination of leg disks	10^{-8} M	20 h	Fristrom et al., 1976
K_c cells	Elongation of cells	10^{-8} M	2–3 days	Cherbas et al., 1980
Drosophila fat body	Induction of protein P_1 mRNA by fat body from ECD 1 mutant	10^{-8} M	3 h	Sommé-Martin et al., 1987

Three main processes have been used to create a carboxylic function:

(1) Conversion of the 6-keto group into a carboxymethoxime (CMO) derivative (Borst and O'Connor, 1972, 1974; Porcheron et al., 1976; Maroy et al., 1977)—

$$\text{ecdysone } C\text{-}6=O + NH_2-O-CH_2-COOH \longrightarrow$$
$$\text{ecdysone } C\text{-}6=N-O-CH_2-COOH$$

(2) Succinylation of the secondary alcohols at positions 2, 3, or 22 (Lauer et al., 1974; De Reggi et al., 1975; Horn et al., 1976)—

$$\text{ecdysone } C\text{-}2-OH + (CH_2)_2(CO)_2O \longrightarrow$$
$$\text{ecdysone } C\text{-}2-0-CO-CH_2-CH_2-COOH$$

(3) Introduction of a carboxylic function at position C-26 by oxidation of inokosterone (Spindler et al., 1978)—

$$\text{inokosterone } C\text{-}26-OH \longrightarrow \text{inokosterone } C\text{-}26-OOH$$

The coupling is done after activation of the carboxylic function by any of the classical methods involving a chloroformate or a carbodiimide:

(1) Chloroformate (Borst and O'Connor, 1974; De Reggi et al., 1975; Porcheron et al., 1976)—

$$\text{ecdysone}-COOH + R'-O-CClO \longrightarrow$$
$$R'-O-CO-O-OC-\text{ecdysone} + HCl$$
$$\longrightarrow \text{protein}-NH-CO-\text{ecdysone} + CO_2 + R-OH$$

(2) Carbodiimide (Lauer et al., 1974; Horn et al., 1976; Spindler et al., 1978)—

$$\text{ecdysone}-COOH + \text{protein}-NH_2 + R-N=C=N-R' \longrightarrow$$
$$\text{protein}-NH-CO-\text{ecdysone} + R-NH-CO-NH-R'$$

A reversed approach was developed by Hung et al. (1980) in which 20-hydroxyecdysone was coupled to previously succinylated thyroglobulin with 1-ethyl-3-(dimethylaminopropyl)carbodiimide in the presence of the acylation catalyst 4-dimethylaminopyridine.

The number of ecdysone molecules coupled to protein varies according to the protein: 20 when albumin was used as a carrier, and 150 when thyroglobulin was used. The protein–ecdysone complex appears to be strongly immunogenic, and a production of high titers of antibodies is generally elicited after the first injection. The titers of antibodies reach, in the better cases, between 1/20,000 (Porcheron et al., 1976) and 1/40,000 (De Reggi et al., 1975). The dissociation constants (K_d) of anti-ecdysone antibodies vary in the range of 5×10^{-10} to 5×10^{-9} M.

The specificities of antibodies are not complete, and generally antibodies recognize both ecdysone and 20-hydroxyecdysone, as well as a series of closely related ecdysteroids (Table 13.4). Even the monoclonal antibodies for 20-hydroxyecdysone from Immunotech (Marseilles, France) recognize ecdysone with a cross-reactivity factor of 10 (Connat, 1987).

TABLE 13.4. Specificities of Antisera DUL-1 (Obtained Against Ecdysone) and DBL-1 (Obtained Against 20-Hydroxyecdysone) Detected as Cross-Reaction Factors of Various Ecdysteroids with Ecdysone in RIA[a]

	Cross-reaction factor	
Ecdysteriod	*DUL-1*	*DBL-1*
Ecdysone	1	1
2-Deoxyecdysone	1.6	4
Makisterone A	2.1	2.9
3-Epiecdysone	3.5	1.8
3-Dehydroecdysone	3.8	17
20-Hydroxyecdysone	47	2.8
3-Epi-20-hydroxyecdysone	120	3
2-Deoxy-20-hydroxyecdysone	140	30
26-Hydroxyecdysone	360	6.8
Polypodine B	200	45
20,26-Dihydroxyecdysone	560	21
Muristerone A	620	350
Inokosterone	1000	9
2,14,22,25-Tetradeoxyecdysone	8200	∞
2,22,25-Trideoxyecdysone	∞	∞
Ponasterone A	∞	3.3
Cyasterone	∞	11
Poststerone	∞	18

SOURCE: Reum et al. (1981).

[a] Fifty percent displacement of [^{3}H]ecdysone was observed at a concentration of 13 nM (DUL-1) and 24 nM (DBL-1).

Immunoassay requires tracers. In RIA, tracers are radiolabeled with high specific activity. Two kinds of analogues are used: [23,24,^{3}H]ecdysone at 50–80 Ci/mmol, and iodinated analogues at a maximum specific activity of 2000 Ci/mmol. The later are prepared from tyrosylated derivatives: the iodinated analogues are obtained by a linkage of the modified hapten via its carboxylic function to a phenolic derivative in the same way as it was linked to the carrier protein. The iodination is done according to the chloramine T (ChT) method (Hunter and Greenwood, 1962):

$$HO-\Phi-(CH_2)_2-NH-OC-R+2Na^{125}I \xrightarrow{ChT}$$
$$HO-{}^{125}I_2\Phi-(CH_2)_2-NH-OC-R$$

Owing to their higher specific activity and their closer resemblance with the immunogen, iodinated derivatives allow the use of higher dilutions of antisera and therefore increase the sensitivity of the assay ($\times 4$) (Hirn and Delaage, 1980).

The RIA has several disadvantages caused by the utilization of a radioisotope, necessitating special precautions. Recently, new immunoassays have been developed: (1) chemiluminescence immunoassay (CIA), which used ecdysone-6-carboxymethoxime aminobutylethylisoluminol (ABEI) to form a chemiluminescent tracer (Reum et al., 1984; and (2) enzyme immunoassay (EIA), with an ecdysone derivative coupled to acetylcholinesterase (Porcheron et al., 1989).

After incubation of antibodies, analogue, and biological extract (or calibration references), the separation of free and bound fractions must be performed. This separation can be done in various ways: (1) absorption by dextran-coated charcoal of the free fraction; (2) rinsing of the free fractions when the antibodies have been previously coated onto the incubation tube (Porcheron et al., 1989); (3) precipitation of the antibody–ecdysteroid complex by ammonium sulfate or by propanol-2; or (4) precipitation of the antibodies by a second antibody directed against the first antibody (Lazarovici et al., 1983; Walgraeve et al., 1986) or by protein A from *Straphylococcus aureus* (Warren et al., 1986). Alternatively, it is possible to use an equilibrium dialysis method (De Reggi et al., 1975) (Table 13.5).

The lower limit of sensitivity can be estimated, in the best cases, as 4×10^{-10} mol (1.6 pg) (Hirn and Delaage, 1980), but in fact the sensitivity currently achieved by most ecdysteroid radioimmunoassays is rather in the range of 2–20×10^{-9} mol (10–100 pg).

13.6.   Methods for Extraction and Separation

13.6.1.   *Extraction Procedures*

Sample preparation is classically performed by solvent extraction, generally alcohol (ethanol or methanol) extraction, followed by solvent partition(s) and chromatography (Morgan and Poole, 1976; Lafont et al., 1980b; Hoffmann and Hétru, 1983; Horn and Bergamasco, 1985).

The first step is homogenization of biological sample with an organic solvent. Next, the extraction solvent is concentrated or evaporated to dryness. After that, partitions are often used to eliminate very apolar or very polar molecules from samples. Coarse chromatographic procedures (e.g., preparative silica TLC, or liquid chromatography with small silica or reversed-phase columns) can be performed before using more efficient techniques.

Recently, new procedures for sample processing were designed that performed extraction and partition at the same time by water/choloroform homogenization. Ecdysteroids dissolved in the water phase are absorbed on C_{18} bonded silica cartridges (Sep-Pak), then eluted with a step gradient of methanol (or acetonitrile) in water (Lafont et al., 1982; see also Watson and Spaziani, 1982; Pimprikar et al., 1984).

Immunoabsorption was used for the analytical-scale purification of ecdysteroids from insects (Reum et al., 1981). Antiecdysone rabbit antibodies were covalently linked to Sepharose 4B and used as immunoabsorbents. These were able to bind ecdysteroids from aqueous crude extracts. Bound ecdysteroids could be eluted quantitatively from the absorbents using 3 *M* sodium trichloroacetate.

13.6.2.   *Separation Procedures*

13.6.2.1.   THIN-LAYER CHROMATOGRAPHY

TLC can be used during sample preparations as for isolation purpose. Most often, TLC is performed on silica gel plates containing a fluorescent additive absorbing UV light at 254 nm with a convenient solvent system like chloroform : methanol (80:20) or cholorform : 96% ethanol (80:20) (Horn and Bergamasco, 1985). Polar ecdysteroids like ecdysonoic acids and phosphate conjugates can be separated using ethyl acetate : ethanol : water (20:80:10). Two-dimensional TLC is also feasi-

TABLE 13.5. Ecdysone Immunoassays

Reference	Hapten	Antiserum dilution	Analogue	Free and bound separate
Borst and O'Connor, 1972, 1974	20-hydroxy-ecdysone–CMO	1/100	[^{3}H]Ecdysone	Precipitation by Ammonium sulfate
Lauer et al., 1974	20-hydroxyecdysone Monohemisuccinate		[^{3}H]Ecdysone	Precipitation by ammonium sulfate
De Reggi et al., 1975	20-Hydroxyecdysone 22-monohemisuccinate	1/40,000	[^{125}I]Succinyl 20-ecdysone tyrosine methyl ester	Equilibrium dialysis
Porcheron et al., 1976	Ecdysone–CMO	1/20,000	[^{125}I]6-Carboxymethyloxime 20-hydroxyecdysone tyramine amide	Charcoal or precipitation by propanol-2
Horn et al., 1976	20-Hydroxyecdysone 22-monohemisuccinate	1/10,000	[^{3}H]Ecdysone	Precipitation by ammonium sulfate
Lazarovici et al., 1983			[^{3}H]Ecdysone	Precipitation of antibody by second antibody
Warren et al., 1986			[^{3}H]Ecdysone	Precipitation with protein A from Staphylococcus aureus

Reference	Hapten	Dilution	Tracer	Separation method
Maroy et al., 1977	Polypodine B–CMO	1/100	[^{3}H]Ecdysone	Precipitation by Ammonium sulfate
Spindler et al., 1978	Inokosterone-26-oic acid	1/5,000	[^{3}H]Ecdysone	Precipitation by ammonium sulfate
Hung et al., 1980	20-Hydroxyecdysone monohemisuccinate		[^{3}H]Ecdysone	Precipitation by ammonium sulfate
Soumoff et al., 1981	20-Hydroxyecdysone 2-monosuccinate	1/20,000	[^{3}H]Ecdysone	Precipitation by ammonium sulfate
Warren et al., 1986			[^{3}H]Ecdysone	Precipitation with protein A from Staphylococcus aureus
Reum et al., 1981	Ecdysone–CMO	1/2,500		
Lazarovici et al., 1983			[^{3}H]Ecdysone	Precipitation of antibody by second antibody
Reum et al., 1984	Ecdysone–CMO		Ecdysone–CMO–ABEI (chemiluninescent immunoassay)	Charcoal absorption of the free
Porcheron et al., 1989	Ecdysone–CMO	1/100,000	Ecdysone–CMO couples to acetyl-chcline esterase	Rinsage of the free (coated antibodies)

ble (Briers and de Loof, 1983). The development of high-performance thin-layer chromatography (HPTLC) has increased the possibilities of TLC (Wilson and Lafont, 1986). Recently, the use of reversed-phase TLC on C-18 coated plates or on silica gel HPLTC plates impregnated either with paraffin (Wilson et al., 1981; 1982c) or coated for ion-pair chromatography (Lewis and Wilson, 1984) has been proposed with methanol : water mixtures for development (Table 13.6).

TABLE 13.6. R_f Values Obtained for Ecdysteroids by Normal-Phase TLC on Silica Gel TLC Plates Using Chloroform : Ethanol (4:1) and by Reversed-Phase TLC on C_{18} Bonded TLC Plates Using Methanol : Water (65:35)

Compound	np-TLC R_f	rp-TLC R_f
22-Isoecdysone	0.13	0.16
20-Hydroxyecdysone	0.15	0.47
Inokosterone	0.17	0.47
Makisterone A	0.20	0.40
Ecdysone	0.21	0.35
Polypodine B	0.22	0.47
Ajugasterone C	0.22	0.36
Muristerone A	0.27	0.37
Dacrysterone	0.27	0.39
2-Deoxy-20-hydroxyecdysone	0.31	0.28
Poststerone	0.32	0.42
Pterosterone	0.32	0.37
Cyasterone	0.33	0.44
2-Deoxyecdysone	0.38	0.19
Ponasterone C	0.38	0.33
Calonysterone	0.42	0.28
Ponasterone A	0.42	0.19
2-Deoxy-3-epiecdysone	0.44	0.15
Kaladasterone	0.49	0.26
20-Hydroxyecdysone 2-cinnamate	0.53	0
Pinnasterol	0.56	0.10
Polypodine B 2-cinnamate	0.56	0
Ponasterone C 2-cinnamate	0.65	0
Acetylpinnasterol	0.68	0
Carpesterol	0.86	0

SOURCE: Wilson (1985).

Ecdysteroids in quantities as low as 500 ng can be clearly distinguished under UV illumination as dark spots on a fluorescent background (Mayer and Svoboda, 1978). Ecdysteroids can also be visualized by spraying the plates with vanillin–sulfuric acid reagent, which gives characteristic colored spots (Horn and Bergamasco, 1985). Similarly, sulfuric acid induces fluorescence of ecdysteroids when excited with light of a suitable wavelength. None of these detection methods is specific for ecdysteroids, but these three methods can be successively performed on the same plate. In this case, a spot that absorbs UV light, fluoresces with sulfuric acid, and reacts with vanillin–sufuric acid has a good probability of being an ecdysteroid, even if other controls are still necessary.

TLC is often used before assays (bioassays or RIA) to separate ecdysteroids from a crude extract. After chromatography, horizontal silica gel bands are scraped and eluted with methanol, then assayed. TLC is also widely used for metabolic studies of radiolabeled ecdysteroids. The radioactivity can be analyzed either with a TLC scanner (Wilson and Lafont, 1986), by fluorography (Lehmann and Koolman, 1986), or by liquid scintillation counting after elution.

13.6.2.2.   DROPLET COUNTERCURRENT CHROMATOGRAPHY

DCCC is an efficient method for the preparative separation of ecdysteroids owing to its high sample capacity (several grams) and because it requires only 1–2 litres of solvent per analysis (Kubo and Hanke, 1986). Separation in DCCC is accomplished by pumping droplets of a mobile phase through columns of stationary liquid phase. Solutes are separated according to their partition coefficients between the two immiscible liquid phases. Ecdysteroids from crude extracts are generally separated by DCCC with a solvent system of chloroform : methanol : water (13:7:4, v/v/v) in the ascending method. Effluent can be monitored by UV detection (Kubo et al., 1984, 1985a) (Fig. 13.7).

13.6.2.3.   GAS–LIQUID CHROMATOGRAPHY

Ecdysteroids are nonvolatile compounds, which can be gas chromatographed only after being converted into trimethylsilyl (TMS) ethers (Katz and Lensky, 1970). GLC analysis underwent a marked period of development during the 1970–1980 decade, but in recent years has had much more limited use.

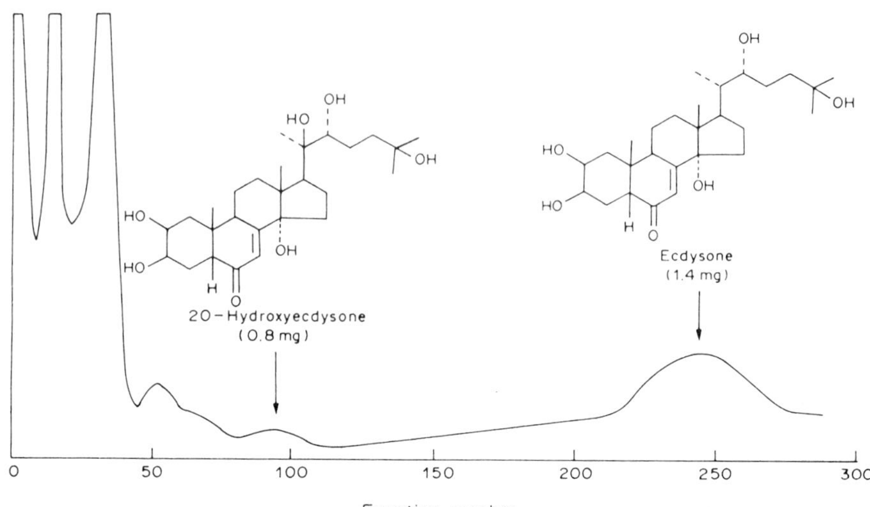

FIGURE 13.7. Analysis by DCCC of a crude extract from *Bombyx mori* pupae. (After Kubo et al., 1985a.)

TMS ether formation first used bis(trimethylsilyl)acetamide, which was later replaced by TMSI (trimethylsilylimidazole) (Ikekawa et al., 1972; Miyazaki et al., 1973) or BSTFA [*N*,*O*-bis(trimethylsily)trifluoroacetamide] (see Morgan and Wilson, 1980). For some specific detection procedures (see below: ECD), HFB–TMS (heptafluorobutyryl–TMS) derivatives are used (see Koreeda and Teicher, 1977). Several silylation procedures have been compared as to their ability to fully silylate both secondary and tertiary hydroxyl groups (Morgan and Poole, 1976; Bielby et al., 1980, 1986). In fact the tertiary 14α-OH group is little reactive, and its trimethylsilylation requires either high temperatures (140°C) or long reaction times (overnight). In many cases it is advisable to purify derivatives prior to injection on the GLC column. Removal of excess reagents or their byproducts can be easily achieved by partioning between acetonitrile and hexane: the apolar TMS ethers are found in the hexane phase. Alternatively, TMS ethers can be purified by TLC on silica plates (Morgan and Woodbridge, 1971) or HPLC (Lafont et al., 1980b).

GLC analysis requires high temperature (280°C) and can be performed either on packed or capillary columns (Evershed et al., 1987). A phytoecdysteroid not present in animals (e.g., cyasterone) can be used as internal standard for quantification (Ikekawa et al., 1972; Wilson et al., 1982a). Detection can be performed in several ways: one can use a universal flame-ionization detector (FID), with a rather low sensitivity, or (better yet) an electron-capture detector (ECD), which allows detection in the picogram range (Poole and Morgan, 1975; Poole et al., 1975); alternatively, GLC can be interfaced with a mass spectrometer, giving both high selectivity and high sensitivity. In the mass fragmentographic mode, 100-pg amounts are quantified with good accuracy (Evershed et al., 1987). Individual ecdysteroids are characterized by a combination of their retention time and the mass of one (single-ion monitoring, or SIM) or several (multiple-ion detection, or MID) of their characteristic fragments (Morgan and Woodbridge, 1974; Lafont et al., 1980b). Repeated scanning of masses over a wide range is also feasible yielding complete mass spectra of the different peaks.

In fact, GLC/FID and GLC/ECD are apparently no longer used, and only GLC/MS is presently used for specific purposes, i.e., to identify ecdysteroids present in low amounts in new biological materials when the available material does not allow a more complete identification using NMR techniques. GLC has, in fact, been almost totally replaced by liquid chromatographic techniques, which are easier to perform and allow easy recovery of injected compounds for further analyses, e.g., RIA. GLC apparatus might also be interfaced with a radioactivity monitor for metabolic studies, but to our knowledge this possibility has not as yet been investigated.

13.6.2.4. HIGH-PERFORMANCE LIQUID CHROMATOGRAPHY

High-performance liquid chromatography (HPLC) of ecdysteroids was introduced by Hori in 1969. Since that time this technique has been considerably improved, and presently it is the most efficient and widely used separation procedure. This subject has been extensively reviewed (see Schooley and Nakanishi, 1973; Lafont et al., 1980b, 1981; Rees and Isaac, 1985; Horn and Bergamasco, 1985; Lafont, 1987).

Classically, several chromatographic modes can be distinguished, including (1) normal-phase systems (silica or polar-bonded phases), (2) reversed-phase systems (most users), and (3) ion-exchange or ion-pairing systems (in the case of ionic metabolites only). The choice of which chromatographic system to use depends on the polarity of the ecdysteroids to be separated (Table 13.7).

Many types of detectors can be used. The presence of a 7-ene-6-one chromophore allows easy detection of nanogram amounts (of pure compounds) at 254 nm (fixed-wavelength UV detectors) or, better yet,

TABLE 13.7. Selected Examples of Chromatograph Systems Commonly Used for Analysis of Ecdysteroids[a]

Class of ecdysteroids	Type of chromatography	Reference
Polar	*Reverse-phase chromatography*	
	Methanol : water	Hétru et al., 1985
	Methanol : sodium acetate	Isaac et al., 1983
	Acetonitrile : Tris-HClO$_4$	Lafont et al., 1980a,b
		Modde et al., 1984
	Acetonitrile : Tris-HCl	Hétru et al., 1985
	Acetonitrile : sodium citrate	Modde et al., 1984
	Ion-pair chromatography	
	Acetonitrile : cetrimide phosphate	Lafont et al., 1980b
	Methanol : Tetrabutylammonium	Isaac et al., 1983
		Scalia and Morgan, 1985
	Ion-exchange chromatography	
	Ammonium acetate (Partisil SAX)[b]	Rees and Isaac, 1985
Medium	*Reverse-phase chromatography*	
	Ethanol : water	Hori, 1969
	Methanol : water	Schooley et al., 1972;
		Schooley and Nakanishi, 1973

	Acetonitrile/water	Holman and Meola, 1978
		Wilson et al., 1982b
	Acetonitrile : 0.1% trifluoroacetic acid	Nirdé et al., 1983
	Acetonitrile : buffer	Lafont et al., 1979
	Tetrahydrofuran : water	Wilson et al., 1982b
	Dioxan : water	Wilson et al., 1982b
	Normal-phase chromatography	
	Silica—dichloromethane : isopropanol : water	Lafont et al., 1979, 1980b, 1981
	isooctane : isopropranol : water	Beydon et al., 1987b
	hexane : ethanol : methanol : acetonitrile	Moribayashi et al., 1985
	Diol—dichloromethane : isopropanol : water	Lafont et al., 1981
	APS—dichloromethane : isopropanol : methanol	Dinan et al., 1981
Apolar	*Reverse-phase chromatography*	
	Methanol : Tris–$HClO_4$	Diehl et al., 1985
	Methanol	Kubo et al., 1987
	Acetonitrile : isopropanol	C. Blais, personal communication
	Normal-phase chromatography	
	Silica—dichloromethane : isopropanol : water	Diehl et al., 1985

[a] The use of gradient elution allows a single system to separate the three classes of ecdysteroids in a single run.

[b] SAX = strong anion exchanger.

[c] APS = aminopropyl silane.

242 nm (spectrophotometers). Diode-array detectors yield more information from the chromatogram, as they continuously record the absorbance over a chosen wavelength range and directly give the UV spectrum of every peak; this is an efficient way to assess compound identity and purity (Beydon et al., 1987a; Lafont, 1987). Fluorescence detection can be used, provided that suitable derivatives have been prepared before injection (Kubo and Komatsu, 1986). HPLC can be interfaced with a mass spectrometer, and this approach has been applied to many compounds including insect juvenile hormones (Mauchamp et al., 1981). To our knowledge, however, there is no successful report as to the application of this method to ecdysteroids. Quantitative analyses using HPLC are feasible using a suitable phytoecdysteroid as internal standard (Lafont et al., 1982; Wilson et al., 1982a).

On-line radioactivity monitoring appears to be a very powerful method for metabolic studies, as it permits an *exact* check for comigration with a reference compound (this is not possible with off-line procedures when very large fractions are collected).

In many cases, HPLC is in fact used as a separation method essentially coupled with an RIA on collected fractions. This will be examplified in the next section.

13.6.2.5. SUPERCRITICAL FLUID CHROMATOGRAPHY

SFC is a new technique that combines characteristics of both GLC and HPLC. The mobile phase is a supercritical fluid (e.g., liquid carbon dioxide—possibly modified by a small percentage of methanol) and the column is either a packed HPLC column [e.g., ODS-bonded silica] or a capillary glass column (similar to a capillary GLC column coated with a suitable organic film). Owing to the low viscosity of the mobile phase, diffusion coefficients are much higher (100 times) than in normal fluids and as a consequence large flow rates may be used with a high efficiency, which results in very short analysis times. In addition, many types of detectors can be used—either those employed with HPLC or those with GLC (FID). Interfacing with an IR spectrometer and MS is also quite feasible.

Only preliminary experiments have been performed with ecdysteroids (Morgan et al., 1988; Raynor et al., 1988) demonstrating the feasibility of this method, and efficient separations have been obtained with reference compounds (Fig. 13.8). Application to biological sam-

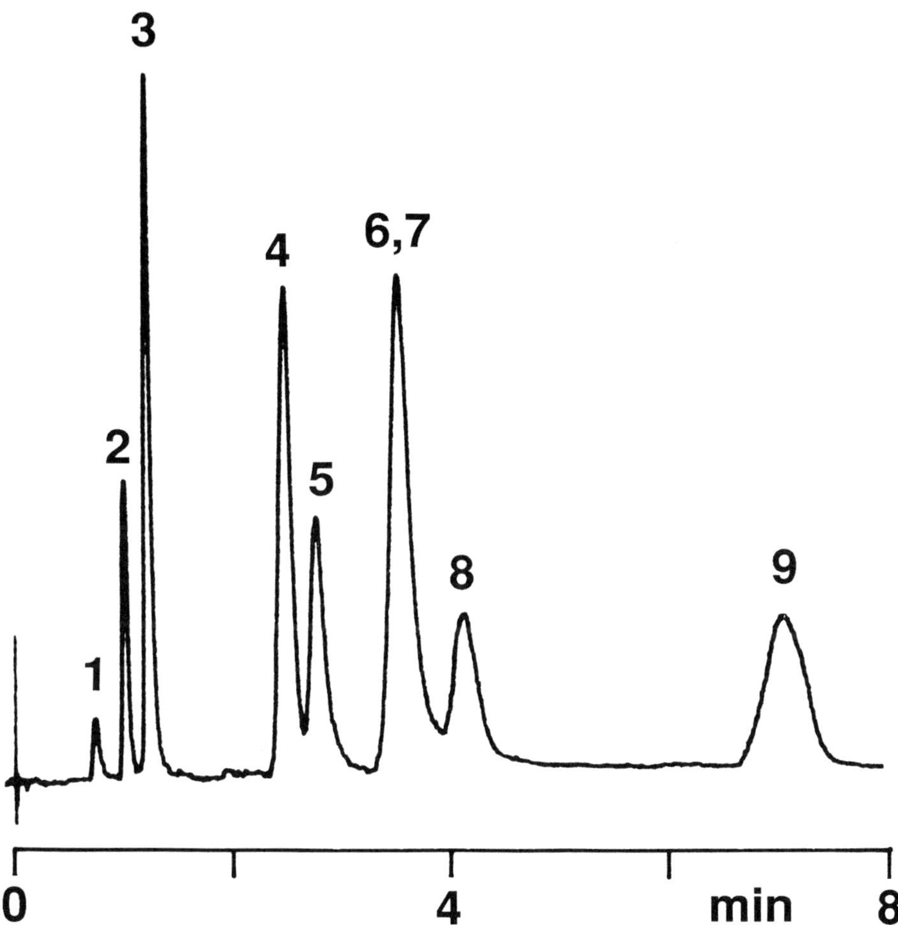

FIGURE 13.8. Separation of ecdysteroids by supercritical fluid chromatography (SFC). (Courtesy of Dr. I. D. Wilson.)

ples remains to be performed in order to further assess the interest of this new technique.

13.6.2.6.   HIGH-VOLTAGE PAPER ELECTROPHORESIS

Paper electrophoresis can be used for the separation of ionic conjugates or ecdysonoic acids from nonionic parent compounds, which do not move (see, e.g., Lafont et al., 1983). Although efficient, this method has had few applications as of yet.

13.7. Practice of Ecdysteroid Analysis

Some examples have been selected to show how various authors have used the methods described above in actual cases. The first five examples illustrate isolation/identification of ecdysteroids; the last two concern ecdysteroid quantification.

13.7.1. *Identification of Ecdysteroids Whose Reference is Available*

Makisterone A has been identified as the major ecdysteroid from pupae of the honeybee *Apis mellifera* (Feldlaufer et al., 1985; Feldlaufer and Svoboda, 1986).

Previous studies had demonstrated that the honeybee is unable to dealkylate C_{28} and C_{29} plant sterols to cholesterol. As a consequence, 24-alkyl sterols account for almost 99% of total sterols. Since it is generally accepted that neutral sterols serve as precursors for MHs, honeybee pupae were examined for ecdysteroids. Makisterone A, a C_{28} 20-hydroxyecdysone analogue, was identified as the major ecdysteroid.

Its identification was facilitated by the mere fact that makisterone A standard was commercially available. Pupae (855 g) were homogenized in methanol, then methanol : water. The supernatants were combined and dried *in vacuo*. Partitions were performed to remove apolar lipids (*n*-hexane : methanol : water) and polar materials (*n*-butanol : water).

After drying the residue was placed on a silica column (Florisil, 30 g, 60–100 mesh) in methanol : chloroform (5:95) and eluted with a step gradient of methanol in chloroform. Portions of column fraction containing the pupal ecdysteroids were fractionated by reversed-phase HPLC (C_8 column with methanol : water (35:65)) and analyzed by RIA. An immunoreactive UV-absorbing fraction matching the retention time of authentic makisterone A was observed (Fig. 13.9). This fraction (yielding ~ 111 μg) was subsequently chromatographed by silica HPLC (silica column with methylene chloride : propanol-2 : water (125:30:2)). A single peak comigrating with reference makisterone A was again observed. Its identity was further assessed by mass spectrometry and ^{1}H-NMR.

Finally, it was determined that [^{3}H]campesterol (a C_{28} sterol) was converted by honeybee pupae to a compound comigrating with makisterone A both in reversed-phase and silica HPLC (Feldlaufer and Svoboda, 1986).

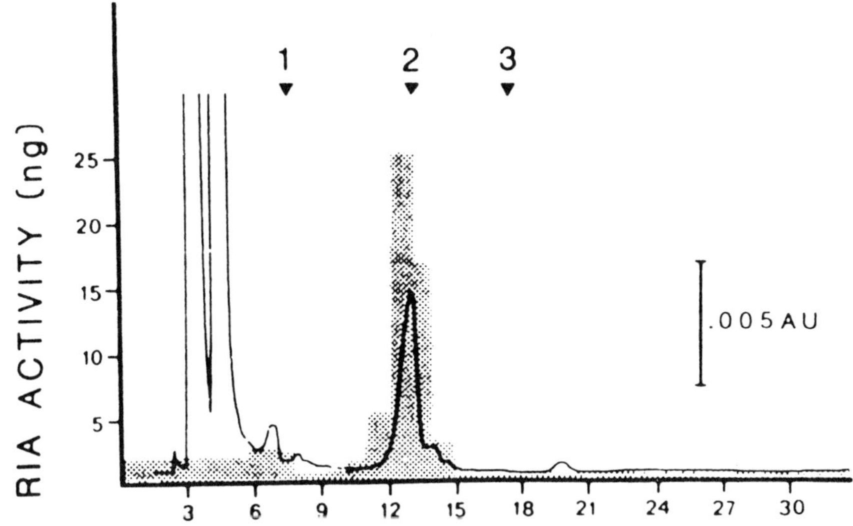

FIGURE 13.9. Analysis of ecdysteroid fraction from honeybee pupae by rpHPLC. UV 254-nm trace and RIA activity expressed as ecdysone equivalents (shaded areas). *Conditions:* C_8 column. Mobile phase: methanol:water (35:65). Flow rate: 1 ml/min. Column temperature: 33°C. Arrows: elution times of 20-hydroxyecdysone (1), of makisterone A (2), and of ecdysone (3). (After Feldlaufer et al., 1985.)

13.7.2. *Identification of New Ecdysteroids from Rich Materials*

Major ecdysteroids have been isolated from the pycnogonid *Pycnogonum litorale* (Bückmann et al., 1986).
Preliminary RIA tests provided evidence that *P. litorale* contains extremely high levels of ecdysteroids (up to 150-ng ecdysone equi./mg dry weight). About 3000 adult animals of both sexes (63 g dry weight) were homogenized in absolute methanol then centrifuged. After evaporation, the dry residue was partitioned between water and chloroform. The ecdysteroids dissolved in the aqueous extract were absorbed onto a 10-ml column filled with preparative C_{18} bonded silica phase (recovered from Sep-Pak cartridges). The column was rinsed with 50 ml of 20% aqueous methanol; then the ecdysteroids were eluted with 100% methanol and evaporated to dryness.

An aliquot of the ecdysteroid-enriched fraction was first analyzed by reversed-phase HPLC (Fig. 13.10); then the whole sample was fractioned on a semipreparative column. The "ecdysteroids" (in fact the major UV-absorbing peaks) were collected and absorbed on Sep-Pak

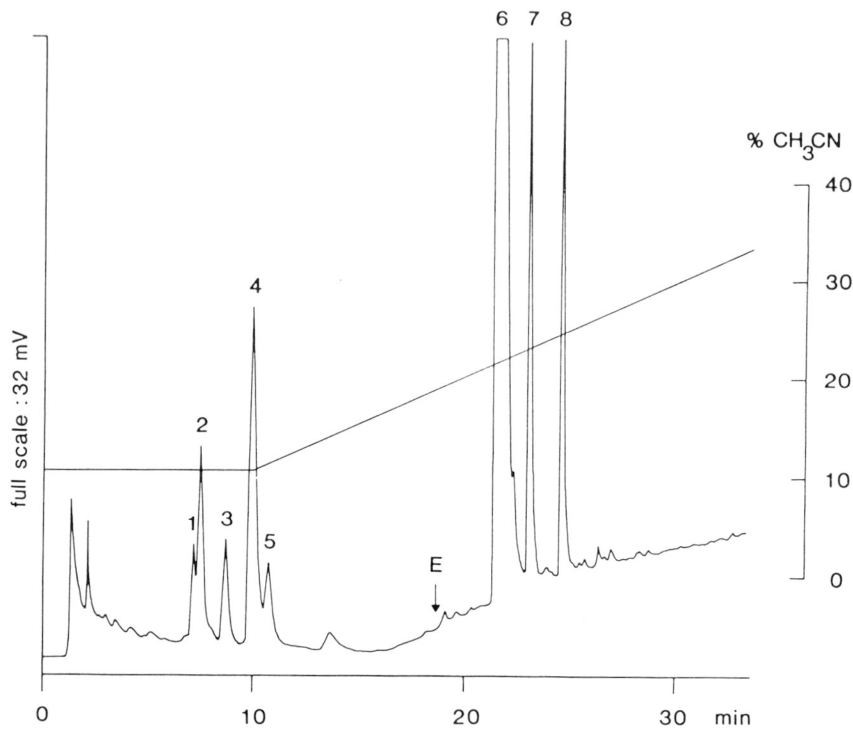

FIGURE 13.10. HPLC analysis of *Pycnogonum* ecdysteroids on a RP-18 Supersphere column. Isocratic conditions of 16% acetonitrile in 0.1% trifluoroacetic acid (TFA) for 10 min followed by a linear gradient 16–40% acetonitrile in 0.1% TFA in 25 min, at 1 ml/min flow rate; UV detection at 254 nm. Compounds 1–8: major ecdysteroids; arrow indicates the retention time of ecdysone (E) standard. (After Bückmann et al., 1986.)

cartridges (after removal of acetonitrile by evaporation under reduced pressure), rinsed with water, and eluted with methanol. They were further purified by HPLC on a silica column (Zorbax-Sil) with dichloromethane : propanol-2 : water (125:25:2).

Ecdysteroids were isolated and identified by a combination of mass spectrometry (CID) and ^{1}H-NMR: one of the compounds was 20-hydroxyecdysone; two further ecdysteroids showed no hydroxyl group at C-22 (22-deoxy-20,26-dihydroxyecdysone and 22-deoxy-20-hydroxyecdysone). Other compounds were esters of ecdysteroids with acetic acid (20,26-dihydroxyecdysone 22-acetate and 20-hydroxyecdysone 22-acetate) or with glycolic acid (ecdysone 22-glycolate and 20-hydroxyecdysone 22-glycolate).

13.7.3. *Identification of Ecdysone Metabolites* *Produced* In Vivo

20-Hydroxyecdysone 3-acetate and 20-hydroxyecdysone 3-acetate 2-phosphate have been isolated from *Locusta migratoria* feces (Modde et al., 1984).

[^{3}H]20-Hydroxyecdysone was injected into last-nymphal instar locusts. The feces of the injected insects were collected over a 24-h period. They were extracted with chloroform : water, and the ecdysteroids contained in the water phase were absorbed on a C_{18} Sep-Pak cartridge, then eluted with 60% aqueous methanol, and then 100% methanol. Samples were analyzed by reverse-phase HPLC. In HPLC at least eight labeled peaks were observed. Some of them disappeared if extracts were incubated with *Helix pomatia* digestive juice: they corresponded to hydrolyzable conjugates. One of these peaks ("peak 4" in Fig. 13.11), the major conjugate, gave 20-hydroxyecdysone 3-acetate during a 1-h incubation with *H. pomatia* juice. It was obvious from reversed-phase HPLC that this double-conjugate was an anion, because decreasing the pH of the mobile phase increased the retention time. Further investigations were performed by double-labeling experiments with [^{3}H]20-hydroxyecdysone and either [^{32}P]phosphate or [^{35}S]sulfate. Only [^{32}P]phosphate gave positive results (Fig. 13.11). It was inferred that "peak 4" was a phosphate ester.

The position of phosphate conjugation remained to be determined. The most probable sites for this conjugation were the still free secondary hydroxyl groups at C-2 or C-22. The phosphate conjugate was acetylated, then briefly treated with *H. pomatia* juice to remove the phosphate group, and this yielded a 3,22-diacetate. It was deduced that peak 4 was most probably conjugated with phosphate at C-2; 800 μg of this compound were further isolated from feces of male adults of *L. migratoria* that received daily injections of 10 μg of 20-hydroxyecdysone (plus [^{3}H]20-hydroxyecdysone). The pure compound was analyzed by negative-ion fast-atom bombardment (FAB$^-$) mass spectrometry. The spectrum obtained confirmed that peak 4 was a 20-hydroxyecdysone acetophosphate, with both acetate and phosphate on the nucleus, thus supporting the proposed structure (now confirmed by NMR data; J.-P. Girault and R. Lafont, unpublished work).

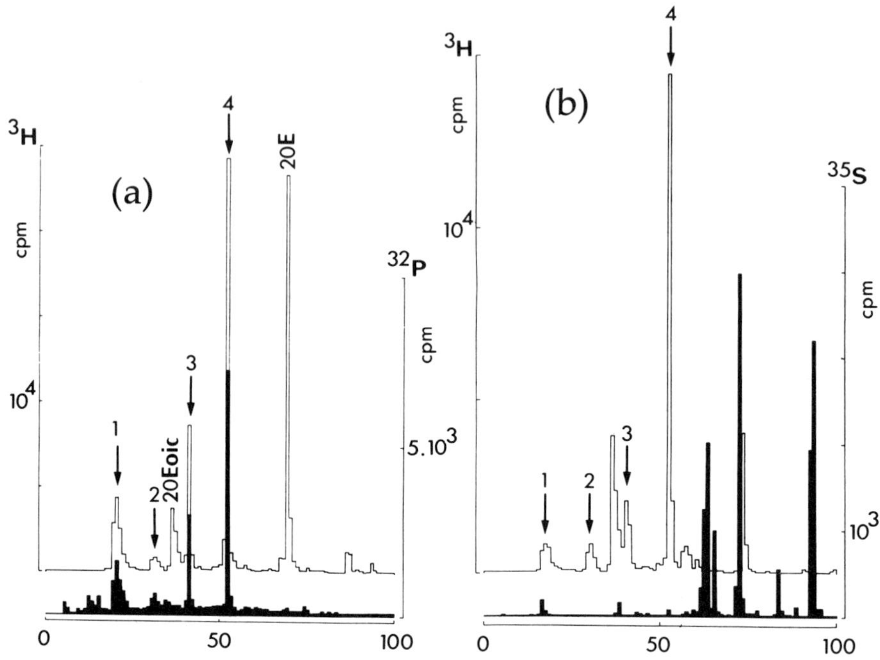

FIGURE 13.11. HPLC analysis of metabolites in feces of male adult locusts injected with [^{3}H] + 20-hydroxyecdysone and sodium [^{32}P]phosphate (*a*), and injected with [^{3}H] = 20-hydroxyecdysone and sodium [^{35}S]sulfate (*b*). *Operating conditions:* RP Ultrasphere ODS column, gradient elution from 8% to 40% acetonitrile in tris/perchloric acid buffer (20 m*M*, pH 7.5) for 60 min at 1 ml/min; 0.4 min fractions were collected and counted for radioactivity after addition of 2 ml of scintillation cocktail. Arrows: major 20-hydroxyecdysone conjugates (20Eoic = 20-hydroxyecdysonoic acid).

13.7.4. *Identification of Ecdysone Metabolites Produced* In Vitro

3-Dehydro-20-hydroxyecdysone and 3-epi-20-hydroxyecdysone have been produced by cytosolic preparations from *Pieris brassicae* incubated with 20-hydroxyecdysone (Blais and Lafont, 1984).

Animals were homogenized in 50-m*M* potassium phosphate buffer (pH 7.2). Cytosolic preparations (100,000*g* supernatants) were incubated with 20-hydroxyecdysone ($10^{-5}M$) plus [^{3}H]20-hydroxyecdysone for 1 h. Two major metabolites were separated by HPLC on a reversed-phase column and a silica column. Both compounds were further characterized by acetonide formation and mass spectrometry. The position of the keto group at C-3 for 3-dehydro-20-hydroxyecdysone was confirmed by ^{1}H-NMR analysis in D_2O (Table 13.8).

13.7.5. Utilization of Cholesterol Labeling for the Isolation of Ecdysteroids

26-Hydroxyecdysone conjugates from eggs of the tobacco hornworm, *Manduca sexta*, have been isolated and identified (Thompson et al., 1987b).

The method used for the isolation of ecdysteroids was particularly inventive in that labeled cholesterol was used to spot ecdysteroids in HPLC analyses. The large amounts of ecdysteroids in *M. sexta* eggs facilitated their ensuing identification.

[^{14}C]Cholesterol was injected into female pupae before eclosion to label ecdysteroids. Eggs or ovaries from labeled animals were homogenized in methanol then methanol-water 70%. After evaporation, a butanol/water partition was performed. Ecdysteroid conjugates were in the aqueous phase. The conjugates redissolved in H_2SO_4 ($10^{-2}N$) were absorbed onto a column of Amberlite XAD-2 (reverse-phase packing). The column was rinsed with water and the conjugates were eluted with 95% ethanol. The extract was further fractionated on a C_{18} Sep-Pak cartridge. The ecdysteroid conjugates were separated by RP-HPLC on a C_{18} µBondapak column by isocratic elution with 30% methanol in 0.025 M NaH_2PO_4 buffer. The labeled peaks indicated putative ecdysteroids (Fig. 13.12). After collection via RP-HPLC from pooled material, the conjugates were finally purified by desalting on C_{18} Sep-Pak cartridges.

Maturing eggs of the tobacco hornworm contain at least three ecdysteroid conjugates, two of which had been previously identified as 26-hydroxyecdysone 26-phosphate (Thompson et al., 1985b) and 26-hydroxyecdysone 22-glucoside (Thompson et al., 1987a), only the third had to be characterized. This compound exhibited an absorbance maximum at 240 nm, it was hydrolysable by *Helix pomatia* juice and yielded 26-hydroxyecdysone. Further identification was performed by FAB$^-$ mass spectrometry, which confirmed the assigned molecular weight of 560 for the ecdysteroid conjugate, suggesting a sulfate or phosphate of 26-hydroxyecdysone. The very strong peaks in the spectrum at m/z-79 and m/z-97, corresponding to PO_3^- and $H_2PO_4^-$ ions, respectively, established that the conjugate was indeed a phosphate. ^{1}H-NMR gave evidence for a 2-phosphate ester.

TABLE 13.8. Physicochemical Characterization of 20-Hydroxyecdysone Metabolites as Compared to 20-Hydroxyecdysone

	Mass spectrum[a] [value in mass units m/e (corresponding ion)	Acetonide derivatives[b] (relative retention time)	Chemical shifts of proton signals (∞) in ppm[c]		
			H-2α	H-3	H-22β
20-hydroxy ecdysone	481 (MH$^+$), 463 (MH$^+$–H$_2$O, 427 (MH$^+$–3 H$_2$O), 363 [MH+–(C-22–C-27) 345 (363–H$_2$O)	20-Hydroxyecdysone (1) 20,22-Monoacetonide (0.42) 2,3-Monoacetonide (0.18) Diacetonide (0.1)	3.89	4.04	3.44
3-Epi-20-hydroxy-ecdysone	481, 463, 445, 427 363, 345.	3-Epi-20-hydroxyecdysone (0.9) 20,22-Monoacetonide (0.34)	3.76[d]	3.50[d]	3.43[d]
3-Dehydro 20-hydro xyecdysone	479, 461, 443, 425 361, 343,	3-Dehydro-20-hydroxy-ecdysone (0.28) 20,22-Monoacetonide (0.12)	4.45	—	3.44

SOURCE: Modified from Blais and Lafont (1984).

[a] Chemical ionization/desoprtion mass spectrometry with ammonia as reagent gas.

[b] Separations were performed by Sil–HPLC with methylene chloride : propanol-2 : water (125:25:2)

[c] NMR with a Bruker WM 250 apparatus, in deuterochloroform.

[d] From Beydon et al. (1987b).

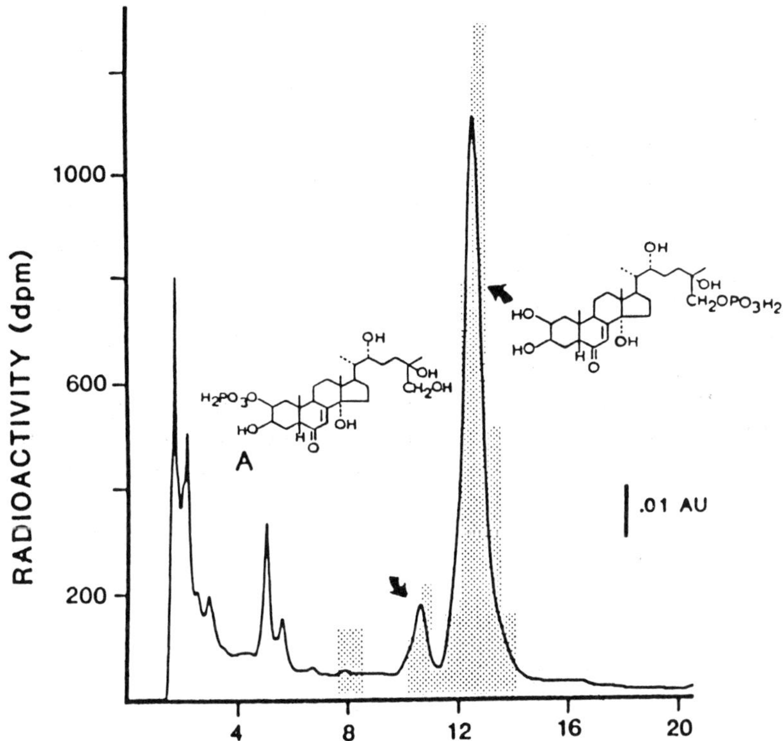

FIGURE 13.12. Reversed-phase (rp) HPLC analysis of partially purified ecdysteroid conjugates from ovaries of day-4 *Manduca sexta*. *Operating conditions:* rp IBM C_8 column; isocratic elution with 26% methanol in 0.03 M NaH_2PO_4 buffer (pH 5) at a flow rate of 1 ml/min; detection at UV 254 nm (trace). Shaded areas indicate radioactivity. (After Thompson et al., 1987b.)

13.7.6. *Quantification of Ecdysteroids by Reversed-Phase HPLC/RIA*

Warren et al. (1986) utilized two complementary antisera for quantification of ecdysteroids by RP-HPLC/RIA.

In this example, the authors describe the use of HPLC techniques for the separation of ecdysteroids in animal extracts using RIA for their detection. The identification of the major ecdysteroids had been previously performed, and RIA calibration curves had been established.

Samples were thawed and homogenized in acetronitrile, then centrifuged. The pellets were reextracted with aqueous acetronitrile 50%, and the supernatants were pooled. Apolar compounds were eliminated from samples by a partition against hexane. The extract was further purified on a C_{18} Sep-Pak cartridge, then injected into RP-HPLC with the gradient mode. Recovery of ecdysteroids was monitored by RIA of aliquots.

The technique described by Warren et al. is of special interest owing to the use of two nonspecific, complementary antisera to identify and quantify ecdysteroids. The H-22 antiserum (from D. H. S. Horn) is able to detect metabolites with hydroxyls at C-20 and C-26 in addition to conjugates at C-22 and C-26 hydroxyls with high cross-reactivities factors. The H-2 antiserum from J. D. O'Connor better recognizes some ecdysteroids resulting from metabolism of the A-ring, such as 3-epimers or conjugates at C-3 and C-2.

After HPLC separation, ecdysteroids were quantified by RIA of collected fractions. The concentration for each ecdysteroid was calculated according to its cross-reactivity factor. The advantages of the tech-

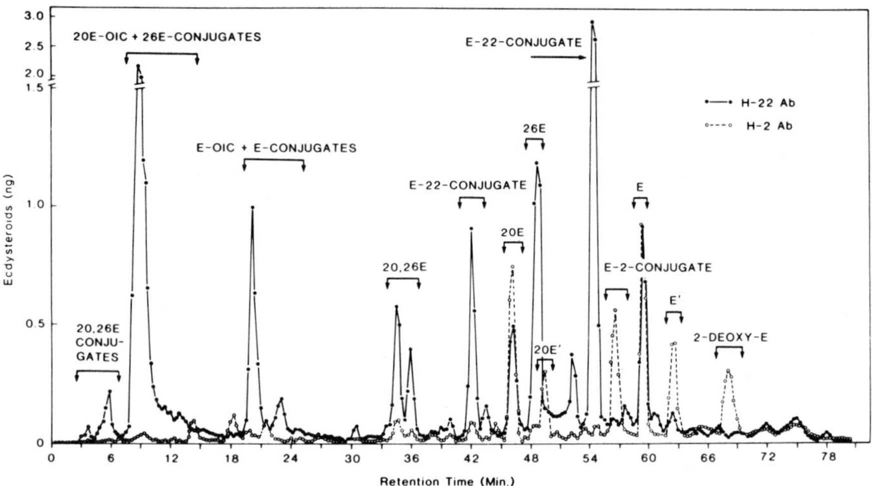

FIGURE 13.13. HPLC/RIA analysis of ecdysteroids from the gut of day-14 female pupa *Manduca sexta*. *Operating conditions:* RP µBondapak column; gradient elution from 100% solvent A (5% acetonitrile : 95% 20 mM tris/ perchlorate buffer, pH 7.5) to a mixture of 85% solvent A 15% solvent B (95% acetonitrile : 5% tris/perchlorate buffer) in 60 min, followed by 5-min isocratic 15%B : 85%A; then a 10-min cleansing shift to 100% B. Flow rate: 1 ml/min. Collected fractions were assayed by RIA using two different antisera (H-2Ab and H-22Ab). *Abbreviations:* E = ecdysone; 20E = 20-hydroxyecdysone; 20Eoic = 20-hydroxyecdysonoic acid; Eoic = ecdysonoic acid; 20,26E = 20,26-dihydroxyecdysone; 26E = 26-hydroxyecdysone; 20E' = 3-epi-20-hydroxy-ecdysone; E' = 3-epiecdysone; 2-deoxy-E = 2-deoxyecdysone. (After Warren et al., 1986.)

nique are apparent in Fig. 13.13, showing the RIA analysis of RP-HPLC fractions of a gut extract from day-14 developing female *Manduca* adults. Three ecdysone conjugates of diverse polarities (23 min, 42 min, and 54.5 min) are detected readily by H-22 antibodies, indicating the involvement of the C-22 hydroxyl with conjugating moieties. A C-2 ecdysone conjugate eluting at 56.5 min is detected differentially by H-2 antibodies, as are 3-epi-20-hydroxyecdysone (50 min) and 3-epiecdysone (62.5 min), and 2-deoxyecdysone (68 min).

13.7.7. Quantification of Ecdysteroids After Long-Term Cholesterol Labeling

The use of long-term radiolabeling with cholesterol seems to be a powerful tool for investigating ecdysteroids originated in cholesterol (Beydon et al., 1981; Dinan and Rees, 1981; Beydon and Lafont, 1987) for the following reasons. (1) It is easy to separate ecdysteroids from unconverted cholesterol during extraction, and in RP-HPLC the labeled peaks are ecdysteroids; however, this may not be valid (a) when ecdysteroids are too apolar like ecdysteroid fatty acid esters and (b) in adults of some species that synthesize noticeable amounts of glucoside derivative(s) of steroids whose polarity is similar to ecdysteroid ones (Thompson et al., 1985a). (2) All ecdysteroids are observed and quantified easily at the same time, assuming they have the same specific activity as their parent compound (cholesterol). The validity of the latter assumption was established by detailed experiments (see the discussion in Beydon and Lafont, 1987). The incorporation rates of labeled cholesterol into ecdysteroids are low, owing to the respective sizes of ecdysteroid and cholesterol pools, but they can nevertheless be measured accurately, provided that sufficient radioactivity was injected.

[^{3}H]Cholesterol was injected into animals at the beginning of the studied instar (Beydon et al., 1981; Beydon and Lafont, 1987). Animals taken at various times of development were extracted by chloroform : water (1:1). After centrifugation, the chloroform phase and pellet were reextracted with water. After centrifugation, the pellet was further extracted with acetone : ethanol (1:1). After centrifugation, the supernatant was added to the chloroform phase. Nearly all unconverted labeled cholesterol was in this mixture. Ecdysteroids in the water phase were absorbed in a C_{18} Sep-Pak cartridge, then eluted with methanol. After evaporation, the residue was dissolved with injection buffer and analyzed by RP-HPLC. Fractions were collected, then mixed with a scintillation cocktail and counted for their radio-

activity (Fig. 13.14). In RP-HPLC analyses of [³H]cholesterol metabolites in *Pieris brassicae,* all labeled peaks were ecdysteroids the chemical
nature of which was previously determined (Lafont et al., 1983; Beydon et al., 1987b).

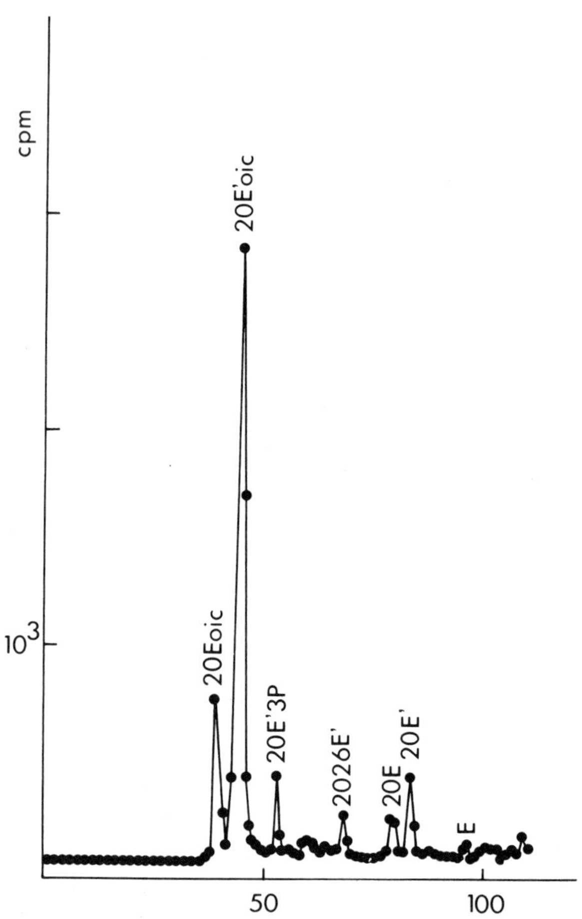

FIGURE 13.14. RP-HPLC analysis of [³H]ecdysteroids from pharate pupae of
Pieris brassicae. [³H]Ecdysteroids were originated in the conversion of
[³H]cholesterol injected into day-1 last-instar larvae. *Operating conditions:* RP
Zorbax ODS column; gradient elution from 8% to 40% acetonitrile in 20 m*M*
potassium citrate buffer (pH 6.6) in 60 min at a flow rate of 1 ml/min; 0.4-min
fractions were collected and counted for radioactivity. *Abbreviations:*
E = ecdysone;   20E = 20-hydroxyecdysone;   20Eoic = 20-hydroxyecdysonoic
acid; 20E'oic = 3-epi-20-hydroxyecdysone acid; 20,26E' = 3-epi-20,26-dihydroxy-
ecdysone; 20E' = 3-epi-20-hydroxyecdysone; 20E'3P = 3-epi-20-hydroxyecdysone
3-phosphate.

13.8. Summary

The presently available methods for ecdysteroid analysis have reached a fair degree of sophistication, and they can solve most identification problems, even if small amounts only (i.e., less than 0.1 mg) of material are available. They have established the complexity of the ecdysteroid family, which by no means has been fully characterized. Thus, within the last 10 years, the number of identified zoo-ecdysteroids from arthropods has increased by five times (Fig. 13.15), and it may be expected that the present list (Table 13.1) is far from definitive. As a consequence of the complex pattern of ecdysteroids present in a single animal at a given time (see Fig. 13.13) and of the biological significance of all these compounds [active hormone(s) and inactivation/storage metabolites], it becomes meaningless to determine "hormone levels" by a single RIA of a crude extract and it is necessary to determine by specific methods the concentrations of individual ecdysteroids. Of course, if the biological significance of the various ecdysteroids is established in a given species, it may become sufficient to measure the concentration of only biologically active compounds.

In every case, this necessitates the use of chromatographic methods for the separation of complex ecdysteroid mixtures. In many cases,

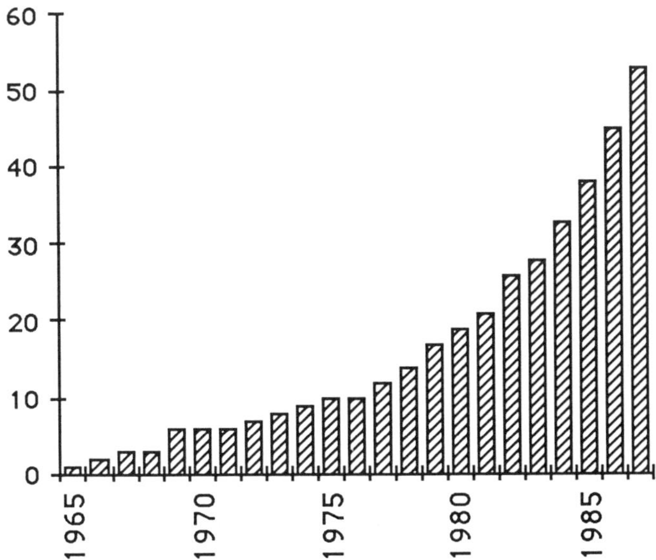

FIGURE 13.15. Evolution of the number of ecdysteroids isolated from arthropods.

TLC methods are not able, unless optimized, to resolve all compounds. GLC techniques are not suited for polar metabolites (ecdysonoic acids or phosphate conjugates), and this explains the increasing popularity of HPLC techniques. They offer a wide range of procedures (see Table 13.7) that can be used to solve any problem, and they are efficient enough to be used without special optimization. It must, however, be kept in mind that (at least as a control) several different systems should be combined to identify a given compound.

Regarding detection and quantification problems, off-line procedures using an immunoassay (probably EIA will become the most popular system) are well suited for such tasks. HPLC/UV works with a weaker sensitivity, but this method may suffice if large concentrations are present. HPLC with fluorescent detection ought to be further studied in order to ascertain whether it can work at low concentrations and maybe compete with HPLC/IA.

Acknowledgments

Research in our laboratory was supported by MEN and CNRS (UA 686). We are indebted to Dr. N. Morin for mass spectral analyses, to Professor J.-P. Girault for NMR analyses, and to M. Andrianjafintrimo for technical assistance in HPLC studies. We also thank Dr. I. D. Wilson for providing various unpublished information.

References

Bergamasco, R. and D. H. S. Horn. 1980. The biological activities of ecdysteroids and ecdysteroid analogues. Pp. 299–324 *in* J. A. Hoffmann (ed.), *Progress in Ecdysone Research.* Elsevier/North-Holland Publ., Amsterdam and New York.

Beydon, P. and R. Lafont. 1987. Long-term cholesterol labeling as a convenient means for measuring ecdysteroid production and catabolism *in vivo:* application to the last larval instar of *Pieris brassicae.* Arch. Insect Biochem. Physiol. 5: 139–154.

Beydon, P., J.-P. Girault, and R. Lafont. 1987b. Ecdysone metabolism in *Pieris brassicae* during the feeding last larval instar. Arch. Insect Biochem. Physiol. 4: 139–149.

Beydon, P., J. Claret, P. Porcheron, and R. Lafont. 1981. Biosynthesis and inactivation of ecdysone during the pupal–adult development of the cabbage butterfly, *Pieris brassicae.* Steroids 38: 633–650.

Beydon, P., A. Fabre, J. Goichon, and R. Lafont. 1987a. Use of a diode-array detector for HPLC analysis of ecdysteroids in biological extracts. Communication presented at the 8th Ecdysone Workshop, Marburg, W. Germany.

Bielby, C. R., E. D. Morgan, and I. D. Wilson. 1986. Gas chromatography of ecdysteroids as their trimethylsilyl esthers. J. Chromatogr. 351: 57–64.

Bielby, C. R., A. R. Gande, E. D. Morgan, and I. D. Wilson. 1980. Practical aspects of the preparation and chromatography of the trimethylsilyl esters of ecdysteroids. J. Chromatogr. 194: 43–53.

Blais, C. and R. Lafont. 1984. Ecdysteroid metabolism by soluble enzymes from an insect: metabolic relationship between 3β-hydroxy-, 3α-hydroxy- and 3-oxo-ecdysteroids. Hoppe-Seyler's Z. Physiol. Chem. 365: 809–818.

Borst, D. W. and J. D. O'Connor. 1972. Arthropod molting hormone: radioimmune assay. Science (Wash., DC) 178: 418–419.

Borst, D. W. and J. D. O'Connor. 1974. Trace analysis of ecdysones by gas–liquid chromatography, radioimmunoassay and bioassay. Steroids 24: 637–656.

Briers, T., and A. De Loof. 1983. Distribution and metabolism of ecdysteroids in the adult yellow mealworm beetle, *Tenebrio molitor*. Insect Biochem. 13: 513–522.

Bückmann, D., G. Starnecker, K.-H. Tomaschko, E. Wilhelm, R. Lafont, and J.-P. Girault. 1986. Isolation and identification of major ecdysteroids from the pycnogonid *Pycnogonum littorale* (Ström) (Arthropoda, Pantopoda). J. Comp. Physiol. 156B: 759–765.

Butenandt, A., and P. Karlson. 1954. Über die Isolierung eines Metamorphosephormons der Insekten in kristallisierter Form. Z. Naturforsch. 9B: 389–391.

Cherbas, L., C. D. Yonge, P. Cherbas, and C. M. Williams. 1980. The morphological response of Kc–H cells to ecdysteroids: hormonal specificity. Wilhelm Roux's Arch. Dev. Biol. 189: 1–15.

Connat, J. L. 1987. Aspects endocrinologiques de la physiologie du développement et de la reproduction chez les tiques. Thesis, University of Dijon, France.

Crosby, T., R. P. Evershed, D. Lewis, K. P. Wigglesworth, and H. H. Rees. 1986. Identification of ecdysone 22-long-chain fatty acyl esters in newly-laid eggs of the cattle tick *Boophilus microplus*. Biochem. J. 240: 131–138.

Delaage, M. A., M. H. Hirn, and M. L. De Reggi. 1982. Radioimmunoassay of ecdysteroids. Methods in Enzymol. 84: 350–358.

De Reggi, M. L., M. H. Hirn, and M. A. Delaage. 1975. Radioimmunoassay of ecdysone: an application of *Drosophila* larvae and pupae. Biochem. Biophys. Res. Commun. 66: 1307–1315.

Diehl, P. A., J.-L. Connat, J.-P. Girault, and R. Lafont. 1985. A new class of apolar ecdysteroid conjugates: esters of 20-hydroxyecdysone with long-chain fatty acids in ticks. Int. J. Invertebr. Reprod. Dev. 8: 1–13.

Dinan, L. N. and H. H. Rees. 1978. Preparation of 3-epi-ecdysone and 3-epi-

20-hydroxyecdysone. Steroids 32: 629–638.

Dinan, L. N. and H. H. Rees. 1981. Incorporation *in vivo* of [4-^{14}C]cholesterol into the conjugated ecdysteroids in ovaries and eggs of *Schistocerca gregaria*. Insect Biochem. 11: 255–265.

Dinan, L. N., P. L. Donnahey, H. H. Rees, and T. W. Goodwin. 1981. High-performance liquid chromatography of ecdysteroids and their 3-epi, 3-dehydro and 26-hydroxy derivatives. J. Chromatogr. 205: 139–145.

Evershed, R. P., J. G. Mercer, and H. H. Rees. 1987. Capillary gas chromatography–mass spectrometry of ecdysteroids. J. Chromatogr. 390: 357–369.

Faux, S., D. H. S. Horn, E. J. Middleton, H. M. Fales, and M. E. Lowe. 1969. Moulting hormones of a crab during ecdysis. Chem. Commun. 4: 175–176.

Feldlaufer, M. F. and J. A. Svoboda. 1986. Makisterone A: a 28-carbon insect ecdysteroid. Insect Biochem. 16: 45–48.

Feldlaufer, M. F., A. W. Herbert, Jr., J. A. Svoboda, M. J. Thompson, and W. R. Lusby. 1985. Makisterone A: the major ecdysteroid from the pupa of the honey bee. Insect Biochem. 15: 597–600.

Fraenkel, G. 1935. A hormone causing pupation in the blowfly *Calliphora erythrocephala*. Proc. R. Soc. Lond. 118B: 1–12.

Fraenkel, G., and J. Zdarek. 1970. The evaluation of the ''*Calliphora* test'' as an assay for ecdysone. Biol. Bull. (Woods Hole) 139: 138–150.

Fristom, J. W. and M. A. Yund. 1976. Characteristics of the actions of ecdysones on *Drosophila* imaginal discs, cultured *in vitro*. Pp. 161–178 *in* K. Maramorosch (ed.), *Invertebrate Tissue Culture: Research Applications*. Academic Press, Orlando, Florida.

Galbraith, M. N., and D. H. S. Horn. 1969. Insect moulting hormones: crustecdysone (20-hydroxyecdysone) from *Podocarpus elatus*. Aust. J. Chem. 22: 1045–1057.

Galbraith, M. N., D. H. S. Horn, E. J. Middleton, and R. J. Hackney. 1968. Structure of deoxycrustecdysone, a second crustacean moulting hormone. J. Chem. Soc. Chem. Commun. pp. 83–85.

Gilgan, M. W. and M. E. Zinck. 1972. Estimation of ecdysterone from sulphuric acid–induced fluorescence. Steroids 20: 95–104.

Girault, J.-P. and R. Lafont. 1988. The complete ^{1}H-NMR assignment of ecdysone and 20-hydroxyecdysone. J. Insect Physiol. 34:701–706.

Hampshire, F. and D. H. S. Horn. 1966. Structure of crustecdysone, a crustacean moulting hormone. J. Chem. Soc. Chem. Commun. pp. 37–38.

Hétru, C., M. Lagueux, B. Luu and J. A. Hoffmann. 1978. Adult ovaries of *Locusta migratoria* contain the sequence of biosynthetic intermediates for ecdysone. Life Sci. 22: 2141–2154.

Hétru, C., B. Luu, and J. A. Hoffmann. 1985. Ecdysone conjugates: isolation and identification. Methods Enzymol. 111: 411–419.

Hétru, C., D. Fraisse, J.-C. Tabet, and B. Luu. 1984. Confirmation, par spectrométrie de masse en mode FAB, de l'identification des esters d'AMP d'ecdystéroïdes dans les oeufs du criquet migrateur. C. R. Acad. Sci. Paris [Sér II] 299: 429–432.

Hirn, M. L. and M. A. Delaage. 1980. Radioimmunological approaches to the quantification of ecdysteroids. Pp. 69–82 *in* J. A. Hoffmann (ed.), *Progress in Ecdysone Research*. Elsevier/North-Holland Publ., Amsterdam and New York.

Hocks, P. and R. Wiechert. 1966. 20-Hydroxy-ecdyson, isoliert aus Insekten. Tetrahedron Lett. pp. 2989–2993.

Hoffmann, J. A. and C. Hétru. 1983. Ecdysone. Pp. 65–88 *in* R. G. H. Downer and H. Laufer (eds.), *Invertebrate Endocrinology*, Vol. 1: *Endocrinology of Insects*. Liss, New York.

Hoffmeister, H. and H. F. Grützmacher. 1966. Zur Chemie des Ecdysterons. Tetrahedron Lett. pp. 4017–4023.

Holman, G. M. and R. W. Meola. 1978. A high-performance liquid chromatography method for the purification and analysis of insect ecdysones: application to measurement of ecdysone titres during pupal–adult development of *Heliothis zea*. Insect Biochem. 8: 275–278.

Hori, M. 1969. Automatic column chromatographic method for insect moulting steroids. Steroids 14: 33–45.

Horn, D. H. S. 1971. The ecdysones. Pp. 333–459 *in* M. Jacobson and D. G. Crosby (eds.), *Naturally Occurring Insecticides*. Dekker, New York.

Horn, D. H. S. and R. Bergamasco. 1985. Chemistry of ecdysteroids. Pp. 185–248 *in* G. A. Kerkut and L. I. Gilbert (eds.), *Comprehensive Insect Physiology, Biochemistry and Pharmacology*, Vol. 7. Pergamon Press, Oxford and Elmsford, New York.

Horn, D. H. S., J. S. Wilkie, B. A. Sage, and J. D. O'Connor. 1976. A high-affinity antiserum specific for the ecdysone nucleus. J. Insect Physiol. 22: 901–905.

Huber, R. and W. Hoppe. 1965. Zur Chemie des Ecdysons. VII. Die Kristall- und Molekülstruktur-analyse des Insektenverpuppungshormons Ecdysone mit des automatisierten Faltmolekülmethode. Chem. Ber. 98: 2403–2424.

Hung, D. T., S. A. Benner, and C. M. Williams. 1980. Preparation of an ecdysone immunogen for radioimmunoassay work. J. Biol. Chem. 255: 6047–6048.

Hunter, W. H. and F. C. Greenwood. 1962. Preparation of human growth hormone of high specific activity. Nature (Lond.). 194: 495.

Ikekawa, N., T. Ikeda, E. Mizuno, E. Ohnishi, and S. Sakurai. 1980. Isolation of a new ecdysteroid, 2:22-dideoxy-20-hydroxyecdysone from the ovaries of the silkworm *Bombyx mori*. J. Chem. Soc. Chem. Commun. pp. 448–449.

Ikekawa, N., F. Hattori, J. Rubio-Lightbourn, H. Miyazaki, M. Ishibashi, and C. Mori. 1972. Gas chromatographic separation of phytoecdysones. J. Chromatogr. Sci. 10: 233–242.

Isaac, R. E., H. H. Rees, and T. W. Goodwin. 1981a. Isolation of 2-deoxy-20-hydroxyecdysone and 3-epi-2-deoxyecdysone from eggs of the desert locust, *Schistocerca gregaria*. J. Chem. Soc. Chem. Commun. pp. 418–420.

Isaac, R. E., H. H. Rees, and T. W. Goodwin. 1981b. Isolation of ecdysone 3-acetate as a major ecdysteroid from the developing eggs of the desert locust, *Schistocerca gregaria*. J. Chem. Soc. Chem. Commun. pp. 594–595.

Isaac, R. E., N. P. Milner, and H. H. Rees. 1982b. High-performance liquid chromatography of ecdysteroids and ecdsteroid-22-phosphates. J. Chromatogr. 246: 317–322.

Isaac, R. E., N. P. Milner, and H. H. Rees. 1983. Identification of ecdysonoic acid and 20-hydroxyecdysonoic acid isolated from developing eggs of *Schistocerca gregaria* and pupae of *Spodoptera littoralis*. Biochem. J. 213: 261–265.

Isaac, R. E., H. P. Desmond, and H. H. Rees. 1984. Isolation and identification of 3-acetylecdysone 2-phosphate, a metabolite of ecdysone, from developing eggs of *Schistocerca gregaria*. Biochem. J. 217: 239–243.

Isaac, R. E., M. E. Rose, H. H. Rees, and T. W. Goodwin. 1982a. Identification of ecdysone-22-phosphate and 2-deoxyecdysone-22-phosphate in eggs of the desert locust, *Schistocerca gregaria*, by fast-atom bombardment mass spectrometry and N.M.R. spectroscopy. J. Chem. Soc. Chem. Commun. pp. 249–251.

Jegla, T. C. and J. D. Costlow. 1979. The *Limulus* bioassay for ecdysteroids. Biol. Bull (Woods Hole) 156: 103–114.

Kaplanis, J. N., W. E. Robbins, M. J. Thompson, and S. R. Dutky. 1973. 26-Hydroxyecdysone: new insect hormone from the egg of the tobacco hornworm. Science (Wash., DC) 180: 307–308.

Kaplanis, J. N., M. J. Thompson, S. R. Dutky, and W. E. Robbins. 1979. The ecdysteroids from the tobacco hornworm during pupal–adult development five days after peak titer of molting hormone activity. Steroids 34: 333–345.

Kaplanis, J. N., M. J. Thompson, S. R. Dutky, and W. E. Robbins. 1980. The ecdysteroids from young embryonated eggs of the tobacco hornworm. Steroids 36: 321–336.

Kaplanis, J. N., L. A. Tabor, M. J. Thompson, W. E. Robbins, and T. J. Shortino. 1966. Assay for ecdysone (moulting hormone) activity using the housefly, *Musca domestica*. Steroids 8: 625–631.

Kaplanis, J. N., S. R. Dutky, W. E. Robbins, M. J. Thompson, E. L. Lindquist, D. H. S. Horn, and M. N. Galbraith. 1975. Makisterone A: a 28-carbon hexahydroxy moulting hormone from the embryo of the milkweed bug. Science (Wash., DC) 190: 681–682.

Karlson, P. 1956. Biochemical studies on insect hormones. Vitam. Horm. 14: 227–266.

Karlson, P., H. Bugany, H. Döpp, and G.-A. Hoyer. 1972. 3-Dehydroecdyson, ein Stoffwechselprodukt des Ecdysons bei der Schmeissfliege *Calliphora erythrocephala* Meigen. Hoppe-Selyer's Z. Physiol. Chem. 353: 1610–1614.

Katz, M., and Y. Lensky. 1970. Gas chromatography analysis of ecdysone. Experientia (Basel) 26: 1043.

Kelly, T. J., C. W. Woods, R. E. Redfern, and A. B. Borkovec. 1981. Makisterone A: the molting hormone of larval *Oncopeltus*. J. Exp. Zool. 218: 127–132.

King, D. S. 1972. Metabolism of α-ecdysone and possible immediate precursors by insects *in vivo* and *in vitro*. Gen. Comp. Endocrinol. (Suppl.) 3: 221–227.

Koolman, J. 1980. Analysis of ecdysteroids by fluorometry. Insect Biochem. 10: 381–386.

Koolman, J. and K.-D. Spindler. 1977. Enzymatic and chemical synthesis of 3-dehydroxyecdysone, a metabolite of the moulting hormone of insects. Hoppe-Seyler's Z. Physiol. Chem. 358: 1339–1344.

Koreeda, M., and B. A. Teicher. 1977. Chemical analysis of insect molting hormones. Pp. 207–240 *in* R. B. Turner (ed.), *Analytical Biochemistry of Insects*. Elseview, Amsterdam and New York.

Kubo, I. and F. J. Hanke. 1986. Chemical methods for isolating and identifying phytochemicals biologically active in insects. Pp. 225–250 in J. R. Miller and T. A. Miller (eds.), *Insect–Plant Interactions* (Springer Ser. in Experimental Entomology). Springer-Verlag, Berlin and New York.

Kubo, I., and S. Komatsu. 1986. Micro-analysis of prostaglandins and ecdysteroids in insects by high-performance liquid chromatography and fluorescence detection. J. Chromatogr. 362: 61–70.

Kubo, I., J. A. Klocke, and S. Asano. 1981. Insect ecdysis inhibitors from the East African medicinal plant *Ajuga remota* (Labiatae). Agric. Biol. Chem. 45: 1925–1927.

Kubo, I., A. Matsumoto, and J. J. Ayafor. 1984. Efficient isolation of a large amount of 20-hydroxyecdysone from *Vitex madiensis* (Verbenaceae) by droplet counter-current chromatography. Agric. Biol. Chem. 48: 1683–1684.

Kubo, I., A. Matsumoto, and S. Asano. 1985a. Efficient isolation of ecdysteroids from the silkworm, *Bombyx mori*, by droplet counter-current chromatography. Insect Biochem. 15: 45–47.

Kubo, I., A. Matsumoto, and F. J. Hanke. 1985b. The ^{1}H-NMR assignment of 20-hydroxyecdysone. Agric. Biol. Chem. 49: 243–244.

Kubo, I., S. Komatsu, Y. Asaka, and G. De Boer. 1987. Isolation and identification of apolar metabolites of ingested 20-hydroxyecdysone in frass of *Heliothis virescens* larvae. J. Chem. Ecol. 13: 785–794.

Lachaise, F., M. F. Meister, C. Hétru, and R. Lafont. 1986. Studies on the biosynthesis of ecdysone by the Y-organs of *Carcinus maenas*. Mol. Cell. Endocrinol. 47: 257–267.

Lafont, R. 1987. HPLC analysis of ecdysteroids in plants and animals. In H. Kalasz and E. S. Ettre (eds.), *Proceedings of the Budapest Chromatography Conference*. Akademia Kiado, Budapest. Pp. 1–15.

Lafont, R., G. Sommé-Martin, and J.-C. Chambet. 1979. Separation of ecdysteroids by using high-pressure liquid chromatography on microparticulate supports. J. Chromatogr. 170: 185–194.

Lafont, R., P. Beydon, G. Sommé-Martin, and C. Blais. 1980a. High-performance liquid chromatography of ecdysone metabolites applied to the cabbage butterfly, *Pieris brassicae*. Steroids 36: 633–650.

Lafont, R., G. Sommé-Martin, B. Mauchamp, B. F. Maume, and J.-P. Delbecque. 1980b. Analysis of ecdysteroids by high-performance liquid chromatography and coupled gas–liquid chromatography/mass spectrometry. Pp. 45–68 *in* J. A. Hoffmann (ed.), *Progress in Ecdysone Research.* Elsevier/North-Holland Publ., Amsterdam and New York.

Lafont, R., P. Beydon, B. Mauchamp, G. Sommé-Martin, M. Andrianjafintrimo, and P. Krien. 1981. Recent progress in ecdysteroid analytical methods. Pp. 125–144 *in* F. Sehnal, A. Zabza, J. J. Menn, and B. Cymborowski (eds.), *Regulation of Insect Development and Behaviour.* Wroclaw Technical University Press, Wroclaw, Poland.

Lafont, R., J.-L. Pennetier, M. Andrianjafintrimo, J. Claret, J.-F. Modde, and C. Blais. 1982. Sample processing for high-performance liquid chromatography of ecdysteroids. J. Chromatogr. 236: 137–149.

Lafont, R., C. Blais, P. Beydon, J.-F. Modde, U. Enderle, and J. Koolman. 1983. Conversion of ecdysone and 20-hydroxyecdysone into 20-oic derivatives is a major pathway in larvae and pupae of species from three insect orders. Arch. Insect Biochem. Physiol. 1: 41–58.

Lauer, R. C., P. H. Solomon, K. Nakanishi, and B. F. Erlanger. 1974. Antibodies to the insect moulting hormone β-ecdysone. Experientia (Basel) 30: 560–562.

Lazarovici, P., D. Shapiro, and E. Shaaya. 1983. Determination of ecdysteroid titers by a double-antibody precipitation in the tropical warehouse moth, *Ephestia cautella* (Lepidoptera, Phycitidae). Comp. Biochem. Physiol. 74: 525–529.

Lehmann, M. and J. Koolman. 1986. Influence of forskolin on the metabolism of ecdysone and 20-hydroxyecdysone in isolated fat body of the blowfly, *Calliphora vicina.* Biol. Chem. Hoppe-Seyler 367: 387–394.

Lewis, S. and I. D. Wilson. 1984. Factors affecting the ion-pair reversed-phase thin-layer chromatography of organic acids on paraffin-coated and C_{18} bonded silica gel. J. Chromatogr. 312: 133–140.

Maroy, P., J. Vargha, and K. Horvath. 1977. Rapid heterologous haptene radioimmunoassay for insect moulting hormone. FEBS (Fed. Eur. Biochem. Soc.) Lett. 81: 319–322.

Maroy, P., G. Kaufmann, and A. Dübendorfer. 1988. Embryonic ecdysteroids of *Drosophila melanogaster.* J. Insect Physiol. 34:633–637.

Mauchamp, B., R. Lafont, and P. Krien. 1981. Analysis of juvenile hormones by high-performance liquid chromatography coupled with mass spectrometry. Pp. 21–31 *in* G. E. Pratt and G. T. Brooks (eds.), *Juvenile Hormone Biochemistry.* Elseview/North-Holland Publ., Amsterdam and New York.

Mayer, R. T. and J. A. Svoboda. 1978. Thin-layer chromatographic *in situ* analysis of insect ecdysones via fluorescence-quenching. Steroids 31: 139–150.

McCarthy, J. F. 1979. Ponasterone A: a new ecdysteroid from the embryos and serum of brachyuran crustaceans. Steroids 34: 799–806.

Milner, N. P. and H. H. Rees. 1985. Involvement of 3-dehydroecdysone in the 3-epimerization of ecdysone. Biochem. J. 231: 369–374.

Miyazaki, H., M. Ishibashi, C. Mori, and N. Ikekawa. 1973. Gas phase microanalysis of zooecdysones. Anal. Chem. 45: 1164–1168.

Modde, J.-F., R. Lafont, and J. A. Hoffmann. 1984. Ecdysone metabolism in *Locusta migratoria* larvae and adults. Int. J. Invertebr. Reprod. Dev. 7: 161–183.

Morgan, E. D. and A. P. Woodbridge. 1971. Insect moulting hormones (ecdysones): identification as derivatives by gas chromatography. J. Chem. Soc. Chem. Commun. pp. 475–476.

Morgan, E. D. and A. P. Woodbridge. 1974. Mass spectrometry of insect moulting hormones: trimethylsilyl-methoxime derivatives of ecdysone and 20-hydroxyecdysone. Organ. Mass Spectrom. 9: 102–110.

Morgan, E. D. and Poole. C. F. 1976. The extraction and determination of ecdysones in arthropods. Adv. Insect Physiol. 12: 17–62.

Morgan, E. D. and I. D. Wilson. 1980. Progress in the analysis of ecdysteroids. Pp. 29–43 *in* J. A. Hoffmann (ed.), *Progress in Ecdysone Research.* Elseview/North-Holland Publ., Amsterdam and New York.

Morgan, E. D. and I. D. Wilson. 1988. Methods for separation and physicochemical quantification of ecdysteroids. *In* J. Koolman (ed.), *Ecdysone* 441: 165–169.

Morgan, E. D., S. J. Murphy, D. E. Games, and I. Mylchreest. 1988. Analysis of ecdysteroids by supercritical fluid chromatography. J. Chromatogr. (in press).

Moribayashi, A., H. Kurahashi, and T. Ohtaki. 1985. Comparative studies on ecdysone metabolism between mature larvae and pharate pupae in the fleshfly, *Sarcophaga perigrina*. Arch. Insect Biochem. Physiol. 2: 237–250.

Nakanishi, K. 1971. The ecdysones. Pure Appl. Chem. 25: 167–195.

Nirdé, P., G. Torpier, M. L. De Reggi, and A. Capron. 1983. Ecdysone and 20-hydroxyecdysone: new hormones for the human parasite *Schistosoma mansoni*. FEBS (Fed. Eur. Biochem. Soc.) Lett. 151: 223–227.

Ohnishi, E., T. Mizuno, F. Chatani, N. Ikekawa, and S. Sakurai. 1977. 2-Deoxy-a-ecdysone from ovaries and eggs of the silkworm, *Bombyx mori*. Science (Wash., DC) 197: 66–67.

Pimprikar, G. D., M. J. Coign, H. Sakurai, and J. R. Heitz. 1984. High-performance liquid chromatographic determination of ecdysteroid titers in the house fly. J. Chromatogr. 317: 413–419.

Poole, C. F. and E. D. Morgan. 1975. Structural requirements for the electron-capturing properties of ecdysone. J. Chromatogr. 115: 587–590.

Poole, C. F., E. D. Morgan, and P. M. Bebbington. 1975. Analysis of ecdysone by gas chromatography using electron capture detection. J. Chromatogr. 104: 172–175.

Poole, C. F., S. Singhawangcha, A. Zlatkis, and E. D. Morgan. 1978. Determination of bifunctional compounds. Part III. Polynuclear aromatic boronic acids as selective fluorescent reagents for HPTLC and HPLC. J. High Resolut. Chromatogr. Chromatogr. Commun. 2: 96–97.

Porcheron, P., J. Foucrier, C. Gros., P. Pradelles, P. Cassier, and F. Dray. 1976. Radioimmunoassay of arthropod molting hormone: β-ecdysone antibodies production and ^{125}I-iodinated tracer preparations. FEBS (Fed. Eur. Biochem. Soc.) Lett. 61: 159–162.

Porcheron, P., M. Morinière, and P. Pradelles. 1989. Development of a competitive enzyme immunoassay of ecdysteroids using acetylcholinesterase as label. Insect Biochem. 19: 117–122.

Raynor, M. W., J. P. Kithinji, I. K. Barker, K. D. Bartle, and I. D. Wilson. 1988. Supercritical fluid chromatography of ecdysteroids. J. Chromatogr. 436: 497.

Redfern, C. P. F. 1984. Evidence for the presence of makisterone A in *Drosophila* larvae and the secretion of 20-deoxymakisterone A by the ring gland. Proc. Nat. Acad. Sci. USA 81: 5643–5647.

Rees, H. H. and R. E. Isaac. 1985. Biosynthesis and metabolism of ecdysteroids and methods of isolation and identification of the free and conjugated compounds. Methods Enzymol. 111: 377–410.

Reum, L., D. Haustein, and J. Koolman. 1981. Immunoabsorption as a means for the purification of low molecular weight compounds: isolation of ecdysteroids from insects. Z. Naturforsch. 36D: 790–797.

Reum, L., W. Klinger, and J. Koolman. 1984. A new immunoassay for ecdysteroids based on chemiluminescence. Pp. 249–252 *in* L. J. Kricka, P. E. Stanley, G. H. G. Thorpe, and T. C. Whitehead (eds.), *Analytical Applications of Bioluminescence and Chemiluminescence*. Academic Press, Orlando, Florida.

Richards, G. 1978. The relative biological activities of α- and β-ecdysone and their 3-dehydro derivatives in the chromosome puffing assay. J. Insect Physiol. 24: 329–335.

Robinson, P. D., E. D. Morgan, I. D. Wilson, and R. Lafont. 1987. The metabolism of ingested and injected ecdysone by final instar larvae of *Heliothis armigera*. Physiol. Entomol. 12: 321–330.

Romer, F. 1979. Ecdysteroids in snails. Naturwissenschaften 66: 471–472.

Sannasi. A. and P. Karlson. 1974. Metabolism of ecdysone: Phosphate and sulphate esters as conjugates of ecdysone in *Calliphora vicina*. Zool. Jahrb. Abt. Allg. Zool. Physiol. Tiere 78: 378–386.

Sato, Y., M. Sakai, S. Imai, and S. Fujioka. 1968. Ecdysone activity of plant-originated molting hormones applied on the body surface of lepidopterous larvae. Appl. Entomol. Zool. 3: 39–51.

Scalia, S. and E. D. Morgan. 1985. Simultaneous determination of free and conjugated ecdysteroids by liquid chromatography. J. Chromatogr. 346: 301–308.

Schooley, D. A. and K. Nakanishi. 1973. Application of high-pressure liquid chromatography to the separation of insect molting hormones. Pp. 37–54 *in* E. Heftmann (ed.), *Modern Methods of Steroid Analysis*. Academic Press, Orlando, Florida.

Schooley, D. A., G. Weiss, and K. Nakanishi. 1972. A simple and general extraction procedure for phytoecdysones based on reversed-phase absorption chromatography. Steroids 19: 377–382.

Sommé-Martin, G., J. Colardeau, and R. Lafont. 1987. Induction of P_1 gene in *Drosophila melanogaster*: regulation by various ecdysteroids. Communication presented at the 8th Ecdysone Workshop, Marburg, W. Germany.

Soumoff, C., D. H. S. Horn, and J. D. O'Connor. 1984. Production of a new antiserum to arthropod molting hormone and comparison with two other antisera. J. Steroid Biochem. 14: 429–435.

Spindler, K.-D., C. Beckers, U. Gröschel-Stewart, and H. Emmerich. 1978. A radioimmunoassay for arthropod moulting hormones, introducing a novel method of immunogen coupling. Hoppe-Seyler's Z. Physiol. Chem. 359: 1269–1275.

Takemoto, T., S. Ogawa, N. Nishimoto, K.-Y. Yen, K. Abe, T. Sato, K. Ogawa, and M. Takahashi. 1967. The isolation of ecdysterone from the radix of *Achyranthes obtusifolia* Lam. Yakugaku Zasshi 87: 1521–1524.

Thomson, J. A., F. P. Imray, and D. H. S. Horn. 1970. An improved *Calliphora* bioassay for insect moulting hormones. Aust. J. Exp. Biol. 48: 321–328.

Thompson, M. J., J. N. Kaplanis, W. E. Robbins, and R. T. Yamamoto. 1967. 20,26-Dihydroxyecdysone, a new steroid with moulting hormone activity from the tobacco hornworm *Manduca sexta* (Johannson). J. Chem. Soc. Chem. Commun. pp. 650–653.

Thompson, M. J., J. A. Svoboda, H. H. Rees, and K. R. Wilzer. 1987b. Isolation and identification of 26-hydroxyecdysone 2-phosphate: an ecdysteroid conjugate of eggs and ovaries of the tobacco hornworm, *Manduca sexta*. Arch. Insect Biochem. Physiol. 4: 183–190.

Thompson, M. J., J. N. Kaplanis, W. E. Robbins, S. R. Dutky, and H. N. Nigg. 1974. 3-Epi-20-hydroxyecdysone from meconium of the tobacco hornworm. Steroids 24: 359–366.

Thompson, M. J., G. F. Weirich, H. H. Rees, J. A. Svoboda, M. F. Feldlaufer, and K. R. Wilzer. 1985b. New ecdysteroid conjugate: isolation and identification of 26-hydroxyecdysone 26-phosphate from eggs of the tobacco hornworm, *Manduca sexta* (L.). Arch. Insect Biochem. Physiol. 2: 227–236.

Thompson, M. J., J. A. Svoboda, W. R. Lusby, H. H. Rees, J. E. Oliver, G. F. Weirich and K. R. Wilzer. 1985a. The conversion of [^{14}C]cholesterol to 5-[^{14}C[pregnen-3β,20β-diol glucoside in the tobacco hornworm, *Manduca sexta*. J. Biol. Chem. 29: 15410–15412.

Thompson, M. J., M. F. Feldlaufer, R. Lozano, H. H. Rees, W. R. Lusby, J. A. Svoboda, and K. R. Wilzer, Jr. 1987a. Metabolism of 26-[^{14}C]hydroxyecdysone 26-phosphate in the tobacco hornworm, *Manduca sexta* L., to a new ecdysteroid conjugate: 26-[^{14}C]hydroxyecdysone 22-glucoside. Arch. Insect Biochem. Physiol. 4: 1–15.

Touchstone, J. C. 1986. Ecdysteroids. Pp. 119–135 *in CRC Handbook of Chromatography:Steroids*. CRC Press, Boca Raton, Florida.

Tsoupras, G. 1982. Identification d'esters adénosinemonophosphoriques d'ecdystéroïdes dans les oeufs d'un insecte, *Locusta migratoria*: recherches sur la destinée de ces molécules au cours du développement embryonnaire. Thesis, University Louis Pasteur, Strasbourg, France.

Tsoupras, G., B. Luu, and J. A. Hoffmann. 1982b. Isolation and identification of three ecdysteroid conjugates with a C-20 hydroxy group in eggs of

Locusta migratoria. Steroids 40: 551–560.

Tsoupras, G., B. Luu, and J. A. Hoffmann. 1983. A cytokinin (isopentenyl-adenosyl-mononucleotide) linked to ecdysone in newly-laid eggs of *Locusta migratoria.* Science (Wash., DC) 220: 507–509.

Tsoupras, G., C. Hétru, B. Luu, M. Lagueux, E. Constantin, and J. A. Hoffmann. 1982a. The major conjugates of ecdysteroids in young eggs and in embryos of *Locusta migratoria.* Tetrahedron Lett. 23: 2045–2048.

Walgraeve, H., E. Van Beek, G. Criel, G. Van Brussel, and A. De Leenheer. 1986. Comparison of three separation systems for the radioimmunoassay of ecdysteroids in *Artemia* (Crustacea: Branchiopoda). Insect Biochem. 16: 41–44.

Warren, J. T., W. Smith, and L. I. Gilbert. 1984. Simplification of the ecdysteroid radioimmunoassay by the use of the protein A from *Staphylococcus aureus.* Experientia (Basel) 40: 393–394.

Warren, J. T., B. Steiner, A. Dorn, M. Pak, and L. I. Gilbert. 1986. Metabolism of ecdysteroids during the embryogenesis of *Manduca sexta.* J. Liquid Chromatogr. 9: 1759–1782.

Watson, R. D. and E. Spaziani. 1982. Rapid isolation of ecdysteroids from crustacean tissues and culture media using Sep-Pak C-18 cartridges. J. Liquid Chromatogr. 5: 525–535.

Weirich, G. F., M. J. Thompson, and J. A. Svoboda. 1986. In vitro ecdysteroid conjugation by enzymes of *Manduca sexta* midgut cytosol. Arch. Insect Biochem. Physiol. 3: 109–126.

Wilson, I. D. 1985. Thin-layer chromatography of ecdysteroids. J. Chromatogr. 318: 373–377.

Wilson, I. D. and R. Lafont. 1986. Thin-layer chromatography and high-performance thin-layer chromatography of [^{3}H] metabolites of 20-hydroxyecdysone. Insect Biochem. 16: 33–40.

Wilson, I. D., S. Scalia, and E. D. Morgan. 1981. Reversed-phase thin-layer chromatography for the separation and analysis of ecdysteroids. J. Chromatogr. 212: 211–219.

Wilson, I. D., C. R. Bielby, and E. D. Morgan. 1982a. Evaluation of some phytoecdysteroids as internal standards for the chromatographic analysis of ecdysone and 20-hydroxyecdysone in arthropods. J. Chromatogr. 236: 224–229.

Wilson, I. D., C. R. Bielby, and E. D. Morgan. 1982b. Selective effects of mobile and stationary phases in reversed-phase high-performance liquid chromatography of ecdysteroids. J. Chromatogr. 238: 97–102.

Wilson, I. D., C. R. Bielby, and E. D. Morgan. 1982c. Studies on the reversed-phase thin-layer chromatography of ecdysteroids on C_{12} bonded and paraffin-coated silica. J. Chromatogr. 242: 202–206.

Part I Taxonomic Index

Acanthonyx lunulatus, 304, 306
acari, 61, 63, 65, 66, 69, 70
acarina, 6, 8, 12, 13
Acheta domesticus, 135, 142–145, 156, 197, 423
acorn barnacle, 42
Aedes aegypti, 140, 142, 349, 430, 435
Aedes caspius, 430
Aedes detritus, 430
Amblyomma hebraeum, 12, 64, 67, 68, 77, 242–245, 287
Amblyomma variegatum, 243, 245
amblypygi, 8
Amblyseius brassili, 67, 68
American cockroach, 342
American dog tick, 43, 142, 167
amphipod, 248, 259, 279, 298, 300
Anastrepha suspensa, 405, 437
Adroctonus australis, 238
annelid, 6, 16, 17, 19, 22
annelida, 16, 17, 363
Anocentor nitens, 306
Anomuran, 40
Anostraca, 7
Antheraea pernyi, 144
Antheraea polyphemus, 43, 406
aphid, 15, 38
Aphis fabae, 437
Apis mellifera, 198, 401, 427, 437, 490
Apterygota, 9, 279
apterygote, 10, 289
Aquatic Chelicerata, 4, 6, 11, 16, 17, 24, 25
Arachnid, 69, 78, 237, 241, 249, 287
Arachnida, 8, 63, 69, 70
Araneae, 8, 61, 64, 68, 70
Araneida, 8, 12
Araneus, 239
Araneus cornutus, 302
Argas arboreus, 77
Argas habraeum, 66, 77
Argas walkerae, 67, 68
Argasidae, 67, 68, 241, 242
Argasis, 301
Armadillidum vulgare, 258, 300

Artemia salina, 7, 245, 246
arthropod, 10, 15–23, 76, 79, 127, 142, 231–235, 237, 238, 240, 254, 259–262, 277, 289, 342–344, 363, 394, 457, 458, 501
Asellus aquaticus, 258
Astacidea, 249
Astacus astacus, 249
Astacus leptodactylus, 256, 303
Attacus atlas, 435

Balanus amphitrite, 246, 247
Balanus balanoides, 7, 299
Balanus balanus, 12, 24
Balanus eburneus, 247, 303
Balanus galeatus, 42, 78, 79
Balanus nubilus, 47
barnacles, 47, 78, 246–248, 303, 304
Biosteres longicaudatus, 437
Blaberus discoidalis, 423
Blatta orientalis, 435
Blattaria, 15
Blattella germanica, 10, 342, 420, 435
Blattodea, 91
blow fly, 130, 374, 375
blue crab, 40, 252
Bombus hypnorum, 198, 430
Bombus terrestris, 423, 430
Bombyx, 15
Bombyx mori, 152, 153, 183, 327, 331, 332, 423, 429, 431, 460
Boophilus decoloratus, 68, 77
Boophilus microplus, 12, 63–69, 244, 245, 283, 461
brachyuran crabs, 250–252, 254
branchiopod, 246
brine shrimp, 245
Buthus occitanus, 238

cabbage looper, 132, 135, 137, 141, 147, 148, 156
Callinectes, 254
Callinectes sapidus, 252, 254, 292, 461, 463
Calliphora, 203, 231, 232, 474, 475

Calliphora stygia, 463
Calliphora vicina, 332, 341, 345, 346,
    350, 354, 365, 374, 375, 376, 437,
    461
Calliphora vomitaria, 423
Calpodes ethlius, 195
Cancer, 255, 256
Cancer antennarius, 255, 256, 299, 305
Cancer borealis, 47
Cancer irroratus, 47
Carausius, 23
Carausius morosus, 201
Carcinus, 249, 257, 287, 294, 300, 308
Carcinus maenas, 44, 47–49, 75, 249,
    251, 252, 254, 282, 286, 294, 300,
    460
cecropia moth, 391
centipede, 234, 260–262
Cerebratulus marginatus, 21
Cerura vinula, 206
Chelicerata, 61
chelicerate, 4, 75, 231, 261, 262, 277,
    278, 287, 289, 292, 295, 296, 299
Chelonus, 163, 437
Chilo, 474, 475
Chilo suppressalis, 206
Chilopoda, 4, 5, 9, 278
Chironomus thummi, 216, 380
Chironumus tentans, 364, 374, 376
Cirripedia, 5, 7, 24, 78
clawed lobsters, 249, 251
cockroach, 135, 140, 143, 145, 157
Coelenterate, 19
Coleoptera, 13, 135, 423, 435
Collembola, 6, 7, 9, 10, 279
Colorado potato beetle, 142, 145, 157
conifer, 279, 284, 282, 285, 290, 297,
    300, 307
crab, 47–49, 75, 78, 142, 248, 249,
    251, 252, 254–256
Crangon, 256
Crangon vulgaris, 231
crayfish, 40, 44, 47, 52, 248–250,
    252–257, 297, 298, 327
Crustacea, 4, 5, 7, 8, 10–13, 17, 18,
    23, 24, 63, 64, 69, 79, 363, 398
Crustacean, 4, 6, 8–10, 14, 75, 76–79,
    231–234, 236, 246, 248–251, 255–
    259, 261, 262, 277–279, 285, 287,
    289, 290, 293, 294, 296, 299, 307,

    342, 393, 398, 420, 433
Ctenolepisma, 195
Culex pigiens, 423
Culex quinquefasciatus, 142
Cyperus iria (plant), 420, 432
cyprid, 42, 78, 79, 246

Danaus, 203
Danaus plexippus, 139, 435
Daphnia magna, 43
decapod, 248, 249, 251, 255, 256,
    258, 277, 279, 285, 294
Decapoda, 5, 7, 11, 14, 75, 79
Dermacentor albipictus, 66, 68
Dermacentor variabilis, 66–68, 76, 142,
    242, 243
Dermanyssidae, 67
Dermanyssus gallinae, 65, 66
Dermaptera, 86
desert locust, 331
Diabrotica undecempunctata, 435
Diatraea grandiosella, 145, 206, 435
Diatraea saccharalis, 148, 153
Dictyoptera, 13, 423, 435
Diplopoda, 4, 5, 9, 260, 278
Diploptera punctata, 87, 94, 95, 97–
    101, 114, 143, 135, 140, 157, 204,
    348, 349, 423, 431, 432, 435, 438
Diplura, 9
Diptera, 13, 86, 135, 140, 165, 423,
    435, 437
Drosophila, 366, 376, 474, 475
Drosophila hydei, 135, 139, 437
Drosophila melanogaster, 195, 341,
    349, 365, 374, 377, 420, 422, 423,
    433, 435, 461, 463
Dugesiella, 239
Dugesiella hentzi, 302
Dysdercus, 209
Dysdercus fasciatus, 13, 24, 423

Eliminus modestus, 42
Embioptera, 86
Entomostraca, 5, 7, 11, 12, 24
Entomostracan, 17
Ephemeroptera, 10
Ephestia, 193
Ephestia cautella, 138
Epilachna varivestis, 435
Eriochier, 255

European corn borer, 145

fiddler crab, 48, 49, 249, 251
flatworm, 19
flesh fly, 130, 131, 379
fly, 231, 232
fruit fly, 135

Galleria mellonella, 92, 134, 135, 139,
   152, 156, 157, 159, 161–165, 188,
   395, 406–408, 411, 423, 429, 435,
   435
Gammarus fossarum, 259
Gammarus pulex, 259
Gastrimargus africanus, 432, 435
Gecarcinus, 254, 256, 257
Gecarcinus lateralis, 252, 254, 255,
   286, 308, 460
German cockroach, 10, 342
Glossina, 23
Glossina morsitans, 204
glowworm, 15
Gomphocerus, 203
Gomphocerus rufus, 109, 111
green crab, 48, 49
Gryllus bimaculatus, 432

Haemaphysalis cancinna, 76
Hanseniella ivorensis, 9, 261
harvestman, 8, 237, 241
Heliothis armigera, 461
Heliothis virescens, 132, 148, 152, 153,
   343, 421, 423, 435, 462
Heliothis zea, 148, 152, 153
Helix pomatia, 286, 351, 493, 495
Helleria brevicornis, 258
Hemigrapsus nudus, 249, 256
Hemiptera, 13, 386, 423, 437
Hemipteran, 10, 231
Heterometrus swammerdami, 8
Hexapoda, 13
Hippodamia convergens, 208
Hirudinea, 20
Homarus americanus, 42, 44, 47, 48,
   51, 52, 63, 249, 251, 254, 297, 398,
   424
Homarus gammarus, 279
Homoptera, 368, 437
honey bee, 490, 491
horseshoe crab, 6, 233, 235, 236, 250

house cricket, 135, 142–145, 156, 164
house fly, 130, 131, 144, 237, 341
Hyalomma dromedarii, 67, 68, 242,
   244, 245
Hyalophora cecropia, 38, 85, 87, 88,
   144, 184, 391, 400, 401, 406–408,
   410, 421–423, 431, 435
Hyalophora gloveri, 157, 158, 435
Hydra, 19
Hymenoptera, 13, 423, 437
Hypantria cunea, 132

Idotea balthica, 258
Insect, 75, 77, 231, 233, 234, 249,
   261–277, 279, 285, 287; as *apter-
   ygote*, 4, 5, 6, 16; as *pterygote*, 4, 5,
   16; *social-*, 10, 15
Insecta, 8, 9, 11, 13, 15, 16, 24, 127,
   279, 363
Invertebrate, 18
Ips confusus, 207
Isopod, 249, 258, 259, 279
Isopoda, 5, 14
Isoptera, 13, 423, 437
Ixodes ricinus, 77
ixodid, 301
Ixodiadae, 67, 68, 242

Jasus, 252
Jasus lalandei, 4, 232, 251, 327, 460

Kalotermes flavicollis, 198
king crab, 5

Labidura riparia, 427
Labinia marginata, 185
Lampyris noctiluca, 13, 15
land crab, 252
Leiurus quinquestriatus, 12, 64, 238
Lepidoptera, 13, 38, 109, 129, 132,
   134, 135, 139, 140, 146, 156, 157,
   165–167, 423, 435
Lepidopteran, 350, 394, 433
Lepisma, 185
Leptinotarsa decemlineata, 105, 110,
   142, 145, 157, 158, 195, 401, 415,
   416, 423, 427, 430, 435
Leucania separata, 199
Leucophaea maderae, 110, 111, 202,
   423, 435

Libinia dubia, 41
Libinia emarginata, 12, 41, 44–52, 75, 142, 398, 420, 424
Ligia, 258
Ligia oceanica, 7, 258, 279
Limulidae, 289
Limulus, 4, 11, 12, 22, 78, 233–236, 262, 278, 289, 298, 299, 474, 475
Limulus polyphemus, 6, 232, 239, 302, 305
Lithobiomorpha, 9
Lithobius, 9, 13, 260
Lithobius forficatus, 9, 260, 303, 305
lobopod, 9, 16, 17
lobster, 40, 42, 43, 47, 48, 79, 232, 248, 250, 252, 254, 279, 294, 298, 299
locust, 10, 15, 143, 380, 493, 494
Locusta migratoria, 131, 143, 145, 161, 196, 285, 331–335, 337–341, 344–346, 348, 350, 353, 403, 416, 418, 420, 422, 423, 427, 431, 433, 435, 460, 461–463, 493
Locusta migratoria migratorioides, 98, 431

Macrobrachium rosenbergi, 14
Macrotermes, 199
Macrotermes subhyalinus, 423, 437
Maia squinado, 279
Malacostraca, 5, 7, 11, 12, 24, 248
Malacostracan, 7, 14, 248
Mammalian, 249
Mamestra brassicae, 138
Mandibulates, 231
Manduca sexta, 85, 86, 88, 89, 91–98, 101, 103–106, 108, 111, 113, 114, 131–140, 143–146, 148–150, 152–156, 158, 159, 161–164, 167, 188, 327, 331–334, 338, 340–342, 344–351, 354, 356, 392, 393, 396, 397, 401, 409, 411, 415, 418, 421–423, 430, 431, 433, 435, 438, 460, 461, 462, 495, 497, 498, 499
mealworm, 135, 395
Megoura viciae, 420, 437
Melanoplus, 203
Melanoplus sanguinipes, 423, 435
Melipona quadrifasciata, 437
Melolontha melolontha, 411, 435

Merostomata, 8
migratory locust, 331
millipede, 260
mite, 8, 76
mollusca, 17, 363
monarch butterfly, 139
mosquito, 140, 164, 166, 430
moth, 43, 474, 475
mud crab, 40
Musca, 237, 260
Musca domestica, 130, 341, 411, 415, 423, 437
myriapod, 4, 6, 231, 261, 262, 277, 278, 285, 289, 292, 295, 299, 342
Myriapoda, 4, 5, 9, 11, 13, 16, 24, 61

Nasutermes, 199
Nauphoeta cinerea, 13, 24, 143, 145, 196, 348, 349, 393, 394, 398, 401, 411, 418, 420, 423, 424, 431, 435, 438
Nauplii, 42, 246
Nematode, 6, 16, 20, 342
Nemertinea, 21, 23
Nephtys, 20, 22
Nereis, 20
Nezara viridula, 423
Niphargus virei, 259
Nonlepidopteran, 393, 394

Oligochaeta, 20
Oncopeltus, 474
Oncopeltus fasciatus, 188, 405, 408, 409, 423, 437, 463
Onychophora, 16
Opilio ravennae, 241
opilionid, 6
Opilionida, 8
Orchestia cavinmana, 249, 286
Orchestia gammarella, 7, 14, 259, 279, 298
Orchestia gammarellus, 43
Orconectes, 256, 257, 303, 305
Orconectes immunis, 255, 256
Orconectes limosus, 249, 253–257, 304
Orconectes obscurus, 250, 294, 300
Orconectes sanborni, 254
Orconectes virilis, 250, 257
Ornithodoros moubata, 12, 63–66, 69, 75, 242–245, 285, 287, 288, 296,

342, 462
Ornithodoros parkeri, 65, 66, 142, 241, 242, 245
Ornithodoros procinus, 66, 77, 241, 242, 296
Ornithodoros turicata, 8
Orthoptera, 13, 140, 423, 435
Ostrinia nubilalis, 145, 159
Ovalipes ocellatus, 47

Pachygrapsus, 255, 256
Pachygrapsus crassipes, 252, 284
Pachygrapsus marmoratus, 294
Palaemon, 305
Palaemon elegans, 309
Palaemon serratus, 253, 279
Palaemonetes, 300
Palaemonetes pugia, 249, 295
Palinura, 250
Palinurus, 380
Palinurus longipes, 298
Panorpa, 202
Panulirus longipes, 250
Parascaris equorum, 342
Pauropoda, 9, 260
Pedipalpida, 8
Penaeus setiferus, 44
Penaeus vannamei, 47
Peracarida, 300
Periplaneta, 203, 380
Periplaneta americana, 13, 24, 105, 110, 342, 343, 421–423, 435
Periplaneta fuliginosum, 423
Phalangida, 18
Phormia regina, 130
Phytoseiidae, 67
Phytoseiulus persimilis, 67
Pieris, 193
Pieris brassicae, 295, 332–334, 337, 338, 345, 346, 349, 351, 401, 403, 412, 435, 461, 463, 494, 500
Pieris rapae, 153
Pisaura mirabilis, 12, 239, 240
Pisauridae, 67
Platyhelminth, 21
Platyhelminthes, 17, 19, 20, 363
Platynereis dumerilii, 18
Podocarpus narkaii, 279
Podogona, 8
Polychaeta, 20

Polychaete, 18, 19, 22
Porcellio, 258
Porcellio dilatatus, 258, 286
Porcellionid, 259
Portunus trituberculatus, 306
prawn, 41, 300
Proasellus cavaticus, 258
Procambarus, 250, 297, 302
Procambarus bartonni, 49
Procambarus clarkii, 40, 47, 152, 297
Procambarus simulans, 300
Protostomia, 363
Protura, 9
Pseudoscorpionida, 8
Psocoptera, 86
Pterygota, 9, 279
Pycnogonid, 8, 237, 278, 491
Pycnogonida, 7
Pycnogonum literale, 7, 239, 460, 461, 491, 492
Pyrrhocoris apterus, 188, 408

rat, 378
Reticulitermes flavipes, 111
Rhipicephalus appendiculatus, 306
Rhipicelphalus sanguineus, 68
Rhithropanopeus harrisii, 42, 43, 78, 251
Rhodnius, 380, 407
Rhodnius prolixus, 183, 407, 408, 427
ribbon worm, 21
Ricinulei, 8
Roscius, 23

Samia cynthia, 400, 423, 435
Sarcophaga bullata, 130, 415, 437
Sarcophaga peregrina, 379
Saturniidae, 391
Scaptotrigona, 198
Scaptotrigona postica, 437
Schistocerca gregaria, 87, 197, 331, 332, 334–341, 343, 345, 346, 349, 352–354, 421, 423, 427, 435, 460, 461–463
Schistocerca nitens, 411
Schistocerca vaga, 132, 421, 423, 435
scorpion, 8, 13, 64, 234, 237, 238, 278, 342
Scorpionida, 8, 11, 13, 61, 63, 64, 69
Scutigera, 9

Scutigeromorpha, 9
Scylla serrata, 424, 433
sea spider, 6, 237
Semibalanus balanoides, 246, 248
Sesarma, 249
Sesarma reticulatum, 249, 254
shrimp, 40, 231, 248, 249, 253, 279, 295, 300
silk moth, 38
silkworm, 10, 251, 277, 327
snail, 472
Solenopsis invicta, 208
Solifugae, 8
Solpugida, 8
southern armyworm, 165
southwestern cornborer, 145
Sphaeroma, 258
Sphaeroma serratum, 258
spider, 8, 63, 69, 234, 237, 239, 240, 278, 299, 432
spider crab, 41, 45, 75, 279
spiny lobster, 4, 232, 248, 250, 251
Spodoptera eridania, 165
Spodoptera littoralis, 138, 332, 345, 346, 349–351, 461
Spodoptera litura, 199
Staphylococcus aureus, 478
sugarcane borer, 148, 153
sun spider, 8
swamp crayfish,
Symphyla, 5, 9

Taeniopoda eques, 435
Tarantulas, 239
Tardigrada, 6
Taxus, 260
Teleogryllus commodus, 423, 435
Tenebrio, 380
Tenebrio molitor, 69, 135, 342, 395, 406–408, 411, 421, 423, 435
Tegenaria ferruginea, 288
termite, 38, 472
Terrestrial Chelicerata, 5, 8, 9, 12, 16, 18, 24
Tetranychidae, 67
Tetranychus persimilis, 69
Tetranychus urticae, 67, 68
Thermobia domestica, 185, 411, 427, 437
Thysanura, 6, 7, 9, 10, 279, 437
tick, 4, 6, 8, 9, 69, 75–79, 142, 234, 237, 241–245, 257, 262, 278, 279, 285, 287, 296, 297, 301, 342, 343
tobacco budworm, 343
tobacco hornworm, 131–137, 143, 144, 152, 156, 164, 327, 333, 495
Triatoma, 201
Trichoplusia ni, 107, 134, 135, 137–141, 147–149, 152–154, 156, 164, 194, 435
trilobite, 6, 17
Trogoderma, 206
tsetse fly, 380
Turbellaria, 19

Uca, 256, 302
Uca pugilator, 47–50, 52, 251, 254
Uca pugnax, 47
Uniramian, 16, 260
uropod, 295
Uropygi, 8

Vanessa io, 411
Vertebrate, 277, 370

water bears, 6
whip scorpion, 8
white shrimp, 44

Xenopus laevis, 378
Xiphosura, 6

Yellow mealworm, 342

Zoea, 251
Zoeae, 251
Zootermopsis anguisticollis, 423

Part I Subject Index

abnormal development, 235, 239
accelerated molt, 235, 249, 250, 258
accessory sex glands (ASG), 393,
    433
    biosynthetic activity, 87, 88
    JH methyltransferase, 88, 433
    morphology, 88
acetylation, 336, 337, 339, 341, 468
acetylcholine esterase, 151, 152
2-Acetylecdysone, 336, 337
    22-phosphate, 336, 337
3-Acetylecdysone, 336, 337, 339,
    340, 463
    2-phosphate, 336, 337, 339, 340,
    463
    22-phosphate, 336, 337
3-Acetyl-20-hydroxyecdysone, 330,
    336, 338–340, 463
    2-phosphate, 330, 336, 338–340,
    463
    22-phosphate, 339, 340, 463
acron, 19
activator, 183
acyclic sesquiterpenoid ester. *See*
    methyl farnesoate
acylation, 336, 342, 353
allatostatins (ASF), 98, 99–102
    factors involved in release, 99–102
    isolation of, 98
allatropins (ATF), 96, 97, 99–102
    factors involved in release, 99–102
    isolation of, 97
    source, 96
alumina, 403
p-aminoglutethimide, 348
ammonia, 412, 465
anamorphic growth, 4
androgenic gland, 12, 14
    in Crustacea, 14, 75, 306
    inhibition of, 14
    in Insecta, 15
androgenic hormone (AH), 14, 15
    in Crustacea, 14, 54
    and its effects, 14
    in Insecta, 15

antennalectomy, 261
antibody, 389, 425, 429, 431, 477
antiecdysteroid, 371
antigen, 425
apolar conjugates, 235, 238, 245,
    260, 262, 285–287, 342, 343
apolar metabolites, 238, 244, 257,
    259, 286, 287, 342
apolysis, 3, 4, 230, 231, 235, 239,
    242, 244, 246, 247, 250, 260, 303
archicephalon, 19
arthropod, 75, 277, 286, 295, 457
    brain evolution in, 19, 20
    evolution of, 15
    land invasion by, 15
    monophyletic origin, 15
    origin from lobopod ancestor, 15,
    16, 20
    phylogeny, 20
    polyphyletic origin, 15
arylphorin, 103–105, 114, 379
Ashburner model, 377
assay, 473. *See also* house fly
asymmetry in rates of hormone
    biosynthesis, 422
ATP, 352; ecdysteroid phospho-
    transferase, 347, 351, 352, 354
autoradiography, 374, 375

bioassay, 231, 236, 406, 416, 473
brassinosteroid, 371
n-butanol, 232, 248

CA. *see* corpora allata
calcium, 100
    role in CA activity, 100
capillary GC, 405, 411, 412, 416, 418
carbon monoxide, 348, 350
cardiac stimulation, 22
cartridges, 403, 479
caste determination, 111, 199
celite, 248, 400
cell death, 193
cerebral gland, 13

chemical ionization (CI) mass spectrometry, 410, 412, 465
chloroform, 232, 233, 243, 467, 479, 490, 491, 499
cholesterol, 245, 256, 262, 282, 328, 336, 341, 343, 346, 347, 495, 499
chromatin, 365, 371
circulatory system, 21, 22
CM-Sephadex, 232
collar gland, 6, 17
color regulation, 22
column chromatography, 401
composite cuticle, 187
conjugation, 328, 335, 336, 353
    phospho-, 336, 351–353
corpora allata (CA), 12, 13, 127, 135–138, 158, 183, 421, 422, 424, 430, 431, 433
    from embryos, 87
    in vitro incubation, 87, 392, 421, 422, 431, 433
    and JH synthesis, 128, 135–138, 158
    and JH acid synthesis, 88, 89, 128, 135–138, 430, 431, 433
    morphology of, 86, 87
    ontogeny of, 86
    ultrastructure of, 86
    and volume-activity measurement, 87
corpora cardiaca (CC), 12, 13, 93–95, 97, 98, 100–102, 209, 396
corticosteroid-binding globulin, 112, 113
crustacean morphotypes, 40–41
    regulation by eyestalk, 41, 294
    and role of methyl farnesoate, 51, 398
crustaceans, 433;
    and effects of JH on larval development and metamorphosis, 42, 76
    and effects of JH on reproduction, 43, 142
    postmetamorphic morphogenesis, 42
    and role of the eyestalk on larval development and metamorphosis, 40
    and role of the eyestalk on reproduction, 41

Crustecdysone, 232, 278, 460
cuticle, 3, 190, 231, 235, 239, 240, 244, 250, 253, 256, 258–262
    deposition by epidermis, 3
cuticular proteins, 190
cyasterone, 484
cyclodiene, 130
cyclohexane, 248
cytochrome P-450, 348, 350, 354

DCCC (droplet countercurrent chromatography), 482, 483
dealation, 209
3-Dehydro-2-deoxyecdysone, 340
3-Dehydroecdysone, 287, 350, 461, 468
3-Dehydroecdysteroid, 333, 464, 470
2-Deoxyecdysone, 285, 329, 331, 332, 334, 337, 339, 340, 458, 460, 468, 477
    22-adenosinemonophosphate, 339, 463
    2-hydroxylase, 347, 350, 354
    22-phosphate, 330, 336, 337, 339, 343, 462
25-Deoxyecdysone, 458
2-Deoxy-20-hydroxyecdysone, 329, 331, 332, 338, 458, 460, 468, 478
    22-phosphate, 336, 338, 462
deuteromethanol, 402
deutocerebrum, 20
developmental program, 190, 201, 216, 235
DFP (O, O-diisopropyl phosphorofluoridate), 146–148, 153
di- and triglycerides, 403
2,3-Diacetylecdysone, 330, 337, 339, 340
    22-phosphate, 330, 337, 339, 340
2,3-Diacetyl-20-hydroxyecdysone, 339
    22-phosphate, 339
diapause, 66–68, 77, 145, 146, 199, 206, 237, 299
diapause hormone, 200, 206
dichloromethane, 404
22,25-Dideoxyecdysone, 284, 285, 341
differentiation, 3, 199, 204
    cellular, 2

(2*E*,6*E*,10-alpha-*cis*)-
 Dihomofarnesoic acid, 421
20,26-Dihydroxyecdysone, 287, 327,
 329, 331–334, 339–341, 346, 347,
 353, 459, 460
disposable cartridges, 394, 401, 403
distal organ, 23
DNA binding domain, 211
DNTFP (3-[(*E*)-4,8-dimethyl-3,7-
 nonadienylthio]-1,1,1-trifluoro-2-
 propanone, 147, 151
domain, 370
DOPA decarboxylase, 364
dose response, 234, 242, 247
dose-dependent, 242, 251, 303

EC (see ecdysone)
 α-ecdysone, 4, 136–138, 231, 277
 β-ecdysone, 4, 231, 232, 278
ecdysial glands, 6, 207, 210, 261
ecdysis, 132–138, 290
ecdysone, 231–235, 237, 238, 241–
 246, 248, 249, 251, 255, 257, 259–
 262, 277, 278, 282, 286, 327, 329,
 331–334, 339–344, 346–348, 350–
 353, 458, 466, 468, 477
 3-acetate, 330, 336, 337, 340, 343,
  461
 22-adenosinemonophosphate,
  339, 463
 22, 25-dideoxyecdysone, 242
 3-epimerase, 350, 354
 25-β-D-glucopyranoside, 342
 20-monooxygenase, 347–349, 354
 oxidase, 347, 350, 351, 354
 2-phosphate, 336, 337, 468
 3-phosphate, 330, 337, 339, 340
 22-phosphate, 330, 336, 337, 339,
  343, 344, 352, 462, 468
ecdysonoic acid, 283–287, 329, 331,
 333, 334, 339, 341, 345, 346, 353,
 459, 470, 479
ecdysteroid, 235–246, 248–262, 278,
 326–360, 424, 429, 430, 455–512
 3α, 327, 333, 335, 351, 353
 3β, 327, 335, 336, 351, 353
 acetate, 342, 459, 468, 469
 acetylation of, 468, 469
 acid, 328, 329, 333, 336, 340, 344–
  347, 353, 354
 action on metamorphosis, 136–138

action on molt-cycle, reproduction
 and diapause, 203, 296
action on target tissues, 30, 302
action on wandering behavior,
 136–138
acyl esters of, 342, 457, 459, 499
in Annelida, 6, 17
apomorphic or newly evolved
 form, 18
in apterygote insects, 6, 235, 236,
 281
in Aquatic Chelicerata, 6
bioassays, 231–233, 236, 327, 473
biological activity, 327, 333, 352,
 353, 370, 473
28-C compounds of, 463
cholesterol labeling of, 245, 256,
 336
in Collembola, 6, 281
conjugate, 328, 330, 331, 335, 336,
 339, 340, 342–344, 346, 351–353,
 497
in Crustacea, 4–6, 234, 245, 246,
 248, 257–259, 278
3-dehydroecdysteroids, 457, 461
25-deoxyecdysone, 4, 279, 282,
 283, 458, 468
2-deoxy-20-hydroxyecdysone, 4,
 5, 232, 251, 460, 466
double conjugates of, 463, 493
droplet countercurrent chromatography (DCCC), 482, 483
α-ecdysone, 231
endogenous levels and their variations, 134–138, 292
3-epiecdysteroids, 351
extraction, 231–234, 236–239, 241–
 243, 245, 246, 248, 251, 252, 258–
 261, 479
formation of acetonides of, 470,
 494
formation of fluorescent derivates
 of, 471
gas-liquid chromatography (GLC),
 237, 483
glucoside, 335, 341, 342, 457
glucuronide, 335
high-performance thin-layer
 chromatography (HPTLC), 482
high-performance liquid chromatography (HPLC), 233, 485, 497

ecdysteroid (*continued*)
high-voltage paper electrophoresis, 489
hydrolysis of conjugates of, 335, 336, 339, 343, 351, 353, 472
20-hydroxyecdysone, 4, 101–102, 234, 236, 237, 460
26-hydroxyecdysone, 460, 468, 478
20-hydroxyecdysone 3-acetate 2-phosphate, 463
26-hydroxyecdysteroids, 460
immunoassay, 474
in trilobites, 6
in Insecta, 9, 10, 134–138
inokosterone, 6, 252, 278, 297, 461, 476
interaction with nerve cells, 380
IR spectroscopy, 251, 464
makisterone A, 231, 252, 278, 457
mass spectrometry, 231, 465, 495
metabolism, 234, 235, 238, 242, 245, 258–260, 262, 285, 326–360
metabolites, esterase-labile, 257, 259, 342, 343
method of investigation
in Mollusca, 17
in Myriapoda, 9, 234, 260, 261, 278
in Nematoda, 6, 17
NMR spectroscopy, 234, 465
occurrence, 363
phosphate, 336, 339, 343, 344, 354, 459, 479
22-phosphate phosphohydrolase, 347, 352, 354
phytoecdysteroids, 457
in Platyhelminthes, 17
plesiomorphic or ancestral form, 17
ponasterone A, 6, 236, 241, 252, 279, 281–284, 286, 458
precursors of, 245, 252, 256, 282, 460
radioimmunassay, 231, 233, 236, 238, 240, 243, 245, 254–256, 260, 261, 278
receptors, 245, 249, 257
regulation of CA (corpora allata) activity, 101
side-chain cleavage product of, 283, 286, 463

sources (glands), 4, 231, 241, 254–256, 258–260, 262
sulfate, 335
supercritical fluid chromatography (SFC), 488, 489
synthesis, 231, 241, 245, 248, 254, 255, 256, 261, 282
in Terrestrial Chelicerata, 8, 234, 237–239, 241, 278
thin layer chromatography (TLC), 232, 236
in Thysanura, 6, 281
in trilobites, 6
UV spectroscopy, 252, 464
zooecdysteroids, 457
eclosion hormone, 138
egg-development neurosecretory hormone, 142
eggs, 143–145, 164, 289, 495
electron impact (EI) mass spectrometry, 410, 465
embryogenesis, 143–145, 155, 156, 164, 333, 335, 336, 339–341, 343, 344, 346, 352, 354
embryotoxic effects, 197
3-Epi-2-deoxyecdysone, 329, 334–337, 339, 340, 461, 499
3-phosphate, 330, 335–337, 340, 353
3-Epiecdysone, 327, 329, 333, 334, 337, 350–353, 461, 468
3-phosphate, 330, 337, 463
3-Epiecdysonoic acid, 333, 334, 341, 345, 346
epidermal cell, 188, 231, 235, 244
epidermal growth, 3
3-Epi-20-hydroxyecdysone, 327, 329, 332–334, 338, 346, 353, 461, 478, 494, 499
3-phosphate, 330, 338
3-Epi-26-hydroxyecdysone, 329, 334, 335, 341, 346, 353, 461
3-Epi-20-hydroxyecdysonoic acid, 329, 333–335, 341, 346, 347
3-Epi-20,26-dihydroxyecdysone, 327, 329, 332–334, 341, 346, 353, 461
epigamy, 18
epitoky, 18
epimerization, 327, 328, 333–335, 346, 347, 351, 353

epimorphic growth, 3
10,11-epoxidase, 128, 132, 135, 142, 143, 154–156, 163–165
epoxide hydrolase, 128, 132, 135, 142, 143, 154–156, 163–165
EPPAT (*O*-ethyl-S-phynylphosphoramidothioate), 134, 137, 138, 152
esterase labile, 242, 257
esterase-resistant, 259
ethanol, 232, 233
ethyl [10-^{3}H]farnesoate, 420
Evolution, 1, 4, 18, 21, 24, 185, 201
  of arthropods, 1, 185
  of arthropod brain and its neurosecretory cells, 18
  or hormones of IS, 18
  of JH or its counterparts, 24, 185, 201
  of molting hormones or ecdysteroids, 16
  of morphogenetic hormones, 4
  of neurohemal organs, 21
exogenous ecdysteroids, 235, 237, 239, 241, 242, 245, 246, 249–251, 259, 261, 295
exuviation factor, 258
eyestalk ablation, 249, 255
  effects on crustacean morphogenesis, 40
  effects on mandibular organ structure, 44
eyestalk factors, 52
  effects on mandibular organ, 52
  extracts, 52
  PDH (pigment-dispersing hormone), 52
  RPCH (red pigment-concentrating hormone), 52

F-19 NMR, 151
(2*E*,6*E*) [^{3}H] farnesoic acid, 421, 424
farnesol, 408, 420
farnesylacetone, 14, 54, 75
Fat body, 108, 111, 135, 154–156, 164, 185, 195, 348–350, 379, 495
fatty acids, 409
ferredoxin, 348
fly bioassay, 231, 232, 237, 260
free ecdysteroids, 236, 237, 239, 240, 244

functions of JHBP (JH-binding protein), 112, 113, 127, 128, 146–150

GABA (γ-aminobutyric acid)-receptors, 379
gas chromatography (GC), 241, 243, 245, 248, 252, 405, 410, 416, 422, 483
gastrolith, 250, 253, 255
GC. *See* gas chromatography
GC with electron capture detection (GC/ECD), 243, 403, 412, 414–146, 435, 473, 485
GC/MS (gas chromatography/mass spectrometry), 46–48, 243, 252, 402, 410, 412, 415, 416, 418–420, 422, 424, 432, 435, 473, 485
  chemical ionization mode, 410, 412, 435
  radioisotope dilution method, 410, 411
gene, 216, 365, 371, 377
  activation of, 216, 365
  ecdysteroid dependent, 371, 377
genetic switch, 160–162, 183, 215
giant chromosomes, 364, 374, 376
glucosylation, 336, 341, 353
α-glucuronidase, 351, 352
glyceride esters of fatty acids, 401
golden oil, 184, 400
grace's medium, 421
growth, 3, 4
  anamorphic, 4
  in arthropods, 3
  teloblastic, 4

HC10$_4$ (perchloric acid), 412, 416
head gland, 6, 17
hemimetabola, 140, 183, 186
hemocyanin, 22
hemolymph, 26, 128, 129, 231, 233, 242, 261, 392, 400, 420
hemolymph ecdysteroid, 233, 236, 238, 245, 252, 261, 293
hemolymph high-affinity, high-molecular-weight JH-binding protein (JHBP), 109, 112
hemolymph protein, 195, 252, 379
heptafluorobutyryl iodide, 418

heteromorphic growth, 187
hexahydrofarnesylacetone, 14
hexane, 232, 243
high-performance liquid chromatography (HPLC), 45–46, 233, 234, 236–239, 243–246, 252, 389, 396, 399, 403, 404, 412, 415, 422, 425, 429, 431, 457, 469, 471, 485
histogenesis, 3
HMG (3-hydroxy-3methylglutaryl), 90
HMG-CoA reductase, 238
hog liver esterase, 238
holometabola, 135, 183, 187
homogenizer, 400
homomevalonate, 92
hormone inactivation products, 238, 257, 262
hormone(s), 232
    analogues, 365, 367
    as cofactors of enzymes, 363
    controlling cell permeability, 216, 363, 380
hormones of inhibitory system (IS), 10, 11
    apomorphic or newly evolved form, 24
    in Aquatic Chelicerata, 11
    in Crustacea, 12
    in Insecta, 13
    in Myriapoda, 13
    plesiomorphic form, 24
    in Terrestrial Chelicerata, 12
HPLC. *See* high-performance liquid chromatography
HPLC/MS, 234, 412
human serum albumin, 425, 426
hydrolysis, 328, 335, 336, 339, 341, 343, 344, 351–353, 472, 473
hydroprene, 42
20-Hydroxyecdysone, 231–233, 235–254, 256, 257, 259–262, 279, 283, 286, 287, 327, 329, 331–334, 338–344, 346, 347, 350–353, 457, 458, 460, 468, 496
    2-acetate, 330, 338
    3-acetate, 338, 339, 461
    20-Hydroxyecdysone 3-acetate 2-phosphate, 463, 493
    26-Hydroxyecdysone 2-

phosphate, 462
    22-linoleate, 330, 343, 462
    22-oleate, 343, 462
    22-palmitate, 343, 462
    22-phosphate, 330, 336, 338, 339, 462
    22-stearate, 343, 462
26-Hydroxyecdysone, 329, 331-335, 338-342, 344, 347, 353, 359, 460
    22-glucoside, 330, 341, 342, 344, 462
    2-phosphate, 330, 338, 340, 341, 344, 462
    26-phosphate, 330, 331, 335, 338, 340, 341, 344, 462
20-Hydroxyecdysonoic acid, 329, 331, 333–335, 339, 345–347, 353, 459
hydroxylase, 256, 347–350
hydroxylation, 231, 257, 261, 328, 331, 332, 353
    C-2, 331, 332, 353
    C-20, 327, 331–333, 348, 353
    C-26, 327, 331–333, 344, 353

imaginal disks, 191, 366, 368, 393, 433
    biosynthetic activity, 88, 89, 138
    JH acid methyltransferase (JHAMT), 88, 89, 393, 433
imaginal primordia, 191
immunoassay, 473
immunohistochemistry, 374, 376
induction of JH esterase, 142, 143, 145, 146, 157–163, 194
inhibitory system. *See* hormones of inhibitory system
inokosterone, 61, 251, 252, 278, 297, 461, 476
insect growth regulators (IGRs), 392
insecticyanin, 210
integument, 188, 253, 257, 261
internal standard, 410, 412, 414, 416, 418, 420, 438
isoamylacetate, 243
isobutane, 412, 465

JH *See* juvenile hormone
JH acid methyltransferase (JHAMT), 431

JH and JHAs, 61;
   effect on diapause, 66–68, 145,
   146, 206
   effect on egg development, 43, 66,
   143–145, 197
   effect in JH esterase, 157–159
   effect on metamorphosis, 138
   effect on reproduction, 43, 65–68,
   139–143, 201
   effect on wandering behavior,
   135–137
JH binding proteins (JHBP), 102,
   113, 127, 128, 146–150
JH biosynthesis, 392, 421, 430
   higher homologue synthesis, 92
   in vitro, 143, 421, 430, 431, 433
   JH III synthesis, 89, 92
JH-deficient black mutant, 409
JH enantiomer binding, 107, 108,
   110, 111
JH 10,11-diol, 396, 397, 402, 412,
   425, 426
JH II 10,11-diol; *threo*-glycol;
   erythro-glycol, 397
JH-like compounds, 49–51, 61
   action on molt, reproduction and
   diapause, 65–67, 76, 77
   methods of investigation, 45
juvabione, 408
juvenile hormone (JH), 183, 397
   (+)- and (−)-JH I, 395, 396
   10R,11S absolute configuration,
   396
   acid methyltransferase (JHAMT),
   88, 89, 92, 431
   α, β-unsaturated ester moiety, 431
   analogues, 185, 206, 216, 392, 403,
   409, 427
   antibodies, 389, 425, 429, 431
   B$_3$ (Methyl-(2E)-6,7; 10,11,
   diepoxy-3,7,11-trimethyl-2-
   dodecadienoate, 423, 433
   binding proteins (JHBP), 102, 113,
   127, 128, 146–150, 211
   binding proteins (JHBP), 102, 113,
   127, 128, 146–150, 211
   binding studies, 110, 111, 211
   10,11 diol-*bis*-heptafluorobutyrate,
   413–416
   10,11 diol-*bis*-trifluoroacetate,
   412–415
   bioassay, 188, 389, 394, 406
   chemistry, 184, 395
   cross-contamination, 432
   cross-reactivity in RIA, 424, 429
   detection of, 389, 404, 409, 412,
   414, 420, 422, 427, 435
   11-deuteromethoxy-10-OH, 417–
   419
   effect on crustaceans, 42, 43, 78
   effect on cuticle, 187, 192
   effects on egg development, 43,
   143–145, 202, 407
   effect on egg viabillity, 143–145,
   197, 407
   enantiomerically pure, 438
   enzyme-linked immunoassay, 425
   ethyl-branched JHs, 92, 111, 394
   ethyl ester analogues of, 414, 416,
   417
   evolution of, 181, 201
   extraction methods, 389, 400, 401
   extraction and purification of, 389,
   399, 402, 421
   feeding and topical bioassay for,
   409
   functional role, 127, 132–146,
   157–162
   gas chromatography/mass spectrometry, 389, 393, 394, 402, 410,
   411, 416, 418, 424, 432, 435
   gonadotropic activity, 127, 139–
   143, 145, 146, 201, 394, 407, 408,
   433
   hemolymph titer, 127, 128, 136,
   412, 415, 429
   homosesquiterpenoid precursors
   of, 92, 421, 422
   in vitro technique, 389, 392, 398,
   421, 433
   interendocrine control, 85, 101,
   102
   internal standard, 400, 402, 414,
   415, 417, 426
   iso-JH 0, d$_3$-methoxydrin derivative, 397
   [^{3}H]Iso JH II ethylester, 410, 417,
   418, 420
   isolation of, 394, 401
   JH acid, 87–89, 91, 92, 128, 131–

JH acid (continued)
138, 205, 391, 393, 398, 401
JH I, 13, 184, 391–394, 396, 406–
408, 410–412, 415, 421, 423, 425,
430, 431, 435
JH II, 13, 185, 391–394, 396, 406,
407, 410–412, 415, 423, 425, 430,
431, 435
JH II and III 10,11-diols, 412, 415,
420
JH III (methyl 10,11-
epoxyfarnesoate, 13, 184, 391–396,
404, 406, 407, 411, 412, 415, 416,
419, 423, 425, 430, 431, 435; absolute configuration in, 396
JH 0, 13, 184, 391, 393, 397, 398,
407, 415, 418, 435
limit of detection, 410, 411, 412,
415, 416, 418, 430
mass spectral (MS) analysis, 409,
410, 416
4Me-JH I, 11, 13, 391, 393, 397,
398, 407, 418, 435
11-methoxy-10-(2,4-
dichlorobenzoate), 414, 415
11-methoxy-10-
heptafluorobutyrate, 414, 416
11-methoxy-10-
nonafluorohexyldimethylsilyl, 417,
419
11-methoxy-10-
pentaflurorphenoxyacetate, 413–
415
methoxyhydrin derivatives, 397,
402, 413–415, 419
methyl 10,11-epoxy-7-ethyl-3,11-
dimethyl-2,6-tridecadienoate, 395
methyl (2E,6E,10-cis)-10,11-
epoxy-3,7,11-trimethyl-2,6-tridec-
adienoate, 395
4-methyl JH I (4ME-JH I). See
under juvenile hormone, 4Me-JH I
microderivatization techniques,
413, 414, 417
modulation during egg development, 143–145, 164, 429
morphogenetic effect of, 187, 394,
407, 408
neck ligation bioassay for, 409
11-nonafluorooctoxy-10-

heptafluorobutyrate, 417, 418
11-nonafluorohexoxy-10-OH, 413,
416, 417
nuclear magnetic resonance
(NMR), 395, 431
occurrence in other arthropods,
43, 127, 128, 142
oxirane ring, 397, 410, 415
physicochemical methods, 390,
392, 409, 416, 427, 430, 431
presence in crustaceans, 43, 53,
142, 185
n-propyl ester analogue of [³H] JH
I, 415, 420
qualitative and quantitative measurements, 394, 409, 416, 417, 429,
431, 432, 435
quantification of, 400, 404, 406,
409, 417, 421, 424, 427, 429, 432,
435
10R,11S absolute configuration,
396
radiochemical assay (RCA) for in
vitro JH biosynthesis, 87, 143, 421,
430, 433
radioimmunoassay (RIA), 63, 64,
87, 389, 392, 394, 403, 404, 422,
424, 427, 430, 431, 433, 438
regulation of biosynthesis, 93,
102, 210, 422
role in insects, 127, 132–146, 157–
163, 183
role in metamorphosis, 135–138,
185
source of the methylester, 91, 92,
392, 421
specific antibody, 424
structure, 127, 129, 389
structure and stereochemistry of
the JHs and JH acids, 127, 129,
389, 395, 396
10-trans JH II, 414
[³H]2E,6E,10-trans JH 0, 415
transmethylation, 420
11-tridecafluorononoxy-10-OH
derivatives, 418
wax tests for, 406–409
juvenile hormone acids, 128, 136–
138, 401, 420–422, 424, 430, 431,
433. See also specific JH acids

juvenile hormone analogues (JHAs),
135, 140, 201, 406
juvenile hormone esterase, 195, 210,
215
  characterization, 146–150, 153, 154,
  156
  definition of, 146–150
  distribution and source, 135, 139,
  154–156, 195
  inhibitor studies, 134, 137, 138,
  140, 146–148, 150–154
  occurrence, 198
  properties of purified enzyme,
  153, 154
  substrate specificity, 129, 146, 148,
  149
juvenile hormone esterase inhibitors, 134, 137, 138, 140, 146–148,
  150–153
  importance, 134, 137, 138, 140,
  146–148, 152, 153
  NMR studies, 151
  organophosphates, 134, 137, 138,
  146–148, 151–153
  structure-activity studies, 150–153
  trifluoromethylketones, 147, 150–
  153
juvenile hormone esterase
  purification, 157–162, 195, 210
  affinity chromatography, 152, 153
  classical methods, 152
juvenile hormone esterase regulation, 157–162, 195, 210
  by head factor, 160–162
  by JH, 157–159, 195, 210
juvenile hormone esterase regulation by environmental factors,
  139–143, 145, 146, 162, 163, 183
  feeding, 162
  light cycle, 145, 146
  mating, 139–143
  nutrition, 163
  parasitism, 163
  stress, 162, 183
juvenile hormone metabolism, 128–
  132, 135, 139, 142, 143–146, 154–
  156, 162–165, 206
  assay methods, 129, 130
  conjugation reactions, 128, 129,
  131, 132

  epoxide hydration, 128, 132, 135,
  142, 143, 154–156, 163–165
  ester hydrolysis, 128–130, 131,
  150, 154–156, 162, 163
  in vitro pathways, 128, 129
  in vivo pathways, 131, 132
  occurrence in other arthropods,
  142, 143
  oxidation reactions, 130–132
  role in diapause, 145, 146, 206
  role in early larval instar development, 132–134
  role in embryogenesis, 143–145
  role in metamorphosis, 134, 138
  role in migration, 139
  role in pupal development, 139
  role in reproduction, 139, 143
  role in ticks, 142, 143
juvenile hormone-like compounds,
  35–61
  in Arachnida, 61
  in Crustacea, 35–68
juvenoids, 392

kaladasterone, 365, 367

larval hemolymph protein, 195, 215
LC (liquid chromatography). *See*
  HPLC
LC/MS (liquid chromatography/
  mass spectrometry). *See* HPLC/
  MS
limb regeneration, 254, 309
limulus bioassay, 236, 233
lipids, 232, 243, 399, 400, 405, 424,
  426
liquid scintillation counting, 238, 421
long-chain fatty acis, 235, 243
low-affinity JH-protein associations,
  103, 105
  lipophorin JH binding, 104, 105
  storage protein JH binding, 103,
  104
lymphatic tissue, 9

makisterone A, 231, 252, 278, 329,
  367, 368, 457, 473, 477, 490
mandibular gland (organ), 12, 44,
  75, 185, 393, 433

function in reproduction, 44, 49,
50, 75, 142
regulation, 51–53, 75, 398
secretory products, 44, 45–47, 398,
424, 433
as source of methyll farnesoate
(MF), 12, 45–47, 75, 142, 185, 202,
398
structure, 44
Manduca sexta (d$_O$) fifth stadium,
135–137
fifth-stadium larvae, 134, 138
prepupal stage, 138, 430
mass fragmentography, 234
mass spectrometry (MS), 234, 237,
395, 396, 397, 416, 419, 465, 490
medusae, 19
α'Me OTFP [3-(2-nonethio)-1,1,1-
trifluoro-2-propanone], 147, 151
mesonotum, 406
metabolites, 234, 426
metamorphosis, 3, 4, 377
effects of JH in regulation in crustaceans, 42, 43, 78, 79
effects of methyl farnesoate in
regulation in crustaceans, 49–51,
75
regulation in insects, 134, 138,
183, 185, 216
methane, 412, 465
methanol, 232, 233, 243, 413
methanol layers, 232
methionine, 392, 420
methionine (^{14}C-labeled
methionine), 396
methoprene, 140
(R)-α-methoxy-α-
trifluoromethylphenylacetyle
chloride, 396
methyl (2E,6E)-3,7,11-trimethyl-
2,6,10-dodecatrienoate. *See also*
methyl farnesoate
methyl farnesoate (MF), 12, 13, 45–
54, 75, 197, 389, 393, 395, 398,
410, 420, 422, 423, 424, 433
hemolymph levels, 47, 48, 398,
420
methods of measurement, 45–48,
398, 420
regulation by eyestalk factors,
51–53, 75

role in development, 51
role in reproduction, 49, 50, 142
synthesis by manidubular organ,
45–47, 75, 142, 185, 398
methyl transferase (see also
JHAMT), 431
metyrapone, 348, 350
mevalonate, 90, 91
MF (methyl farnesoate), 45–54, 142.
See also methyl farnesoate
Mg^{2+}, 352
MH (molt hormone), 231, 233, 235,
237, 236, 241, 243–145, 248, 249,
252, 257, 258, 261, 262
migratory behavior, 139
mitochondrial proliferation, 195
mitosis, 3
mode of action of allato-active factors, 99–101
calcium, 99, 100
cyclic AMP, 100, 101
molt cycle, 189, 234, 235, 244–246,
248, 250, 252–255, 257–260, 288
molt-inducing, 234, 237, 259
molt-related, 244–246
molt-specific genes, 245
molting, 231, 235, 237, 239–242, 244,
249, 250, 253, 254, 258–260, 262
control of, 3, 11, 231, 235
molting glands, 6–9, 231, 249
in Aquatic Chelicerata, 6
in Crustacea, 7, 8, 282, 294
in Insecta, 9
in *Limulus polyphemus*, 7
in Myriapoda, 9
in *Pycnogonum litorale*, 7
molting hormone, 4, 191, 217, 231,
326–360, 474
active form, 368
activity, 239, 327, 333
molting pad, 194
molting response, 231, 235
monoclonal antibodies, 110, 438
monooxygenase, 128, 132, 155, 156,
347–349, 354
morphogenesis, 3
definition of, 3
regulation by environmental factors, 3
morphogenetic hormones, 4, 15,
185, 199

of arthropods, 4
evolution of, 4
evolutionary aspects of, 4
role in caste differentiation, 15, 199
role in color change, 15
role in eye morphogenesis, 15
role in morphogenesis, 4, 185
role in regeneration, 15
role in sclerotization, 15
role in silk gland differentiation, 15
role in spermatogenesis, 15
role in wing polymorphism, 15
their glands, 4
motor neurons, 19
MS. *See* GC/MS, mass spectrometry
muristerone, 365, 367, 477

Na^+/K^+ ATPase, 198
NADPH (nicotinamide adenine dinucleotide phosphate), 130, 350, 351
NADPH-cytochrome P-450 reductase, 348
NaF, 352
neotenin, 13
nephrocytes, 9
as sources of ecdysteroids, 9
α-naphthyl acetate esterase, 134, 140, 143, 146–149, 151–153, 160, 161
neurohemal organ, 10, 13, 22
evolution of, 21
primitive, 21
Neuroparsin A, 99
Neuroparsin B, 99
neurosecretory cells (NSC), 96, 97, 207
new cuticle formation, 189, 231, 235, 239, 242, 246, 250, 259, 260, 262
NMR. *See* nuclear magnetic resonance
normal-phase HPLC, 233, 403, 485
nuclear magnetic resonance, 151, 234, 237, 395, 431, 465, 490

oenocytes, 6, 241
as sources of ecdysteroids, 6, 241, 286

oestradiol binding proteins, 364
20-OHE. *See* 20-hydroxyecdysone
olive oil, 409
oocytes, 378
oogenesis, 203, 237
oothecin, 205
organ culture, 233, 241, 255, 256
organophosphate inhibitors, 134, 137, 138, 146–148, 151–153
OTFP (3-octylthio-1,1,1-trifluoro-2-propanone), 147, 150, 151, 153
ovary, 260
as source of ecydsteroids, 260
oxidation, 128–132, 155, 156, 331, 348
3-Oxoecdysone, 350
3-Oxoecdysteroid, 351
3α-reductase, 350, 351, 344
3β-reductase, 350, 351, 354

paper factor, 197, 408
paraffin oil, 406
paraffin wax, 406
paraoxon (*O,O*-diethyl-*O*-*p*-nitrophenyl phospate), 147, 152
patency, 204
pattern formation, 3, 15
perisympathetic organ, 23
phagocytosis, 194
phosphatase, 343, 352, 353
acid, 351, 352
phosphorylation, 90, 336, 337, 339, 341, 350, 379
photoaffinity labeling, 106, 110, 111, 367, 376
piperonyl butoxide, 350
PO_4^{3-}, 352
polar lipids, 232
polar metabolites, 238, 257, 335
polarimetry, 395
polarity rule, 107
polymorphism, 199, 200
ponasterone A, 236, 241, 252, 279, 281, 284, 286, 341, 365, 367, 458, 460, 477
precocious molting, 246, 250, 251, 258, 261
prohormone, 138, 393, 398
prostomium, 20
protein kinase C, 205

prothoracic glands, 6, 200, 347, 348, 350
protocerebrum, 20
pseudergates, 195
PTTH (prothoracicotropic hormone), 137, 157, 206
puff, 216, 290, 337, 364, 377
  early, 377
  formation, 364, 377
  induction, 216, 377
  intermolt, 290, 377
  late, 337

queen bees, 199

radioimmunoassay (RIA), 63, 64, 231, 233, 234, 403, 404, 422, 424–427, 429, 430, 435, 474
radiolabeled propionate, 424
receptor, 210
  "activation," 364
  affinity, 365, 366
  affinity labeling, 367
  "cytoplasmic", 374
  definition, 365
  dissociation rate, 365, 372
  genes, 370
  heterogeneity, 368
  intracellular localization, 364, 374
  JH, 210
  labeling, 365, 367
  measurement, 372
  purification, 368
  stability, 365
  "translocation," 365, 374
regenerating limb buds, 251
regeneration, 308
reversed-phase HPLC, 46, 47, 233, 243, 259, 485
RIA. See radioimmunoassay
ring (Weismann's) gland, 10, 433
rostral gland, 3

S-adenosyl-L-methionine (SAM), 45, 91, 92
S-methyl radiolabeled methionine, 91, 92, 392, 420
D-saccharic acid 1,4-lactone, 352
salivary glands, 366, 377
Schneider's organ, 13

selected ion monitoring mode (GC/MS), 402, 416, 418, 419
sensory neuron, 19
SEP-PAK cartridge, 233, 234, 244, 245, 403, 479, 498
sephadex G-25, 352
sepharose-S(CH2)₄SCH₂COCF₃, 152
silica gel, 232, 239, 248, 251, 260, 394
silicic acid, 232, 243
sinus gland, 10, 12
solitary phase, 196
spinning behavior, 194
steroid-secreting cells, 249
steroids, 17
  as bioregulators, 17
  as universal biomolecules, 17
sterols, 403
storage protein receptor, 379. See also Arylphorin
structure-function, 150–153
styrene oxide, 130
subesophageal ganglion, 20, 160, 161
sulfatase, 131, 351, 352
supermolting, 237, 241, 242, 296
supraesophageal ganglion, 20, 22

teloblastic growth, 4
Tenebrio topical and injection tests, 406, 407
tetrocerebrum, 20
TFD (1,1,1-trifluoro-3Z,4-methyl-dodecene-2-one), 147, 151
TFT (1,1,1-trifluorotetradecan-2-one), 147, 150
thin-layer chromatography, 129, 130, 398, 401, 402, 415, 421, 479
TLC. See thin-layer chromatography
TMIS (N-trimethylsilylimidazole), 243, 248, 252
trans-acting factors, 214
transcription, 212, 364
transcriptional control, 206
transesterification, 420
trifluoroacetic anhydride, 412
trifluoromethylketones, 147, 150–153
trihomosesquiterpenoid, 397
trihomosesquiterpenols, 17
trilobitomorpha (trilobites), 20
trimethylsilylimidazole, 243, 248, 252

UV absorbance, 251, 404, 431, 464

ventral gland, 9, 10
vertebrate steroid(s), 112, 113, 211
  receptors, 211, 370, 378
vitellogenesis, 139, 143, 202, 203,
  254, 363
vitellogenin, 202, 203, 211

wandering, 194, 415, 430

worker bees, 199

X-organ, 12, 13, 18

Y-organ, 6, 8, 17, 249, 252, 254–256,
  258–262
  biosynthesis, 248, 252, 282
  secretion, 252, 256
Y-organectomy, 254, 255, 259
Y-organless, 249, 255

PART II

Embryonic and Postembronic Sources

Embryonic Sources

1. Embryonic Sources of Morphogenetic Hormones in Arthropods, August Dorn (80 pp.)

Postembryonic Sources

2. Morphology, Histology, and Ultrastructure of JH-Producing Glands in Insecta, Pierre Cassier (115 pp., 5 figures, 35 photos, 1 table)
3. Morphology, Histology, and Ultrastructure of Cephalic Neuronemal Organs and Their Roles in Morphogenetic Processes in Myriapoda, Michel Descamps, François Sahli, Catherine Jamault-Navarro, J. Caplet (70 pp., 10 figures, 5 photos, 1 table)
4. Morphology, Histology. and Ultrastructure of the Ecdysial Gland (Y-organ) in Crustacea, Eugene Spaziani (41 pp., 2 figures, 5 photos)
5. Ecdysial Glands and Ecdysteroids in Terrestrial Chelicerata, C. Juberthie and J.C. Boneric (42 pp., 9 figures, of which, 3 are photos)
6. Morphology, Histology, and Ultrastructure of the Ecdysial Glands in Myriapoda, Gerhard Seifert (48 pp., 5 figures, 11 photos)
7. Anatomy, Histology, Ultrastructure and Functions of the Prothoractic or Ecdysial Glands in Insects. J. Beaulaton (155 pp., 5 figures, 15 photos, 2 tables)
8. Morphology, Histology, Ultrastructure, and Organogenesis of the Androgenic Gland in Crustacea, Genevieve G. Payen (42 pp., 5 figures, 4 photos)
9. Morphology, Histology, and Ultrastructure of the Maxillary Gland in Crustaceans: Their Probable Function in Morphogenesis, Gertrude W. Hinsch (23 pp., 1 figure, 10 photos)

10. Morphology, Histology, and Ultrastructure of the
     Mandibular Gland in Crustacea, Gertrue W. Hirsch
     (23 pp., 9 photos)

TAXONOMIC INDEX

SUBJECT INDEX

PART III

Roles in Histogenesis, Organogenesis, and Morphogenesis

In Embryogenesis

1. Role of Gradient Factor in Arthropod Morphogenesis, V.J.A. Novak (68) + 12
2. Roles of Juvenile Hormone and Molting Hormone in Embryogenesis of Arthropods, A. Dorn (50)
3. Roles of Morphogenetic Hormones in Embryonic Cuticle Deposition in Arthropods, G. Sbrenna (68) + 7
4. Roles of Morphogenetic Hormones in Embryonic Diapause, O. Yamashita and K. Suzuki (113) + 6

In Postembryonic Events

5. Roles of Morphogenetic Hormones in the Metamorphosis of Arthropods Other Than Insects, K. D. Spindler (39) + 3
6. Roles of Molting Hormone in Sclerotization in Insects, C. E. Sekeris (148) + 2
7. Roles of Morphogenetic Hormones in Morphological Color Changes in Arthropods, A. Bouthier and P. Noel (89) + 5
8. Termite Polymorphism and Morphogenetic Hormones, Ch. Noirot and C. Bordereau (50) + 4
9. Roles of Morphogenetic Hormones in Caste Polymorphism in Sting Bees, H. Rembold (33) + 9
10. Roles of Morphogenetic Hormones in Caste Polymorphism in Stingless Bees, H.V.W. Velthuis and M.J. Sommeijar (56) + 4
11. Roles of Morphogenetic Hormones in Caste Polymorphism Bumble Bees, P. F. Roseler (25) + 4
12. Role of Morphogenetic Hormones in Caste Polymorphism Ants, L. Passera and J. P. Suzzoni (60) + 10
13. Morphogenetic Hormones and Phase Polymorphism in Locusts, M. P. Pener (96) + 2

14. Role of Androgenic Gland Hormone in Determining the Sexual Characters in Crustacea, G. G. Payen (35) + 4

15. Juvenile Hormone and Aphid Polymorphism, T. H. Mittler (41)

16. Role of Morphogenetic Hormones in the Morphogenesis of Eyes in Insects, M. Mouze (38) + 9

17. Roles of Morphogenetic Hormones in Regeneration, D. Bulliere and F. Bulliere (47) + 14

18. The Oenocytes of Insects: Differentiation, Changes During Molting, and Their Possible Involvement in the Secretion of Molting Hormone, F. Romer (35) + 13

19. Role of Morphogenetic Hormones in Spermatogenesis in Myriapoda, M. Descamps (42) + 24

20. Hormonal Control of Early Metamorphosis in Flies, J. Zdarej and P. Sivasubramanian (30) + 9

TAXONOMIC INDEX

SUBJECT INDEX